Site-specific Cancer Nanotheranostics

This book addresses various aspects of recent progress in the development of all types of tumor-responsive and site-specific delivery platforms together with the range of related chemistries. It is a timely volume as global research in oncology is moving towards more biochemical signal–based and cell environment–inspired therapeutic strategies. Furthermore, the book emphasizes the need to explore various characterization techniques for specific stimuli-responsive nanoplatforms and toxicological and regulatory aspects.

Features

- Focused, comprehensive volume providing a different angle on theranostics in cancer intervention.
- Targets not only the multifunctionality of nanosystems but provides a comprehensive overview of niche technical aspects.
- Given the chemistries presented here, it provides both early-career and experienced readers with strategies from leading authorities in the field globally.

Site-specific Cancer Nanotheranostics
A Microenvironment-responsive Approach

Edited by
Yashwant V. Pathak, Jayvadan K. Patel,
Namdev L. Dhas and Vipul P. Patel

CRC Press is an imprint of the
Taylor & Francis Group, an **informa** business

First edition published 2024
by CRC Press
6000 Broken Sound Parkway NW, Suite 300, Boca Raton, FL 33487–2742

and by CRC Press
4 Park Square, Milton Park, Abingdon, Oxon, OX14 4RN

© 2024 selection and editorial matter, Yashwant V. Pathak, Jayvadan K. Patel, Namdev L. Dhas and Vipul P. Patel; individual chapters, the contributors

CRC Press is an imprint of Taylor & Francis Group, LLC

Reasonable efforts have been made to publish reliable data and information, but the author and publisher cannot assume responsibility for the validity of all materials or the consequences of their use. The authors and publishers have attempted to trace the copyright holders of all material reproduced in this publication and apologize to copyright holders if permission to publish in this form has not been obtained. If any copyright material has not been acknowledged please write and let us know so we may rectify in any future reprint.

Except as permitted under U.S. Copyright Law, no part of this book may be reprinted, reproduced, transmitted, or utilized in any form by any electronic, mechanical, or other means, now known or hereafter invented, including photocopying, microfilming, and recording, or in any information storage or retrieval system, without written permission from the publishers.

For permission to photocopy or use material electronically from this work, access www.copyright.com or contact the Copyright Clearance Center, Inc. (CCC), 222 Rosewood Drive, Danvers, MA 01923, 978–750–8400. For works that are not available on CCC please contact mpkbookspermissions@tandf.co.uk

Trademark notice: Product or corporate names may be trademarks or registered trademarks and are used only for identification and explanation without intent to infringe.

Library of Congress Cataloging-in-Publication Data
Names: Pathak, Yashwant V., editor. | Patel, Jayvadan K., editor. | Dhas, Namdev L., editor. | Patel, Vipul P., editor.
Title: Site-specific cancer nanotheranostics : a microenvironment-responsive approach / edited by Yashwant V. Pathak, Jayvadan K. Patel, Namdev L. Dhas, Vipul P. Patel.
Description: First edition. | Boca Raton : CRC Press, 2024. | Includes bibliographical references and index.
Identifiers: LCCN 2023014950 (print) | LCCN 2023014951 (ebook) | ISBN 9781032434827 (hardback) | ISBN 9781032437606 (paperback) | ISBN 9781003368731 (ebook)
Subjects: MESH: Neoplasms—drug therapy | Theranostic Nanomedicine—methods | Tumor Microenvironment—drug effects | Nanoparticle Drug Delivery System—pharmacology
Classification: LCC RC271.C5 (print) | LCC RC271.C5 (ebook) | NLM QZ 267 | DDC 616.99/4061—dc23/eng/20230621
LC record available at https://lccn.loc.gov/2023014950
LC ebook record available at https://lccn.loc.gov/2023014951

ISBN: 978-1-032-43482-7 (hbk)
ISBN: 978-1-032-43760-6 (pbk)
ISBN: 978-1-003-36873-1 (ebk)

DOI: 10.1201/9781003368731

Typeset in Times
by Apex CoVantage, LLC

Family is not only an important thing. It's everything. This book is dedicated to my parents, Kantilal and Kamuben; my wife Sneha; my son Shubh; my nephews Sai and Yakin; lovely Hely; and Kiva, without whom this book would not have been possible.

—Jayvadan K. Patel

"Being a family means you are a part of something very wonderful. It means you will love and be loved for the rest of your life". This book is dedicated to my parents Laxmanrao and Parvati; my wife Ritu; and my daughter Shivanya; my grandfather Baburao and grandmother Laxmibai; my uncles Ramrao and Vitthalrao; my beloved sisters Sita, Neeta and Mukta; and the whole Dhas family, without whom this book would not have been possible.

—Dr. Namdev L. Dhas

This book is dedicated to my beloved parents; my wife Falguni; my kids Gati and Megh; my friends; and Hon. Amitdada Kolhe sir (managing trustee, Sanjivani Rural Education Society, Kopargaon, Maharashtra, India), without whom this book would not have been possible.

—Dr. Vipul P. Patel

Dedicated to the memories of my academic mentors who shaped and encouraged me all through my career: Prof Ramlal Nikore (Nagpur University), Prof A K Dorle (Nagpur University), Prof Gregory Gregoriadis (Royal Free Hospital, London), Prof Simon Benita (Hebrew University of Jerusalem, Israel and Prof Robert Levy (University of Michigan, Ann Arbor, Michigan)

—Dr. Yashwant V. Pathak

Contents

Foreword xi
Preface...xiii
Editors xv
List of Contributors xvii

Chapter 1 Nanoparticles and the Tumor Microenvironment: Challenges and Opportunities 1

    Philemon N. Ubanako, Samson A. Adeyemi, Pavan Walvekar and Yahya E. Choonara

Chapter 2 Materials and Chemistries of Tumor Microenvironment–Responsive Delivery Platforms... 21

    Hillary Mndlovu and Yahya E. Choonara

Chapter 3 pH-Responsive Delivery Nanoplatforms in Cancer Theranostics 38

    Himanshu Paliwal, Bindiya Chauhan, Devesh U Kapoor, Bhavik H Satani, Shivani Patel, Yashwant V. Pathak and Bhupendra G. Prajapati

Chapter 4 Enzyme-Responsive Delivery Nanoplatforms in Cancer Theranostics...... 54

    Jigar Vyas, Isha Shah and Bhupendra G. Prajapati

Chapter 5 Redox-Responsive Delivery Nanoplatforms in Cancer Theranostics 77

    Pavan Walvekar, Nombeko Sikhosana, Samson A. Adeyemi, Philemon N. Ubanako and Yahya E. Choonara

Chapter 6 ROS-Responsive Delivery Nanoplatforms in Cancer Theranostics 98

    Naina Rajak, Praveen Kumar, Namdev L. Dhas, Yashasvi Singh and Neha Garg

Chapter 7 Hypoxia-Responsive Delivery Nanoplatforms in Cancer Theranostics 114

    Priyanka Mishra, Yamini B Tripathi, Namdev L. Dhas and Neha Garg

Chapter 8 Dual/Multi-Responsive Delivery Nanoplatforms in Cancer Theranostics 120

    Kunj Vyas and Mayur M Patel

Chapter 9 Temperature-Responsive Delivery Nanoplatforms in Cancer Theranostics... 142

    Mduduzi N. Sithole and Yahya E. Choonara

Chapter 10 Light-Responsive Delivery Nanoplatforms in Cancer Theranostics 155

Atul Garkal, Anam Sami, Jahanvi Patel, Lajja Patel, Namdev L. Dhas and Tejal Mehta

Chapter 11 Magnetic Field-Responsive Delivery Nanoplatforms in Cancer Theranostics.. 170

Manish P. Patel, Rutvi V. Patel, Mehul R. Chorawala, Avinash K. Khadela, Sandip P. Dholakia and Jayvadan K. Patel

Chapter 12 Ultrasound-Responsive Delivery Nanoplatforms in Cancer Theranostics 184

Ruchi Tiwari, Vaseem Ahmad Ansari, Juhi Mishra and Namdev L. Dhas

Chapter 13 Exogenous Dual/Multi-Responsive Delivery Nanoplatforms in Cancer Theranostics.. 208

Mershen Govender and Yahya E. Choonara

Chapter 14 Aptamer-Based Targeting of Nanoplatforms for Cancer Theranostics 225

Gaurav Tiwari, Ancha Kishore Babu, Charul Khatri and Namdev L. Dhas

Chapter 15 Peptide-Based Targeting of Nanoplatforms for Cancer Theranostics 245

Samson A. Adeyemi, Philemon N. Ubanako, Pavan Walvekar and Yahya E. Choonara

Chapter 16 Antibody-Based Targeting of Nanoplatforms for Cancer Theranostics 275

Malvin Ofosu-Boateng, Seth Kwabena Amponsah and Benedicta Obenewaa Dankyi

Chapter 17 Carbohydrate-Based Targeting of Nanoplatforms for Cancer Theranostics.. 286

Emmanuel Boadi Amoafo, Malvin Ofosu-Boateng, Kwasi Agyei Bugyei and Seth Kwabena Amponsah

Chapter 18 Subcellular Targeting of Nanoparticles for Cancer Theranostics 296

Vivek Patel, Ajay J. Khopade and Jayvadan K. Patel

Chapter 19 Characterization Techniques for Stimuli-Responsive Delivery Nanoplatforms in Cancer Treatment 322

Deepak Kulkarni, Dipak Gadade, Harshad Kapare, Namdev L. Dhas and Mayuri Ban

Chapter 20 Toxicological Aspects of Tumour Microenvironment-Responsive Nanoplatforms 339

Ofosua Adi-Dako, Benoit Banga N'guessan, Joseph Adusei-Sarkodie, Nii Hutton Mills and Doris Kumadoh

Chapter 21 Commercialization Challenges of Tumor Microenvironment–Responsive Nanoplatforms 352

Snigdha Das Mandal, Devanshu J Patel, Surjyanarayan Mandal and Jayvadan K. Patel

Index 363

Foreword

Nanotechnology is developing quickly; therefore innovations from yesterday might not be relevant today. To prevent study duplication, it is crucial for scientists to communicate clearly. The work of assembling the most important developments into a practical format is extremely difficult because translational research on customized site-specific and microenvironment-responsive nanoplatforms for biomedical applications encompasses a number of disciplines.

In addition to providing a missing bridge in the literature, this book, written by some of the most knowledgeable professionals on the subject, also successfully spans the gap between all the various subfields in this discipline.

It offers a whole range of services, from therapeutic applications to diagnostics in cancer treatment. In addition to introducing fundamental sciences, basic research and therapeutic approaches resulting from it, it also discusses recent advancements. Most importantly, it anticipates the difficulties and chances that lie ahead. For scientists working in the field and for young researchers interested in site-specific and microenvironment-responsive nanoplatforms for biomedical or more specifically theranostic nanoplatforms for cancer treatment, this book is a toolbox of answers and concepts.

Contributions to this book include innovative ideas like focusing on malignant cells and using the tumor microenvironment (TME) as one of the stimuli to demonstrate the increased efficacy of therapeutic moiety-loaded nanoplatforms for the treatment and detection of cancer. The chapters give an in-depth analysis of the most recent advancements in the field of theoretical and experimental methods for designing innovative nanoplatforms and their theranostic uses in the treatment of cancer.

The expertise of the editors and their interest in the field have made it possible to fully comprehend the research issues discussed in this book, which features excellent contributions from top global experts in the field. This book will soon become a standard reference for researchers at various levels, including undergraduate, graduate and postgraduate students as well as working scientists in the fields of pharmacy, pharmaceutics, chemistry, biochemistry, pharmacology, material science, biology and clinical research, as well as in the interdisciplinary fields of nanotechnology. Also, it will be highly sought after for professionals focusing on advancements in cancer therapies and medical diagnostics. This book will be helpful to investors, policymakers, regulators, funding organizations and college professors who are interested in nanotechnology.

Lt. Gen. (Dr.) M. D. Venkatesh
Vice-Chancellor
Manipal Academy of Higher Education (MAHE),
Manipal-576104
Udupi, Karnataka State,
India

Preface

Cancer is an epidemic burden and remains prevalent worldwide. The significant mortality rates due to cancer-related deaths are because of later diagnosis and complicated treatment protocols. In the global scenario, lack of healthcare is another major problem, which contributes to the high mortality rates. The current therapeutic modalities are of very high price and the techniques to detect cancer are expensive and beyond the reach of majority populations.

Alternative approaches are required to improve detection and enhance the efficacy of current treatment options as well as to explore new avenues for diagnosis of cancer and concurrently treatment. Nanotheranostic platforms, with biocompatible nanoparticles to explore applications for both diagnostics and targeted delivery, are a new a potential approach for cancer management in recent years. Enormous development of nanomaterials and major innovations in biomedical engineering techniques have led to rapid expansion of the nanotheranostics field, allowing for effective, sensitive and specific diagnosis; real-time monitoring of drug delivery; and enhanced treatment efficacies across various malignancies.

A nanotheranostic platform approach potentially provides a personalized and targeted approach to cancer therapy, wherein nanoparticles, which can be designed to detect specific biomarkers of the target malignant region, allow real-time monitoring or visualization of the target and finally deliver therapeutic options in a more precise manner. In recent years, nanosensor and nanomedicine technologies and products have experienced great strides in development and have paved the way for promising means of nanotheranostics implementation in cancer management.

A microenvironment-responsive approach is a recent development in cancer treatment, and it is helping us understand how cancer treatment can be modified and revolutionized without significant side effects of anti-cancer drugs. Tumor microenvironment is created by dominating tumor induced interactions and tumor itself. Although various immune effector cells are recruited to the tumor site, their anti-tumor functions are downregulated, largely in response to tumor-derived signals. Infiltrates of inflammatory cells present in human tumors are chronic in nature and are enriched in regulatory T cells (T_{reg}) as well as myeloid suppressor cells (MSCs). Immune cells in the tumor microenvironment not only fail to exercise anti-tumor effector functions, but they are co-opted to promote tumor growth. Sustained activation of the NF-κB pathway in the tumor milieu represents one mechanism that appears to favor tumor survival and drive abortive activation of immune cells (Whiteside, 2008).

This book addresses a microenvironment-responsive approach for the development of site-specific cancer theranostics. This edited book presents 21 chapters written by leading authors who are contributing to this area of research in their respective institutions. The first chapter covers nanoparticles and tumor microenvironment challenges and opportunities. Chapters 2 to 6 are dedicated to materials and chemistries of tumor microenvironment, pH-responsive delivery for cancer theranostics, enzyme-responsive nanoplatforms, redox-responsive nanoplatforms and reactive oxygen species-responsive nanoplatforms.

Chapters 7 to 12 cover hypoxia-responsive delivery platforms, dual/multiresponsive delivery systems, temperature-responsive delivery, light-responsive delivery, magnetic field delivery and ultrasound-responsive delivery to the cancer tumor microenvironment.

Chapter 13 addresses exogenous dual and multiresponsive delivery platforms and the use of different substrates such as aptamer based, peptide, antibody, carbohydrate and subcellular-targeting nanoparticle systems for cancer theranostics.

Chapters 20 and 21 cover the toxicological aspects of tumor environment–responsive nanoplatforms. Chapter 21 discusses the commercialization challenges of tumor microenvironment-responsive nanoplatforms.

We owe a lot to all the chapter authors and wish to express our sincere gratitude for their contributions. Without their support; this book would not have seen the market.

The Springer staff has always supported our book proposal, and we are highly indebted to Carolyn Spenser, other desk editors, copyright editors and printing press supporters who have made this book possible with their sincere efforts.

Our families always have to share their time when we embark on such a tedious journey as editing a book. We definitely owe a lot to our respective families.

We are glad to see the book and share with all the scientists, academicians and industry people working in this field. We have tried our level best to make it excellent. If you find any errors, please do share with the editors, and we will try to rectify them in the next edition.

With best wishes to our readers, and we look forward to your constructive and critical comments.

Yashwant V. Pathak, Jayvadan K. Patel, Namdev L. Dhas and Vipul P. Patel

REFERENCE

Whiteside TL. The tumor microenvironment and its role in promoting tumor growth, Oncogene. 2008;27(45):5904–5912.

Editors

Dr. Yashwant V. Pathak completed his Ph.D. in pharmaceutical technology from Nagpur University, India and EMBA and MS in Conflict Management from Sullivan University, USA. He is the Associate Dean for the Faculty Affairs at Taneja College of Pharmacy, University of South Florida, Tampa, Florida. With extensive experience in academia and industry, he has over 350 research publications, abstracts, chapters and reviews; and over 70 books in nanotechnology and drug delivery systems; nutraceuticals; conflict management and several in cultural studies. His areas of research include drug delivery systems and nanotechnology applications for pharmaceuticals and nutraceuticals. He has traveled extensively to over 80 countries to network with scientific experts and is actively involved with many pharmacy colleges in different countries. He is Adjunct Professor at Faculty of Pharmacy, Airlangga University, Surabaya, Indonesia.

Dr. Jayvadan K. Patel is a formulation scientist at Aavis Pharmaceuticals, USA and Professor Emeritus, Faculty of Pharmacy, Sankalchand Patel University, India. He has more than 26 years of research, academic and industry experience. Dr. Patel has expertise in design and conducts pre-formulation, formulation, process, stability and container closure development studies for drug products. He conducted research in the areas of formulation modifications and potential new product development. He has published more than 270 research and review papers in international and national journals. He has co-authored 20 books and contributed 116 book chapters to books published by well-reputed publishers. His papers have been cited more than 5500 times, with more than 30 papers getting more than 50 citations each. He has a 35 h-index and 125 i10-index to his credit. He has guided 106 M.Pharm. students and mentored 46 Ph.D. scholars. He is already decorated with a dozen awards and prizes both national and international. He was the recipient of the very prestigious AICTE-Visvesvaraya Best Teachers Award-2020 by the All India Council for Technical Education, government of India, and the APTI-Young Pharmacy Teacher Award (2014) by the Association of Pharmaceutical Teachers of India. He is a reviewer of more than 75 and an editorial board member of 20 reputed scientific journals. He has completed 12 industry and government-sponsored research projects.

Dr. Namdev L. Dhas is an assistant professor in the Department of Pharmaceutics, Manipal College of Pharmaceutical Sciences, Manipal Academy of Higher Education (MAHE), Manipal, Udupi, Karnataka, India. He earned his Ph.D. from the Institute of Pharmacy, Nirma University, Ahmedabad, Gujarat, in 2021. His areas of research include material science and engineering, multimodal therapeutic platforms and inorganic nanocomposites for cancer therapy. The main focus is different synthesized metal-based nanoplatforms and how their properties affect the efficiency of therapy and diagnosis. He has 31 publications in the form of several research and review articles in international journals. Out of 31 publications, 2 articles were published in the journal *Coordination Chemistry Reviews* with an impact factor of 24.883, 6 articles have an 11.467 impact factor (*Journal of Controlled Release*) and 1 article has 10.723 (*Carbohydrate Polymers*). As of now, he has achieved a cumulative impact factor of more than 180 by publishing articles in several national and international journals. His scientific articles have been cited more than 500 times. Recently, he received a research grant from SRG-SERB in August 2022. He was recently awarded Excellent Researcher in Pharmaceutical Science by the Indian Drug Manufacturing Association (IDMA) and Healthcopeia in July 2022. He also received the Young Researcher Award-2020 organized by Institute of Scholars, Bangalore, India.

Dr. Vipul P. Patel is a director and professor at SRES Sanjivani College of Pharmaceutical Education and Research (An Autonomous Institute), Kopargaon, Maharashtra. He has 22 years of teaching and research experience in the fields of pharmaceutical technology and biotechnology. He has guided more than 60 M.Pharm. students and 10 Ph.D. students. He has received more than 6.0 crore rupees grant from various funding agencies like AICTE, DBT, GUJCOST, RGSTC and DDU-GKY (government of India) for research, institutional development and rural skill development purposes. He has published more than 70 research publications in various reputed national and international journals and 4 book chapters in internationally published books, with 1 book authored. He has a 13 h-index and 18 i10-index, with more than 500 citations of published papers to his credit He had published two Indian patents in the area of anticancer drug discovery and tablets. He had completed more than 12 lakhs rupees of consultancy projects in the area of in vitro cytotoxic cell lines. He has also completed an industry project sponsored by SciTech Specialties Pvt, Ltd., Sinnar, Nashik, Maharashtra, on effervescent tablets. He is a professional member of various pharmaceutical associations, including IPA, GSPC, Pharmacy Council of India (as an elected inspector) and Fellow in Institute Chemist India. He also received the Best Teacher Award 2020 (InSc-Bangalore, approved by Ministry of HRD, New Delhi). His core areas of research are novel drug delivery systems for probiotics, anticancer drug discovery and cytotoxicity and bioavailability enhancement.

Contributors

Vaseem Ahmad Ansari
Department of Pharmaceutics,
Integral University,
Lucknow, Uttar Pradesh, India

Seth Kwabena Amponsah
Department of Medical Pharmacology,
University of Ghana Medical School,
Accra, Ghana

Ofosua Adi-Dako
Department of Pharmaceutics and
  Microbiology, School of Pharmacy
University of Ghana, Legon, Accra, Ghana

Joseph Adusei-Sarkodie
Department of Pharmacognosy and Herbal
  Medicine, School of Pharmacy,
University of Ghana,
Legon, Accra, Ghana

Samson A. Adeyemi
Wits Advanced Drug Delivery Platform,
  Department of Pharmacy and
  Pharmacology, Faculty of Health Sciences,
University of the Witwatersrand,
Johannesburg, South Africa

Emmanuel Boadi Amoafo
Department of Pharmaceutical Sciences,
North Dakota State University,
North Dakota, USA

Selorme Adukpo
Department of Pharmaceutics and
  Microbiology, School of Pharmacy,
University of Ghana,
Legon, Accra, Ghana

Kwasi Agyei Bugyei
Department of Medical Pharmacology,
University of Ghana Medical School,
Accra, Ghana

Mayuri Ban
Department of Pharmaceutical Chemistry,
  Srinath College of Pharmacy,
Aurangabad, Maharashtra, India

Ancha Kishore Babu
KPJ Healthcare University, Persiaran seriemas,
Nilai, Negeri Sembilan, Malaysia

Bindiya Chauhan
School of Pharmacy,
Parul University,
Vadodara, Gujarat, India

Mehul R. Chorawala
L. M. College of Pharmacy,
Ahmedabad, Gujarat, India

Yahya E. Choonara
Wits Advanced Drug Delivery Platform,
  Department of Pharmacy and
  Pharmacology, Faculty of Health Sciences,
University of the Witwatersrand,
Johannesburg, South Africa

Benedicta Obenewaa Dankyi
Family Health University College,
Accra, Ghana

Namdev L. Dhas
Department of Pharmaceutics, Manipal College
  of Pharmaceutical Sciences,
Manipal Academy of Higher Education
  (MAHE),
Manipal, Karnataka state, India

Sandip P. Dholakia
L. M. College of Pharmacy,
Ahmedabad, Gujarat, India

Mershen Govender
Wits Advanced Drug Delivery Platform
  Research Unit, Department of Pharmacy
  and Pharmacology, School of Therapeutic
  Science, Faculty of Health Sciences,
University of the Witwatersrand,
Johannesburg, South Africa

Neha Garg
Department of Medicinal Chemistry, Institute
  of Medical Sciences,
Banaras Hindu University,
Varanasi, Uttar Pradesh, India

Atul Garkal
Department of Pharmaceutics,
Institute of Pharmacy,
Nirma University,
Ahmedabad, Gujarat, India

Dipak Gadade
Delhi Skill and Entrepreneurship University,
IIT Dwaraka Campus,
New Delhi, India

Devesh U. Kapoor
Dr. Dayaram Patel Pharmacy College, Sardar
Baug, Station Road,
Bardoli, Gujarat, India

Avinash K. Khadela
L. M. College of Pharmacy,
Ahmedabad, Gujarat, India

Charul Khatri
Pranveer Singh Institute of Technology
(Pharmacy), Bhauti,
Kanpur, Uttar Pradesh, India

Praveen Kumar
Department of Medicinal Chemistry, Institute
of Medical Sciences,
Banaras Hindu University,
Varanasi, Uttar Pradesh, India

Ajay J. Khopade
Sun Pharmaceutical Industries Ltd.,
Vadodara, Gujarat, India

Deepak Kulkarni
Department of Pharmaceutics, Srinath College
of Pharmacy,
Aurangabad, Maharashtra, India.

Harshad Kapare
Department of Pharmaceutics,
Dr. D.Y. Patil Institute of Pharmaceutical
Sciences and Research, Pune, Maharashtra,
India

Doris Kumadoh
Department of Pharmaceutics, Centre for Plant
Medicine Research, Mampong-Akuapem,
Ghana

Hillary Mndlovu
Wits Advanced Drug Delivery Platform,
Department of Pharmacy and
Pharmacology, Faculty of Health Sciences,
University of the Witwatersrand,
Johannesburg, South Africa

Surjyanarayan Mandal
R&D Department Themis Medicare Ltd,
Uttarakhand, India

Snigdha Das Mandal
Department of pharmacology, Parul Institute of
Pharmacy and Research,
Parul University,
Vadodara, Gujarat, India

Tejal Mehta
Department of Pharmaceutics, Institute of
Pharmacy,
Nirma University,
Ahmedabad, Gujarat, India

Nii Hutton Mills
Department of Pharmaceutics and
Microbiology, School of Pharmacy,
University of Ghana,
Legon, Accra, Ghana

Priyanka Mishra
Department of Medicinal Chemistry, Institute
of Medical Sciences,
Banaras Hindu University,
Varanasi, Uttar Pradesh, India

Juhi Mishra
Pranveer Singh Institute of Technology
(Pharmacy), Bhauti,
Kanpur, Uttar Pradesh, India

Benoit Banga N'guessan
Department of Pharmacology and Toxicology,
School of Pharmacy,
University of Ghana,
Legon, Accra, Ghana

Malvin Ofosu-Boateng
Department of Pharmaceutical Sciences,
University of Tennessee Health Science Centre,
Tennessee, USA

Contributors

Himanshu Paliwal
Drug Delivery System Excellence Center and
 Department of Pharmaceutical Technology,
 Faculty of Pharmaceutical Sciences,
Prince of Songkla University,
Hat Yai, Songkhla, Thailand

Vivek Patel
Sun Pharmaceutical Industries Ltd.,
Vadodara, Gujarat, India

Devanshu J. Patel
Department of Pharmacology, Parul Institute of
 Medical Science and Research,
Parul University,
Vadodara, Gujarat, India

Jahanvi Patel
Department of Pharmaceutics, Institute of
 Pharmacy,
Nirma University,
Ahmedabad, Gujarat, India

Lajja Patel
Department of Pharmaceutics, Institute of
 Pharmacy,
Nirma University,
Ahmedabad, Gujarat, India

Manish P. Patel
L. M. College of Pharmacy,
Ahmedabad, Gujarat, India

Jayvadan K. Patel
Formulation Scientist, Aavis Pharmaceuticals,
 USA and
Professor Emeritus, Faculty of Pharmacy,
Sankalchand Patel University,
Visnagar, Gujarat, India

Rutvi V. Patel
L. M. College of Pharmacy,
Ahmedabad, Gujarat, India

Yashwant V. Pathak
College of Pharmacy, University of South
 Florida, United States and
Adjunct Professor at Faculty of Pharmacy,
Airlangga University,
Surabaya, Indonesia

Shivani Patel
School of Pharmacy,
Parul University,
Vadodara, Gujarat, India

Mayur M. Patel
Department of Pharmaceutics, Institute of
 Pharmacy,
Nirma University,
Ahmedabad, Gujarat, India

Bhupendra G. Prajapati
Dept. of Pharmaceutics and Pharmaceutical
 Technology, Faculty of Pharmacy, Shree
 S.K. Patel College of Pharmaceutical
 Education & Research,
Ganpat University,
Mehsana, Gujarat, India

Naina Rajak
Department of Medicinal Chemistry, Institute
 of Medical Sciences,
Banaras Hindu University,
Varanasi-221005, Uttar Pradesh,
 India

Anam Sami
Department of Pharmaceutics, Institute of
 Pharmacy,
Nirma University,
Ahmedabad, Gujarat, India

Bhavik H. Satani
Maliba Pharmacy College and Vedanta
 Ayurvedic Hospital,
Uka Tarsadia University,
Tarsadi, Bardoli, Gujarat,
 India

Isha Shah
Sigma Institute of Pharmacy,
Vadodara, Gujarat, India

Nombeko Sikhosana
Wits Advanced Drug Delivery Platform,
 Department of Pharmacy and
 Pharmacology, Faculty of Health
 Sciences,
University of the Witwatersrand,
Johannesburg, South Africa

Mduduzi N. Sithole
Wits Advanced Drug Delivery Platform
  Research Unit, Department of Pharmacy
  and Pharmacology, School of Therapeutic
  Science, Faculty of Health Sciences,
University of the Witwatersrand,
Johannesburg, South Africa

Yashasvi Singh
Department of Urology, 3rd floor, CSSB,
  Institute of Medical Sciences
Banaras Hindu University,
Varanasi, Uttar Pradesh, India

Yamini B. Tripathi
Department of Medicinal Chemistry, Institute
  of Medical Sciences,
Banaras Hindu University,
Varanasi, Uttar Pradesh, India

Ruchi Tiwari
Pranveer Singh Institute of Technology
  (Pharmacy),
Kanpur, Uttar Pradesh, India

Gaurav Tiwari
Pranveer Singh Institute of Technology
  (Pharmacy), Bhauti,
Kanpur, Uttar Pradesh, India

Philemon N. Ubanako
Wits Advanced Drug Delivery Platform,
  Department of Pharmacy and
  Pharmacology, Faculty of Health
  Sciences,
University of the Witwatersrand,
Johannesburg,
  South Africa

Jigar Vyas
Sigma Institute of Pharmacy,
Vadodara, Gujarat,
  India

Kunj Vyas
Department of Pharmaceutics, Institute of
  Pharmacy,
Nirma University,
Ahmedabad, Gujarat,
  India

Pavan Walvekar
Wits Advanced Drug Delivery Platform,
  Department of Pharmacy and
  Pharmacology, Faculty of Health
  Sciences,
University of the Witwatersrand,
Johannesburg,
  South Africa

1 Nanoparticles and the Tumor Microenvironment
Challenges and Opportunities

*Philemon N. Ubanako, Samson A. Adeyemi,
Pavan Walvekar and Yahya E. Choonara*

CONTENTS

1.1 Overview of Cancer and Importance of the Tumor Microenvironment in Resistance and Therapy 1
1.2 How the TME Influences Tumor Progression 3
    1.2.1 Tumor Vasculature and Tumor Endothelial Cells 4
    1.2.2 Cancer-Associated Fibroblasts 5
    1.2.3 Tumor-Associated Macrophages and Other Immune Cells 6
    1.2.4 The Extracellular Matrix 7
    1.2.5 Deregulated pH of the TME/Metabolic Alterations in the Tumor Microenvironment 8
    1.2.6 Hypoxia in the TME 9
1.3 Strategies to Target the TME 10
    1.3.1 Anticancer Nanoparticles Targeting the TME 10
1.4 Challenges of Nanoparticle-Based Therapies 12
    1.4.1 The Challenge of Tumor Heterogeneity 12
    1.4.2 Nanoparticle-Induced Aberrant Gene Expression 12
    1.4.3 Inadvertent Epigenetic and Post-Translational Modifications 13
    1.4.4 Reproducible Nanoparticle Synthesis 14
    1.4.5 Nanoparticle Screening Strategies 14
    1.4.6 Scalable Manufacturing 14
1.5 Conclusions 15
References 15

1.1 OVERVIEW OF CANCER AND IMPORTANCE OF THE TUMOR MICROENVIRONMENT IN RESISTANCE AND THERAPY

The global fight against cancer remains a challenging endeavor for researchers and clinicians. In 2020, there were an estimated 19.3 million new cases of cancer worldwide and over 10 million cancer deaths, according to GLOBOCAN (1). Cancer cases are predicted to increase to over 20 million per year by 2025 (2). Effective cancer therapies should be able to thwart the proliferation of rogue cells and prevent cancer resurgence while sparing normal cells. Although chemotherapy is the gold standard for cancer treatment, its efficacy is constrained primarily by tumor microenvironment (TME) heterogeneity, multidrug resistance, toxicity to normal cells, and poor drug distribution to the tumor (3). The key challenge in standard chemotherapy is to deliver a high enough drug concentration to eradicate the tumor without inflicting adverse effects on nearby normal tissues (4). Clinical studies have indicated that chemotherapeutic drugs such as taxanes and doxorubicin are only modestly effective, partly due to their insufficient penetration or delivery to tumor cells (5, 6).

Resistance to traditional chemotherapeutic treatments and targeted therapies is still a major concern despite the progress in anticancer therapies for a variety of malignancies. A seminal study in non-small cell lung cancer treatment showed that cytotoxic chemotherapy with predominantly used treatment modalities in randomly tested patients had reached its therapeutic plateau (7). This sparked interest and ushered in molecularly targeted therapeutics, which rely on specific molecular alterations in cancer-associated proteins or other molecules. Cancer treatment has since then evolved into molecular profiling of tumors to discover potentially druggable molecular targets on tumors and on specific components of the tumor microenvironment (2).

Treatment failure and tumor progression are largely instigated by innate or acquired drug resistance, which results because of mechanisms associated with specific populations of cancer cells or with various components of the TME (8). Physical barriers produced by these TME characteristics reduce the possibility of a chemotherapeutic drugs traversing tumor blood vessels to target cancer cells. To circumvent these challenges, there has been a growing tendency to use nanotechnology for enhanced drug delivery.

Nanotechnology provides beneficial effects such as longer drug exposure, targeted delivery, improved solubility, and a higher therapeutic index in cancer treatment (9). New methods for cancer treatment and diagnostics have been developed recently due to advances in nanotechnology. When compared to conventional chemotherapeutic drugs, nanoparticle-based drugs possess enhanced stability, biocompatibility, targeting ability, and an enhanced permeability and retention (EPR) effect (10). Due to the EPR effect, nanoparticles preferentially accumulate more extensively in tumor tissue than in healthy tissue, which encourages the extravasation of nanochemotherapeutics (10). However, aberrant tumor vasculature, dense stroma and high interstitial pressure structure restrict the homogenous intra-tumoral distribution and limit the diagnostic and therapeutic efficacy of nanotherapeutics (11).

Additionally, the uneven biodistribution of nanoparticle-based formulations in healthy stromal cells disrupts the normal interaction between the tumor and stroma (12). The transvascular and interstitial transport of nanotherapeutics, which is highly impacted by various components of the tumor microenvironment, is crucial for the efficient delivery of drugs to solid tumors. The poor delivery of nanoparticles into solid tumors is crucial to the development of cancer nanomedicine. This has detrimental effects on the cost, manufacture, toxicity, imaging, and therapeutic efficacy on cancer nanomedicine (13).

Overall, the biodistribution of nanochemotherapeutics and their biological effects are influenced by the tumor microenvironment. Transport of nanoparticles and macromolecules into and within solid tumors is hampered by several tumor-specific pathophysiological characteristics. First, in contrast to blood vessels within normal tissues, tumor vasculature is structurally abnormal and functionally compromised (14). Blood vessels in the TME show increased production of proangiogenic growth factors, such as VEGFR-2 (vascular endothelial growth factor receptor-2), which results in the creation of immature arteries unable to maintain appropriate perfusion of tumor tissues, is the direct cause of this increased permeability (14, 15). Second, expression levels of cytokines like TGF-1 by tumor and stromal cells stimulates collagen synthesis in the tumor interstitial space, resulting in the development of a fibrous extracellular matrix (ECM) (16). The ECM prevents diffusion and limits the extravasation of nanoparticles into the tumor parenchyma (12). Third, tumor hypoxia results in an abnormal vasculature and areas with inadequate blood circulation and modifies the ECM by rendering it stiff and fibrotic. This hinders the ability of nano-based therapies to access different tumor-related areas (17, 18). Fourth, it has been demonstrated that nanoparticles can be taken up by immune cells, including macrophages and monocytes in the blood, tissues, and tumors, thereby limiting the concentration of nanoparticles that are delivered to cancer cells (19, 20).

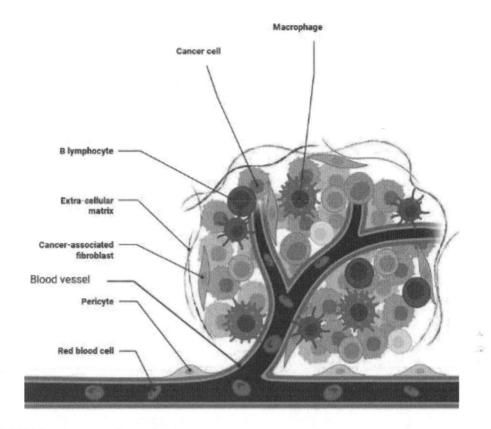

FIGURE 1.1 The tumor microenvironment.

Designing innovative nanomedical technologies to enable more effective and homogeneous delivery into solid tumors has garnered much interest recently. However, approaches that modify the TME to make it more receptive to nanomedicine delivery have received less attention and present a novel strategy to investigate combination approaches for cancer nanomedicine. Researchers have examined methods to modify the TME by normalizing the extracellular matrix and tumor vasculature to treat cancer more effectively (12). The clinical translation of nanomedicine will be accelerated by finding solutions to the nanoparticle delivery challenge.

1.2 HOW THE TME INFLUENCES TUMOR PROGRESSION

Growing evidence alludes to the critical function of the TME in influencing chemotherapeutic resistance, aberrant cell proliferation, metastasis, and tumor progression. In clinical oncology, the TME significantly influences the efficacy of chemotherapeutic treatments (11, 12, 17, 21). To develop successful treatment approaches to combat acquired drug resistance, suppress tumor development, and inhibit metastasis, it is crucial to have a detailed understanding of tumor biology and the mechanisms by which individual components of the TME influence tumor progression. The "seed and soil" theory of carcinogenesis presented by Stephen Paget states that a tumor tends to metastasize to a secondary site in the body that has a similar environment to that of the primary tumor location (22, 23). This theory emphasizes the critical influence of the tumor microenvironment in tumor development and progression. The following section outlines various components of the TME, how they influence cancer development and progression, how they interact with nanoparticles, and the nanotherapeutic strategies that have been harnessed to target these TME components.

1.2.1 TUMOR VASCULATURE AND TUMOR ENDOTHELIAL CELLS

Blood vessels are responsible for transporting nutrients and oxygen to and removing metabolic waste products from tissues all over the body. Tumors are vascularized markedly differently from normal healthy tissues. The atypical properties of tumor blood vessels enable both passive and active targeting by nanoparticles (NPs). In active targeting, NPs bind overexpressed receptor molecules, such as specific integrins or VEGF to enter the tumor microenvironment, while passive nanoparticles rely on their inherent characteristics. Tumor blood vessels are disorganized, not following the hierarchical branching pattern of blood vessels that is seen in normal tissues (24). Additionally, abnormal spaces found between adjacent tumor endothelial cells cause hemorrhage. Most tumor cells are relatively far away from any blood vessels due to the rapid growth of many cancers. By limiting the oxygen supply, this separation from blood arteries causes a hypoxic tumor microenvironment. Next, hypoxia triggers the high expression of VEGF, which results in increased interstitial fluid pressure and increased vessel permeability. Normal blood vessel growth and development is governed by a fine equilibrium between antiangiogenic and proangiogenic molecules. However, the fast-growing tumor blood vessels stimulated by predominantly proangiogenic factors are observed to be unusually fragile, irregularly shaped, dilated, convoluted, and extremely porous (25). Due to this anomaly, the tumor's vascular network is chaotic and convoluted and has a propensity to restrict the blood flow to downstream blood vessels. Additionally, the non-laminar blood flow, porous character, and diverse structure of the blood vessels frequently lead to poor blood distribution in tumors. As a result, poorly perfused tumor areas show hypoxia, low extracellular acidic pH, and high interstitial fluid pressure and are more difficult to access with chemotherapy (26, 27).

1.2.1.1 Targeting the Tumor Vasculature

Overexpression of certain proangiogenic factors, such as VEGF, PDGF, and fibroblast growth factor (FGF), results in disorderly structural development in these newly created tumor blood vessels (28). Hence, strategies to target such factors have been developed to restore tumor vasculature and curb tumor progression.

For example, the monoclonal antibody bevacizumab is an inhibitor of all the isoforms of VEGF-A (29) and has been used for the treatment of several types of cancer (30). However, due to limitations such as poor tumor penetration and the development of drug resistance, nanoparticle strategies have been harnessed to improve its efficacy. In one study, bevacizumab-loaded albumin nanoparticles were synthesized and used to treat colorectal cancer in a xenograft mouse model of colorectal cancer (31). Both forms of bevacizumab—free and nanoencapsulated—caused a 50% reduction in tumor volume when compared to controls. However, micro PET imaging investigation revealed that bevacizumab-loaded nanoparticles (B-NP-PEG) had a higher efficacy than free bevacizumab in terms of reducing the volume of the glycolysis and metabolic tumor. These findings concurred with B-NP-PEG's capacity to boost intratumor bevacizumab levels by around a factor of four when compared to the traditional formulation.

Parallel to this, B-NP-PEG revealed six times less bevacizumab in blood than the antibody's aqueous formulation, indicating a lower likelihood of any potential negative side effects. By further analyzing angiogenesis through CD31 staining, the findings indicated that nanoencapsulated bevacizumab significantly decreased angiogenesis compared with the control (31). Imatinib mesylate (IMA) has been shown to normalize A549-derived tumor vasculature by blocking the signaling of platelet-derived growth factor (PDGF). It is interesting to note that IMA treatment increased the formation of micelles around 23 nm while drastically reducing the accumulation of nanoparticles with a size of approximately 110 nm. Additionally, IMA treatment reduced the number of NPs that were distributed inside tumors while increasing the number of micelles that had a more uniform distribution. The anti-tumor efficacy investigation also proved that the efficacy of paclitaxel-laden nanomicelles might be greatly enhanced by IMA pretreatment. By normalizing the tumor vasculature, the endothelial gaps could be minimized, which could restrict the shedding of malignant tumor cells into blood vessels and limit the potential for metastasis (32).

1.2.2 Cancer-Associated Fibroblasts

Quiescent fibroblasts are resilient cells that are ubiquitously present throughout the body. They can become stimulated to differentiate into myofibroblasts, also known as cancer-associated fibroblasts (CAFs). CAFs are a heterogenous population of cells of tumor stroma that play a significant role in cancer development through the regulation of chemoresistance, angiogenesis, metabolism, and metastasis (33, 34). CAFs are also known to provide structural support for tumor cells and, depending on the context, can contribute to either pro- or antitumorigenic roles.

CAFs are also responsible for producing most of the components of the ECM and are known to build and maintain the various components of the tumor microenvironment and secrete enzymes responsible for remodeling the ECM (36).

CAFs can promote tumorigenesis, invasion, and metastasis by releasing cytokines and growth factors into the bloodstream. Furthermore, CAFs were reported to induce resistance to anticancer treatments on cancer cells and support the TME's immunosuppressive milieu. Researchers further found that growth factors like TGF-1 induced the CAF-mediated epithelial-to-mesenchymal transition in bladder cancer cells (37). By increasing the production of the soluble cytokine macrophage colony-stimulating factor, researchers demonstrated that CAF polarized the tumor-promoting M2 macrophage phenotype, which in turn promoted the growth, invasion, and metastasis of PDAC cells (38).

1.2.2.1 Targeting CAFs

Several strategies have been developed to target CAFs and CAF-activated molecular signaling pathways to curb tumor progression. The presence of a CAF-rich tumor microenvironment has been suggested to induce chemoresistance to abraxane in pancreatic ductal adenocarcinoma (PDAC) (39) When compared to gemcitabine alone, a nanoparticle albumin-bound paclitaxel (abraxane) and gemcitabine combination therapy decreases the amount of CAFs in PDAC while enhancing response rate, disease-free survival, and overall survival (40). In a recent study, researchers found that ovarian cancer and TME cells activated ovarian CAFs, while 20 nm gold nanoparticles (GNPs) prevented CAF activation, as shown by alterations in cell migration, morphology, and expressed molecular markers. The GNPs achieved their therapeutic effects by changing the expression or secretion of several proteins released by ovarian cancer cells and TME cells involved in fibroblast activation or inactivation, such as PDGF, TGF-1, TSP1, and uPA (41). Compelling evidence has indicated that near-infrared photothermal induced therapy (NIR-PIT) targeting cancer cells induces a phenomenon known as the super-enhanced permeation and retention (SUPR) effect, a markedly enhanced form of the EPR effect that encompasses the death of vascular smooth muscle cells and pericytes, resulting in widespread extravasation of nanotherapeutics into tumors (42).

Researchers examined both in vitro and in vivo whether CAF-PIT treatment could eliminate the stimulation of CAFs and overcome drug resistance to 5-fluorouracil in esophageal cancer. Cancer cells developed chemoradiotherapy resistance through the activation of CAF in vitro and resistance to 5-fluorouracil (FU) in CAF-co-inoculated murine tumor models. In vitro, CAF stimulation enhanced cancer cell invasion/migration and stemness, which could be reversed by removing CAF stimulus. In vitro, CAFs-PIT showed a very selective effect on CAFs. The combination of 5-FU with NIR-PIT successfully produced a tumor decrease of 70.9%, but 5-FU alone only managed to reduce tumors by 13.3%, as shown by CAF removal by CAFs-PIT in vivo. This suggests that 5-FU sensitivity was restored, and CAFs-PIT may be able to overcome therapeutic resistance by eliminating CAFs.

Fibroblast activation protein (FAP) is a type II integral membrane serine protease that is mainly expressed in 90% of human malignancies and is highly expressed by CAFs (43). To enhance CAF targeting properties, researchers developed photosensitive ZnF16Pc-encapsulated nanoparticle cages based on photoimmunotherapy (nano-PIT) coupled to an anti-FAP single-chain variable fragments (anti-FAP-scFv) ligand (44). The nano-PIT selectively eliminated CAFs, resulting in dramatically increased T cell infiltration and effective tumor suppression (44). In another study in which bilateral 4T1 mammary carcinoma tumor models were treated with photodynamic nano-PIT therapy, irradiated tumors displayed a considerably lower expression of positive SMA and collagen staining two days following photodynamic therapy when compared to the unirradiated tumors (45). Additionally, it was shown that the side tumors that had been exposed to radiation had a greater accumulation of macromolecules and nanoparticles (45). These findings showed that nano-PIT-anti-FAP ligand-mediated elimination of CAFs may result in the degradation of the ECM, allowing for deeper penetration of nanoparticles.

1.2.3 Tumor-Associated Macrophages and Other Immune Cells

Macrophages are a diverse group of resident and recruited immune cells that participate in hemostasis and host defense. Through their normal functions, macrophages should decrease tumors. Nevertheless, immune reprogramming of the tumor has shown that tumor-associated macrophages (TAMs) promote cancer development, drive tumorigenesis, and are a poor prognostic marker in many tumor types (46, 47). Macrophages that infiltrate tumors are known as TAMs and are derived from the differentiation of monocytes and recruited to tumor tissue in response to chemokine expression (48).

TAMs constitute an essential part of the TME and are known to influence tumor initiation and malignant progression through enhancing tumor growth, angiogenesis, immune suppression, metastasis, and chemoresistance (49, 50). A high population of TAMs is associated with poor prognosis in many types of cancers (51). Mounting evidence shows that several of the clinically approved chemotherapy drugs have specialized properties that influence the recruitment, polarization, and tumor-promoting functions of macrophages in tumor microenvironments (52–54). Macrophages are known to primarily differentiate into two main subsets that are known as the classically activated, pro-inflammatory M1 or the alternatively activated, anti-inflammatory M2 macrophages (55).

In carcinogenesis, TAMs are largely polarized into the tumor-promoting M2 macrophage phenotype, which has less cytotoxic effects than M1 macrophages. Moreover, M2 macrophages use a multiplicity of mediators to promote increased cancer cell proliferation, migration, immunosuppression, angiogenesis, and resistance to chemotherapy (55–59). Tumor angiogenesis is a hallmark for carcinogenesis (60). Mounting evidence shows that tumor angiogenesis can be promoted by TAM-secreted proteins such as VEGF, MMPs, interleukin-8, plasmin, urokinase-type plasminogen activator, thymidine phosphorylates, biosynthesis class F protein, basic fibroblast growth factor (bFGF), and phosphatidylinositol glycan (61–63).

TAMs have also been proven to stimulate metastasis by inducing epithelial-to-mesenchymal transition (EMT) of tumor cells through the activation of the EMT signaling pathway and remodeling of ECM in the TME. TAMs induced EMT by secreting proteins such as transforming growth factor-β (TGF-β) and tumor necrosis factor alpha (TNF-α) (62). Proteases such as matrix metalloproteinases (MMPs) and cathepsins are known to be involved in degrading the protein component of the extracellular matrix, thereby inducing tumor EMT, invasion, and metastasis and developing chemoresistance (62). IL-4 is an anti-inflammatory cytokine reported to stimulate the proteolytic action of cathepsin and enhance pancreatic and breast cancer proliferation and invasion mediated by Toll-like receptor 2 signaling (52). TAMs also induce immunosuppression in renal cell carcinoma by the expression of cytokines, which include TGF-β, IL-6, TNF-α, and IL-1β (64), while other TAM-induced proinflammatory cytokines such as IL-23, IL-17, and IL-6 stimulate tumor-mediated inflammation, which promotes tumor proliferation (65, 66).

1.2.3.1 Targeting TAMs

Cancer nanotherapy employs two primary strategies to target TAMs: first, transforming M2 to M1 TAMs to boost antitumor immune protection and second, eliminating M2 TAMs. By applying the first strategy, researchers employed hyaluronic acid (HA) to alter manganese dioxide (MnO_2) nanoparticles, hence converting the pro-tumorigenic M2 TAMs into anti-tumorigenic M1 TAMs. This improved the capacity of MnO_2 to decrease hypoxia and suppress chemoresistance in 4T1 breast cancer murine models (67). Moreover, optical imaging using noninvasive near-infrared fluorescence showed excellent biodistribution of the MnO_2 nanoparticles to the tumor site.

Because M2 TAMs exhibit elevated levels of the mannose receptor, researchers developed a system of mannose-altered nanoparticles that can be shed from PEG to specifically target M2 TAMs. Mannose-modified poly (lactic-co-glycolic acid) (PLGA) NPs were conjugated to sheddable PEG using an acid-responsive linker. Acid-responsive PEG was shed in the acidic TME to expose PEG and mannose-conjugated PLGA, enabling the M2 TAMs to ingest the particles but blocking normal macrophage absorption in the mononuclear phagocyte cellular system organs due to their neutral pH (68).

1.2.4 THE EXTRACELLULAR MATRIX

Tumor invasion and metastasis, a critical hallmark of cancer progression, is a multistep process that involves proteolytic breakdown of extracellular matrix, modification of the interactions between cells and the ECM, and cancer cell migration through the basement membrane. This proteolytic cleavage is highly stimulated by matrix metalloproteinases that are usually highly expressed in the cancer-associated ECM (69). The ECM is a complex structural network consisting of several macromolecules; primarily fibrous proteins and proteoglycans, arranged in a cell- or tissue-selective manner (70). The intricate combination of components of the ECM such as collagens, proteoglycans, elastin, and cell-binding glycoproteins influence the structural and mechanical properties of tissues.

Although many stromal and tumor cell types can synthesize ECM components, CAFs are primarily responsible for the synthesis, secretion, assembly, and modification of the composition and structural organization of the ECM (36). The tumor-associated ECM usually exhibits an altered structural architecture and deregulated post-translational modifications of ECM-associated proteins such as collagens and matrix metalloproteinases. The ECM also secretes a plethora of bioactive molecules that influence organ development, proliferation, differentiation, migration, tissue homeostasis, inflammation, and various disease processes (71).

The tumor ECM shows enhanced stiffness and density and is constantly remodeled by stromal and tumor cells through the synthesis and degradation of ECM proteins (36, 72). Each component in the ECM plays a role in tumor progression. Overall, the stiffness of tumor-associated ECM is known to be a barrier to effective delivery of drugs to tumors. Collagen, the most abundant protein in the ECM, interacts with several proteins in the ECM such as matrix metalloproteinases, fibronectin, hyaluronic acid, and laminin to promote angiogenesis, proliferation, invasion, and metastasis in ovarian cancer (73).

1.2.4.1 Targeting the ECM

The ECM promotes carcinogenesis through several mechanisms and is pivotal for the initiation of metastasis, which accounts for about 90% of cancer-related mortality. Hence, targeting the ECM is a crucial strategy to inhibit cancer progression. An essential constituent of the tumor ECM, laminin, can be targeted using laminin-mimicking molecules. In one study, researchers developed a laminin-mimicking peptide to construct an artificial ECM in order to inhibit tumor invasion (74). They showed that the laminin-mimicking peptide could self-assemble through hydrophobic interactions to form nanoparticles. Nanoparticles could be converted into nanofibers after binding to laminin and integrin receptors on tumor cells (74). The synthetic ECM laminin NP mimics inhibited metastasis in a number of solid tumor models due to their long retention times and ability to accumulate at tumor sites for three days (74). These findings suggest that biomimetic nanoparticles provide creative approaches that can fortify the ECM by controlling metastasis and adhesion.

Mechanical insights into the composition and architecture of the normal ECM are crucial and have inspired innovative strategies of targeting the tumor ECM. The stiffness of the ECM, which is associated with cancer progression, can be induced by high activity of lysyl oxidase (LOX) and increased integrin signaling mediated by collagen-altering enzymes such as LOX, P4HA1, P4HA2, and PLOD2 in the TME (75, 76). In one study, researchers monitored the assembly of collagen in a nano-based 3D matrix using advanced microscopy (77). The authors identified a strategy by which the natural alignment of endogenous collagen could be modified by LOXL2 antibodies without changing the composition of the ECM. They showed that the LOXL2 antibody-mediated modification of collagen morphology abrogated the migration, adhesion, and invasion of bladder cancer cells and further decreased tumor growth in a breast cancer mouse model (77). Their findings suggested that nanoparticle-based modulators of collagen fibril alignment that regulate the ECM assembly can inhibit the invasion of tumor cells.

A similar study showed that breast cancer tumor growth could be inhibited by modifying the ECM using lysyl oxidase-targeting poly(d,l-lactide-co-glycolide) nanoparticles. The researchers demonstrated that the synthesized poly(d,l-lactide-co-glycolide) nanoparticles coated with LOX-inhibiting antibodies (LOXAbNPs) suppressed the growth of breast cancer cells in vitro and in mouse xenograft models more efficiently than soluble anti-LOX antibody (78). They further proved that the anticancer properties of the LOXAbNPs correlated with their ability to impede in vivo collagen cross-linking. Moreover, these findings established that these tumor-targeting LOXAbNPs can incorporate imaging agents to visualize TME for cancer diagnostics (78).

Cyclooxygenase-2 (COX-2) is closely associated with ECM formation and tumor-associated angiogenesis, which are pivotal cancer hallmarks. In one study, the authors used celecoxib, a COX-2 inhibitor, to improve the delivery of paclitaxel-loaded nanomicelles to tumors in a nude mouse tumor model and enhance its therapeutic efficacy. Celecoxib modulated the TME by reducing cancer-associated fibroblasts and disrupting fibronectin bundles, thereby distorting the ECM, and by normalizing the tumor vasculature and improving tumor perfusion. Interestingly, Celecoxib significantly increased the accumulation and penetration of 22-nm micelles in vivo (not 100-nm nanoparticles) in tumors and enhanced the efficacy of paclitaxel-encapsulated nanomicelles in murine tumor models (79).

1.2.5 Deregulated pH of the TME/Metabolic Alterations in the Tumor Microenvironment

Due to the high biosynthetic and bioenergetic demand of tumors, energy metabolism of normal cells differs markedly from that of tumor cells. While most normal cells (except cells and skeletal myocytes during high-intensity physical activity) utilize mainly oxidative phosphorylation to produce energy, the high metabolic requirement in most malignant cells stimulates high glucose uptake and employs enhanced glycolysis, which produces considerable amounts of lactate even in the presence of ample oxygen supply. This bioenergetic phenotype is known as the Warburg effect or aerobic glycolysis (80). The increased generation of lactic acid coupled with reduced vascular perfusion leads to an acidic tumor microenvironment with pH values of 6.5–6.9 compared with pH 7.2–7.4 in non-cancerous tissues (81). Moreover, the hydration of surplus CO_2 characteristic of the TME generates carbonic anhydrase, a rich source of hydrogen ions, which further acidifies the TME (82).

The acidic TME has been shown to contribute to tumor progression by inducing metastasis through the degradation of the ECM. In another study performed in nude mice, extracellular acidification was demonstrated to enhance metastasis of melanoma cells (83). Further evidence has been provided that the acidic TME influences the polarization of macrophages from the M1 to the M2 tumor-promoting phenotype, which in turn induced the proliferation of prostate cancer cells (84). Tumor acidity has also been shown to be involved in resistance to chemotherapy and some targeted therapies (85, 86). Tumor acidity can be targeted utilizing pH-sensitive drug delivery systems, such as drugs that interfere with the generation of H+ or bicarbonate transporters.

1.2.5.1 Targeting TME Acidity

Targeting tumor acidity is one of the strategies of passive targeting of nanotherapeutics to the tumor microenvironment. pH-responsive nanomaterials are relatively stable in the blood and normal tissue but are unstable in acidic tumor microenvironment. This leads to changes in the extent of ionization in pH-sensitive groups, thereby inducing changes in the nanomaterials that release targeted anti-cancer therapeutics (87). When the pH in the tumor microenvironment decreases, pH-responsive polymer nanoparticles carrying anti-tumor therapeutics can alter their structures or characteristics, offering precise targeted tumor therapy.

In one study, researchers developed doxorubicin and verapamil-loaded liposomal nanoparticles conjugated with Her-2 antibody and incorporating the pH-sensitive compound malachite green carbinol base as an anti-breast cancer therapeutic strategy. Malachite green carbinol base was converted into a carbocationic molecule when exposed to an acidic environment, which triggered the liposome to release doxorubicin only at the intended target site (88). A P-gp inhibitor, Verapamil, was included to reverse the effects of multi-drug resistance in both in vitro and in vivo breast cancer models, which further enhanced the anticancer activity of the formulation (88).

In another study, the investigators designed DOX-loaded liposomes that were pH responsive and coated with glycol chitosan, demonstrating how the cationic character of the liposomes changes from their anionic state at physiological pH to their cationic state in an acidic extracellular TME. In animals bearing the T6–17 tumor, this charge difference facilitated the cellular internalization of liposomes, thereby increasing their anti-cancer effect (89).

1.2.6 Hypoxia in the TME

A complex vascular network is responsible for the delivery of oxygen and nutrients to normal and tumor cells. Since the diffusion range of oxygen from blood vessels to cells is limited to 100 to 200 μM, tumor cells must generate more blood vessels to be able to exceed this size limit (90). Therefore, the disparity between rapidly proliferating tumor cells and adequate blood supply causes the TME to become increasingly hypoxic. Hypoxia therefore results due to increased oxygen demand from proliferating tumor cells, which exceeds its supply (91).

Tumor hypoxia inhibits apoptosis, stimulates immunosuppression, and supports epithelial-to-mesenchymal transition, thereby enhancing invasion and metastasis (92). Hypoxia also develops an oxygen gradient within the tumor, which promotes tumor heterogeneity and fosters a more malignant cancer phenotype. Moreover, it has been demonstrated that hypoxia triggers the activity of hypoxia inducible factor 1-alpha (HIF-1-alpha), which in turn mediates the upregulation of drug efflux transporter proteins (93, 94). HIF-1 has been found to be overexpressed in a variety of human malignancies, and targeting it has been demonstrated to be a potential anticancer therapeutic strategy (93, 95).

1.2.6.1 Targeting Hypoxia

Using nanoparticles to suppress hypoxia has also been extensively investigated. One method to prevent a hypoxic environment is to knock down the HIF-1 gene. Numerous studies have demonstrated the potency of HIF-1 siRNA-containing complex nanosystems in overcoming cancer treatment resistance (96, 97). Hypoxia-sensitive NPs can provide superior platforms for combination therapy, such as PEGylated nanoparticles for targeting tumor hypoxia (98, 99).

Additionally, the transcriptional activity of the key regulator of hypoxia, HIF-1-α, depends on heat shock protein 90 (HSP90); therefore, inhibiting HSP90 can similarly suppress the expression

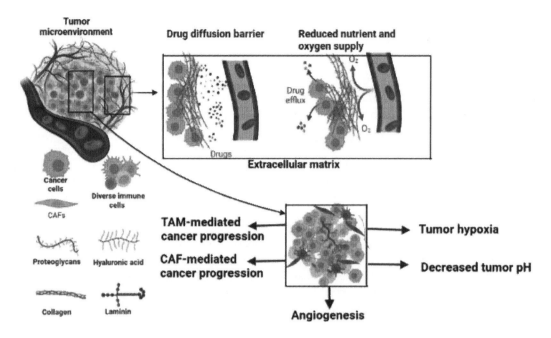

FIGURE 1.2 Roles of the TME in tumorigenesis and response to cancer therapy.

of HIF-1 (100). It has been demonstrated that the HSP90 inhibitor in 17AAG-loaded photo-therapeutic nanoporphyrin NPs significantly improves the treatment of bladder cancer (101).

1.3 STRATEGIES TO TARGET THE TME

The tumor microenvironment possesses malignant hallmarks that have been targeted for anticancer therapy using genetic and pharmacological methods. NPs can be targeted actively or passively to tumor cells to improve their therapeutic potential and mitigate systemic toxicity. To achieve passive targeting, NPs are designed to target tumor and TME vulnerabilities such as low pH and hypoxia and enhanced permeability and retention effect (which utilizes superior blood vessel permeability and reduced lymphatic drainage of cancer cells). Active targeting exploits several types of receptors such as folate, transferrin, epidermal growth factor (EGFR), and glycoproteins, which are usually overexpressed on many types of cancer cells (102).

Therapeutic approaches employed to target the TME include both genetic and pharmacological methods. Genetic approaches include gene editing or gene knockdown technologies that can reprogram the TME by targeting specific upregulated proteins in cellular components of the TME. Pharmacological methods used in this review will focus on how components of the TME have been successfully targeted using nanotherapeutics as anticancer therapeutic strategies. Two primary strategies are generally employed to target the TME. First, the TME could be primed or sensitized to enable more efficient uptake of nanochemotherapeutics. Second, nanocarriers could be targeted to cancer cells based on the expression of specific receptors or enzymes.

1.3.1 ANTICANCER NANOPARTICLES TARGETING THE TME

The small size of nanomaterials coupled with varied methods of synthesis confers advantageous characteristics such as high surface-area-to-volume ratio, targeted delivery, multi-functionality, controlled release, enhanced EPR effect, and diverse surface modification potential (103). Their small sizes afford them the ability to access remote areas of the body such as the brain and

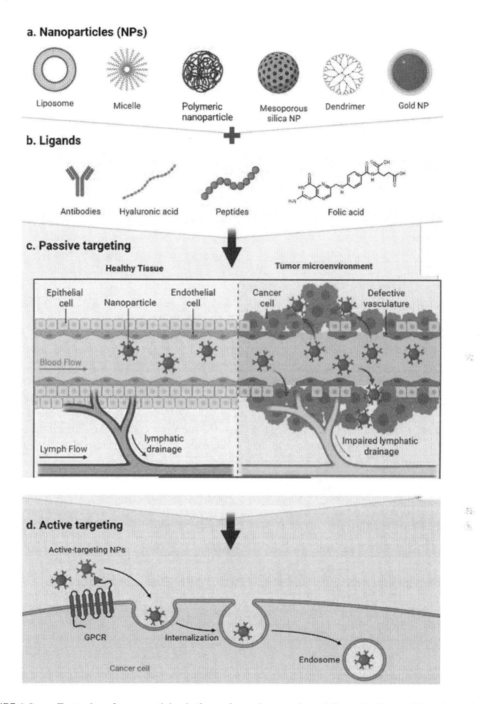

FIGURE 1.3 a. Examples of nanoparticle platforms for anticancer drug delivery. b. Types of ligands used to functionalize nanoparticles for targeted anticancer drug delivery. c. An example of active targeting, in which a ligand-decorated nanoparticle is targeted at a G-protein coupled receptor.

certain types of solid tumors, making them suitable for biomedical applications such as diagnosis and therapy (104). Nanoparticles display numerous physicochemical properties such as composition, size, shape, structure, and surface chemistry that are responsible for their unique interactions with different components of the TME. The sizes, shapes, and surface properties of therapeutic nanoparticles are often critical because these factors significantly influence the efficiency of the transport of nanotherapeutics and, consequently, determine the therapeutic potential of the drugs (105, 106).

To deliver drugs efficiently and achieve the EPR effect, NPs that range from 10 to 100 nm in diameter are typically regarded as acceptable. Smaller NPs from 1–2 nm are known to readily leak out from blood vessels and are toxic to normal cells, while those less than 10 nm in diameter are easily filtered by the kidneys (107), whereas particles above 100 nm are more likely to be eliminated from circulation by phagocytic cells (108). Moreover, the surface characteristics of NPs can influence their bioavailability, biocompatibility, and half-life. Nanoparticles coated with hydrophilic materials such as polyethylene glycol possess improved biocompatibility and colloidal stability in biological fluids and reduce opsonization and hence prevent clearance by the immune system (109, 110). Nanoparticles are usually modified to become hydrophilic, thereby extending the duration that drugs remain in circulation and enhancing tumor penetration and accumulation (106, 109). The combination of varied nanoparticle properties determines how therapeutically effective they are in the treatment of cancer.

The Food and Drug Administration (FDA) has approved a few nanoparticle drugs that have been developed for the treatment of multiple types of solid tumors, including non-small cell lung cancer (NSCLC), breast cancer, and metastatic pancreatic adenocarcinoma. They include liposomal daunorubicin (DaunoXome)(111), PEGylated liposomal doxorubicin (Doxil)(112), paclitaxel albumin-bound particles (Abraxane) (113), liposomal irinotecan (Onivyde) (114), and a liposome-encapsulated doxorubicin citrate (Myocet) (115).

Some severe side effects have been noted in some of these nanoformulations. For example, although Onivyde showed improved therapeutic index/elevated biodistribution of irinotecan and increased pharmacokinetic qualities, the formulation was known to cause serious adverse effects, including neutropenia and gastrointestinal toxicity that resulted in severe diarrhea (116). Although the introduction of these nano-therapeutics has enhanced treatment responsiveness, these treatments' potential benefits are rarely fully achieved. Given the variability of the components that make up the TME, nanoparticles may not be as effective as expected because of the lowered tumor absorption of these substances (117).

1.4 CHALLENGES OF NANOPARTICLE-BASED THERAPIES

Although nanotherapeutics are generally targeted actively or passively to the tumor or tumor microenvironment, several inadvertent effects could impact their toxicity and efficacy.

1.4.1 The Challenge of Tumor Heterogeneity

Tumor heterogeneity refers to variations between cancer cells inside a single tumor, variations between a primary and a secondary tumor, or variations between tumors of the same type in different people. These variations could be caused by differential gene expression of malignant cells of the tumor as some cancer cells may harbor mutations that are not present other cells of the same tumor. The degree of tumor heterogeneity may have a significant impact on the diagnosis and treatment of cancer and its sensitivity or resistance to therapy. Phenomena that distinguish a tumor microenvironment from normal tissue such as the EPR effect are also known to be non-universal. There are also varying degrees of expression of specific cell surface receptors on various cellular components of the tumor microenvironment that can be targeted. This prompts the need for the development of more personalized TME nanotherapies that target specific subsets of cancer patients.

1.4.2 Nanoparticle-Induced Aberrant Gene Expression

There is a paucity of data about the effects and mechanisms of NP-related characteristics on cellular behavior due to the diversity of NP size, shape, and surface characteristics. Cell–nanoparticle interactions, which rely on the physicochemical properties of nanoparticles and unique cellular contexts, are critical in nanomedicine because they determine both the efficacy and toxicity of the nanomaterials. To understand cell–nanoparticle interactions in living systems, we require knowledge regarding the dynamic activity of nanoparticles during cellular uptake, their intracellular trafficking, and mutual interactions with cell organelles (104, 118).

The effect of several nanoparticle characteristics on cellular structure and function has also been investigated in vitro with a validated multiparametric high-content imaging method (119). In one research study, the authors used high-content imaging and gene expression analysis to investigate cell–nanoparticle interactions that encompassed nanoparticle uptake, gene expression analysis, cell viability and morphology, autophagy, Reactive oxygen species (ROS) induction, mitochondrial health, and cell cycle progression in mouse-derived multipotent neural stem cells (C17.2) and human umbilical cord-derived vascular endothelial cells (HUVEC) (119). The underlying mechanisms were then clarified by in-depth gene expression research and high-content cell cycle phase analyses. The results showed a substantial relationship between membrane damage and the induction of autophagy and the degree of NP surface hydrophobicity, which is stronger than the impact of surface charge for charges between 50 and +20 mV (119). Similar studies could follow the same procedure to elucidate the underlying mechanisms of cell–nanoparticle interactions and even further enhance the method by assessing extra parameters, such as NP size, shape, and specific surface functionalization. This could offer a widely applicable format for researching cell–NP interactions under incredibly repeatable circumstances.

Specifically, they evaluated the effect of hydrophobicity and surface charge of gold nanoparticles on cellular structure and function. An obvious challenge for such an interesting study would be the difficulty of replicating the methodology in an in vivo setting: a problem that is common in much in vitro nanotoxicology. Moreover, lack of long-term research in the existing methodology is a concern that is common to in vitro nanotoxicology. The exponential dilution of nanoparticles during cell division in rapidly growing cells renders the investigating of long-term effects in vitro challenging (120). The complexity, heterogeneity, and dynamism of the TME further requires that the data be interpreted cautiously.

To evaluate the pharmacokinetics, biodistribution, efficacy, and safety of NPs to gauge their in vivo performance, the use of animal models is compulsory. Various animal models such as patient-derived xenografts, genetically modified mice, and cell line-based xenografts are currently used to evaluate nanoparticles. Interestingly, EPR is typically less consistent in cancer patients than it is in animal models, and no single model can entirely replicate all features of cancer.

1.4.3 INADVERTENT EPIGENETIC AND POST-TRANSLATIONAL MODIFICATIONS

The challenges of using nanochemotherapeutics include their ability to induce inadvertent genetic and epigenetic changes, some of which are responsible for the induction of toxicity. Some studies have proposed that nanotoxicity results from aberrant gene expression of genes involved in major cellular pathways (106), while other studies have shown that inadvertent epigenetic modifications can be induced by nanoparticle drug formulations (107). Nanoparticles are known to induce inadvertent epigenetic modifications such as histone modifications, DNA methylation, and miRNA-induced control of gene expression, which do interfere with biological pathways such as inflammation, oxidative stress, and MAPK pathway activation (121).

ROS production and oxidative stress may be driven by NP-mediated epigenetic alterations. Recent studies showed that the promoters of critical regulators of oxidative stress such as glutathione reductase and glutathione peroxidase were hypermethylated and hypomethylated, respectively, in BALB/c mice treated with gold NPs (122). Although nanoparticle-mediated genetic and epigenetic modifications could influence their systemic toxicity and efficacy, little is known about how NPs affect epigenetic modifications or whether their physicochemical characteristics influence epigenetic effects and how they do so.

1.4.4 Reproducible Nanoparticle Synthesis

The development of successful therapeutic nanoparticles is contingent on the identification of ideal physiochemical properties of their constituent materials. There is currently a significant amount of knowledge on the various factors that can lead to efficient cell targeting, immune evasion, tumor extravasation, and diffusion-controlled drug release and internalization (123). Nevertheless, due to the difficulty of efficiently synthesizing NP libraries with distinctive traits, systematic parallel screening of the plethora of nanoparticle properties remains challenging. Recently, microfluidic systems have garnered significant attention due to the rapid nanoparticle synthesis and self-assembly which yield nanoparticles with narrower size distribution, amendable physicochemical characteristics, and higher batch-to-batch reproducibility when compared to traditional bulk techniques, which typically form nanoparticles with high polydispersity indices (124, 125). In addition, the fabrication of highly dispersed NPs with highly uniform particle replication in terms of size, shape, chemical composition, surface characteristics, and drug loading has been made possible by an innovative technology known as particle replication in non-wetting template (PRINT) (126–128).

1.4.5 Nanoparticle Screening Strategies

Cancer cells do not proliferate and metastasize in isolation but through their dynamic interactions with cellular and non-cellular components of the TME, which include fibroblasts, macrophages, endothelial cells, and the ECM (8, 11, 12, 79). One of the strategies by which cancer cells demonstrate their ingenuity is by harnessing complex molecular and cellular pathways of the tumor microenvironment to support and promote their growth, metastasis, and resistance to therapy (79).

Profound insights into the mechanisms of cancer biology have been elucidated in 2D cell culture models that have laid the groundwork for further research. Nevertheless, the physiological relevance of the findings obtained from 2D cell culture models should be cautiously interpreted because molecular and cellular interactions are not fully recapitulated, which restricts our understanding of the interactions between cells and the TME. To understand the complex interactions between tumor cells and the TME, 2D in vitro methods are insufficient. More physiologically relevant models such as 3D culture platforms, animal tumor models, organoids, and lab-on-chip methods have been utilized to better simulate the TME and elucidate the mechanisms by which tumor cells interact with the TME (80, 81).

Tumor-on-a-chip technology is a cutting-edge approach that combines microfabrication, microfluidics, biomaterials research, and tissue engineering. It provides new prospects for creating and using functioning 3D in vitro human tumor models for oncology, immunotherapy, and drug discovery research. These microdevices allow researchers to conduct tightly regulated microscale studies of the interactions between tumor cells, immune cells, and cells in the TME, which are not possible with basic 2D cell cultures or model animals (81). To evaluate the therapeutic potential of NPs, it appears relevant to compare the behavior of NPs in biomimetic microdevices with NPs in animal models.

1.4.6 Scalable Manufacturing

Generally, when NP formation requires numerous steps or sophisticated technology, large-scale and reproducible synthesis becomes more challenging. Optimizing formulation parameters or changing formulation techniques is almost always required when transitioning from the laboratory to the clinic; therefore, early NP design and engineering should take scale-up aspects into account (128).

Liposomes and polymeric systems are simple NPs with desirable physicochemical features containing small-molecule APIs. These NPs can be easily scaled up using manufacturing unit operations that are easily accessible and often utilized in the pharmaceutical sector (123). However, scaling up more complicated nanomedicines that incorporate biological components or biological targeting ligands could present new problems and necessitate modifying current unit processes or creating brand-new production techniques. Such nanomedicines are composed of a mixture of two or more therapies and are created through layer-by-layer assembly or feature numerous functional units, such as theranostics or multistage systems (123).

1.5 CONCLUSIONS

The mechanism by which cancers become resistant to conventional treatments has garnered significant attention. Tumor progression, invasion, metastasis, and resistance to therapy have been strongly linked to the tumor microenvironment. A new era in anticancer therapeutics has been ushered in employing nanotechnology, in which numerous forms of NPs, including inorganic, organic, and hybrid NPs, including diverse targeting strategies have been extensively exploited. In comparison to conventional therapies, NP-based drug delivery methods have better stability, biocompatibility, pharmacokinetics, and tumor targeting abilities. Due to these benefits, nanopharmaceuticals are frequently used in chemotherapy, targeted therapeutics, gene therapy, radiation, and hyperthermia.

Additionally, nano-based delivery systems possess efficient platforms for combination cancer therapy, which assists in suppressing drug resistance mechanisms that include the overexpression of efflux transporters, dysfunctional apoptotic signaling pathways, and hypoxic and low-pH tumor microenvironments. The EPR effect is recognized to be dynamic, complex, and heterogeneous within the same tumor and in different tumors, allowing cancer nanotherapeutics to accumulate in solid tumors. The development of personalized nanomedicine could be sparked by the identification of EPR effect-associated biomarkers, allowing for targeted nanotherapeutics to cancer patients most likely to benefit from it. To fully harness the potential of targeting the tumor microenvironment in cancer therapy, unravelling the complexities of how biological processes are involved in the systemic delivery of nanotherapeutics to solid tumors is necessary. These processes include the interaction of nanoparticles with proteins, blood, and blood vessels; extravasation to and interaction with the perivascular TME; penetration of tumor tissue; NP uptake by tumor cells; and their intracellular trafficking. Moreover, optimizing and standardizing controllable and reproducible nanoparticle synthesis, scalable manufacturing, and nanoparticle screening methods are critical for clinical translation of nanoparticle-based drug delivery systems to the tumor microenvironment.

REFERENCES

1. Sung H, Ferlay J, Siegel RL, Laversanne M, Soerjomataram I, Jemal A, et al. Global cancer statistics 2020: GLOBOCAN estimates of incidence and mortality worldwide for 36 cancers in 185 countries. CA: A Cancer J Clin. 2021;71(3):209–249.
2. Zugazagoitia J, Guedes C, Ponce S, Ferrer I, Molina-Pinelo S, Paz-Ares L. Current challenges in cancer treatment. Clin Ther. 2016 Jul 1;38(7):1551–1566.
3. Król M, Pawłowski KM, Majchrzak K, Szyszko K, Motyl T. Why chemotherapy can fail? Pol J Vet Sci. 2010;13(2):399–406.
4. Muchmore JH, Wanebo HJ. Regional chemotherapy: Overview. Surg Oncol Clin North Am. 2008 Oct 1;17(4):709–730.
5. Kyle AH, Huxham LA, Yeoman DM, Minchinton AI. Limited tissue penetration of taxanes: A mechanism for resistance in solid tumors. Clin Cancer Res. 2007 May 1;13(9):2804–2810.
6. Primeau AJ, Rendon A, Hedley D, Lilge L, Tannock IF. The distribution of the anticancer drug doxorubicin in relation to blood vessels in solid tumors. Clin Cancer Res. 2005 Dec 15;11(24 Pt 1):8782–8788.
7. Schiller JH, Harrington D, Belani CP, Langer C, Sandler A, Krook J, et al. Comparison of four chemotherapy regimens for advanced non–small-cell lung cancer. New England J Med. 2002 Jan 10;346(2):92–98.
8. Yang M, Li J, Gu P, Fan X. The application of nanoparticles in cancer immunotherapy: Targeting tumor microenvironment. Bioact Mater. 2021 Jul 1;6(7):1973–1987.
9. Boyce MW, LaBonia GJ, Hummon AB, Lockett MR. Assessing chemotherapeutic effectiveness using a paper-based tumor model. Analyst. 2017 Jul 24;142(15):2819–2827.

10. Maeda H. Toward a full understanding of the EPR effect in primary and metastatic tumors as well as issues related to its heterogeneity. Adv Drug Deliv Rev. 2015 Aug 30;91:3–6.
11. Fernandes C, Suares D, Yergeri MC. Tumor microenvironment targeted nanotherapy. Front Pharmacol. 2018;9. [Internet] [cited 2022 Nov 16] Available from: www.frontiersin.org/articles/10.3389/fphar.2018.01230.
12. Stylianopoulos T, Munn LL, Jain RK. Reengineering the physical microenvironment of tumors to improve drug delivery and efficacy: From mathematical modeling to bench to bedside. Trends Cancer. 2018 Apr;4(4):292–319.
13. Wilhelm S, Tavares AJ, Dai Q, Ohta S, Audet J, Dvorak HF, et al. Analysis of nanoparticle delivery to tumours. Nat Rev Mater. 2016 Apr 26;1(5):1–12.
14. Jain RK. Normalization of tumor vasculature: An emerging concept in antiangiogenic therapy. Science. 2005 Jan 7;307(5706):58–62.
15. Huang Y, Kim BYS, Chan CK, Hahn SM, Weissman IL, Jiang W. Improving immune-vascular crosstalk for cancer immunotherapy. Nat Rev Immunol. 2018 Mar;18(3):195–203.
16. Wei Y, Kim TJ, Peng DH, Duan D, Gibbons DL, Yamauchi M, et al. Fibroblast-specific inhibition of TGF-β1 signaling attenuates lung and tumor fibrosis. J Clin Invest. 2017 Oct 2;127(10):3675–3688.
17. Jing X, Yang F, Shao C, Wei K, Xie M, Shen H, et al. Role of hypoxia in cancer therapy by regulating the tumor microenvironment. Mol Cancer. 2019 Nov 11;18(1):157.
18. Gilkes DM, Semenza GL, Wirtz D. Hypoxia and the extracellular matrix: Drivers of tumour metastasis. Nat Rev Cancer. 2014 Jun;14(6):430–439.
19. Qie Y, Yuan H, von Roemeling CA, Chen Y, Liu X, Shih KD, et al. Surface modification of nanoparticles enables selective evasion of phagocytic clearance by distinct macrophage phenotypes. Sci Rep. 2016 May 19;6(1):26269.
20. Lucas AT, White TF, Deal AM, Herity LB, Song G, Santos CM, et al. Profiling the relationship between tumor-associated macrophages and pharmacokinetics of liposomal agents in preclinical murine models. Nanomed. 2017 Feb;13(2):471–482.
21. Calcinotto A, Filipazzi P, Grioni M, Iero M, De Milito A, Ricupito A, et al. Modulation of microenvironment acidity reverses anergy in human and murine tumor-infiltrating T lymphocytes. Cancer Res. 2012;72(11):2746–2756.
22. Paget S. The distribution of secondary growths in cancer of the breast. The Lancet. 1889 Mar 23;133(3421):571–573.
23. Fidler IJ. The pathogenesis of cancer metastasis: The "seed and soil" hypothesis revisited. Nat Rev Cancer. 2003 Jun;3(6):453–458.
24. McDonald DM, Choyke PL. Imaging of angiogenesis: From microscope to clinic. Nat Med. 2003 Jun;9(6):713–725.
25. Geevarghese A, Herman IM. Pericyte-endothelial crosstalk: Implications and opportunities for advanced cellular therapies. Transl Res. 2014 Apr;163(4):296–306.
26. Jain RK. Normalizing tumor microenvironment to treat cancer: Bench to bedside to biomarkers. J Clin Oncol. 2013 Jun 10;31(17):2205–2218.
27. Belli C, Trapani D, Viale G, D'Amico P, Duso BA, Della Vigna P, et al. Targeting the microenvironment in solid tumors. Cancer Treat Rev. 2018 Apr;65:22–32.
28. Goel S, Duda DG, Xu L, Munn LL, Boucher Y, Fukumura D, et al. Normalization of the vasculature for treatment of cancer and other diseases. Physiol Rev. 2011 Jul;91(3):1071–1121.
29. Ellis LM. Mechanisms of action of bevacizumab as a component of therapy for metastatic colorectal cancer. Semin Oncol. 2006 Oct 1;33:S1–7.
30. Garcia J, Hurwitz HI, Sandler AB, Miles D, Coleman RL, Deurloo R, et al. Bevacizumab (Avastin®) in cancer treatment: A review of 15 years of clinical experience and future outlook. Cancer Treat Rev. 2020 Jun 1;86:102017.
31. Luis de Redín I, Expósito F, Agüeros M, Collantes M, Peñuelas I, Allemandi D, et al. In vivo efficacy of bevacizumab-loaded albumin nanoparticles in the treatment of colorectal cancer. Drug Deliv and Transl Res. 2020 Jun 1;10(3):635–645.
32. Zhang B, Hu Y, Pang Z. Modulating the tumor microenvironment to enhance tumor nanomedicine delivery. Front Pharmacol. 2017 Dec 22;8:952.
33. Sahai E, Astsaturov I, Cukierman E, DeNardo DG, Egeblad M, Evans RM, et al. A framework for advancing our understanding of cancer-associated fibroblasts. Nat Rev Cancer. 2020 Mar;20(3):174–186.
34. Kalluri R. The biology and function of fibroblasts in cancer. Nat Rev Cancer. 2016 Sep;16(9):582–598.
35. Liu T, Han C, Wang S, Fang P, Ma Z, Xu L, et al. Cancer-associated fibroblasts: An emerging target of anti-cancer immunotherapy. J Hematol Oncol. 2019 Aug 28;12(1):86.

36. Liu T, Zhou L, Li D, Andl T, Zhang Y. Cancer-associated fibroblasts build and secure the tumor microenvironment. Front Cell Dev Biol. 2019;7:60.
37. Zhuang J, Lu Q, Shen B, Huang X, Shen L, Zheng X, et al. TGFβ1 secreted by cancer-associated fibroblasts induces epithelial-mesenchymal transition of bladder cancer cells through lncRNA-ZEB2NAT. Sci Rep. 2015 Jul 8;5:11924.
38. Zhang A, Qian Y, Ye Z, Chen H, Xie H, Zhou L, et al. Cancer-associated fibroblasts promote M2 polarization of macrophages in pancreatic ductal adenocarcinoma. Cancer Med. 2017 Feb;6(2):463–470.
39. Alvarez R, Musteanu M, Garcia-Garcia E, Lopez-Casas PP, Megias D, Guerra C, et al. Stromal disrupting effects of nab-paclitaxel in pancreatic cancer. Br J Cancer. 2013 Aug;109(4):926–933.
40. Von Hoff DD, Ervin T, Arena FP, Chiorean EG, Infante J, Moore M, et al. Increased survival in pancreatic cancer with nab-paclitaxel plus gemcitabine. New England J Med. 2013 Oct 31;369(18):1691–1703.
41. Zhang Y, Elechalawar CK, Hossen MN, Francek ER, Dey A, Wilhelm S, et al. Gold nanoparticles inhibit activation of cancer-associated fibroblasts by disrupting communication from tumor and microenvironmental cells. Bioact Mater. 2021 Feb;6(2):326–332.
42. Sano K, Nakajima T, Choyke PL, Kobayashi H. Markedly enhanced permeability and retention effects induced by photo-immunotherapy of tumors. ACS Nano. 2013 Jan 22;7(1):717–724.
43. Garin-Chesa P, Old LJ, Rettig WJ. Cell surface glycoprotein of reactive stromal fibroblasts as a potential antibody target in human epithelial cancers. Proc Natl Acad Sci U S A. 1990 Sep;87(18):7235–7239.
44. Zhen Z, Tang W, Wang M, Zhou S, Wang H, Wu Z, et al. Protein nanocage mediated fibroblast-activation protein targeted photoimmunotherapy to enhance cytotoxic t cell infiltration and tumor control. Nano Lett. 2017 Feb 8;17(2):862–869.
45. Li L, Zhou S, Lv N, Zhen Z, Liu T, Gao S, et al. Photosensitizer-encapsulated ferritins mediate photodynamic therapy against cancer-associated fibroblasts and improve tumor accumulation of nanoparticles. Mol Pharm. 2018 Aug 6;15(8):3595–3599.
46. Zhang M, He Y, Sun X, Li Q, Wang W, Zhao A, et al. A high M1/M2 ratio of tumor-associated macrophages is associated with extended survival in ovarian cancer patients. J Ovarian Res. 2014 Feb 8;7:19.
47. Ries CH, Cannarile MA, Hoves S, Benz J, Wartha K, Runza V, et al. Targeting tumor-associated macrophages with anti-CSF-1R antibody reveals a strategy for cancer therapy. Cancer Cell. 2014 Jun 16;25(6):846–859.
48. Jinushi M, Komohara Y. Tumor-associated macrophages as an emerging target against tumors: Creating a new path from bench to bedside. Biochim Biophys Acta (BBA)—Rev Cancer. 2015 Apr 1;1855(2):123–130.
49. Qian BZ, Pollard JW. Macrophage diversity enhances tumor progression and metastasis. Cell. 2010 Apr 2;141(1):39–51.
50. Allavena P, Sica A, Garlanda C, Mantovani A. The yin-yang of tumor-associated macrophages in neoplastic progression and immune surveillance. Immunol Rev. 2008;222(1):155–161.
51. Zhang Q wen, Liu L, Gong C yang, Shi H shan, Zeng Y hui, Wang X ze, et al. Prognostic significance of tumor-associated macrophages in solid tumor: A meta-analysis of the literature. PLoS One. 2012 Dec 28;7(12):e50946.
52. Gocheva V, Wang HW, Gadea BB, Shree T, Hunter KE, Garfall AL, et al. IL-4 induces cathepsin protease activity in tumor-associated macrophages to promote cancer growth and invasion. Genes Dev. 2010 Feb 1;24(3):241–255.
53. Shree T, Olson OC, Elie BT, Kester JC, Garfall AL, Simpson K, et al. Macrophages and cathepsin proteases blunt chemotherapeutic response in breast cancer. Genes Dev. 2011 Dec 1;25(23):2465–2479.
54. Su P, Jiang L, Zhang Y, Yu T, Kang W, Liu Y, et al. Crosstalk between tumor-associated macrophages and tumor cells promotes chemoresistance via CXCL5/PI3K/AKT/mTOR pathway in gastric cancer. Cancer Cell Int. 2022 Sep 23;22(1):290.
55. Gordon S. Alternative activation of macrophages. Nat Rev Immunol. 2003 Jan;3(1):23–35.
56. Suarez-Lopez L, Sriram G, Kong YW, Morandell S, Merrick KA, Hernandez Y, et al. MK2 contributes to tumor progression by promoting M2 macrophage polarization and tumor angiogenesis. Proc Natl Acad Sci. 2018 May;115(18):E4236–4244.
57. Solinas G, Germano G, Mantovani A, Allavena P. Tumor-associated macrophages (TAM) as major players of the cancer-related inflammation. J Leukoc Biol. 2009;86(5):1065–1073.
58. Tu D, Dou J, Wang M, Zhuang H, Zhang X. M2 macrophages contribute to cell proliferation and migration of breast cancer. Cell Biol Int. 2021 Apr;45(4):831–838.
59. Wang N, Liang H, Zen K. Molecular mechanisms that influence the macrophage m1-m2 polarization balance. Front Immunol. 2014;5:614.
60. Hanahan D, Weinberg RA. Hallmarks of cancer: The next generation. Cell. 144(5):646–674.

61. Pepper MS. Role of the matrix metalloproteinase and plasminogen activator-plasmin systems in angiogenesis. Arterioscler Thromb Vasc Biol. 2001 Jul;21(7):1104–1117.
62. Santoni M, Massari F, Amantini C, Nabissi M, Maines F, Burattini L, et al. Emerging role of tumor-associated macrophages as therapeutic targets in patients with metastatic renal cell carcinoma. Cancer Immunol Immunother. 2013 Dec;62(12):1757–1768.
63. Priceman SJ, Sung JL, Shaposhnik Z, Burton JB, Torres-Collado AX, Moughon DL, et al. Targeting distinct tumor-infiltrating myeloid cells by inhibiting CSF-1 receptor: Combating tumor evasion of antiangiogenic therapy. Blood. 2010 Feb 18;115(7):1461–1471.
64. Ikemoto S, Yoshida N, Narita K, Wada S, Kishimoto T, Sugimura K, et al. Role of tumor-associated macrophages in renal cell carcinoma. Oncol Rep. 2003 Nov 1;10(6):1843–1849.
65. Kong L, Zhou Y, Bu H, Lv T, Shi Y, Yang J. Deletion of interleukin-6 in monocytes/macrophages suppresses the initiation of hepatocellular carcinoma in mice. J Exp Clin Cancer Res. 2016 Sep 2;35(1):131.
66. Grivennikov SI, Wang K, Mucida D, Stewart CA, Schnabl B, Jauch D, et al. Adenoma-linked barrier defects and microbial products drive IL-23/IL-17-mediated tumour growth. Nature. 2012 Nov;491(7423):254–258.
67. Song M, Liu T, Shi C, Zhang X, Chen X. Bioconjugated manganese dioxide nanoparticles enhance chemotherapy response by priming tumor-associated macrophages toward M1-like phenotype and attenuating tumor hypoxia. ACS Nano. 2016 Jan 1;10(1):633.
68. Zhu S, Niu M, O'Mary H, Cui Z. Targeting of tumor-associated macrophages made possible by PEG-sheddable, mannose-modified nanoparticles. Mol Pharm. 2013 Sep 3;10(9):3525–3530.
69. Konstantinopoulos PA, Karamouzis MV, Papatsoris AG, Papavassiliou AG. Matrix metalloproteinase inhibitors as anticancer agents. Int J Biochem Cell Biol. 2008 Jun 1;40(6):1156–1168.
70. Hynes RO, Naba A. Overview of the matrisome—an inventory of extracellular matrix constituents and functions. Cold Spring Harb Perspect Biol. 2012 Jan;4(1):a004903.
71. Lu P, Takai K, Weaver VM, Werb Z. Extracellular matrix degradation and remodeling in development and disease. Cold Spring Harb Perspect Biol. 2011 Dec;3(12):a005058.
72. Lu P, Weaver VM, Werb Z. The extracellular matrix: A dynamic niche in cancer progression. J Cell Biol. 2012 Feb 20;196(4):395–406.
73. Cho A, Howell VM, Colvin EK. The extracellular matrix in epithelial ovarian cancer—a piece of a puzzle. Front Oncol. 2015 Nov 2;5:245.
74. Guo Z, Hu K, Sun J, Zhang T, Zhang Q, Song L, et al. Fabrication of hydrogel with cell adhesive micropatterns for mimicking the oriented tumor-associated extracellular matrix. ACS Appl Mater Interfaces. 2014 Jul 23;6(14):10963–10968.
75. Holback H, Yeo Y. Intratumoral drug delivery with nanoparticulate carriers. Pharm Res. 2011 Aug 1;28(8):1819–1830.
76. Han W, Chen S, Yuan W, Fan Q, Tian J, Wang X, et al. Oriented collagen fibers direct tumor cell intravasation. Proc Natl Acad Sci. 2016 Oct 4;113(40):11208–11213.
77. Grossman M, Ben-Chetrit N, Zhuravlev A, Afik R, Bassat E, Solomonov I, et al. Tumor cell invasion can be blocked by modulators of collagen fibril alignment that control assembly of the extracellular matrix. Cancer Res. 2016 Jul 14;76(14):4249–4258.
78. Kanapathipillai M, Mammoto A, Mammoto T, Kang JH, Jiang E, Ghosh K, et al. Inhibition of mammary tumor growth using Lysyl oxidase-targeting nanoparticles to modify extracellular matrix. Nano Lett. 2012 Jun 13;12(6):3213–3217.
79. Zhang B, Jin K, Jiang T, Wang L, Shen S, Luo Z, et al. Celecoxib normalizes the tumor microenvironment and enhances small nanotherapeutics delivery to A549 tumors in nude mice. Sci Rep. 2017 Aug 30;7(1):10071.
80. DeBerardinis RJ, Lum JJ, Hatzivassiliou G, Thompson CB. The biology of cancer: Metabolic reprogramming fuels cell growth and proliferation. Cell Metab. 2008 Jan;7(1):11–20.
81. Estrella V, Chen T, Lloyd M, Wojtkowiak J, Cornnell HH, Ibrahim-Hashim A, et al. Acidity generated by the tumor microenvironment drives local invasion. Cancer Res. 2013 Jan 3;73(5):1524–1535.
82. Kato Y, Ozawa S, Miyamoto C, Maehata Y, Suzuki A, Maeda T, et al. Acidic extracellular microenvironment and cancer. Cancer Cell Int. 2013 Sep 3;13(1):89.
83. Rofstad EK, Mathiesen B, Kindem K, Galappathi K. Acidic extracellular pH promotes experimental metastasis of human melanoma cells in athymic nude mice. Cancer Res. 2006 Jul 3;66(13):6699–6707.
84. El-Kenawi A, Gatenbee C, Robertson-Tessi M, Bravo R, Dhillon J, Balagurunathan Y, et al. Acidity promotes tumour progression by altering macrophage phenotype in prostate cancer. Br J Cancer. 2019 Oct;121(7):556–566.

85. De Milito A, Fais S. Tumor acidity, chemoresistance and proton pump inhibitors. Future Oncol. 2005 Dec;1(6):779–786.
86. Apicella M, Giannoni E, Fiore S, Ferrari KJ, Fernández-Pérez D, Isella C, et al. Increased lactate secretion by cancer cells sustains non-cell-autonomous adaptive resistance to MET and EGFR targeted therapies. Cell Metab. 2018 Dec 4;28(6):848–865.e6.
87. Asad MI, Khan D, Rehman AU, Elaissari A, Ahmed N. Development and in vitro/in vivo evaluation of pH-sensitive polymeric nanoparticles loaded hydrogel for the management of psoriasis. Nanomater (Basel). 2021 Dec 17;11(12):3433.
88. Liu Y, Li LL, Qi GB, Chen XG, Wang H. Dynamic disordering of liposomal cocktails and the spatio-temporal favorable release of cargoes to circumvent drug resistance. Biomater. 2014 Mar 1;35(10):3406–3415.
89. Yan L, Crayton SH, Thawani JP, Amirshaghaghi A, Tsourkas A, Cheng Z. A pH-responsive drug-delivery platform based on glycol chitosan–coated liposomes. Small. 2015;11(37):4870–4874.
90. Carmeliet P, Jain RK. Principles and mechanisms of vessel normalization for cancer and other angiogenic diseases. Nat Rev Drug Discov. 2011 Jun;10(6):417–427.
91. Dewhirst MW, Cao Y, Moeller B. Cycling hypoxia and free radicals regulate angiogenesis and radiotherapy response. Nat Rev Cancer. 2008 Jun;8(6):425–437.
92. Casazza A, Di Conza G, Wenes M, Finisguerra V, Deschoemaeker S, Mazzone M. Tumor stroma: A complexity dictated by the hypoxic tumor microenvironment. Oncogene. 2014 Apr;33(14):1743–1754.
93. Vadde R, Vemula S, Jinka R, Merchant N, Bramhachari PV, Nagaraju GP. Role of hypoxia-inducible factors (HIF) in the maintenance of stemness and malignancy of colorectal cancer. Crit Rev Oncol/Hematol. 2017 May 1;113:22–27.
94. Xia S, Yu S ying, Yuan X lin, Xu S peng. Effects of hypoxia on expression of P-glycoprotein and multidrug resistance protein in human lung adenocarcinoma A549 cell line. Zhonghua Yi Xue Za Zhi. 2004 Apr 1;84(8):663–666.
95. Reddy KR, Guan Y, Qin G, Zhou Z, Jing N. Combined treatment targeting HIF-1α and Stat3 is a potent strategy for prostate cancer therapy. The Prostate. 2011;71(16):1796–1809.
96. Luan X, Guan YY, Liu HJ, Lu Q, Zhao M, Sun D, et al. A tumor vascular-targeted interlocking trimodal nanosystem that induces and exploits hypoxia. Adv Sci. 2018;5(8):1800034.
97. Hajizadeh F, Moghadaszadeh Ardebili S, Baghi Moornani M, Masjedi A, Atyabi F, Kiani M, et al. Silencing of HIF-1α/CD73 axis by siRNA-loaded TAT-chitosan-spion nanoparticles robustly blocks cancer cell progression. Eur J Pharmacol. 2020 Sep 5;882:173235.
98. Joshi U, Filipczak N, Khan MM, Attia SA, Torchilin V. Hypoxia-sensitive micellar nanoparticles for co-delivery of siRNA and chemotherapeutics to overcome multi-drug resistance in tumor cells. Int J Pharm. 2020 Nov 30;590:119915.
99. Hao D, Meng Q, Jiang B, Lu S, Xiang X, Pei Q, et al. Hypoxia-activated PEGylated paclitaxel prodrug nanoparticles for potentiated chemotherapy. ACS Nano. 2022 Sep 27;16(9):14693–14702.
100. Semenza GL. Evaluation of HIF-1 inhibitors as anticancer agents. Drug Discovery Today. 2007 Oct 1;12(19):853–859.
101. Long Q, Lin T Yin, Huang Y, Li X, Ma A Hong, Zhang H, et al. Imaging-guided photo-therapeutic nanoporphyrin synergized HSP90 inhibitor in patient-derived xenograft bladder cancer model. Nanomed. 2018 Apr;14(3):789–799.
102. Yao Y, Zhou Y, Liu L, Xu Y, Chen Q, Wang Y, et al. Nanoparticle-based drug delivery in cancer therapy and its role in overcoming drug resistance. Front Mol Biosci. 2020;7. [Internet] [cited 2023 Jan 17] Available from: www.frontiersin.org/articles/10.3389/fmolb.2020.00193.
103. Garbayo E, Estella-Hermoso de Mendoza A, Blanco-Prieto MJ. Diagnostic and therapeutic uses of nanomaterials in the brain. Curr Med Chem. 2014;21(36):4100–4131.
104. Hemmerich PH, Mikecz AH von. Defining the subcellular interface of nanoparticles by live-cell imaging. PLoS One. 2013 Apr 26;8(4):e62018.
105. Verma A, Stellacci F. Effect of surface properties on nanoparticle-cell interactions. Small. 2010 Jan;6(1):12–21.
106. Albanese A, Tang PS, Chan WCW. The effect of nanoparticle size, shape, and surface chemistry on biological systems. Annu Rev Biomed Eng. 2012;14(1):1–16.
107. Venturoli D, Rippe B. Ficoll and dextran vs. globular proteins as probes for testing glomerular permselectivity: Effects of molecular size, shape, charge, and deformability. Am J Physiol Renal Physiol. 2005 Apr;288(4):F605–613.
108. Decuzzi P, Pasqualini R, Arap W, Ferrari M. Intravascular delivery of particulate systems: Does geometry really matter? Pharm Res. 2009 Jan;26(1):235–243.

109. Yang Q, Jones SW, Parker CL, Zamboni WC, Bear JE, Lai SK. Evading immune cell uptake and clearance requires PEG grafting at densities substantially exceeding the minimum for brush conformation. Mol Pharm. 2014 Apr 7;11(4):1250–1258.
110. Zhang Z, Lin M. Fast loading of PEG—SH on CTAB-protected gold nanorods. RSC Adv. 2014;4(34):17760–17777.
111. Kaposi's sarcoma: DaunoXome approved. AIDS Treat News. 1996 May 3;(no 246):3–4.
112. Sonneveld P, Hajek R, Nagler A, Spencer A, Bladé J, Robak T, et al. Combined pegylated liposomal doxorubicin and bortezomib is highly effective in patients with recurrent or refractory multiple myeloma who received prior thalidomide/lenalidomide therapy. Cancer. 2008 Apr 1;112(7):1529–1537.
113. Miele E, Spinelli GP, Miele E, Tomao F, Tomao S. Albumin-bound formulation of paclitaxel (Abraxane® ABI-007) in the treatment of breast cancer. Int J Nanomed. 2009;4:99–105.
114. Passero FC, Grapsa D, Syrigos KN, Saif MW. The safety and efficacy of Onivyde (irinotecan liposome injection) for the treatment of metastatic pancreatic cancer following gemcitabine-based therapy. Expert Rev Anticancer Ther. 2016 Jul 2;16(7):697–703.
115. Batist G, Barton J, Chaikin P, Swenson C, Welles L. Myocet (liposome-encapsulated doxorubicin citrate): A new approach in breast cancer therapy. Expert Opin Pharmacother. 2002 Dec 1;3(12):1739–1751.
116. Wang-Gillam A, Li CP, Bodoky G, Dean A, Shan YS, Jameson G, et al. Nanoliposomal irinotecan with fluorouracil and folinic acid in metastatic pancreatic cancer after previous gemcitabine-based therapy (NAPOLI-1): A global, randomised, open-label, phase 3 trial. The Lancet. 2016 Feb 6;387(10018):545–557.
117. Moody AS, Dayton PA, Zamboni WC. Imaging methods to evaluate tumor microenvironment factors affecting nanoparticle drug delivery and antitumor response. Cancer Drug Resistance. 2021 Jun 19;4(2):382–413.
118. De Jong WH, Borm PJ. Drug delivery and nanoparticles: Applications and hazards. Int J Nanomed. 2008 Jun;3(2):133–149.
119. Manshian BB, Moyano DF, Corthout N, Munck S, Himmelreich U, Rotello VM, et al. High-content imaging and gene expression analysis to study cell–nanomaterial interactions: The effect of surface hydrophobicity. Biomater. 2014 Dec 1;35(37):9941–9950.
120. Summers HD, Rees P, Holton MD, Rowan Brown M, Chappell SC, Smith PJ, et al. Statistical analysis of nanoparticle dosing in a dynamic cellular system. Nat Nanotech. 2011 Mar;6(3):170–174.
121. Zhang W, Liu S, Han D, He Z. Engineered nanoparticle-induced epigenetic changes: An important consideration in nanomedicine. Acta Biomaterialia. 2020 Nov 1;117:93–107.
122. Tabish AM, Poels K, Byun HM, luyts K, Baccarelli AA, Martens J, et al. Changes in DNA methylation in mouse lungs after a single intra-tracheal administration of nanomaterials. PLoS One. 2017 Jan 12;12(1):e0169886.
123. Shi J, Kantoff PW, Wooster R, Farokhzad OC. Cancer nanomedicine: Progress, challenges and opportunities. Nat Rev Cancer. 2017 Jan;17(1):20–37.
124. Valencia PM, Basto PA, Zhang L, Rhee M, Langer R, Farokhzad OC, et al. Single-step assembly of homogenous lipid-polymeric and lipid-quantum dot nanoparticles enabled by microfluidic rapid mixing. ACS Nano. 2010 Mar 23;4(3):1671–1679.
125. Rhee M, Valencia PM, Rodriguez MI, Langer R, Farokhzad OC, Karnik R. Synthesis of size-tunable polymeric nanoparticles enabled by 3D hydrodynamic flow focusing in single-layer microchannels. Adv Mater. 2011;23(12):H79–83.
126. Gratton SEA, Pohlhaus PD, Lee J, Guo J, Cho MJ, Desimone JM. Nanofabricated particles for engineered drug therapies: A preliminary biodistribution study of PRINT nanoparticles. J Control Release. 2007 Aug 16;121(1–2):10–18.
127. Beletskii A, Galloway A, Rele S, Stone M, Malinoski F. Engineered PRINT(®) nanoparticles for controlled delivery of antigens and immunostimulants. Hum Vaccin Immunother. 2014;10(7):1908–1913.
128. Xu J, Wong DHC, Byrne JD, Chen K, Bowerman C, DeSimone JM. Future of the particle replication in nonwetting templates (PRINT) technology. Angew Chem Int Ed Engl. 2013 Jun 24;52(26):6580–6589.

2 Materials and Chemistries of Tumor Microenvironment–Responsive Delivery Platforms

Hillary Mndlovu and Yahya E. Choonara

CONTENTS

2.1 Introduction 21
2.2 Physical and Chemical Attributes of TME..22
2.3 TME Physical and Chemical Attribute–Inspired Drug Delivery Systems ...23
2.4 Stimuli-Sensitive Moieties in DDS Targeting TME.....................................24
    2.4.1 pH-Responsive Moieties in SRDDS 24
    2.4.2 Redox-Induced Drug Delivery Materials 31
    2.4.3 Moieties in DDSs that Facilitate Enzyme-Induced Drug Delivery ...31
    2.4.4 ROS-Induced Drug Release Moieties...33
2.5 Conclusion 33
References...34

2.1 INTRODUCTION

Tumours have a high level of heterogeneity from the patient, tissue, and cellular to molecular levels. This multi-scale heterogeneity limits the treatment of many classes of cancer. Amongst the major concerns in cancer therapy is the indiscriminate killing or removal of all cells within a given tumour volume, which may leave tumour cells for relapse [1]. Receptor-targeting therapies also lead to off-target effects and toxicity to normal cells given that the targeted receptors are also expressed in normal cells [1, 2]. Targeted blockage of receptors also leads to cancers adapting to therapies and also patients developing adverse immune-related events [3, 4]. The cancer cell's adaptability during cancer progression proves to be a bigger concern in cancer therapy.

    The changes in cancer progression include induced hypoxia suitable for tumour angiogenesis, inadequate nutrient transport changes, tumour energy metabolism, and extracellular metabolite lactic acid reduces pH of the tumor microenvironment (TME) [5]. These TME changes along with dense extracellular matrix (ECM) in cancer-associated fibroblasts (CAFs) are among the major obstacles for drugs to reach the target site and induce therapeutic effects [1, 5, 6]. However, the physical and chemical properties of the tumour cells and their environment are more consistent across different tumours than their biological markers counterparts. This gives insight into the design of effective target-specific drug delivery systems (DDSs) that can respond to the stimuli/properties of the tumour microenvironment.

    TME-responsive DDS are intelligent formulations designed to release bioactives stimulated by the changing tumour environments. The microenvironmental stimuli may include shrinking or expanding of the tumour, varying surface charge, and regulation of other physiochemical properties [7]. The tumour microenvironment stimuli highlighted here can be used to develop intelligent formulations that can directly or indirectly respond to one or more of those microenvironment stimuli [7, 8]. Conventional cancer therapies present limitations such as insufficient therapeutic effect, dose-limiting toxicity, and inability to overcome the highly adaptive and aggressive nature of tumour cells [9, 10]. Other imitations of conventional cancer therapies include poor tumoral penetration and uncontrolled drug release [7].

In an attempt to tackle the limitations, smart nano-enabled stimuli-responsive drug delivery platforms have been investigated to increase drug potency, reach target tumour cells, and avoid drug leakage [10, 11]. The concept behind developing a stimuli-responsive drug delivery platform is simple: to have the DDS act as a switch to open or close the release of drugs in response to endogenous, exogenous, or multi-stimuli such as pH, enzyme and redox changes, temperature, light, ultrasound, and magnetic field [10].

The design of these DDSs makes use of tumour-sensing moieties that could be absorbed or chemically react with the microenvironment to release drugs at target sites [12, 13]. Bond cleavage, disassembly, or cap removal are the notable ways in which drugs are released to the targeted site [7]. These stimuli-responsive drug delivery system (SRDDS) design strategies leverage the properties of the TME to use macromolecules that can increase drug retention in the tumour site and provide great tumour cell selectivity and co-delivery of drugs, thus inducing optimal antitumor efficacy while limiting adverse reactions [1, 14, 15]. Herein, stimuli-sensitive moieties/materials in SRDDSs will be outlined, along with their interactions/chemistries with the tumour microenvironment. This section will not focus on polymers, detailed fabrication approaches, or formulation types. This section will focus on specific moieties such as cleavable bonds, enzyme-substrate/cleavage sites, functional groups, and atoms which are targeted by TME stimuli to induce drug release. The mechanism with which these moieties facilitate drug delivery will also be highlighted.

2.2 PHYSICAL AND CHEMICAL ATTRIBUTES OF TME

Physical and chemical changes in growing tumours are interconnected, and the changes can induce one or more stimuli to be detected by SRDDSs. The high cellular proliferation rate in tumours leads to a depleted oxygen supply, which forces tumours to adapt to survive in a low O_2 environment [1, 16]. Those two interconnected physical (tumour size) and chemical (low/absent O_2) alterations result in the activation of hypoxia-inducible factor-1 (HIF-1), which regulates cell functions such as apoptosis, cell cycle arrest, angiogenesis, glycolysis, and adaptation to acidic environments [17]. Thus, extracellular acidity is the primary characteristic of the TME which can be targeted by SRDDS. The affected cell functions lead to deformed tissues, that is, tissue stiffening. Tissue stiffening in liver, breast, and prostate tumours is increased by the deposition of highly crosslinked collagen in the ECM [18–21]. Tissue stiffening leads to increased tissue stress and interstitial fluid pressure, which in turn affect cell behaviour such as gene expression, cell proliferation, invasiveness, and ECM organisation [21–24].

Cellular electrical properties are also affected in tumours. The electrical effect in cancer cells includes a high intracellular concentration of positively charged sodium ions and a high concentration of negatively charged glycocalyx on the cell membrane of tumour cells compared to normal cells [1, 25]. These altered ions and charges lead to low resting transmembrane potential (TMP) in cancer cells [26, 27]. The altered electron and proton transport across the cell membrane leads to the accumulation of negative charges on the cell surface which cause further changes in electrical properties of the ECM [26]. The changes in electrical properties of the ECM affect the overall TME electrical properties such as higher electrical conductivity and permittivity within the tumours [28–30]. The high electrical properties of the TME can be used to induce the release of drugs, and electroporation is one of the approaches that can be employed to target the altered electrical properties of the TME [25].

Normal tissues and tumours have been noted to have the following differences: PH changes from 7.4 to 4.5–6.8 [31, 32], distinct hyperthermia ranging from 37 to 42°C induced by intrinsic pyrogenic substances secreted by tumour cells [33], production of hyperactive reactive oxygen species (ROS) [3 µM, (in normal human plasma) to 50–100 µM (extracellular)] in tumours leading to redox imbalance [34, 35], overexpression of certain enzymes (fibroblast-activation protein-a (FAP-a) and matrix

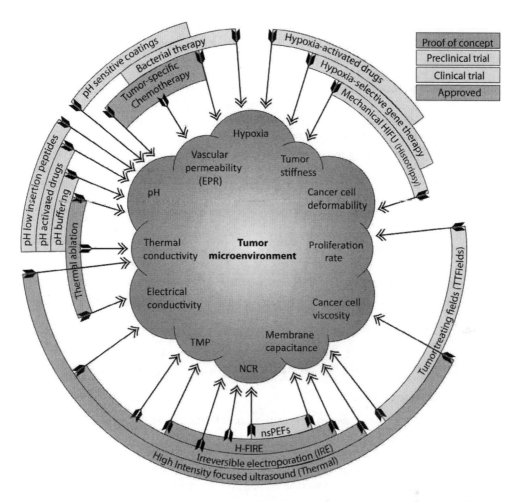

FIGURE 2.1 Stimuli-responsive drug delivery systems targeting TME. The DDSs are colour coded according to their stage of development from proof of concept to being approved for use as cancer therapies.

metalloproteinases (MMPs)) and upregulation of legumain and cathepsin [31, 36, 37], and glutathione (GSH) concentrations decreased from 140 nmol/g tissue in normal tissues to approximately 90 nmol/g tissue (extracellular) in tumours [38]. Those specific physical and chemical changes lead to the design and development of several DDSs targeting the TME (Figure 2.1).

2.3 TME PHYSICAL AND CHEMICAL ATTRIBUTE–INSPIRED DRUG DELIVERY SYSTEMS

There are several therapies targeting the TME. Drug delivery systems can be degloved to respond to different attributes of the TME. The TME attributes can be used as stimuli to trigger drug delivery. Various approaches targeting TME are highlighted in Figure 2.1. The most common stimulus that is targeted in TME is pH variation. The moieties/structures responsible for drug-induced drug lease upon TME exposure will be discussed in detail compared to other stimuli. The pH-sensitive DDS make use of molecules that can be cleaved, protonated, or undergo a pH-induced conformational change to release drugs. Some of these molecules include low pH insertion peptides (pHLIPs), which can undergo membrane-associated folding when exposed to low pH conditions, thereby releasing the bound drug molecules [39, 40]. Drugs can be shielded from the environment to prevent drug leakage and only release at a target site in response to pH as a stimulus [41]. Enzyme-sensitive DDSs are exceptional candidates to be used along with pH-sensitive moieties. Combining enzyme-sensitive materials with pH-sensitive moieties allows for a much more specific targeted drug delivery system. The pH-induced release would target the low pH of TME, whereas substrates with specific enzyme binding sites will direct drug release to specified overexpressed proteins/enzymes. One example of a dual stimuli-sensitive DDS is the conjugation of 6-maleimidocaproicacid and the amine group of gemcitabine to produce amide bond-bearing cathepsin B-sensitive gemcitabine prodrugs. In vivo studies indicated that albumin was capable of binding with the maleimide groups of the linker, which allowed for the transportation of the prodrug to the tumour [42]. The overexpressed cathepsin B and the acidic microenvironment in the tumour lead to the rapid release of gemcitabine owing to the degradation of the amide-containing linker [42]. There are other stimuli targeting DDSs such as those that target low oxygen concentrations in tumours (hypoxia) [43, 44], pore size in tumour tissues [45], and ATP content [7, 46] to name a few. The notable moieties that respond and are reduced to cytotoxic radicals under hypoxia include nitro groups, aromatic N-oxides, aliphatic N-oxides, and transition metals [44]. Moieties that affect permeability and retention of drug molecules include angiotensin, nitric oxide (NO), and vascular endothelial growth factor [1]. ATP aptamers possess a high number of "guanine-cytosine" residues and negatively charged surfaces which enable multiple attachments of anticancer agents, linkers, and cationic polymeric materials [7, 47, 48]. This section will not focus on formulation types that target TME. However, it will focus on the moieties in the stimuli-responsive drug delivery system that allow for stimuli-induced drug release.

2.4 STIMULI-SENSITIVE MOIETIES IN DDS TARGETING TME

2.4.1 pH-Responsive Moieties in SRDDS

pH changes in tumours have garnered more attention in DDSs targeting the lowered pH condition of the TME. Drug release from stimuli-response drug delivery systems proceeds via bond cleavage, protonation, and gas generation [7]. The main mechanism for acid-responsive DDSs is bond cleavage, and this is due to the presence of acid-labile linkers and carriers in the materials. The notable labile acid moieties that are commonly employed in developing pH-responsive DDS include hydrazine [49], acetal [50], imines [51], and metal-organic frameworks (MOFs) [52]. These moieties are hydrolysed in an low-pH environment (pH 4.5–6.5) to release drugs in the targeted tumour site.

2.4.1.1 Metal-Organic Complexes in pH-Responsive DDSs

Metal ions interact with organic ligands to form metal-organic complexes with different structural arrangements such as one-, two-, or three-dimensional structures [7, 52]. One example of such materials is the zeolitic imid-azolate framework-8 (ZIF-8), which was reported to have acceptable biocompatibility [7]. The developed fluorescein-incorporated-ZIF-8 acidic responsive nanospheres exhibited exceptional acid environment responsive behaviour [53]. The approach employed in developing this pH-responsive system made use of ZIF-8, which absorbs, protects, and shields the small molecules and only releases them upon exposure to the low pH of the environment (Figure 2.2A). The fluorescein-incorporated-ZIF-8 nanospheres displayed high stability at pH 7.4 and 50% fluorescein release after 1 hour incubation in an acidic buffered environment (pH 6) [52]. Another MOF development approach can make use of a carboxyl group coordinated with Zn^{2+} ions to enable the initial self-assembly of MOFs [53]. The Zn^{2+} ions are displaced by introducing a different organic linker to form a more stable coordinated structure [53]. Zheng et al. employed this approach, and the resultant nanospheres displayed high doxorubicin (DOX) release (>95%) in acidic environments (pH 5.0–6.0), whereas low concentrations were released at pH 7.4 to 6.5 [53]. Ion (III) can also act as a pH-responsive moiety. A study by Kim et al. highlighted the use of Fe^{III} and 3, 4-dihydroxyphenylalanine as pH-responsive materials in a mussel adhesive protein-based DDS for DOX delivery [54]. The approach employed in these polymeric nanoparticles follows the common

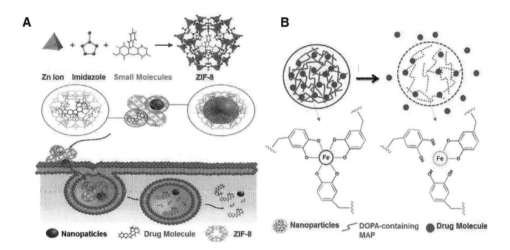

FIGURE 2.2 Mechanism of drug encapsulation and release from pH-responsive DDSs. (A) ZIF-8 frameworks encapsulating small drug molecules and pH-induced drug release at TME (B) Metal coordination as a drug encapsulation technique and pH-induced drug release (reproduced with permission from [7]).

pH-accelerated hydrolysis mechanism where the pH-dependent Fe^{III}–DOPA complexes allowed for the drug to diffuse through crosslinked polymeric networks which were loosened by the low-pH environment (Figure 2.2B) [54]. Another targeted polymeric group in metal coordination is the amino group. A report highlighted the development of pH-responsive platinum coordinated on the amino group on $Fe_3O_4Gd_2O_3$ surface to release cisplatin in acidic conditions [54].

2.4.1.2 Organic Linkers in pH-Responsive DDS

The organic drug-polymer bond linkers in pH-responsive DDSs make use of hydrazone and acetals [50, 55, 56]. The development of pH-responsive polymer–drug conjugates and their properties with a specific focus on organic linkers are well documented [57]. Herein, the materials and chemistries of pH-responsive will be outlined briefly. Similarly to MOFs, the mode of drug release in organic material–linked formulation is by hydrolysis, which is accelerated by the acidic environment (Table 2.1). The acute responsive nature of the hydrozone bond makes it one of the highly used chemical bonds in developing pH-responsive DDS. The hydrazone-polymer conjugates are prepared by modifying the carboxylic group of the polymers with hydrazines. Drugs can also be conjugated with hydrazines. However, not all antitumour drugs have keto groups or active functional groups to facilitate a pH-sensitive linker for drug delivery. Therefore, further functionalisation is needed. The carboxylic group containing drugs can be esterified to obtain a keto site capable of forming a hydrozone [57]. Other targeted linkers such as acetals, *cis*-aconityl, imines, and β-thiopropionate are also employed in developing pH-sensitive DDSs, and their modes of drug release in the acidic environment are highlighted in Table 2.1.

2.4.1.3 Protonation-Induced Drug Release in pH-Responsive DDS

2.4.1.3.1 Functionalisation of Polymeric Materials with pH-Sensitive Moieties

Functionalisation of polymeric materials with carboxylic groups and amines allows for the development of pH-dependent drug release. Drug release proceeds through the protonation of those functional groups in an acidic environment. Polymers can be anionic or cationic depending on their surface charge at pH 7.4. Examples of pH-sensitive materials with acute drug release in an acidic environment include formulations such as the amphiphilic PEG-b- poly[2-(diisopropylamino)ethyl methacrylate]-(PDPA) copolymer and poly[2-(diisopropylamino)ethyl

TABLE 2.1
Stimuli-Responsive Materials/Moieties and Mechanisms of Drug Release

Stimuli	Materials	Responsive Mechanism	Ref.
pH	Hydrazine	Bond cleavage under acidic environment	[49, 57]
	Acetal	Bond cleavage	[50, 57]
	Imines	Bond cleavage	[51, 57]
	Shift base	Bond cleavage	[52, 57]
	B-Thiopropionate	Bond cleavage	[57]

Coordination metal–organic frameworks	bond cleavage		[53, 54]
Redox	Sulphide	GSH cleavage site and ROS	[7, 58–60]

(Continued)

TABLE 2.1 *(Continued)*
Stimuli-Responsive Materials/Moieties and Mechanisms of Drug Release

Stimuli	Materials	Responsive Mechanism	Ref.
	Selenium	GSH cleavage site and ROS	[7, 61, 62]
	Telenium	GSH cleavage site and ROS	[7, 63, 64]
	Boronic ester/acid	ROS	[7, 65, 66]

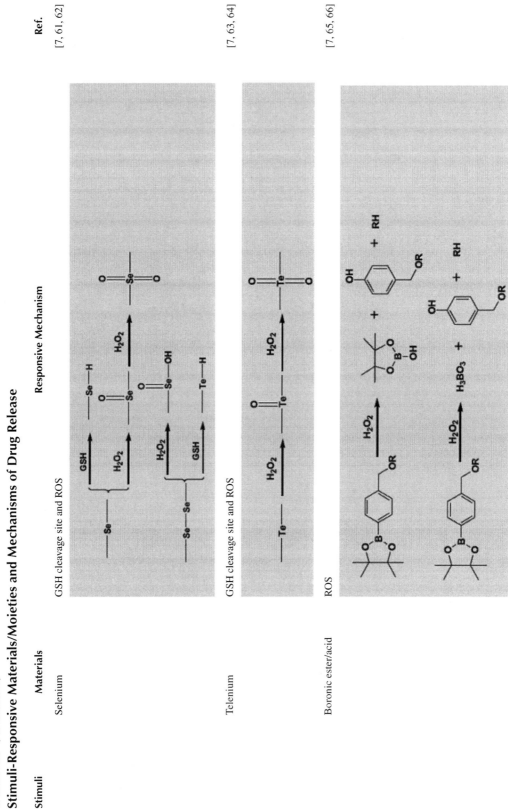

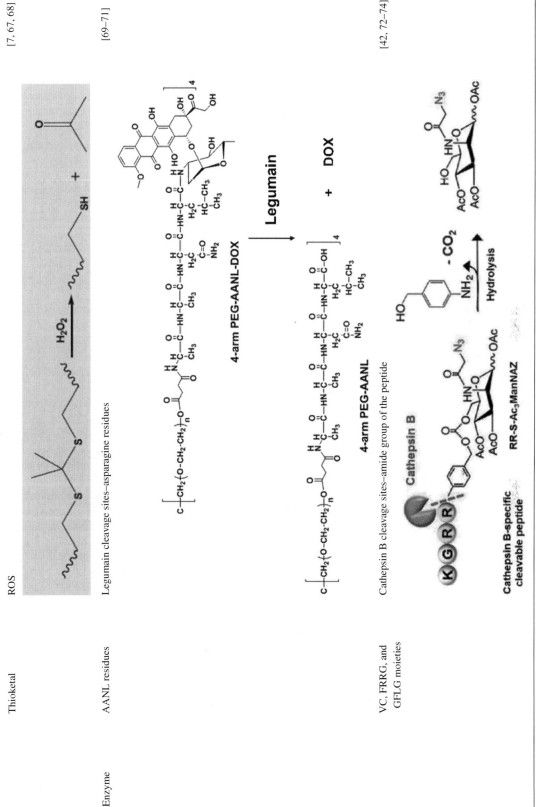

TABLE 2.1 (Continued)
Stimuli-Responsive Materials/Moieties and Mechanisms of Drug Release

Stimuli	Materials	Responsive Mechanism	Ref.
Hypoxia	Nitro groups, aromatic N-oxides, aliphatic N-oxides, and transition metals	Activating nitrogen mustard via nitroreduction as an electronic switch to activate a reactive centre	[44]
Enhanced permeability and retention	Angiotensin, nitric oxide (NO) and VEGF	Diaminofluoresceins (DAFs) reacting with NO to form DAF-FM triazole	[75]

methacrylate]-b-poly[(ethylene glycol) methyl ether methacrylate], to name a few [76, 77]. At pH 7.4 the hydrophobic nature of the PDPA allows for the incorporation of hydrophobic drugs, while in an acidic environment (pH 5), PDPA undergoes the hydrophobicity–hydrophilicity transition, resulting in drug release at the target site [76, 77].

2.4.1.3.2 Protein Amines and Carboxylic Acid Groups as pH-Sensitive Moieties
Proteins can also be used as pH-responsive DDSs due to the presence of both amine and carboxylic groups. Several studies made use of proteins as drug carriers such as ferritin loaded with metal ions, DOX, curcumin, and quercetin [78] and pHLIP [79]. Fe(II)-DOX were loaded in ferritin by partially opening the hydrophobic nanocages of ferritin, and the complex remained stable at pH 6 and started to dissemble in acidic conditions (pH 4.0–5.0) [78]. The disassembly was due to the metal-protein complexes disassembling in an acidic environment [78]. Similar to the mode of drug release in ferritin, pHLIP-based nanosystems also released drugs after undergoing protonation in an acidic environment [79, 80]. Cheng et al. highlighted the formation of a GSH-responsive disulphide bridge in the pHLIP nanosystem [80]. The disulphide bond cleavage in the nanosystem was facilitated by GSH which led to the release of nucleic acid analogues [80]. Several pH-sensitive DDS make use of the protonation mechanism to release drugs. Some of these pH-sensitive DDS include poly(L-histidine)-based pH-sensitive micelles with an amine group becoming protonated during drug release [81] and partial hybridisation and protonation of cytosine leading to disassembly of the gold clusters, thus releasing DOX from oligodeoxynucleotide nanoparticles [82]. In addition to the carboxylic acid and amines that can be targeted when developing pH-sensitive DDSs, inorganic materials such as calcium carbonate [83], zinc oxide [84, 85], and manganese dioxide [86] may be employed in developing pH-sensitive DDS. This is because these inorganic materials can dissolve as nontoxic ions under acidic conditions while remaining stable/relatively insoluble at pH 7.4, thus making them potential pH-sensitive moieties in DDSs [7].

2.4.2 Redox-Induced Drug Delivery Materials

HCO_3^--containing materials and GSH-based nanosystems offer great potential to be used for pH-responsive DDS. A study by Liu et al. highlighted the encapsulating of DOX and HCO_3^- in the lysosomal system [87]. The liposomal system released DOX upon the reaction of HCO_3^- with the acidic environment to produce carbon dioxide [87]. GSH is more highly concentrated (four-fold) in cancer cells than in normal cells [88]. GSH-containing nanosystems are capable of undergoing a redox reaction due to the presence of these functional moieties: disulphide bonds, diselenide/ditelluride moieties, thioether/selenide/tellurium groups, metal-thiol-based linkers, and ferrocenium [7]. The redox reaction mechanism of these moieties is summarised in Table 3.1. Briefly, the disulphide moieties can be incorporated into polymeric material through L-cystine [89], dithiodiglycolic acid [90], and pyridyl disulphide chemistry [91]. The redox drug release mechanism of disulphide-based nanosystems follows the GSH-triggered thiol-disulphide intracellular exchange and S-S dissolution from the nanosystem's disulphide bridges [92]. Ditelluride/diselenide bonds are also moieties that can be explored when developing GSH-sensitive DDSs. Polymeric materials containing the ditelluride/diselenide moieties can be cleaved through the oxidation process to produce selenic/telluric acid or the reduction process, which results in selenol/tellurol formation in different redox environments [93, 94].

2.4.3 Moieties in DDSs that Facilitate Enzyme-Induced Drug Delivery

Tumours are associated with upregulated enzymes which allow them to survive and grow. The upregulated extracellular enzymes include matrix metalloproteinases and hyaluronidases (Has)

[95, 96]. Enzyme-sensitive DDSs have been developed with enzymatic substrates that are specifically recognised and degraded by overexpressed enzymes [7]. This substrate-specific enzyme recognition makes the overexpressed enzymes suitable triggers of drug delivery systems. Herein, examples and chemistries of enzyme-triggered DDRS will be highlighted.

2.4.3.1 Legumain-Induced Drug Release

Peptides are normally designed as substrates recognised by specific enzymes. Enzymes can recognise and cleave specific residues. Legumain, an asparaginyl endopeptidase, has high specificity on asparagine or aspartic acid residues [69]. Legumain targets and cleaves the asparagine/aspartic acid residues in DDSs, thus inducing the release of drug molecules [69–71]. One example of such a DDS is the development of a prodrug copolymer (4-arm PEG-AANL-DOX) by using alanine-alanine-asparagine-leucine (AANL) peptide as the polymer-to-drug link [71]. The prodrug micelle accumulated in the tumour tissue and released DOX upon cleaving the asparagine residues. One limitation of this approach is that, under physiological conditions, legumain is expressed in spleen, kidney, and placenta, however at lower concentration, mainly in its inactive form. An additional moiety would have to be introduced to the DDS to induce a therapeutic effect. A simple conjugation of hyaluronic acid with DOX can improve the target specificity of the DDS given that HA can bind to CD44 receptors overexpressed on cancer cells, while AANL cleavage by legumain can induce drug release [69]. In another study, a legumain-responsive drug delivery system was developed for dual two-step drug release. Ly6c$^+$ inflammatory monocytes were incorporated in alanine-alanine-asparagine-lysine (AANK)–linked nanoparticles containing mertansine and poly(styrene-co-maleic anhydride) (SMA) [97]. The two-step drug release was achieved by sensing the metastasis site, which induced the differentiation of monocytes into macrophages while cleaving the AANK moieties and facilitated the release of SMA at the target site [97].

2.4.3.2 Cathepsin B–Induced Drug Release Moieties

Valine-citrulline (VC), phenylalanine-arginine-arginine-glycine (FRRG), and glycine-phenylalanine-leucine-glycine (GFLG) residues are specifically recognised and cleaved by cathepsin B [42, 72–74]. DDSs linked by these peptides release drugs upon degradation of the amide group of the peptides by the overexpressed cathepsin B and the acidic environment [42]. Cathepsin B enzymatic activity is optimal in an acidic environment (pH 5–6) [98, 99]. This implies that the peptide substrates will not be degraded in the bloodstream; thus drug release can be targeted to a tumour environment with acidic pH conditions. Some of the strategic design of cathepsin B–sensitive DDSs includes the development of a GFLG-containing multifunctional rotaxane structure to act as an open/close switch for delivery of mesoporous silica nanoparticles (MSNs), conjugation of GPLG with paclitaxel (PTX)-PEGylated peptide dendrimer for PTX delivery, and developing bio-orthogonal click-molecule-conjugated nanoparticles with lysine-glycine-arginine-arginine (KGRR) caged metabolic precursors for exogenous generation of glycans (with azide groups) on the cancer cell membrane [59, 74, 100].

2.4.3.3 Lysozyme-Induced Drug Release

Substrate specificity of enzymes is the major mechanism by which drug release is induced in enzymatic-responsive drug delivery systems. DDSs can be designed with substrates that can be recognised and cleaved by enzymes such as esterase, amylases, β-D-galactosidase, and amidases [101–103]. One example of such a DDS is the crosslinking of alkyne-terminated surfactant with esterified diazido [103]. The alkyne-terminated surfactant was attached to the thiolated DNA molecules which would be released upon esterase cleavage of the esterified diazido [103]. The released nucleic acids sequence cleaved transcription factor-associated mRNA in inflammation pathways [103].

2.4.4 ROS-Induced Drug Release Moieties

ROSs such as H_2O_2, $O^{2\bullet}$, $\bullet OH$, $ONOO^-$, and Ocl^- are highly involved in several cellular metabolic pathways [104]. High concentrations of H_2O_2 can trigger drug release. DDSs can be designed to respond to H_2O_2 concentrations. The design of H_2O_2-sensitive DDSs can be achieved by developing chemical structures with S, Se, and Te atoms or the Se-Se bond. The formed C-S and C-Se bonds were targeted redox reactions by ROS such as H_2O_2 [61]. Examples of such DDSs are PEG-PU_x-PEG triblock polymers, with x being one of the previous atoms or bonds [61]; conjugation of 6-maleimidocaproic acid to paclitaxel using a mono-sulfide linker [105]; and development of sulphide-containing poly (propylene sulfide) nanoparticles [60]. Another strategy for the development of ROS-sensitive DDSs is the use of photodynamic therapy (PDT) as a cytotoxic ROS generator and ROS-responsive carriers such as the atoms/bonds mentioned previously. Irradiation of IR780 (pro-PDT) resulted in the production of ROS, which led to the oxidation of ROS-sensitive thioether chains along with the nanoparticle's acid-labile β-carboxylic amide pendants under acidic TME [106].

Bond cleavage is another mechanism with which drug release can be achieved. Se–Se crosslinks can be cleaved by ROS in the process of drug release. The Se–Se cleavage mechanism by ROS was highlighted in a study where tellurium-containing molecules and phospholipid nanostructures were exposed to H_2O_2 [64]. The study noted a hydrophobic-to-hydrophilic transition and the formation of an oxygen atom in the presence of H_2O_2 [64]. The common bonds that can be cleaved to trigger drug release are mainly aryl boronic acid and poly (thioketal). DDSs with aryl boronic acid and poly (thioketal) moieties can be cleaved by ROS [65, 66]. One approach to exploring these moieties in developing ROS-sensitive DDSs is to modify the amino group of amino acid residues with aryl boronic ester to affect protein/protein activity [66]. The protein/peptide activity can be restored by exposing the aryl boronic ester groups of the DDS to H_2O_2 [66]. Another example of ROS cleavable bonds in DDSs is the development of poly(thioketal)-based poly-prodrugs, which can release mitoxantrone upon thioketal bond cleavage by H_2O_2 [68].

2.5 CONCLUSION

The current chapter highlighted stimuli-responsive moieties in DDSs. SRDDS design strategies, which included incorporation/conjugation or functionalisation using specific stimuli-sensitive moieties such as chemical bonds, atoms, functional groups, and amino acid residues, were highlighted. Different polymeric or protein-based drug delivery systems can be formulated. However, the most important part of SRDDSs is the use of one or more of the stimuli-sensitive moieties highlighted previously. Stimuli-sensitive moieties respond to the physical and chemical attributes of the TME. TME physical attributes such as temperature and tumour size can be used to induce drug release. However, they present a few limitations such as drug leakage and off-targeted drug delivery. Enzyme-sensitive DDSs have a similar limitation where substrates bind to either normal or tumour cells due to the presence of the same enzymes. A DDS which responds to the acidic pH of the TME proved to release drugs at targeted sites. The target specificity can be improved by combing moieties that can respond to two stimuli, that is, DRSs with moieties that respond to the acidity of the TME and also have moieties that act as enzyme cleavage sites. In summary, pH-responsive drug delivery systems have moieties such as hydrazine, acetal, imines, and coordination metal-organic frameworks. These moieties can be hydrolysed or protonated to induce drug release. Enzyme-sensitive DDSs have an enzymatic cleavage site, which is mostly a specific amino acid sequence that is cleaved by the enzyme at normal or specific pH conditions. The stimuli-sensitive moieties covered in this section can be used in developing smart stimuli-sensitive drug delivery systems to target different physical and chemical attributes of TME.

REFERENCES

1. Ivey, J.W., et al., *Improving cancer therapies by targeting the physical and chemical hallmarks of the tumor microenvironment.* 2016. **380**(1): p. 330–339.
2. Malik, I.A., et al., *Comparison of changes in gene expression of transferrin receptor-1 and other iron-regulatory proteins in rat liver and brain during acute-phase response.* 2011. **344**(2): p. 299–312.
3. Tu, L., et al., *Assessment of the expression of the immune checkpoint molecules PD-1, CTLA4, TIM-3 and LAG-3 across different cancers in relation to treatment response, tumor-infiltrating immune cells and survival.* 2020. **147**(2): p. 423–439.
4. Bayraktar, S., S. Batoo, S. Okuno, and S. Glück, *Immunotherapy in breast cancer.* 2019: p. 541–552. https://doi.org/10.4103/jcar.JCar_2_19.
5. Jia, R., et al., *Advances in multiple stimuli-responsive drug-delivery systems for cancer therapy.* 2021. **16**: p. 1525.
6. Karagiannis, G.S., et al., *Cancer-associated fibroblasts drive the progression of metastasis through both paracrine and mechanical pressure on cancer tissue interdigital model of metastasis.* 2012. **10**(11): p. 1403–1418.
7. He, Q., et al., *Tumor microenvironment responsive drug delivery systems.* 2020. **15**(4): p. 416–448.
8. Schattling, P., F.D. Jochum, and P.J.P.C. Theato, *Multi-stimuli responsive polymers—the all-in-one talents.* 2014. **5**(1): p. 25–36.
9. Grasmann, G., A. Mondal, and K.J.I.J.O.M.S. Leithner, *Flexibility and adaptation of cancer cells in a heterogenous metabolic microenvironment.* 2021. **22**(3): p. 1476.
10. Li, L., W.-W. Yang, and D.-G.J.J.O.D.T. Xu, *Stimuli-responsive nanoscale drug delivery systems for cancer therapy.* 2019. **27**(4): p. 423–433.
11. Zhou, M., et al., *The application of stimuli-responsive nanocarriers for targeted drug delivery.* 2017. **17**(20): p. 2319–2334.
12. Ju, C., et al., *Sequential intra-intercellular nanoparticle delivery system for deep tumor penetration.* 2014. **126**(24): p. 6367–6372.
13. Paliwal, S.R., et al., *Hyaluronic acid modified pH-sensitive liposomes for targeted intracellular delivery of doxorubicin.* 2016. **26**(4): p. 276–287.
14. Vicent, M.J., H. Ringsdorf, and R.J.A.D.D.R. Duncan, *Polymer therapeutics: Clinical applications and challenges for development.* 2009. **13**(61): p. 1117–1120.
15. Robey, I.F., et al., *Bicarbonate increases tumor pH and inhibits spontaneous metastases.* 2009. **69**(6): p. 2260–2268.
16. Hockel, M. and P.J.J.O.T.N.C.I. Vaupel, *Tumor hypoxia: Definitions and current clinical, biologic, and molecular aspects.* 2001. **93**(4): p. 266–276.
17. Shannon, A.M., et al., *Tumour hypoxia, chemotherapeutic resistance and hypoxia-related therapies.* 2003. **29**(4): p. 297–307.
18. Masuzaki, R., et al., *Assessing liver tumor stiffness by transient elastography.* 2007. **1**(3): p. 394–397.
19. Paszek, M.J., et al., *Tensional homeostasis and the malignant phenotype.* 2005. **8**(3): p. 241–254.
20. Tuxhorn, J.A., et al., *Reactive stroma in human prostate cancer: Induction of myofibroblast phenotype and extracellular matrix remodeling.* 2002. **8**(9): p. 2912–2923.
21. Krouskop, T.A., et al., *Elastic moduli of breast and prostate tissues under compression.* 1998. **20**(4): p. 260–274.
22. Jain, R.K., J.D. Martin, and T.J.A.R.O.B.E. Stylianopoulos, *The role of mechanical forces in tumor growth and therapy.* 2014. **16**: p. 321.
23. Boucher, Y., L.T. Baxter, and R.K.J.C.r. Jain, *Interstitial pressure gradients in tissue-isolated and subcutaneous tumors: Implications for therapy.* 1990. **50**(15): p. 4478–4484.
24. Less, J.R., et al., *Interstitial hypertension in human breast and colorectal tumors.* 1992. **52**(22): p. 6371–6374.
25. Haltiwanger, S. and R.J.E.-B.T.C.F.B.T.C.A. Sundararajan, *Why electroporation is a useful technique for cancer treatments.* 2014. **3**: p. 103–125.
26. Stern, R., B. Milestone, and R.J.M.H. Gatenby, *Carcinogenesis and the plasma membrane.* 1999. **52**(5): p. 367–372.
27. Sree, V.G., K. Udayakumar, and R.J.I.J.O.B.C. Sundararajan, *Electric field analysis of breast tumor cells.* 2011. **2011**.
28. Laufer, S., et al., *Electrical impedance characterization of normal and cancerous human hepatic tissue.* 2010. **31**(7): p. 995.

29. Michel, E., D. Hernandez, and S.Y.J.M.R.I.M. Lee, *Electrical conductivity and permittivity maps of brain tissues derived from water content based on T1-weighted acquisition.* 2017. **77**(3): p. 1094–1103.
30. Sugitani, T., et al., *Complex permittivities of breast tumor tissues obtained from cancer surgeries.* 2014. **104**(25): p. 253702.
31. Zhang, W., et al., *Tumor microenvironment–responsive peptide-based supramolecular drug delivery system.* 2020. **8**: p. 549.
32. Stubbs, M., P. McSheehy, and J.R.J.A.I.E.R. Griffiths, *Causes and consequences of acidic pH in tumors: A magnetic resonance study.* 1999. **39**: p. 13–30.
33. Danhier, F., O. Feron, and V.J.J.O.C.R. Préat, *To exploit the tumor microenvironment: Passive and active tumor targeting of nanocarriers for anti-cancer drug delivery.* 2010. **148**(2): p. 135–146.
34. Purohit, V., D.M. Simeone, and C.A.J.C. Lyssiotis, *Metabolic regulation of redox balance in cancer.* 2019. **11**(7): p. 955.
35. Montero, D., et al., *Intracellular glutathione pools are heterogeneously concentrated.* 2013. **1**(1): p. 508–513.
36. Ji, T., et al., *Using functional nanomaterials to target and regulate the tumor microenvironment: Diagnostic and therapeutic applications.* 2013. **25**(26): p. 3508–3525.
37. Foekens, J.A., et al., *Prognostic significance of cathepsins B and L in primary human breast cancer.* 1998. **16**(3): p. 1013–1021.
38. Perry, R.R., et al., *Glutathione levels and variability in breast tumors and normal tissue.* 1993. **72**(3): p. 783–787.
39. Andreev, O.A., D.M. Engelman, and Y.K.J.M.M.B. Reshetnyak, *pH-sensitive membrane peptides (pHLIPs) as a novel class of delivery agents.* 2010. **27**(7): p. 341–352.
40. Andreev, O.A., D.M. Engelman, and Y.K.J.F.I.P. Reshetnyak, *Targeting diseased tissues by pHLIP insertion at low cell surface pH.* 2014. **5**: p. 97.
41. Han, L., et al., *Acid active receptor-specific peptide ligand for in vivo tumor-targeted delivery.* 2013. **9**(21): p. 3647–3658.
42. Shim, M.K., et al., *Carrier-free nanoparticles of cathepsin B-cleavable peptide-conjugated doxorubicin prodrug for cancer targeting therapy.* 2019. **294**: p. 376–389.
43. Brown, J.M. and W.R.J.N.R.C. Wilson, *Exploiting tumour hypoxia in cancer treatment.* 2004. **4**(6): p. 437–447.
44. Wilson, W.R. and M.P.J.N.R.C. Hay, *Targeting hypoxia in cancer therapy.* 2011. **11**(6): p. 393–410.
45. Matsumura, Y. and H.J.C.R. Maeda, *A new concept for macromolecular therapeutics in cancer chemotherapy: Mechanism of tumor tropic accumulation of proteins and the antitumor agent smancs.* 1986. **46**(12_Part_1): p. 6387–6392.
46. Traut, T.W.J.M. and C. Biochemistry, *Physiological concentrations of purines and pyrimidines.* 1994. **140**: p. 1–22.
47. Wang, G.-H., et al., *ATP triggered drug release and DNA co-delivery systems based on ATP responsive aptamers and polyethylenimine complexes.* 2016. **4**(21): p. 3832–3841.
48. Zhang, J., et al., *Inhibition of cell proliferation through an ATP-responsive co-delivery system of doxorubicin and Bcl-2 siRNA.* 2017. **12**: p. 4721.
49. Yang, X., et al., *Multifunctional stable and pH-responsive polymer vesicles formed by heterofunctional triblock copolymer for targeted anticancer drug delivery and ultrasensitive MR imaging.* 2010. **4**(11): p. 6805–6817.
50. Schlossbauer, A., et al., *pH-responsive release of acetal-linked melittin from SBA-15 mesoporous silica.* 2011. **50**(30): p. 6828–6830.
51. Yu, G., et al., *A pillar [5] arene-based [2] rotaxane lights up mitochondria.* 2016. **7**(5): p. 3017–3024.
52. Zhuang, J., et al., *Optimized metal–organic-framework nanospheres for drug delivery: Evaluation of small-molecule encapsulation.* 2014. **8**(3): p. 2812–2819.
53. Zheng, H., et al., *One-pot synthesis of metal–organic frameworks with encapsulated target molecules and their applications for controlled drug delivery.* 2016. **138**(3): p. 962–968.
54. Kim, B.J., et al., *Mussel-inspired protein nanoparticles containing iron (III)–DOPA complexes for pH-responsive drug delivery.* 2015. **127**(25): p. 7426–7430.
55. Gu, Y., et al., *Acetal-linked paclitaxel prodrug micellar nanoparticles as a versatile and potent platform for cancer therapy.* 2013. **14**(8): p. 2772–2780.
56. Zou, J., et al., *pH-Sensitive brush polymer-drug conjugates by ring-opening metathesis copolymerization.* 2011. **47**(15): p. 4493–4495.
57. Pang, X., et al., *pH-responsive polymer–drug conjugates: Design and progress.* 2016. **222**: p. 116–129.

58. Liang, Y. and K.L.J.B. Kiick, *Liposome-cross-linked hybrid hydrogels for glutathione-triggered delivery of multiple cargo molecules.* 2016. **17**(2): p. 601–614.
59. Luo, C., et al., *Self-assembled redox dual-responsive prodrug-nanosystem formed by single thioether-bridged paclitaxel-fatty acid conjugate for cancer chemotherapy.* 2016. **16**(9): p. 5401–5408.
60. Poole, K.M., et al., *ROS-responsive microspheres for on demand antioxidant therapy in a model of diabetic peripheral arterial disease.* 2015. **41**: p. 166–175.
61. Xu, H., W. Cao, and X.J.A.O.C.R. Zhang, *Selenium-containing polymers: Promising biomaterials for controlled release and enzyme mimics.* 2013. **46**(7): p. 1647–1658.
62. Shao, D., et al., *Bioinspired diselenide-bridged mesoporous silica nanoparticles for dual-responsive protein delivery.* 2018. **30**(29): p. 1801198.
63. Wang, Y., et al., *Ultrasensitive GSH-responsive ditelluride-containing poly (ether-urethane) nanoparticles for controlled drug release.* 2016. **8**(51): p. 35106–35113.
64. Wang, L., et al., *Ultrasensitive ROS-responsive coassemblies of tellurium-containing molecules and phospholipids.* 2015. **7**(29): p. 16054–16060.
65. Jäger, E., et al., *Fluorescent boronate-based polymer nanoparticles with reactive oxygen species (ROS)-triggered cargo release for drug-delivery applications.* 2016. **8**(13): p. 6958–6963.
66. Wang, M., et al., *Reactive oxygen species-responsive protein modification and its intracellular delivery for targeted cancer therapy.* 2014. **126**(49): p. 13662–13666.
67. Kim, J.S., et al., *ROS-induced biodegradable polythioketal nanoparticles for intracellular delivery of anti-cancer therapeutics.* 2015. **21**: p. 1137–1142.
68. Xu, X., et al., *ROS-responsive polyprodrug nanoparticles for triggered drug delivery and effective cancer therapy.* 2017. **29**(33): p. 1700141.
69. Lin, S., et al., *Targeted delivery of doxorubicin to tumour tissues by a novel legumain sensitive polygonal nanogel.* 2016. **8**(43): p. 18400–18411.
70. Scomparin, A., et al., *Two-step polymer-and liposome-enzyme prodrug therapies for cancer: PDEPT and PELT concepts and future perspectives.* 2017. **118**: p. 52–64.
71. Zhou, H., et al., *Legumain-cleavable 4-arm poly (ethylene glycol)-doxorubicin conjugate for tumor specific delivery and release.* 2017. **54**: p. 227–238.
72. Huang, H., et al., *Fabrication of doxorubicin-loaded ellipsoid micelle based on diblock copolymer with a linkage of enzyme-cleavable peptide.* 2015. **133**: p. 362–369.
73. Han, H., et al., *Dual enzymatic reaction-assisted gemcitabine delivery systems for programmed pancreatic cancer therapy.* 2017. **11**(2): p. 1281–1291.
74. Cheng, Y.-J., et al., *Enzyme-induced and tumor-targeted drug delivery system based on multifunctional mesoporous silica nanoparticles.* 2015. **7**(17): p. 9078–9087.
75. Kojima, H., et al., *Fluorescent indicators for imaging nitric oxide production.* 1999. **38**(21): p. 3209–3212.
76. Wang, Y., et al., *A nanoparticle-based strategy for the imaging of a broad range of tumours by nonlinear amplification of microenvironment signals.* 2014. **13**(2): p. 204–212.
77. Shi, X., et al., *pH-Responsive unimolecular micelles based on amphiphilic star-like copolymers with high drug loading for effective drug delivery and cellular imaging.* 2017. **5**(33): p. 6847–6859.
78. Ahn, B., et al., *Four-fold channel-nicked human ferritin nanocages for active drug loading and pH-responsive drug release.* 2018. **130**(11): p. 2959–2963.
79. Wyatt, L.C., et al., *Applications of pHLIP technology for cancer imaging and therapy.* 2017. **35**(7): p. 653–664.
80. Cheng, C.J., et al., *MicroRNA silencing for cancer therapy targeted to the tumour microenvironment.* 2015. **518**(7537): p. 107–110.
81. Jia, N., et al., *Preparation and evaluation of poly (L-histidine) based pH-sensitive micelles for intracellular delivery of doxorubicin against MCF-7/ADR cells.* 2017. **12**(5): p. 433–441.
82. Kim, J., et al., *Tumor-homing, size-tunable clustered nanoparticles for anticancer therapeutics.* 2014. **8**(9): p. 9358–9367.
83. Zhao, Y., et al., *A preloaded amorphous calcium carbonate/doxorubicin@ silica nanoreactor for ph-responsive delivery of an anticancer drug.* 2015. **54**(3): p. 919–922.
84. Ye, D.-X., et al., *ZnO-based nanoplatforms for labeling and treatment of mouse tumors without detectable toxic side effects.* 2016. **10**(4): p. 4294–4300.
85. Wang, Y., et al., *ZnO-functionalized upconverting nanotheranostic agent: Multi-modality imaging-guided chemotherapy with on-demand drug release triggered by pH.* 2015. **54**(2): p. 536–540.
86. Chen, Y., et al., *Break-up of two-dimensional MnO2 nanosheets promotes ultrasensitive pH-triggered theranostics of cancer.* 2014. **26**(41): p. 7019–7026.

87. Liu, J., et al., *CO 2 gas induced drug release from pH-sensitive liposome to circumvent doxorubicin resistant cells.* 2012. **48**(40): p. 4869–4871.
88. Kuppusamy, P., et al., *Noninvasive imaging of tumor redox status and its modification by tissue glutathione levels.* 2002. **62**(1): p. 307–312.
89. Wu, J., et al., *Hydrophobic cysteine poly (disulfide)-based redox-hypersensitive nanoparticle platform for cancer theranostics.* 2015. **127**(32): p. 9350–9355.
90. Zhang, S., et al., *Self-delivering prodrug-nanoassemblies fabricated by disulfide bond bridged oleate prodrug of docetaxel for breast cancer therapy.* 2017. **24**(1): p. 1460–1469.
91. Bulmus, V., et al., *A new pH-responsive and glutathione-reactive, endosomal membrane-disruptive polymeric carrier for intracellular delivery of biomolecular drugs.* 2003. **93**(2): p. 105–120.
92. Li, S., et al., *Sarcoma-targeting peptide-decorated polypeptide nanogel intracellularly delivers shikonin for upregulated osteosarcoma necroptosis and diminished pulmonary metastasis.* 2018. **8**(5): p. 1361.
93. Cao, W., L. Wang, and H.J.N.T. Xu, *Selenium/tellurium containing polymer materials in nanobiotechnology.* 2015. **10**(6): p. 717–736.
94. Cao, W., L. Wang, and H.J.C.C. Xu, *Coordination responsive tellurium-containing multilayer film for controlled delivery.* 2015. **51**(25): p. 5520–5522.
95. Stern, R.J.H.I.C.B., *Hyaluronidases in cancer biology.* 2008: p. 207–220.
96. McAtee, C.O., J.J. Barycki, and M.A.J.A.I.C.R. Simpson, *Emerging roles for hyaluronidase in cancer metastasis and therapy.* 2014. **123**: p. 1–34.
97. He, X., et al., *Inflammatory monocytes loading protease-sensitive nanoparticles enable lung metastasis targeting and intelligent drug release for anti-metastasis therapy.* 2017. **17**(9): p. 5546–5554.
98. Giusti, I., et al., *Cathepsin B mediates the pH-dependent proinvasive activity of tumor-shed microvesicles.* 2008. **10**(5): p. 481–488.
99. Lee, J.S., et al., *Lysosomally cleavable peptide-containing polymersomes modified with anti-EGFR antibody for systemic cancer chemotherapy.* 2011. **32**(34): p. 9144–9153.
100. Shim, M.K., et al., *Cathepsin B-specific metabolic precursor for* in vivo *tumor-specific fluorescence imaging.* 2016. **128**(47): p. 14918–14923.
101. Bernardos, A., et al., *Enzyme-responsive intracellular controlled release using nanometric silica mesoporous supports capped with "saccharides".* 2010. **4**(11): p. 6353–6368.
102. Mondragón, L., et al., *Enzyme-responsive intracellular-controlled release using silica mesoporous nanoparticles capped with ε-poly-L-lysine.* 2014. **20**(18): p. 5271–5281.
103. Awino, J.K., et al., *Nucleic acid nanocapsules for enzyme-triggered drug release.* 2017. **139**(18): p. 6278–6281.
104. D'Autréaux, B. and M.B.J.N.R.M.C.B. Toledano, *ROS as signalling molecules: Mechanisms that generate specificity in ROS homeostasis.* 2007. **8**(10): p. 813–824.
105. Yang, B., et al., *Polydopamine-modified ROS-responsive prodrug nanoplatform with enhanced stability for precise treatment of breast cancer.* 2019. **9**(16): p. 9260–9269.
106. Deng, H., et al., *Reactive oxygen species activated nanoparticles with tumor acidity internalization for precise anticancer therapy.* 2017. **255**: p. 142–153.

3 pH-Responsive Delivery Nanoplatforms in Cancer Theranostics

Himanshu Paliwal, Bindiya Chauhan, Devesh U Kapoor, Bhavik H Satani, Shivani Patel, Yashwant V. Pathak and Bhupendra G. Prajapati

CONTENTS

3.1 Introduction 38
3.2 Challenges Associated with the Treatment of Cancer..40
3.3 Benefits of pH-Responsive Nanodrug Delivery..41
3.4 Application of Various pH-Responsive Nanoparticles 41
    3.4.1 pH-Responsive Gold Nanoparticles..41
    3.4.2 pH-Responsive Liposomes 42
    3.4.3 pH-Responsive Polymersomes..42
    3.4.4 pH-Responsive Multiple Core Shell Complexes..42
    3.4.5 pH-Responsive Dendrimers..44
    3.4.6 pH-Responsive Nanoemulsions 45
    3.4.7 pH-Responsive Hydrogels..45
    3.4.8 pH-Responsive Inorganic Nanoparticles 47
    3.4.9 pH-Responsive Polymeric Micelles..47
3.5 Conclusions...49
References..49

3.1 INTRODUCTION

It is challenging to provide accurate diagnosis and therapeutically effective treatment of cancerous tumors due to their heterogeneous nature. Due to cancer sub-types, high diversity in inter-patient tumor heterogeneity occurs, along with epigenetic and unique genetics followed by various dynamic factors such as lifestyle, age, environment and medical history (1). In cancer theranostics, many nanomaterials have emerged due to their multifunctional and intrinsic molecular properties with helpful properties in various diagnosis and imaging results in successful therapies. Nanomaterials synthesized by chemical processes have several issues because of their toxicity, efficacy and cost, whereas, over chemically synthesized nanoparticles, other bio-inspired nanoparticles are preferred owing to their low toxicity, easy synthesis and low cost. Across the globe, prodigious nanotechnology progress has prompted many researchers due to the enormous possibilities and their application in developing diagnostic and imaging-based cancer therapy. In cancer therapy, a providential paradigm of nanomedicine has emerged via multimodal theranostics and benefits in diagnosis and therapeutic delivery results because of newly formed nanoparticles.

In recent years, nanocarriers have demonstrated high efficacy, especially in drug delivery systems (DDSs) based on the nanotechnology used for antitumor treatment. Nanocarrier size is

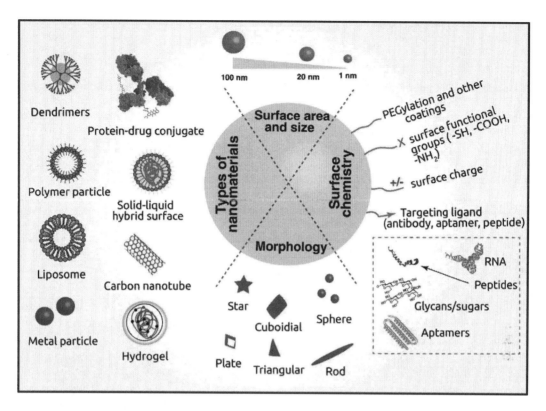

FIGURE 3.1 In cancer therapy, different types of nanomaterials are used for effective therapy (5).

around 100 nm. It's imperative that, based on the drug delivery process, the optimized particle size of nanocarriers plays significant role for therapeutically effective therapy. In the case of pH-responsive nanocarriers, along with programmable size, anti-tumor drugs provide therapeutic high safety and efficacy based on their endogenous stimulus and change in pH (approximate pH for lysosomes to 5.0, endosomes to 5.5, tumor tissue to 6.5). pH-responsive nanocarrier mechanisms are affected by various mechanisms and factors, such as tumor sites, endosomes and lysosomes escaping, lower size for better penetration into tumor parenchyma and disassembly or swelling due to drug release (1, 2).

The development of nanocarriers has shown tremendous progress over the last decade as they have been found to be highly effective drug delivery agents in antitumor therapy. Their versatile surface properties and nanometer size enhance drug pharmacokinetic and pharmacodynamic properties, which helps in providing and developing therapeutically effective therapies with low toxicity. Nanocarriers have been designed in a broad range and tested as micelles, dendrimers, liposomes, star polymers, polymer nanoparticles and various nanoparticles with inorganic compounds such as quantum dots, gold, silica, metal oxide and iron oxide frameworks. Figure 3.1 shows the different types of nanomaterials used for therapeutically effective cancer therapy (3, 4).

There is a need to develop multiple stimuli-responsive nanocarriers to provide efficient DDSs, especially for parenteral (intravenous) drugs, which undergo multiple processes before reaching the site of action for solid tumors. That leads to a reduction in nonspecific interaction with non-targeting sites and cells and adverse effects. By using specially design smart pH-based stimuli-responsive nanocarriers, which increases drug concentration at the site and increases therapeutic efficacy at low doses while reducing side effects (5).

3.2 CHALLENGES ASSOCIATED WITH THE TREATMENT OF CANCER

Per GLOBOCAN data in 2020, 19.3 million new cases and 10 million cancer deaths were estimated. The most frequently diagnosed cancers are colorectal, lung, prostate and breast. Worldwide, lung cancer is one of the leading causes of incident and mortality due to cancer. Cancer treatment in unselected patients has reached its therapeutic plateau due to broad usage of cytotoxic chemotherapies. The paradigm for treating cancer has drastically evolved during the past 15 years as a result of molecular biology information that has been gained. Tumor molecular profiling and predictive molecular target discovery have led to evolution in cancer treatment by genotype-directed precision-based oncology and tumor microenvironment targeting by selective components. Major difficulties associated in cancer treatments are discussed in the following (5–8).

Targeting cancer stem cells is one of the most difficult challenges associated with cancer therapy. Lack of specificity and target-specific drugs are unable to provide long-lasting efficacy against cancer, as this leads to a regression impact on bulk tumors. This fails to eradicate cancer stem cells, and tumor recurrence is commonly observed after discontinuation of drug therapy. There are many barriers to overcome in order to develop highly specifically action on the identification and targeting of cancer stem cells for future therapeutic strategies.

Cancer stem cells have developed drug-resistant properties which induce immunity towards anticancerous drugs. They have developed unique mechanisms to protect themselves against harmful xenobiotic agents. This is a normal stem cell self-renewal process which they undergo and leads to differentiation during the entire individual life span. ATP-binding cassette transporter proteins expressed drug resistance at high levels of stem cells. The most significant relevant example in breast cancer cells in which drug resistance of protein has been observed. Breast cancer resistant protein (BCRP) ABCG2, which belongs to a specific ATP-binding of cassette transporter.

A lack of epigenetic profiling in cancer and the specificity of existing epi-drugs which mainly results due to mutational and other aberrations at the chromosomal level. The primary focus in cancer research is on the identification and determination of genetic anomaly patterns. Meager data are available related to cancer-associated genetic mutations, which causes difficulty in interpretation for enormous genetic deviation in cancer, especially in malignant phenotypes. Numerous studies are in progress and have been conducted with main aim to identify specific epigenetic changes which stimulate cancer development. Examples of gene transcriptional inactivation and hypomethylation lead to genomic instability, which has been observed in cancer studies, like DNA methyl transferase (DNMT), responsible for methylation for CpG dinucleotide.

Diagnosis is difficult in cancer with non-specific symptoms where patients are asymptotic. Esophageal cancer, which is considered one of the most lethal cancers and is difficult to treat. Prostate cancer, mainly found in older people in the over-50 age group, is the most prevalent cancer in males. In early stages of slow growing, it remains symptom free and undetected. With continuous growth, it metastasizes to lymph nodes and bones. It can be detected by biopsy or with a prostrate-specific antigen. Pancreatic cancer is also called a silent disease, as it also non-symptomatic at early stages, and symptoms shown at the lateral stage depend on the cancer stage and location.

In cancer treatment, the unavailability of effective biomarkers is another major hindrance to cancer diagnosis and prognosis. They are important for diagnostic purposes and also have great prognostic value. The right biomarker identification provides the cancer progression stage, and chemotherapeutic drug effects can be easily evaluates to provide specific and accurate cancer treatment. Clinical proteomics have shown promising results which enable early stage detection and cancer disease progression at early stages, identifying effective biomarkers, but this research approach requires more standardization and validation studies.

The limitations of conventional chemotherapeutic agent are due to their toxic effects on all cancer and normal cells. Patient deaths are reported as these therapeutic toxic agents kill rapidly proliferating cancer and normal cells, which results in serious side effects which lead to death. Untargeted radiotherapy also lacks specificity and shows serious toxicity, which also results in deaths in patients. Metastasis is one of the main reasons for difficulties in cancer treatment. Certain cancer asymptomatic nature and lack of diagnosis, without medical intervention, cancer from sites of origin spreads to different body parts. Cancer's first site is the "primary site", and the spread site is the "metastatic or secondary site". Due to the invasion and colonizing ability of cancer, it spreads to different parts in the body. Moreover, certain metastatic events result in too little metastasized cancer, called micrometastases events, which are not easily detectable.

3.3 BENEFITS OF PH-RESPONSIVE NANODRUG DELIVERY

Currently, there are many efficient tumor treatment options, including molecular targeted therapy, radiotherapy, chemotherapy and surgical removal of the tumor. The fields of immunotherapy, gene therapy, photodynamic treatment, photothermal therapy and chemodynamic therapy have all advanced significantly over time. It is well known that the traditional method of administering chemotherapy results in inadequate drug targeting to tumors, low drug use and numerous other issues. Since chemotherapeutic medications are loaded and conveyed on carriers, academics have proposed DDSs. The development of nano-drug carriers that can react to aberrant biochemical indicators in the tumor microenvironment and release medications directly into the tumor site has become a hot topic in tumor therapy research to increase the efficacy of drug delivery techniques. The permeability and retention of medicines in solid tumors can be improved by introducing different stimuli-responsive components into DDSs. Polymers with pH-sensitive properties are frequently employed in stimulus-responsive novel DDSs (9, 10).

3.4 APPLICATION OF VARIOUS PH-RESPONSIVE NANOPARTICLES

3.4.1 pH-Responsive Gold Nanoparticles

Another intriguing type of NPS is gold nanoparticles (Au NPs), which have drawn a lot of interest due to their larger surface area, great drug loading capacity and thiolated compound simplicity in functionalization. Researchers from many teams have studied pH-sensitive Au NPs. For instance, Kumar et al. (9) created pH-responsive Au NPs loaded with doxorubicin (DOX) and embellished them with the brief tripeptide Lys-Phe-Gly (KFG). Glioblastoma (U251), human embryonic kidney, transformed, and cervical carcinoma cell cultures were used to test the generated NPs. According to the MTT assay, less viable cells were seen in the cells exposed to DOX-loaded KFG-Au NPs than in the cells incubated with free DOX. The DOX-KFG-Au NPs internalized HeLa cells more than free DOX, according to the flow cytometry results. Breast cancer (BT-474) cell xenograft nude mice were used for in vivo testing, and the results revealed that the DOX-KFG-AuNP treatment groups had considerably lower tumor sizes than the free DOX treatment groups. A PEG and folic acid (FA) functionalized graphene oxide (GO) embellished with gold nanoparticles (GO-PEG-FA/GN) was created by Samadian et al. (10). Human breast cancer (MCF-7) cells were used to investigate the anti-cancer effectiveness of the newly designed hybrid system, which contained DOX. Due to the weakened stacking and hydrophobic contacts between the drug molecules and the NPs, GO-PEG-FA/GNs demonstrated more drug release at pH 4.0 than that observed at a pH of 7.4, which was indicative of their pH responsiveness. In addition, as compared to free medication, GO-PEG-FA/GNPs were more hazardous to cancer cells. In a different study, Khodashenas et al. (11) looked at the drug delivery of methotrexate (MTX) for the treatment of breast cancer. MTX was added to nanorod-shaped and spherical gelatin-coated NPs. The characterization results demonstrated that the spherical AuNPs' entrapment efficiency was greater than that of Au-nanorods (Au-NRs).

3.4.2 pH-Responsive Liposomes

The hydrophilic heads of the phospholipids are directed toward the aqueous environment, while the hydrophobic tails are directed toward the interior of the bilayer. The unique capacity to encapsulate both hydrophilic (in core) and hydrophobic drugs is a result of liposomes' structure (in bilayer). Because of their non-toxic makeup, biocompatibility, biodegradability and lack of immunogenicity, liposomes rank among the finest DDSs. Because of their amphiphilic properties, phospholipids may encapsulate both hydrophilic and hydrophobic drugs and are similar to natural cell membranes, which enhances efficient cellular uptake (12, 13). Furthermore, adding targeting molecules and/or stealth-imparting polymers to the surface of liposomes is straightforward (such as PEG) in a variety of targeted modalities. The duration of the liposomes circulation in the body is prolonged by PEGylating them, which stops their interaction with RES organs. Because conventional and stealth liposomes cannot be delivered directly to target cells, this fundamental shortcoming led to the development of ligand targeted liposomes. By including one or more ligands that can target receptors and are overexpressed on the surfaces of a particular cell type, such as cancer cells, these liposomes' selective therapeutic efficacy is increased. Liposomes can also release drugs under the control of internal or external cues (14–16). The pH-responsive liposome, which responds to the acidic environment of the tumor's microenvironment to release its contents, is one form of internal stimuli-sensitive liposome that is often used in cancer therapy. The use of neutral lipids is common in pH-responsive liposomes. Negatively charged phospholipids are more likely to fuse with cellular and endosomal membranes in an acidic tumour environment, which causes the release of liposomal contents (17, 18). For example, Zhai et al. employed the pH-sensitive liposome-creating peptide DVar7 (DOPE-DVar7-lip@DOX), which has been the subject of numerous studies focusing on cancer therapy (13), to make DOX-liposomes which are pH responsive. They used flow cytometry and near-infrared (NIR) fluorescence imaging to examine the anti-cancer effectiveness of DOPE-DVar7- lip@DOX in vitro and in vivo. When pH was 5.3 instead of 7.4, DOPE-DVar7-lip@DOX emitted nearly five times as much DOX.

3.4.3 pH-Responsive Polymersomes

These are ideally constructed from biocompatible, biodegradable and amphiphilic polymers (19). They have the potential to be versatile DDSs due to their adaptable membrane formulations, stability in vivo, various physicochemical features, controlled release mechanisms, targeting abilities and capacity to encapsulate a variety of pharmaceuticals, among other things (19, 20). pH-sensitive polymersomes have been developed to quickly respond to even the smallest changes in the pH of the tumor's microenvironment (21, 22). After pH is adjusted, the pendant basic (amine) or acidic (carboxylic acids/sulfonic acids) groups undergo protonation or deprotonation. Since the drugs are released more quickly at the target location as a result of the structural shift, the formation and deformation of polymeric vesicles improves the therapeutic index. However, pH-sensitive polymersomes delay reaction to stimuli, which results in a gradual drug release and ultimately leads to drug resistance in the surrounding cells, which is the greatest obstacle to their utilization. Polymersomes must respond quickly as a result of the decreased pH at sick sites. Further research on pH-responsive polymersomes is required in order to transport, deliver and regulate the release of therapeutic medications to tumor tissue based on the low pH in the area around the tumor tissues (23, 24). Anajafi and Mallik (2015) (22) have added additional commentary on recent developments in polymersomes. Gynecological malignancies may respond to them, but this has not yet been established.

3.4.4 pH-Responsive Multiple Core Shell Complexes

The addition of polyelectrolyte multilayers enables pH-responsive drug encapsulation and release from numerous core-shell nanoparticles (25). Huang and colleagues(25) first created a functionalized core for a mesoporous silica nanoparticles, which was then covered in a number of polyelectrolyte layers. The polyelectrolyte shell was subsequently incorporated with DOX. After 72 hours at pH 5.2, the resulting core-shell nanoparticles with DOX loading showed more than 60% DOX leakage. DOX-loaded nanoparticles were more cytotoxic than free DOX, according to in vitro tests on MCF-7 breast cancer cells. Figure 3.2 depicts the schematic entry and release of drug from mesoporous silica nanoparticles (26). Oligo (ethylene glycol) methyl ether methacrylate (OEGMA), 2-(diisopropylamino) ethyl methacrylate (DPA) and glycidyl methacrylate were combined to create an azide-terminated deblock copolymer by Tian et al. (27). After functionalization with DOTA(Gd) and 4-(prop-2-ynyloxy) benzaldehyde, the resultant copolymer was co-assembled into mixed micelles. Tetrakis[4-(2-mercaptoethoxy) phenyl] ethylene (TPE-4SH) was successfully encapsulated by the GMA moieties present inside the cores, and the ensuing micelles were capable of dual MR and fluorescence imaging. These micelles also have a pH low insertion peptide (pHLIP) surface conjugation, which allowed them to release the cancer drug camptothecin in situ while selectively targeting tumor tissues, as demonstrated by in vivo MR imaging of BALB/c nude mice with tumors. (24). In this synthesis, the Gd-core was first conjugated to the polymer backbone to produce the Gd-core, and then the DOX was enclosed inside the shell that the Gd-core was housed in. One of the molecules in the core was decided to be resorcinol. The ability of 1,3-phenylenediamine to release under pH control was an important factor to consider when selecting the shell unit. In vitro and in vivo research on the novel theranostic nanoparticle UCS-Gd-DOX revealed that the DOX in the shell is effectively and arbitrarily released in tumor acidic environments (pH 5.5). At pH 8.0, 7.0, 6.0, 5.0 and 4.0, respectively, the in-vitro pH-dependent release of DOX was reported to be 5%, 10%, 55%, 75% and 80%. In targeted therapy of tissues harboring cancerous tumors using UCS-Gd-potential DOXs as the pH dropped from 7.0 to 6.0, the medicine released more readily. The pH-switchable MR signal was also visible in in-vitro T1-weighted MR imaging tests that contrasted with resorcinol, one of the UCS-Gd-capacity DOXs. In addition to efficiently reducing the growth of subcutaneous human cervical cancer in mouse xenograft models, the pH-responsive UCS nanoparticle was created to improve the MRI contrast at the tumor site in comparison to other tissues and organs. Theranostic nanoparticles that are Gd conjugated

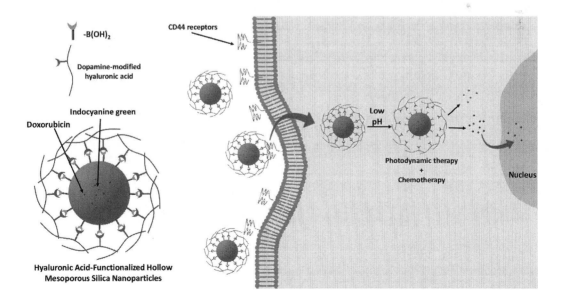

FIGURE 3.2 Pictorial representation of entry and pH-responsive release of anticancer drug from hyaluronic acid-functionalized hollow mesoporous silica nanoparticles (26).

and DOX doped, as well as additional UCS-Gd-DOX applications, are thus all feasible. According to reports, targeted anticancer drug delivery using pH-sensitive magnetic nanoparticles has been developed. A multilayer core-shell nanocarrier with dual responsiveness to pH and magnetism was developed at the start of this decade. When directing the drug-loaded nanocarriers to the sick area, the $Fe_3O_4@SiO_2$ nanoparticles were used as a superparamagnetic core. The mPEG and PBLA segments, on the other hand, functioned as a pH-sheddable hydrophilic core and a hydrophobic inner layer, respectively, to load the drug DOX via hydrophobic interactions. With regard to the targeted intracellular administration of hydrophobic chemotherapeutics in the treatment of cancer, this approach has been shown to be very promising (28, 29). 2017 saw the publication of Karimi and his team's pH-sensitive magnetic nanoparticle method for methotrexate (MTX) targeting of cancers (29).

3.4.5 pH-Responsive Dendrimers

Dendrimers are globular macromolecules and highly branched, with different arms radiating from a central core. Dendrimers are desirable for drug delivery and biological applications due to their high surface functionality, good polydispersity and water solubility. Due in part to the various functional groups on their surface, dendrimers are potential nanocarriers with distinctive physiochemical features that have drawn a lot of interest in studies on cancer therapy (30, 31). pH-responsive PAMAM dendrimers can be formulated by altering the functional groups at the dendrimer surface. Additionally, the anticancer agent is joined to dendrimer-drug conjugates via pH-sensitive links. PAMAM's ability to improve biodistribution of drugs and the EPR effect for targeting tumors make it an advantageous drug carrier (32, 33).

Wang et al. formulated tecto dendrimers (pH responsive) modified with benzimidazole (BMZ). The researchers employed third-generation (G3) PAMAM dendrimers for the core and fifth-generation (G5) PAMAM dendrimers for the shell modified with β-cyclodextrin (BCD). The formulated pH-responsive G3/G5 PAMAM dendrimers were encapsulated with anticancer drug doxorubicin by hydrophobic interactions. Different techniques were used to thoroughly characterize both the loading of the DOX and the synthesis of dendrimer derivatives. The researchers observed that under slightly acidic pH conditions (pH 6), which is analogous to the acidic tumor microenvironment, the encapsulated DOX can be released in a sustained way with a rapid release speed. By using flow cytometry and cancer cell viability assessment, the researchers demonstrated surged intracellular release of DOX and drug-loaded pH-responsive core shell PAMAM dendrimer antitumor activity. The investigators found that pH-responsive core shell tecto dendrimers exhibited significant drug penetration and inhibition of tumor spheroid growth. This attenuation was due to quicker release of DOX in an acidic pH environment. The researchers concluded that pH-sensitive G3/G5 PAMAM tecto dendrimers could be used as intelligent delivery system for different antiproliferative drugs (34).

Mekonnen and his team formulated 4.5-generation PAMAM dendrimers loaded with resveratrol (RSV) and DOX by employing sodium deoxycholate hydrogel (NA-DO-HYD) in a pH tumor microenvironment. By entrapping G4.5 PAMAM-DOX in the RSV-loaded NA-DO HYD, the sequential, regulated and sustained release of RSV and DOX for synergistic anticancer effects was proven. The HeLa cells were employed to assess the synergistic anticancer activity of NA-DO-HYD-SV+G4.5-DOX in Balb/c nude mice. The researchers observed by MTT assay that G4.5 dendrimer and NA-DO-HYD showed no cytotoxicity with highest dose (2 mg mL^{-1}) in HaCaT, HeLa and MDA-MB-231 cells. In the seven days of drug release experiments, the release patterns of DOX and RSV from the NA-DC-HYD-RSV+G4.5-DOX indicated that RSV was released very quickly (71.33 2.19%), followed by DOX (53.48 1.63%). During the course of the 28-day treatment period, a single intratumoral injection of NA-DC-HYD-RSV+G4.5-DOX effectively reduced tumour growth. The primary visceral organs were not histologically harmed by Na-DOC-hyd-RESV+G4.5-DOX. The researchers recapitulated that NA-DO-HYD/4.5 PAMAM dendrimers is a safe, effective and promising combination to deliver RSV and DOX at tumors' pH extracellular microenvironment for cancer treatment (35).

Wang et al. fabricated pH-responsive polyacetal dendrimers (PLDs) functionalized with zwitterionic sulfobetaine (ZSB) by using thiol ene-click and alternant aza-michael addition reactions for anticancer drug delivery. The DOX was loaded with fabricated ZSB-PLD. The researchers observed that ZSB-PLD-DOX showed noteworthy pH responsive drug release and significant antitumor activity (36).

The tripeptide dendrimer-based formulation equipped with polyethylene groups and catechol was investigated for the targeted delivery of bortezomib (BTZ) against metastatic bone cancers. The boronate–catechol linkage with a pH-responsive property, used to load the BTZ on the dendrimers, plays a vital role in the loading and release of BTZ. At pH 7.4, the non-targeted bortezomib nanomedicine had negligible cytotoxicity, but when cyclic RGD (CRGD) moieties were anchored to the dendrimer surface, anticancer activity considerably increased. The researchers observed that breast cancer cells like MDA-MB-231 cells were able to efficiently internalize the bortezomib complex due to the ligand CRGD. In a bone tumor model, the targeted nanomedicine greatly reduced tumor-associated osteolysis and effectively slowed the growth of metastatic bone cancers (37).

Hu et al. also reported the successful delivery of DOX against liver cancer (HepG2 cancer cell lines) by using pH responsive G4 PAMAM dendrimers (G4PD) conjugated with transferrin. The G4PD-TF-DOX exhibited significant antitumor activity and increased apoptotic activity (38).

3.4.6 pH-Responsive Nanoemulsions

Pharmaceutical formulations known as nanoemulsions (NEs) are made up of nanoscale particles. Due to their hydrophobic core, they can encapsulate medications that are only weakly soluble in water. NEs are effectively taken up by tumor cells, which slows the growth of tumors, eliminates toxicity to healthy cells and prevents cancer cells from spreading to other organs. Comparing NEs to conventional drug carriers, their key benefit is that they can be made to specifically target tumor cells while avoiding multidrug resistance (MDR). A novel and promising approach to treating cancer is called NEs. By using a hydrophobic core, it is possible to encapsulate lipophilic medications, solving one of the major issues with cancer treatment medications (39).

Miranda et al. formulated NEs loaded with antitumor drug lapachol (LC). A hot homogenization technique was used to create the NEs, and cryogenic transmission electron microscopy was used to study their morphology (cryo-TEM). By using DLS, the average diameter, polydispersity index and zeta potential were assessed. HPLC was used to gauge encapsulation effectiveness. Additionally, drug release, hemolysis in vitro and short-term storage stability were assessed. Additionally, a breast cancer (4T1) tumor model was used to assess the pharmacokinetic, toxicological, and cytotoxic features of NEs-LC. The NEs-LC physicochemical characterization revealed homogenous stable NEs with a mean diameter of around 171 nm, a zeta potential of roughly 21 mV and an encapsulation level of more than 86%, while the cryo-TEM revealed spherical globules. The encapsulation did not weaken the cytotoxicity activity of LC confirmed by in vitro studies. Additionally, NE-LAP has shown greater anticancer efficacy than free medication, and the therapy had no adverse side effects. As a result, these results imply that NE-LAP can be regarded as a successful cancer therapy method (40).

3.4.7 pH-Responsive Hydrogels

A polymer gel called a pH-sensitive hydrogel changes in volume in response to the ionic strength and pH of the surrounding environment. Many readily hydrolyzable or protonated acids, base groups such as carboxyl groups and amino groups are present in such gels. The external pH has an impact on these groups' ability to dissociate; as a result, when the external pH varies, these groups' level of dissociation changes as well, changing both the internal and external ion concentration. The bond alters the structure of the gel network and the degree of hydrogel swelling by lowering the cross-linking point of the gel network (41, 42). pH-responsive peptide hydrogel is regarded as an appropriate intratumoral injection vehicle for chemotherapy.

Lee et al. combined transferrin, dithiothreitol and an anticancer medication in a salt solvent to fabricate a self-assembling hydrogel. To regulate the concentration of the transferrin protein, dithiothreitol and salt in the solvent, the structural characteristics of the hydrogel were established. The required quantity of anticancer medications was present in the self-assembled hydrogels. With this method, variations in pH and temperature regulate the release ratio and pace of anticancer medications. The researchers observed the evaluated cytotoxicity of the drug-loaded hydrogel, which showed that more than 80% of treated cancer cells were killed with the prepared formulation with 40 h of treatment (43).

For targeted pH-responsive skin cancer therapy, a novel in situ–forming nanocomposite hydrogel (NCH) incorporating PLGA-carboxymethyl chitosan NPs has been developed. Within 5 minutes, this injectable hydrogel, which has a polymer backbone produced from citric acid, gelled and had good swelling (287% of dry weight) and compressive strength (5.26 MPa). The NCH effluents showed pH-dependent drug release and promise therapeutic activity against A431 and G361 skin cancer cells in vitro and were cytocompatible with human dermal fibroblasts. The researchers concluded that the pH-responsive NCH has potential use in the adjuvant treatment of skin cancer (44). The injectable OE peptide hydrogel (pH responsive) delivered the anticancer drug PTX and gemcitabine (GCB) at the tumor site effectively and showed noteworthy antitumor effects (45).

Raza et al. fabricated a paclitaxel (PTX) medication delivery system for targeting tumors utilizing a pH-responsive FER-8 peptide hydrogel (FER-8-PHG). The loading capacity, acid sensitivity, structure, rheology, morphology, drug release, in vitro cytotoxicity and in vivo effectiveness of the pH-sensitive hydrogel system were studied in H22 tumor-bearing mice. At pH 7.4, it was found that the FER-8-PHG fibers had an average size less than 500 nm, which was supported by TEM and DLS analysis. The FER-8-PHG-PTX showed sustained release of PTX at pH 5.5 for almost a week. When compared to free pharmaceuticals, the FER-8-PHG significantly reduced the growth of HepG2 tumor cells and boosted drug accumulation in HepG2 cells, according to in vitro cytotoxicity experiments. Furthermore, FER-8-PHG-PTX dramatically boosted the amount of drugs in tumor tissues and demonstrated sustained retention at the tumor site by intratumoral injection, according to in vivo investigations employing H22-bearing mice. The in vivo anti-tumor evaluation supported pH-sensitive HG-PTX characteristics, which enabled the drug to be activated by the acidic pH environment at tumor sites, provided sustained drug administration and improved tumor inhibition. Finally, FER-8-PHG-PTX offers a desirable approach and prospective platform for effective anti-cancer medication delivery. The carrier can improve therapeutic targeting for the tumor, extend retention, lessen systemic side effects and promote drug accumulation at the tumor site (46).

Cimen et al. fabricated an injectable pH-responsive hydrogel equipped with aldehyde functionalized PEG (PEG-DIBA) and hydrazide functionalized gelatin (ADH-GEL). These hydrogels were loaded with anticancer drug DOX and evaluated against endothelial cell line (HUVEC) and human breast cell line (SVCT). The hybrid hydrogel's measured gelation time was 80 seconds, and the generated hybrid hydrogels showed outstanding injectability and quick self-healing ability. Hybrid hydrogels' gel-sol transition behaviors showed exceptional gelation stability, which is a highly desired property in applications involving controlled drug delivery. The long-term drug release profile of the injectable hybrid ADH-GEL/PEG-DIBA-DOX hydrogels was extremely effective and pH dependent. The hybrid hydrogel's components are all highly biocompatible and even encourage cell growth. Additionally, the MCF-7 human breast cancer cell line and the triple-negative breast cancer cell line were used to test the hydrogels' cytotoxicity (MDA-MB-231). Our findings strongly suggested that this hybrid ADH-GEL/PEG-DIBA-DOX hydrogel, which is injectable, self-healing and pH responsive, has a promising future as a drug carrier for long-term and controlled release applications (47).

3.4.8 pH-Responsive Inorganic Nanoparticles

Inorganic nanoparticles (INPs) exhibit distinctive physical properties, including electronic, optical, magnetic and catalytical, and have good tunable characteristics such as surface, size, shape and morphology. Due to their customizable qualities, including size, surface characteristics and interior cavity dimensions, inorganic nanoparticles are desirable carriers for the delivery of drugs, genes and proteins (48).

Dhavale et al. fabricated magnetic NPs (MNPs) decorated with chitosan and loaded with anticancer drug telmisartan (TSR). The researchers observed that poorly bioavailable antitumor drug TSR therapeutic effectiveness has been reported to be improved through nanoformulation. The TEM, FE-SEM, XRD and BET surface area analyzer and vibrating sample magnetometer were employed to evaluate the prepared CS-MNPs-TSR. TGA and FTIR were used to validate the chitosan coating on MNPs. The TSR was loaded on CS-MNPs by an amide bond between the carboxylic group of TSR and amino group of chitosan. The greater surface area and mesoporous characteristics of chitosan-coated MNPs allowed for higher drug loading. The researchers observed that the prepared formulation was stable at normal physiological conditions; in contrast, release of the drug was higher in the cancerous environment. Studies on the cytotoxicity of PC-3 human prostate cancer cells showed that cell viability was decreased in a dose-dependent manner (49).

3.4.9 pH-Responsive Polymeric Micelles

The spherical supramolecular nanoassemblies known as micelles, which range in size from 20 to 100 nm, have drawn a lot of attention as possible drug nanocarriers because of their distinctive qualities, including high drug-loading capacity, excellent solubility and low toxicity. Individual polymeric molecules, which are synthetic amphiphilic di- or tri-block copolymers made up of hydrophilic and hydrophobic blocks, spontaneously self-assemble to form polymeric micelles. Normally, micelles are structures which are spherical in shape, with a hydrophilic shell and hydrophobic core. By adding "titratable" groups like amines or carboxylic acids to block copolymers or pH-sensitive linkages (hydrazone linkages), pH-responsive polymeric micellar systems can be formulated. Due to their association with aberrant vasculature, nano polymeric micelles of a specific size (30 to 100 nm) can passively aggregate and remain within tumors, generating the desired effects. The polymeric nanomicelles have the potential to increase site-specific drug delivery and also can control the drug release rate in the microenvironment of tumors by moderating their physicochemical properties and chemical structure (50).

Han et al. formulated a polymeric micelle system (pH sensitive) based on MPEG and equipped with chitosan (CTS) for the effective delivery of paclitaxel (PTX) against different breast cancer cell lines. The MPEG-CTS-PTX micelles were prepared by using an ultrasonic probe method. The zeta potential and average particle size of optimized PTX loaded polymeric micelles were +22.8 ± 0.9 mV and 145 ± 3 nm, respectively. The PTX-loaded MPEG-CTS-PTX polymeric micelles were stable under typical physiological settings (pH 7.4), while quick drug release was observed by the researchers in the simulated tumor intracellular microenvironment (pH 5.0), according to an in vitro drug release investigation. The polymer itself was shown to be non-toxic in an in vitro cytotoxicity investigation, and the PTX-loaded micelles displayed superior cytotoxicity and substantial selectivity on tumor cells. The investigators also found that PTX-loaded polymeric micelles could enhance the therapeutic efficiency of PTX and lessen its negative effects confirmed from an in vivo anticancer efficacy investigation. The researchers concluded that MPEG-CTS-PTX polymeric micelles may be promising pH-sensitive nanocarriers for PTX administration, according to all of these findings (51).

Mehnath et al. fabricated pH responsive polymeric micelles conjugated with poly(bis(carboxyphenoxy) phosphazene loaded with anticancer drug PTX. The drug released continuously in the acidic tumor microenvironment, according to in vitro experiments, and had good

TABLE 3.1
List of Studies Based on Delivery of Anticancer Drugs Using a Variety of pH-Responsive Nanoparticles

Formulation	Composition	Anticancer Drug	Cancer Cell Lines	Type of Cancer	References
Hydrogel	Chitosan and pullulan	Doxorubicin (DOX)	HCT116	Colon cancer	(54)
	Chitosan	DOX	MCF-7	Human breast cancer	(55)
	N-carboxyethyl chitosan	DOX	HepG2, L929	Hepatocellular carcinoma	(56)
	Fe_3O_4	Methotrexate (MTX)	MCF-7	Human breast cancer	(57)
	Gelatin/PEG/laponite	DOX	MCF-7 and MDA-MB-231	Human breast cancer and triple negative breast cancer (TNBC)	(47)
	Chitosan-PVA/silica	Cisplatin	HCT-116 and HEK-293	Colon cancer	(58)
	Sodium deoxycholate	DOX and resveratrol (RSV)	MDA-MB-231 and HaCaT	Human breast cancer	(35)
	Protein polysaccharides	Naringenin	Human colorectal cell line	Colorectal cancer	(35)
	Luteolin-PEG	Bortezomib (BTZ)	HCT116	Colorectal cancer	(59)
	Hyaluronic acid	DOX and cisplatin	A549	Lung cancer	(60)
Polymeric nanoparticles/	Bovine serum albumin (BSA)-PLGA NPs	DOX	MCF-7	Human breast cancer	(61)
	Chitosan and glucuronic acid silica NPs	Capecitabine (CCB)	HCT-116	Colorectal cancer	(62)
Inorganic nanoparticles	Polyacrylic acid silica NPs	PTX and arsenic trioxide	MCF-7	Human breast cancer	(63)
	Hyaluronic acid silica NPs	DOX	4T1	Breast cancer	(26)
	Aptamer functionalized silica NPs	DOX	MCF-7	Human breast cancer	(64)
	Chitosan and glucuronic acid silica NPs	5-Flurouracil	HCT-116	Colon cancer	(65)
	Carboxylated chitosan NPs	DOX	HER2	Triple negative breast cancer	(66)
	Silica NPs	DOX	MCF-7	Human breast cancer	(67)
	B-cylcodextrin NPs	5-FU and naproxen	SKBR3 and U87MG	Breast cancer and human glioblastoma	(68)
	Hyaluronic acid, aminated silica NPs	Methotrexate	SMMC-7721	Liver cancer	(69)
	Magnetic NPs	DOX	4T1	Breast cancer	(70)
	Layered double hydroxide magnetic NPs	Lamivudine	A549 and Mel-Rm	Lung and skin cancer	(71)
	Carboxymethyl chitosan amine magnetic NPs	DOX	3T3 and MCF-7	Breast cancer	(72)
	Graphene oxide magnetic NPs	Quercetin	MDA-MB-231	Breast cancer	(73)
	3-aminopropyl triethoxysilane magnetic NPs	5-FU	MCF-7 and HEK293	Breast and kidney cancer	(74)
Nanoemulsions	Water in oil in water (W/O/W)	Quercetin	MCF-7	Human breast cancer	(75)
	Pickering	DOX	4T1	Breast cancer	(76)
	Lipid targeting	DOX	MDA-MB-231	Triple negative breast cancer	(77)
	Chitosan/agarose	5-Fluorouracil/curcumin	MCF-7	Human breast cancer	(78)
	Vitamin E-TPGS	Quercetin	HCT-116 and HT-29	Colon cancer	(79)

biocompatibility. In breast cancer cells, these drug-loaded micelles had a higher potential for cytotoxicity than free PTX. The research findings support the polymeric micelles' positive targeted action and increased anticancer activity, which hold tremendous promise for the safe and efficient treatment of breast cancer cells (52). pH-responsive and CD44-targeting polymer micelles were conjugated with PEG block hydroxyethyl starch, block poly (L-lactic acid) prepared and loaded with natural anticancer drug emodin. The EMO-CD44-PEG-PLA micelles were evaluated against breast cancer cell lines (MDA-MB-231). The DLS and TEM results of the synthesized polymeric EMO-CD44-PEG-PLA micelles revealed an average size of 155.4±0.5 nm. The researchers observed that EMO-CD44-PEG-PLA micelles exhibited significant EMO loading capacity, excellent pH responsiveness and noteworthy thermal stability. An intracellular uptake investigation demonstrates that the increased exposure of CD44p promotes the cellular ingestion of EMO-CD44-PEG-PLA micelles in a significant way. The in vitro studies further demonstrated that EMO-CD44-PEG-PLA micelles had good biocompatibility and anti-tumor properties. EMO-CD44-PEG-PLA polymeric micelles offer a promising method of delivering targeted therapy for breast cancer (53).

3.5 CONCLUSIONS

Several pH-sensitive nanocarriers have been extensively researched in recent years to enhance the targeted and effective delivery to cancer cells. Many researchers have focused upon development and characterization of these nanocarriers, but challenges still exist in designing clinically relevant products. This chapter provides an outline of the reports showing improvement of drug release by specially designed nanoparticle-based carriers made up of either pH-sensitive components or chemically modified. The studies have reported that the tumor microenvironment is usually acidic in nature, which can assist in endogenous release of drugs after stimulation. Although cancer cells show intense metabolism, there is a considerable difference between the pH of tumor and normal cells. This chapter covered applications of several specialized carriers, like polymeric nanoparticles, hydrogels, dendrimers and nanoemulsions, in production of pH-sensitive formulations. The different sections covered important outcomes of some of the recently published studies, along with special emphasis on benefits or limitations of the formulation.

REFERENCES

1. Wu W, Luo L, Wang Y, Wu Q, Dai H-B, Li J-S, et al. Endogenous pH-responsive nanoparticles with programmable size changes for targeted tumor therapy and imaging applications. Theranostics. 2018;8(11):3038–3058.
2. Pramod Kumar EK, Um W, Park JH. Recent developments in pathological pH-responsive polymeric nanobiosensors for cancer theranostics. Frontiers in Bioengineering and Biotechnology. 2020;8.
3. Din FU, Aman W, Ullah I, Qureshi OS, Mustapha O, Shafique S, et al. Effective use of nanocarriers as drug delivery systems for the treatment of selected tumors. International Journal of Nanomedicine. 2017;12:7291–7309.
4. Wei QY, Xu YM, Lau ATY. Recent progress of nanocarrier-based therapy for solid malignancies. Cancers. 2020;12(10).
5. Navya PN, Kaphle A, Srinivas SP, Bhargava SK, Rotello VM, Daima HK. Current trends and challenges in cancer management and therapy using designer nanomaterials. Nano Convergence. 2019;6(1):23.
6. Madamsetty VS, Mukherjee A, Mukherjee S. Recent trends of the bio-inspired nanoparticles in cancer theranostics. Frontiers in Pharmacology. 2019;10.
7. Zugazagoitia J, Guedes C, Ponce S, Ferrer I, Molina-Pinelo S, Paz-Ares L. Current challenges in cancer treatment. Clinical Therapeutics. 2016;38(7):1551–1566.
8. Chakraborty S, Rahman T. The difficulties in cancer treatment. Ecancermedicalscience. 2012;6:ed16.
9. Kumar K, Moitra P, Bashir M, Kondaiah P, Bhattacharya S. Natural tripeptide capped pH-sensitive gold nanoparticles for efficacious doxorubicin delivery both in vitro and in vivo. Nanoscale. 2020;12(2):1067–1074.

10. Samadian H, Mohammad-Rezaei R, Jahanban-Esfahlan R, Massoumi B, Abbasian M, Jafarizad A, et al. A de novo theranostic nanomedicine composed of PEGylated graphene oxide and gold nanoparticles for cancer therapy. Journal of Materials Research. 2020;35(4):430–441.
11. Paroha S, Chandel AKS, Dubey RD. Nanosystems for drug delivery of coenzyme Q10. Environmental Chemistry Letters. 2018;16(1):71–77.
12. Zong W, Shao X, Chai Y, Wang X, Han S, Chu H, et al. Liposomes encapsulating artificial cytosol as drug delivery system. Biophysical Chemistry. 2022;281:106728.
13. AlSawaftah N, Pitt WG, Husseini GA. Dual-targeting and stimuli-triggered liposomal drug delivery in cancer treatment. ACS Pharmacology & Translational Science. 2021;4(3):1028–1049.
14. Awad NS, Paul V, AlSawaftah NM, ter Haar G, Allen TM, Pitt WG, et al. Ultrasound-responsive nanocarriers in cancer treatment: A review. ACS Pharmacology & Translational Science. 2021;4(2):589–612.
15. Abu Lila AS, Ishida T. Liposomal delivery systems: Design optimization and current applications. Biological and Pharmaceutical Bulletin. 2017;40(1):1–10.
16. Yuba E. Development of functional liposomes by modification of stimuli-responsive materials and their biomedical applications. Journal of Materials Chemistry B. 2020;8(6):1093–1107.
17. Zangabad PS, Mirkiani S, Shahsavari S, Masoudi B, Masroor M, Hamed H, et al. Stimulus-responsive liposomes as smart nanoplatforms for drug delivery applications. Nanotechnology Reviews. 2018;7(1):95–122.
18. Zhai L, Luo C, Gao H, Du S, Shi J, Wang F. A dual pH-responsive DOX-encapsulated liposome combined with glucose administration enhanced therapeutic efficacy of chemotherapy for cancer. International Journal of Nanomedicine. 2021;16:3185–3199.
19. AlSawaftah NM, Awad NS, Pitt WG, Husseini GA. pH-Responsive nanocarriers in cancer therapy. Polymers. 2022;14(5):936.
20. Asano I, So S, Lodge TP. Oil-in-oil emulsions stabilized by asymmetric polymersomes formed by AC + BC block polymer co-assembly. Journal of the American Chemical Society. 2016;138(14):4714–4717.
21. Zhao Y, Li X, Zhao X, Yang Y, Li H, Zhou X, et al. Asymmetrical polymer vesicles for drug delivery and other applications. Frontiers in Pharmacology. 2017;8.
22. Anajafi T, Mallik S. Polymersome-based drug-delivery strategies for cancer therapeutics. Therapeutic Delivery. 2015;6(4):521–534.
23. Thambi T, Park JH, Lee DS. Stimuli-responsive polymersomes for cancer therapy. Biomaterials Science. 2016;4(1):55–69.
24. Ray S, Li Z, Hsu C-H, Hwang L-P, Lin Y-C, Chou P-T, et al. Dendrimer- and copolymer-based nanoparticles for magnetic resonance cancer theranostics. Theranostics. 2018;8(22):6322–6349.
25. Huang S, Cheng Z, Chen Y, Liu B, Deng X, Ma Pa, et al. Multifunctional polyelectrolyte multilayers coated onto Gd2O3:Yb3+,Er3+@MSNs can be used as drug carriers and imaging agents. RSC Advances. 2015;5(52):41985–41993.
26. Zhou Y, Chang C, Liu Z, Zhao Q, Xu Q, Li C, et al. Hyaluronic acid-functionalized hollow mesoporous silica nanoparticles as pH-sensitive nanocarriers for cancer chemo-photodynamic therapy. Langmuir. 2021;37(8):2619–2628.
27. Tian S, Liu G, Wang X, Zhang G, Hu J. pH-Responsive tumor-targetable theranostic nanovectors based on core crosslinked (CCL) micelles with fluorescence and magnetic resonance (MR) dual imaging modalities and drug delivery performance. Polymers. 2016;8(6):226.
28. Wang J, Gong C, Wang Y, Wu G. Magnetic and pH sensitive drug delivery system through NCA chemistry for tumor targeting. RSC Advances. 2014;4(31):15856–15862.
29. Karimi Z, Abbasi S, Shokrollahi H, Yousefi G, Fahham M, Karimi L, et al. Pegylated and amphiphilic chitosan coated manganese ferrite nanoparticles for pH-sensitive delivery of methotrexate: Synthesis and characterization. Materials Science and Engineering: C. 2017;71:504–511.
30. Viswanath V, Santhakumar K. Perspectives on dendritic architectures and their biological applications: From core to cell. Journal of Photochemistry and Photobiology B: Biology. 2017;173:61–83.
31. Santos A, Veiga F, Figueiras A. Dendrimers as pharmaceutical excipients: Synthesis, properties, toxicity and biomedical applications. Materials (Basel, Switzerland). 2019;13(1).
32. Bober Z, Bartusik-Aebisher D, Aebisher D. Application of dendrimers in anticancer diagnostics and therapy. Molecules (Basel, Switzerland). 2022;27(10).
33. Zhang M, Zhu J, Zheng Y, Guo R, Wang S, Mignani S, et al. Doxorubicin-conjugated PAMAM dendrimers for pH-responsive drug release and folic acid-targeted cancer therapy. Pharmaceutics. 2018;10(3).

34. Wang J, Li D, Fan Y, Shi M, Yang Y, Wang L, et al. Core–shell tecto dendrimers formed via host–guest supramolecular assembly as pH-responsive intelligent carriers for enhanced anticancer drug delivery. Nanoscale. 2019;11(46):22343–22350.
35. Mekonnen TW, Andrgie AT, Darge HF, Birhan YS, Hanurry EY, Chou H-Y, et al. Bioinspired composite, pH-responsive sodium deoxycholate hydrogel and generation 4.5 poly(amidoamine) dendrimer improves cancer treatment efficacy via doxorubicin and resveratrol co-delivery. Pharmaceutics. 2020;12(11):1069.
36. Wang Y, Huang D, Wang X, Yang F, Shen H, Wu D. Fabrication of zwitterionic and pH-responsive polyacetal dendrimers for anticancer drug delivery. Biomaterials Science. 2019;7(8):3238–3248.
37. Wang M, Cai X, Yang J, Wang C, Tong L, Xiao J, et al. A targeted and pH-responsive bortezomib nanomedicine in the treatment of metastatic bone tumors. ACS Applied Materials & Interfaces. 2018;10(48):41003–41011.
38. Hu Q, Wang Y, Xu L, Chen D, Cheng L. Transferrin conjugated pH- and redox-responsive poly(amidoamine) dendrimer conjugate as an efficient drug delivery carrier for cancer therapy. International Journal of Nanomedicine. 2020;15:2751–2764.
39. Sánchez-López E, Guerra M, Dias-Ferreira J, Lopez-Machado A, Ettcheto M, Cano A, et al. Current applications of nanoemulsions in cancer therapeutics. Nanomaterials. 2019;9(6):821.
40. Mendes Miranda SE, Alcântara Lemos J, Fernandes RS, Silva JO, Ottoni FM, Townsend DM, et al. Enhanced antitumor efficacy of lapachol-loaded nanoemulsion in breast cancer tumor model. Biomedicine & Pharmacotherapy = Biomedecine & Pharmacotherapie. 2021;133:110936.
41. Liu Z, Xu G, Wang C, Li C, Yao P. Shear-responsive injectable supramolecular hydrogel releasing doxorubicin loaded micelles with pH-sensitivity for local tumor chemotherapy. International Journal of Pharmaceutics. 2017;530(1):53–62.
42. Fan D-y, Tian Y, Liu Z-j. Injectable hydrogels for localized cancer therapy. Frontiers in Chemistry. 2019;7.
43. Lee JH, Tachibana T, Yamana K, Kawasaki R, Yabuki A. Simple formation of cancer drug-containing self-assembled hydrogels with temperature and pH-responsive release. Langmuir. 2021;37(38):11269–11275.
44. Gonsalves A, Tambe P, Le D, Thakore D, Wadajkar AS, Yang J, et al. Synthesis and characterization of a novel pH-responsive drug-releasing nanocomposite hydrogel for skin cancer therapy and wound healing. Journal of Materials Chemistry B. 2021;9(46):9533–9546.
45. Liu Y, Ran Y, Ge Y, Raza F, Li S, Zafar H, et al. pH-Sensitive peptide hydrogels as a combination drug delivery system for cancer treatment. Pharmaceutics. 2022;14(3):652.
46. Raza F, Zhu Y, Chen L, You X, Zhang J, Khan A, et al. Paclitaxel-loaded pH responsive hydrogel based on self-assembled peptides for tumor targeting. Biomaterials Science. 2019;7(5):2023–2036.
47. Cimen Z, Babadag S, Odabas S, Altuntas S, Demirel G, Demirel GB. Injectable and self-healable pH-responsive gelatin–PEG/laponite hybrid hydrogels as long-acting implants for local cancer treatment. ACS Applied Polymer Materials. 2021;3(7):3504–3518.
48. Yang K, Zhang S, He J, Nie Z. Polymers and inorganic nanoparticles: A winning combination towards assembled nanostructures for cancer imaging and therapy. Nano Today. 2021;36:101046.
49. Dhavale RP, Dhavale RP, Sahoo SC, Kollu P, Jadhav SU, Patil PS, et al. Chitosan coated magnetic nanoparticles as carriers of anticancer drug telmisartan: pH-responsive controlled drug release and cytotoxicity studies. Journal of Physics and Chemistry of Solids. 2021;148:109749.
50. Manchun S, Dass CR, Sriamornsak P. Targeted therapy for cancer using pH-responsive nanocarrier systems. Life Sciences. 2012;90(11–12):381–387.
51. Han Y, Pan J, Liang N, Gong X, Sun S. A pH-sensitive polymeric micellar system based on chitosan derivative for efficient delivery of paclitaxel. International Journal of Molecular Sciences. 2021;22(13):6659.
52. Mehnath S, Chitra K, Karthikeyan K, Jeyaraj M. Localized delivery of active targeting micelles from nanofibers patch for effective breast cancer therapy. International Journal of Pharmaceutics. 2020;584:119412.
53. Cheng K, Zhou J, Zhao Y, Chen Y, Ming L, Huang D, et al. pH-responsive and CD44-targeting polymer micelles based on CD44p-conjugated amphiphilic block copolymer PEG-b-HES-b-PLA for delivery of emodin to breast cancer cells. Nanotechnology. 2022;33(27):275604.
54. Liang Y, Zhao X, Ma PX, Guo B, Du Y, Han X. pH-responsive injectable hydrogels with mucosal adhesiveness based on chitosan-grafted-dihydrocaffeic acid and oxidized pullulan for localized drug delivery. Journal of Colloid and Interface Science. 2019;536:224–234.

55. Nisar S, Pandit AH, Nadeem M, Pandit AH, Rizvi MMA, Rattan S. γ-Radiation induced L-glutamic acid grafted highly porous, pH-responsive chitosan hydrogel beads: A smart and biocompatible vehicle for controlled anti-cancer drug delivery. International Journal of Biological Macromolecules. 2021;182:37–50.
56. Qu J, Zhao X, Ma PX, Guo B. pH-responsive self-healing injectable hydrogel based on N-carboxyethyl chitosan for hepatocellular carcinoma therapy. Acta Biomaterialia. 2017;58:168–180.
57. Najafipour A, Gharieh A, Fassihi A, Sadeghi-Aliabadi H, Mahdavian AR. MTX-loaded dual thermoresponsive and pH-responsive magnetic hydrogel nanocomposite particles for combined controlled drug delivery and hyperthermia therapy of cancer. Molecular Pharmaceutics. 2021;18(1):275–284.
58. Kouser R, Vashist A, Zafaryab M, Rizvi MA, Ahmad S. pH-responsive biocompatible nanocomposite hydrogels for therapeutic drug delivery. ACS Applied Bio Materials. 2018;1(6):1810–1822.
59. Qing W, Xing X, Feng D, Chen R, Liu Z. Indocyanine green loaded pH-responsive bortezomib supramolecular hydrogel for synergistic chemo-photothermal/photodynamic colorectal cancer therapy. Photodiagnosis and Photodynamic Therapy. 2021;36:102521.
60. Anirudhan TS, Mohan M, Rajeev MR. Modified chitosan-hyaluronic acid based hydrogel for the pH-responsive Co-delivery of cisplatin and doxorubicin. International Journal of Biological Macromolecules. 2022;201:378–388.
61. Palanikumar L, Al-Hosani S, Kalmouni M, Nguyen VP, Ali L, Pasricha R, et al. pH-responsive high stability polymeric nanoparticles for targeted delivery of anticancer therapeutics. Communications Biology. 2020;3(1):95.
62. Narayan R, Gadag S, Cheruku SP, Raichur AM, Day CM, Garg S, et al. Chitosan-glucuronic acid conjugate coated mesoporous silica nanoparticles: A smart pH-responsive and receptor-targeted system for colorectal cancer therapy. Carbohydrate Polymers. 2021;261:117893.
63. Zhang B-b, Chen X-j, Fan X-d, Zhu J-j, Wei Y-h, Zheng H-s, et al. Lipid/PAA-coated mesoporous silica nanoparticles for dual-pH-responsive codelivery of arsenic trioxide/paclitaxel against breast cancer cells. Acta Pharmacologica Sinica. 2021;42(5):832–842.
64. Jin R, Wang J, Gao M, Zhang X. Pollen-like silica nanoparticles as a nanocarrier for tumor targeted and pH-responsive drug delivery. Talanta. 2021;231:122402.
65. Narayan R, Gadag S, Mudakavi RJ, Garg S, Raichur AM, Nayak Y, et al. Mesoporous silica nanoparticles capped with chitosan-glucuronic acid conjugate for pH-responsive targeted delivery of 5-fluorouracil. Journal of Drug Delivery Science and Technology. 2021;63:102472.
66. Lohiya G, Katti DS. Carboxylated chitosan-mediated improved efficacy of mesoporous silica nanoparticle-based targeted drug delivery system for breast cancer therapy. Carbohydrate Polymers. 2022;277:118822.
67. Porrang S, Rahemi N, Davaran S, Mahdavi M, Hassanzadeh B. Synthesis of temperature/pH dual-responsive mesoporous silica nanoparticles by surface modification and radical polymerization for anti-cancer drug delivery. Colloids and Surfaces A: Physicochemical and Engineering Aspects. 2021;623:126719.
68. Beňová E, Hornebecq V, Zeleňák V, Huntošová V, Almáši M, Máčajová M, et al. pH-responsive mesoporous silica drug delivery system, its biocompatibility and co-adsorption/co-release of 5-fluorouracil and naproxen. Applied Surface Science. 2021;561:150011.
69. Rui Q, Yin Z-Z, Cai W, Li J, Wu D, Kong Y. Hyaluronic acid encapsulated aminated mesoporous silica nanoparticles for pH-responsive delivery of methotrexate and release kinetics. Bulletin of the Korean Chemical Society. 2022;43(5):650–657.
70. Pourradi NMA, Babaei H, Hamishehkar H, Baradaran B, Shokouhi-Gogani B, Shanehbandi D, et al. Targeted delivery of doxorubicin by thermo/pH-responsive magnetic nanoparticles in a rat model of breast cancer. Toxicology and Applied Pharmacology. 2022;446:116036.
71. Shahabadi N, Razlansari M, Zhaleh H. In vitro cytotoxicity studies of smart pH-sensitive lamivudine-loaded CaAl-LDH magnetic nanoparticles against Mel-Rm and A-549 cancer cells. Journal of Biomolecular Structure and Dynamics. 2022;40(1):213–225.
72. Obireddy SR, Lai WF. ROS-generating amine-functionalized magnetic nanoparticles coupled with carboxymethyl chitosan for pH-responsive release of doxorubicin. International Journal of Nanomedicine. 2022;17:589–601.
73. Matiyani M, Rana A, Pal M, Rana S, Melkani AB, Sahoo NG. Polymer grafted magnetic graphene oxide as a potential nanocarrier for pH-responsive delivery of sparingly soluble quercetin against breast cancer cells. RSC Advances. 2022;12(5):2574–2588.

74. Tokmedash MA, Seyyedi Zadeh E, Balouchi EN, Salehi Z, Ardestani MS. Synthesis of smart carriers based on tryptophan-functionalized magnetic nanoparticles and its application in 5-fluorouracil delivery. Biomedical Materials. 2022;17(4):045026.
75. Samadi A, Pourmadadi M, Yazdian F, Rashedi H, Navaei-Nigjeh M, Eufrasio-da-silva T. Ameliorating quercetin constraints in cancer therapy with pH-responsive agarose-polyvinylpyrrolidone-hydroxyapatite nanocomposite encapsulated in double nanoemulsion. International Journal of Biological Macromolecules. 2021;182:11–25.
76. Jia L, Pang M, Fan M, Tan X, Wang Y, Huang M, et al. A pH-responsive Pickering nanoemulsion for specified spatial delivery of immune checkpoint inhibitor and chemotherapy agent to tumors. Theranostics. 2020;10(22):9956–9969.
77. Kim B, Hebert JM, Liu D, Auguste DT. A Lipid Targeting, pH-responsive nanoemulsion encapsulating a DNA intercalating agent and HDAC inhibitor reduces TNBC tumor burden. Advanced Therapeutics. 2021;4(3):2000211.
78. Pourmadadi M, Ahmadi M, Abdouss M, Yazdian F, Rashedi H, Navaei-Nigjeh M, et al. The synthesis and characterization of double nanoemulsion for targeted co-delivery of 5-fluorouracil and curcumin using pH-sensitive agarose/chitosan nanocarrier. Journal of Drug Delivery Science and Technology. 2022;70:102849.
79. Enin HAA, Alquthami AF, Alwagdani AM, Yousef LM, Albuqami MS, Alharthi MA, et al. Utilizing TPGS for optimizing quercetin nanoemulsion for colon cancer cells inhibition. Colloids and Interfaces. 2022;6(3):49.

4 Enzyme-Responsive Delivery Nanoplatforms in Cancer Theranostics

Jigar Vyas, Isha Shah and Bhupendra G. Prajapati

CONTENTS

4.1 Introduction 54
    4.1.1 Overview of Cancer 54
    4.1.2 Comparison of Normal Cells and Cancer Cells 55
    4.1.3 Role of Enzymes in the Cellular Cycle 55
    4.1.4 Tumor-Cell Receptors in Cancer 57
4.2 Cancer Theranostics and Nanoplatform 59
    4.2.1 Introduction to Cancer Theranostics 59
    4.2.2 Nanocarriers for Cancer Therapy 59
    4.2.3 Stimuli-Responsive/Smart Drug Delivery Systems for Cancer Therapy 60
4.3 Enzyme-Responsive Nanocarriers for Cancer Therapy 62
    4.3.1 Enzyme-Responsive Controlled Drug Release from Nanocarriers: A General Mechanism 64
    4.3.2 Enzyme-Responsive Nanoparticles 64
    4.3.3 Enzyme-Responsive Liposomes 65
    4.3.4 Other Enzyme-Responsive Nanocarrier Systems 65
    4.3.5 Nanozymes for Cancer Therapy 66
4.4 Enzyme-Responsive Systems for Cancer Diagnosis 67
    4.4.1 Diagnostic Test for Cancer 67
    4.4.2 Stimuli-Responsive Systems for Cancer Theranostics 68
    4.4.3 Enzymatic Probes for Cancer Diagnosis 68
4.5 Current Development and Future Prospects of Enzyme-Responsive Cancer Theranostics 70
    4.5.1 Advantages, Disadvantages, and Applications of Enzyme-Responsive Cancer Theranostics 70
    4.5.2 Current Development of Enzyme-Responsive Cancer Theranostics 71
    4.5.3 Future Prospects of Enzyme-Responsive Cancer Theranostics 71
References 72

4.1 INTRODUCTION

4.1.1 Overview of Cancer

Cancer, also known as malignancy, is defined as abnormal cell multiplication and is one of the major risk factors of mortality worldwide, as well as a major risk to human health (1). Transformed cell refers to the transformation of a normal cell into a malignant cell, and a comparison is shown in Table 4.1. It is a condition with several causes. Unfortunately, it is a tissue-level condition that poses a significant diagnostic barrier, followed by therapy effectiveness. Cancer is triggered by a cascade of gene alterations that disrupt cell function. Furthermore, a carcinogenic toxic agent in the atmosphere affects the cytoplasm and nucleus of cells, causing genetic defects and gene changes (2, 3). Other sources of carcinogenesis include bacteria, viruses, and radiation rays, which contribute to around 7% of all malignancies (4). Cellular networks are generally changed by cancer, and major genes become defective. This disturbance disrupts the cell cycle, leading to uncontrolled proliferation (5). Proto-oncogenes are essential for cellular multiplication and proliferation under normal circumstances; however, when a genetic mutation takes place, they change into oncogenes, which are the most harmful to cell survival (6). Furthermore, the absence of tumor suppressor genes causes unregulated cell division (7).

4.1.2 Comparison of Normal Cells and Cancer Cells

TABLE 4.1
Normal Cells vs Cancer Cells

Normal Cells	Cancer Cells
 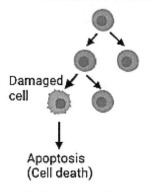	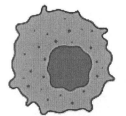 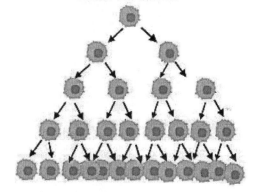
• Regular shape and size	Irregular shape and size
• Controlled growth and cell division	Uncontrolled growth and cell division
• Cause apoptosis	Does not cause apoptosis
• The nucleus is small and light	The nucleus is large and dark
• Cause aerobic glycolysis	Cause lactic acidosis
• RNA and DNA synthesis is normal	Increased synthesis of RNA and DNA
• Catabolism of pyrimidine is normal	Catabolism of pyrimidine is decreased
• The tumor suppressor gene is present	The tumor suppressor gene is lost

4.1.3 Role of Enzymes in the Cellular Cycle

Tumor cells have several genetic and metabolic abnormalities that contribute to their development and recurrence. Cancer and metabolism have complex interrelationships. Cancer cells exhibit metabolic alterations, including increased aerobic glycolysis, mutations in tricarboxylic acid (TCA) cycle metabolic enzymes, and reliance on lipid metabolism, as shown in Figure 4.1 (8).

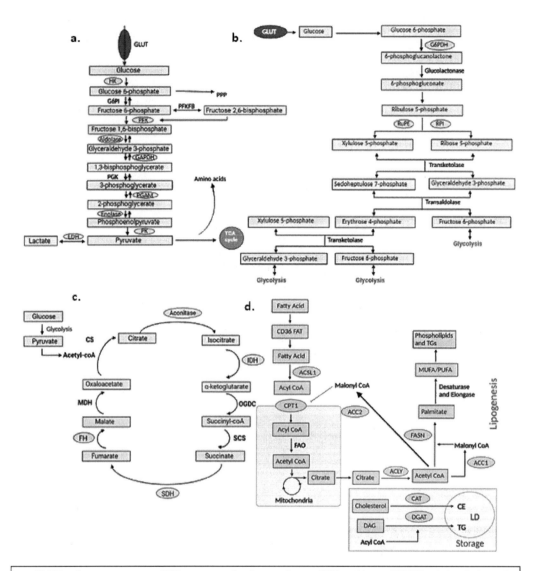

a. **GLUT**- glucose transporter, **PPP**- pentose phosphate pathway, **HK**-hexokinase, **PFKFB**- 6-phosphofructo-2-kinase/fructose 2,6-bisphosphatase, **GAPDH**- glyceraldehyde-3-phosphate dehydrogenase, **PGAM**- phosphoglycerate mutase, **PK**- pyruvate kinase, **LDH**- lactate dehydrogenase, **TCA**- tricarboxylic acid cycle

b. **G6PDH**- glucose 6-phosphate dehydrogenase, **6PGDH**- 6-phosphogluconate dehydrogenase, **RuPE**- ribulose-5-phosphate-3-epimerase, **RPI**- ribose-5-phosphate isomerase

c. **IDH**- isocitrate dehydrogenase, **OGDC**- oxoglutarate dehydrogenase complex, **SCS**- succinyl-CoA synthetase, **SDH**- succinate dehydrogenase, **FH**- fumarate hydratase, **MDH**- malate dehydrogenase, **CS**- citrate synthase

d. **ACSL**- long-chain acyl-CoA synthetase 1, **CPT1**-carnitine palmitoyltransferase 1, **FAO**-fatty acid oxidation, **ACC**- acetyl-CoA carboxylase, **ACLY**-ATP citrate lyase, **FASN**- fatty acid synthase, **TGs**- triglycerol, **MUFA/PUFA**- mono/polyunsaturated fatty acid, **CAT**- carnitine acetyltransferase, **DGAT**- diacylglycerol acyltransferase

FIGURE 4.1 a. Glycolysis pathway, b. pentose phosphate pathway (PPP), c. tricarboxylic acid (TCA) cycle, d. lipid metabolism.

4.1.4 Tumor-Cell Receptors in Cancer

Numerous therapeutically active surface molecules, including those that act as receptors for various ligands, are present on the surface of every cancerous cell, as shown in Figure 4.2. Major histocompatibility complex (MHC) antigens or human leukocyte antigen (HLA), cytokine receptors, cell-adhesion molecules (CAM), Fas/Fas-ligand molecules, growth factor receptors, transferrin, and others are among them. They can alter their expressions, usually to promote the growth and development of tumors. A few are specific to tumor cells and have not developed in normal cells before.

4.1.4.1 Human Leukocyte Antigen

The discovery of tumor-associated antigens (TAAs) sparked expectations that at least some malignancies, particularly lung cancer, may be eradicated by malignant cell T lymphocytes recognizing TAA on specific targets through the MHC restriction mechanism. Antigenic peptides must be adequately expressed on target cells in the MHC antigenic channel, as well as precisely sensitized T cells, for effective anti-tumor cytotoxicity (9). As a result, MHC molecules serve a dual activity as antigenic peptide carriers and lymphocyte receptors. Corrias et al. (10) recently demonstrated that the loss of transporter associated with antigen processing 1 (TAP-1) and TAP-2 mRNA was associated with the deletion or down-regulation of HLA class I antigens on the cell surface in human neuroblastoma cells. Patients with dual HLA expression on their tumor cells seemed to have a longer survival time (11).

4.1.4.2 Cytokine Receptor

Cytokine receptors are another key category of surface molecules on cancer cells. Certain cytokines, including interleukin 2 (IL-2) and IL-6, can operate as cancer growth factors. IL-2 has been demonstrated to serve this function in head and neck cancer, whereas IL-6 acts in multiple myeloma and likely in melanoma (12). However, the specific receptors on the target cells are necessary for cytokine action. IL-2 and IL-6 receptors have been found on cells from various malignancies, both

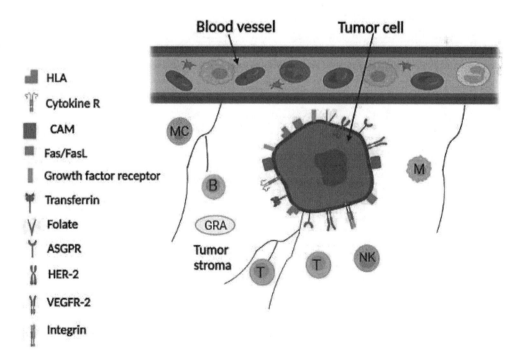

FIGURE 4.2 Interactions between tumor cells and tissue microenvironment.

at the molecular (13) and protein levels (14). On 17 head and neck cancer cell lines, elevated amounts of interleukin 4 (IL-4) receptors were found in vitro and in vivo (15).

4.1.4.3 Cell Adhesion Molecules

Cell-to-cell adhesion and cell-to-extracellular matrix interactions are critical for homeostasis in healthy tissues. Normal epithelial cells, for example, regulate tissue morphology by binding to one another and the extracellular matrix. Cell adhesion molecules are proteins that permit such diverse attachments and are categorized into four forms: integrins, cadherins, selectins, and immunoglobulins. In addition to functioning as physical anchors for the cells, they actively facilitate interactions between the cells and the extracellular environment. Cell adhesion signaling molecular pathways have been widely explored, mainly because mutations and alterations in the expression of these proteins, including cadherins and integrins, are commonly linked with ailments ranging from developmental cognitive impairments to cancer. Furthermore, cell adhesion molecules are required for two key cancer characteristics: anchorage-independent proliferation and loss of cell-to-cell adhesion (16).

4.1.4.4 Fas-Fas Ligand System

The tumor necrosis factor (TNF)-receptor superfamily includes type I and type II members in the form of Fas (CD95/APO-1) and Fas ligand (FasL/CD95L). It is usually recognized that FasL binding to Fas activates the death domain, resulting in cell apoptosis. Tumor cells in cancer may also have Fas in their cytoplasm but only activate FasL. This occurs when FasL on tumor cells interacts with Fas-expressing T lymphocytes, resulting in tumor-infiltrating lymphocyte death. Furthermore, cancer cells may exude FasL, which can interact with Fas on the surface of lymphocytes to trigger apoptosis (17).

4.1.4.5 Growth Factor Receptors

The epidermal growth factor receptor (EGFR), whose stimulation accelerates cancer progression, metastasis, and invasion, and invasion is a major contributor to epithelial malignancies (18). EGFR is a tyrosine kinase receptor that belongs to the ErbB family and sends a signal to cells that induces growth after stimulation via EGFR ligands (such as TGF and EGF). The amount of EGFR ligands in healthy tissue is largely monitored to assure that the kinetics of cell growth exactly indicate the tissues' homeostasis requirements. However, EGFR is usually consistently triggered in cancer due to the continual production of EGFR ligands in a tumour cells (19) or because of a mutation in EGFR that keeps the receptor in a constant activation state (20). The growth and stability of the altered form are dependent on the insulin-like growth factor receptor I (IGF-IR), a transmembrane tyrosine kinase receptor, both in vivo and in vitro. Numerous types of cancer cells have been exhibited to contain a significant amount of IGF-IR (21).

4.1.4.6 Transferrin Receptor

Transferrin receptor 1 (TFR1) is extensively overexpressed in malignancies, and its homeostatic processes for carcinogenesis are complex and frequently linked. TFR1 is overexpressed in tumor cells, primarily to satisfy the iron requirements of tumor cell growth (22). TFR1 is a signaling protein, and Src tyrosine phosphorylation at position 20 promotes anti-apoptosis and increases breast cancer cell survival. TFR1 has also been identified as a mitochondrial regulator that promotes cancer cell proliferation by stimulating the c-Jun N-terminal kinases (JNK) signaling pathway (23).

4.1.4.7 Folate Receptor

The one-carbon metabolism is carried out by the folates. One of the three primary forms of folate transporters is folate receptors (FOLRs). Multiple malignancies have high levels of the folate receptors FOLR1 and FOLR2. FOLR1 overexpression is closely attributed to enhanced tumor growth and poor patient clinical outcome. One-carbon metabolism-independent signaling mechanisms are impacted by FOLR1 (24).

4.1.4.8 Asialoglycoprotein Receptor

Asialoglycoprotein receptor (ASGPR) is a glycoprotein that recognizes substances with terminal galactosyl or acetylaminogalactosyl units (25). To produce a targeting effect, a drug's carrier can be effectively targeted to the hepatic parenchymal cells using galactosyl (Gal) residues or N-acetylaminogalactosyl (GalNAc) residues (26). The amount of ASGPR expression on hepatocellular liver cells is significant to the effectiveness of this method.

4.1.4.9 Human Epidermal Growth Factor Receptor

A number of human malignancies have a pathogenic aspect that is highly affected by the human epidermal growth factor receptor (HER) family of receptors. As a result of receptor dimerization, tyrosine residues in the cytoplasmic region of the receptors are autophosphorylated, resulting in a number of signaling pathways that promote cell growth and the development of tumors. The class consists of four major members: HER-1, HER-2, HER-3, and HER-4, also known as ErbB1, ErbB2, ErbB3, and Erb4, respectively (27). Each of the four HER receptors includes a cysteine-rich exogenous binding affinity site, as well as a transmembrane lipid-soluble region or even an intracellular motif with the catalytic activity of tyrosine kinase (28).

4.1.4.10 Vascular Endothelial Growth Factor Receptor

Both vascular endothelial growth factor (VEGF) and its receptor (VEGFR) have been shown to have key functions in both healthy and pathological oncogenesis, particularly cancer. VEGFR activation can be regulated by ligand binding. A number of downstream enzymatic cascades, including PI3K/AKT/mTOR, RAS/RAF/MEK/ERK, and p38/MAPK, are then triggered as a result of the structural modifications brought on by the ligand in the cytoplasmic region of the VEGFRs (29).

4.1.4.11 Integrin

Integrins are transmembrane receptors that are linked to the biology of both physiological and pathological functions in humans. The development of tumors involves a diverse range of integrins. The integrins exhibited by epithelial cells, such as a61, a64, a2–1, a3–1, and v5, are often recognized in the tumor; however, the level of expression may be varied because many solid tumors are developed from epithelial cells. These integrins normally promote epithelial cell adherence to the basement membrane, but they may also lead to tumor cell migration, proliferation, and survival (30). Integrins are potential targets for cancer treatment because of their function in tumor growth, and various integrin antagonists, like synthetic peptides and antibodies, were employed successfully in the trial for cancer therapies (31).

4.2 CANCER THERANOSTICS AND NANOPLATFORM

4.2.1 Introduction to Cancer Theranostics

Nanotechnology advancements have resulted in tremendous progress in nanomedicine to combat such intricacy. The prospect of precise cancer diagnosis and therapy has resulted in the development of an incredible technology known as "theranostics"; as categorized in Figure 4.3, it is a term that refers to the integrated therapeutic and diagnostic activities performed by a unified system (32). As a result, there is a relationship between diagnosis and therapy that might provide a more personalized therapy plan with the potential to enhance results (33).

4.2.2 Nanocarriers for Cancer Therapy

Various therapies can be used in cancer, as listed in Figure 4.4.

Of the therapies listed, immunotherapy, phototherapy, chemotherapy, and gene therapy utilize novel drug delivery systems where different nano-carriers are used for transporting and targeting

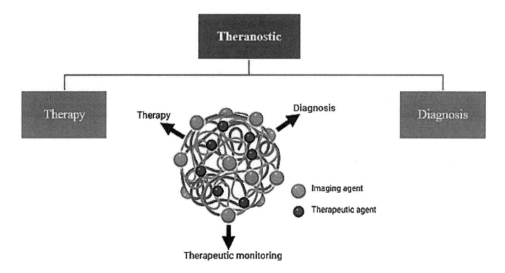

FIGURE 4.3 Theranostics technology comprises two systems, therapy and diagnosis.

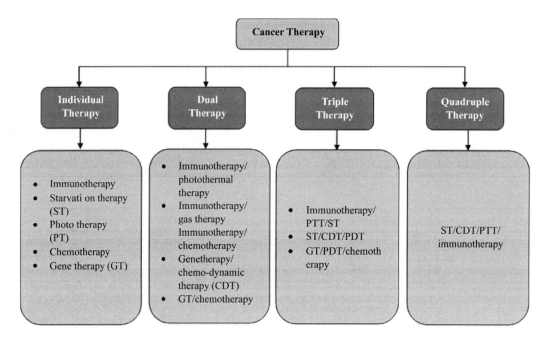

FIGURE 4.4 Different therapies for cancer.

the active pharmaceutical ingredients to the site of action. Many of the nanocarriers have been used in the management of cancer, with the current focus on the stimuli-responsive system, which is one of the major developments in smart drug delivery systems for cancer theranostics.

4.2.3 STIMULI-RESPONSIVE/SMART DRUG DELIVERY SYSTEMS FOR CANCER THERAPY

A completely unforeseen development in biomedical nanotechnology has suddenly transformed traditional drug delivery systems (DDSs) into smart DDS with stimuli-responsive properties. The

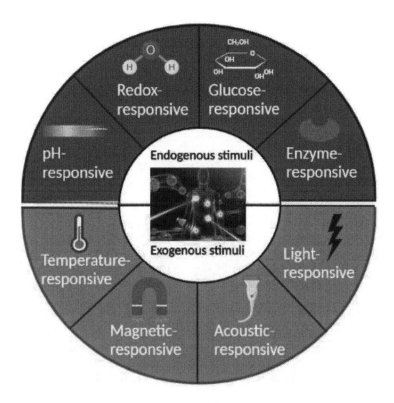

FIGURE 4.5 Classification of exogenous and endogenous stimuli.

term "smart DDS" refers to a method where drugs are only released at the appropriate rates at the action sites, not before they have entered their intended organs or tissue. Using the reaction to certain external or internal triggers, such well-developed nanoplatforms can enhance drug targeting efficacy while lowering payload unwanted effects; both are significant factors in improving patient compliance (34). The classification of the stimuli-responsive system is shown in Figure 4.5.

4.2.3.1 Exogenous Stimuli
Exogenous stimuli are categorized into four types: light, acoustic, magnetic, and temperature-responsive stimuli drug delivery systems.

- *Light-responsive system:* Using light-responsive devices, external light illumination may be employed to reinforce drug release at the intended suitable target. Because light is a promising stimulus with the potential to change the wavelength, power, and damaged regions of irradiation, light-responsive nanocarriers have been widely developed (35). In general, exposure to light irradiation, like ultraviolet-visible radiation and near infrared (NIR), may cause light-sensitive nanocarriers in biological systems to alter directly, for instance, cancer cells or tumors.
- *Acoustic-responsive system:* Ultrasound is widely utilized in therapeutics because of its ability to provide targeted radiation of variable amplitude and frequency with significant spatial and temporal precision (36). Ultrasound, in particular, has been utilized to induce polymer release by triggering their disintegration (37). This breakdown permanently alters the delivery vehicle, affecting both the baseline frequency of molecular release and the release induced along with further ultrasonic exposure.

- *Magnetic-responsive system:* Magnetic-responsive nanocarriers have been developed as a result of magnetic nanocarriers' propensity to cause localized hyperthermia through the effect of an external magnetic field, which may be utilized to induce drug release and tumor removal, as well as their intrinsic magnetic field tropism for tumor targeting (38).
- *Temperature-responsive system:* When compared to other stimuli, temperature is among the simplest and most effective parameters for controlling drug release. Thermo-sensitive nanocarriers are intended to maintain payloads at 37°C and deliver them whenever the temperature rises to 40°C to 45°C (39).

4.2.3.2 Endogenous Stimuli

Endogenous stimuli are categorized into four types: pH, redox, glucose, and enzyme responsive stimuli drug delivery systems.

- *pH-responsive system:* Tumor cells' cytoplasmic pH is substantially lower compared to normal tissues (40). As a result, pH has emerged as an essential trigger for stimuli-responsive drug-loaded nanoplatforms. According to the protonation/ionization process, pH-responsive polymeric nanocarriers with anti-cancer drugs might alter their forms or characteristics in accordance to pH changes in the tumor microenvironment, enabling precisely targeted cancer therapy (41).
- *Redox-responsive system:* As a result of significantly varying reduction potentials and capacities in tumors, redox-responsive nanocarriers have been extensively utilized for the delivery of drugs. For instance, glutathione (GSH) levels in cancer cells are noticeably more than in normal tissues (42).
- *Glucose-responsive system:* Diabetics are at a considerably higher risk of getting numerous forms of cancer, notably tumors of the pancreas, breast, liver, esophagus, and colon. Diabetes and pancreatic cancer both affect the pancreas. Glucose oxidase (GOx), glucose-binding molecules, and phenylboronic acid (PBA) are the most common glucose-sensing components used in the production of chemically synthetic glucose-responsive insulin delivery strategies. The efficacy of formulations to release insulin is affected via glucose-triggered binding-capability alteration, contraction, variation in size of the pore, degradation, and dissolution. (43).
- *Enzyme-responsive system:* Enzyme-responsive nanocarriers have shown to be among the most efficient smart stimuli-responsive nanocarriers. Inflammatory or tumor sites may demonstrate altered expression of particular enzymes, such as phosphatases, glycosidases, and proteases, which can be utilized to localize drug deposition to a specific biological site by enzyme-triggered drug release (44).

4.3 ENZYME-RESPONSIVE NANOCARRIERS FOR CANCER THERAPY

Enzymes are vital for physiological activities. Although some enzymes are synthesized in excess at the tumor site, it is possible to use this system to initiate the delivery of drugs that are responsive to certain enzymes. Numerous enzyme-responsive nanocarriers were developed in order to achieve regulated encapsulated material release in tumor cells, prodrug, or ligand activation, along with morphological alteration, as shown in Figure 4.6 (45). Nanoparticles, liposomes, micelles, dendrimers, and others are among them.

The particular enzyme-triggered payload release facilitates drug delivery to cancerous cells while avoiding carrier exposures throughout the circulation, which may keep bioactive substances active while avoiding adverse effects on normal organs/tissues.

Several approaches have been developed to improve the selectivity of cytotoxic agents by targeting delivery and prodrug activation inside tumor tissue. These approaches are described in Table 4.2.

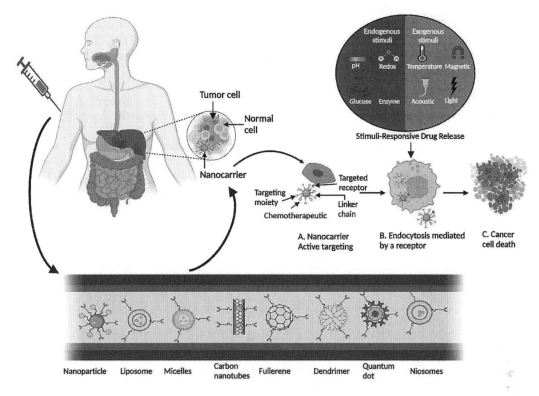

FIGURE 4.6 Nanocarrier systems for cancer theranostics.

TABLE 4.2
Approaches for Targeted Administration and Prodrug Activation

Approach	Mechanism	Example	Ref.
Antibody-directed enzyme prodrug therapy (ADEPT)	Intravenously, a monoclonal antibody (mAb) associated with a drug-activating enzyme (generally an antibody-enzyme fusion protein) adheres to a specific antigen expressed on the tumor surface of the cell, after which a non-toxic prodrug is delivered systemically and is transformed to the chemotherapeutic drug via the pre-targeted enzyme	Folate-carboxypeptidase G2 (CPG2) for metastatic cancer	(46)
Tumor specific antigens/receptors	Drug molecules are conjugated to monoclonal antibodies (mAbs) or ligands for targeted interaction with antigens or receptors located on tumor cell surfaces	Pyrrolobenzodiazepine anti-CD276 for tumors (human colon, ovarian, breast cancer)	(47)
Antibody-targeted, triggered, electrically modified prodrug type strategy (ATTEMPTS)	With the help of an antibody, an inactivated enzymatic drug is delivered to the intended site, and a compound with a greater heparin-binding specificity enables the active form of the enzymatic drug to be released at the specific sites	protein transductiondomains (PTDs) modified TAT-Gel to retained the N-glycosidase was studied for anticancer activity	(48)
Gene-directed enzyme prodrug therapy (GDEPT)	Includes the physical introduction of a gene for a foreign enzyme while leaving non-cancerous cells unaffected. The altered tumour cells produce the enzyme that triggers the systemically supplied non-toxic prodrug	E-coli cytosine deaminase (CD), 5-fluorocytosine (5-FC), several tumor-specific promoters, and virus-based vectors SLN for colorectal cancer	(49)
Polymer-directed enzyme prodrug therapy (PDEPT)	To increase cancer targeting, the polymeric prodrug is first delivered, and then the activating polymer-enzyme conjugate is delivered	5-fluorouracil (5-FU)-chitosan (CTS) (5-FU-CTS-NPs) coupled with the hyaluronidase (HAase) enzyme system for colorectal carcinoma	(50)

4.3.1 Enzyme-Responsive Controlled Drug Release from Nanocarriers: A General Mechanism

Enzyme-stimulus nanocarriers are designed to "orient" the body's systemic circulation at the location of the tumor and then "activate" in the cancerous intercellular environment. When nanocarriers are exposed to enzymes, they exhibit physical and chemical modifications that cause drug release at the appropriate site. The enzyme-triggered activation of nanocarriers, therefore, provides increased internalization and cellular binding, rapid release of drugs at the targeted site, and efficient drug perfusion through tumor mass (51), as shown in Figure 4.8. The structure of nanocarriers is typically composed of four components (core, crown, ligand, and linker), and the breakdown of either component might eventually lead to the disruption of the nanocarrier structure, accompanied by the delivery of encapsulated drugs.

In enzyme-responsive cancer theranostics, the nanocarrier system is used in four different ways, nanocarrier with enzyme-responsive core, crown, linker, and ligand, as shown in Figure 4.9. In enzyme-responsive cancer theranostics, nanoparticles and liposomes are the most often utilized nanocarrier systems; however, other nanocarrier systems are also employed but have not been extensively explored. Examples of all these systems are shown in the following tables.

4.3.2 Enzyme-Responsive Nanoparticles

The delivery of drugs to the lymphatic system, lung, brain, liver, spleen, and other organs with long-term circulation and regulated release profiles has been investigated using a variety of formulations based on enzyme-responsive nanoparticles (NPs). Targeted modifications in the macro-scale structure of NPs in the presence of certain enzymes in the tumor microenvironment generally result in the intended controlled release of drugs. Examples are shown in Table 4.3.

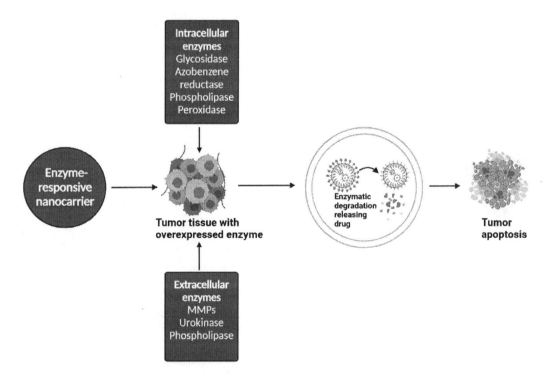

FIGURE 4.7 Enzyme-responsive controlled drug release from nanocarrier mechanism.

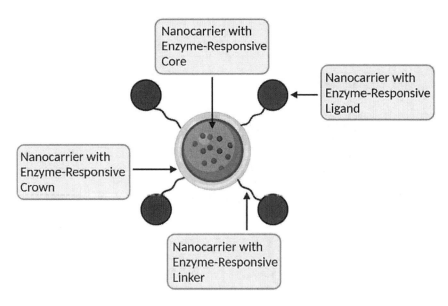

FIGURE 4.8 Types of enzyme-responsive nanocarriers.

TABLE 4.3
Enzyme-Responsive Nanoparticles with Examples

Types of enzymes	Examples	Ref.
Proteolytic enzymes	PEG-coated APNPs composed of an "enzyme responsive core" made of peptides and self-assembled into therapeutic peptides which were activated by proteases	(52)
Hyaluronidase (HAase)	Doxorubicin hydrochloride (Dox) mesoporous silica nanoparticle (MSN)-based targeted delivery encapsulated using biotin-modified hyaluronic acid (HA) with "HAase-responsive crowns"	(53)
Matrix metalloproteinase(MMP)-9 and cathepsin B	PEG-coated gemcitabine (GEM) silica and calcium carbonate-based NPs utilizing two consecutive "enzyme-responsive linkers": peptide GGPLGVRGK for MMP and GFLG for cathepsin B	(54)
Hyaluronidase (HAase)/ transglutaminase	Enzyme-responsive ligand-mediated NPs were formed using hyaluronic acid (HA) as "ligand" and transglutaminase as a supporting carrier which acts as a cross-linker for HA after internalization of nanocarrier	(55)

4.3.3 ENZYME-RESPONSIVE LIPOSOMES

Liposomes are a vesicular system–containing lipid bilayer structure that can encapsulate and transport both hydrophilic and lipophilic materials to the intended site. A morphological change in the lipid bilayer, dissociation of a lipopeptide or lipopolymer included in the bilayer, a loss of a shielding polymer from the surface and an enhancement in cellular uptake, and stimulation of a prodrug in the liposomes are all types of destabilization processes used by enzyme-responsive liposomes to deliver their payload on exposure with the enzyme (56). Examples are shown in Table 4.4.

4.3.4 OTHER ENZYME-RESPONSIVE NANOCARRIER SYSTEMS

In addition to nanoparticles and liposomes, additional carriers such as micelles, carbon nanotubes, fullerene, graphene quantum dots (GQDs), and niosomes have also been employed, as they are

TABLE 4.4
Enzyme-Responsive Liposomes with Examples

Enzyme Types	Example	Ref.
Matrix metalloproteinases (MMP2)	Doxorubicin/tariquidar-coloaded and peptide-functionalized liposomes (DT-pLip) demonstrated greater effectiveness in drug-resistant triple-negative breast cancer (TNBC) tumor growth inhibition	(57)
Secreted phospholipase A2	Doxorubicin (Dox)-loaded sPLA2 responsive liposomes have been formed utilizing 1,2-distearyl-sn-glycero-3-phosphoethanolamine (DSPE), cholesterol, 1,2-distearyl-sn-glycero-3-phosphocholine (DSPC), and DSPE-PEG2000 for treatment of prostate cancer	(58)
Prostate-specific antigen (PSA)	Three domains were formed in siRNA-loaded liposomes containing a peptide, including the PSA-cleavable HSSKYQ, cell-penetrating polyarginine (GR8GC), the polyanionic DGGDGGDGGDGG wherein folate adheres to the prostate-specific membrane antigen (PSMA), a glycoprotein moiety upregulated in prostate cancer cells	(59)
Elastase	DOPE liposomes conjugated to N-acetyl-Ala-Ala is developed for the cytoplasmic delivery of drugs in tumors of the skin, breast, and lungs	(60)
Urokinase plasminogen activator (uPA)	Anti-mitotic N-alkylisatin (N-AI)-loaded liposomes functionalized with the uPA/uPAR targeting ligand, plasminogen activator inhibitor type 2 (PAI-2/SerpinB2) were formed for the treatment of triple-negative breast cancer	(61)
Cathepsin B	Doxorubicin (Dox) liposomes comprising the oligopeptide "GLFG" were formed to exert an effective anticancer effect on Hep G2 cells	(62)

TABLE 4.5
Other Enzyme-Responsive Nanocarriers with Examples

Enzyme Types	Example	Ref.
Matrix metalloproteinases (MMP2)	A ligand-coupled polymeric micelle was designed, and it significantly increased cellular absorption in prostate cancer cells	(63)
Carboxylesterase	SWNTs were designed to bind with the EGFR inhibitor cetuximab and the topoisomerase I inhibitor SN38 for the effective therapy of EGFR over-expressing colorectal cancer	(64)
–	Encapsulated gadofullerene was produced and subsequently functionalized with the cytokine interleukin-13, which is upregulated in human glioblastoma cell lines. To improve water solubility and tumor absorption, the molecule was then functionalized with amine groups, which may sustain a positive charge at physiological pH	(65)
Cathepsin B	PEGylated lysine peptide dendrimer-gemcitabine conjugate was developed and has considerable significance as an antitumor drug in the treatment of breast cancer	(66)
Cathepsin D	Doxorubicin (DOX) graphene quantum dots (GQDs)	(67)
–	Oxaliplatin-niosomes were developed for the treatment of colorectal cancer and can stimulate apoptosis of cancerous cells via various mechanisms, such as assisting in the destabilization and downregulation of the mitochondrial respiratory complex or inducing DNA damage or oxidation of protein, lipid, and enzyme, resulting in cellular damage	(68)

effective in stabilizing and transporting various enzymes to different body organs for various diseases including cancer. However, research on them is still very limited. Examples are shown in Table 4.5.

4.3.5 Nanozymes for Cancer Therapy

Nanozymes, nanomaterials with innate enzyme-like properties, have gained great attention in cancer therapy (69). Furthermore, nanozymes demonstrated great imaging capability in fluorescence imaging in the second near IR range and MRI for in-vivo visualization testing. The approach of combination therapy has shown a significant therapeutic impact on cancer (70). Nanozymes are categorized into four subtypes: peroxidase (POD), oxidase (OXD), catalase (CAT), and superoxide dismutase (SOD).

The anti-tumor properties of cancer treatment modalities that typically rely on O_2 levels, including photodynamic therapy (PDT), sonodynamic therapy (SDT), photothermal therapy (PTT), chemotherapy, and radiotherapy (RT), can be enhanced by the use of nanozymes (71). Furthermore, O_2 consumption and tumor vasoconstriction might increase hypoxia and restrict the efficacy of the aforementioned cancer treatments, resulting in positive feedback (72). Nanozymes are increasingly being employed to improve the efficacy of these treatments.

4.4 ENZYME-RESPONSIVE SYSTEMS FOR CANCER DIAGNOSIS

Enzymes, which are a crucial key element of bio-nanotechnology, possess superior biorecognition capabilities along with excellent catalytic characteristics. Often, altered enzyme expression identified in cancer provides several alternatives for developing nanocarriers with enzyme-labile linkage (73). Typically, cancer has four stages. The actual stage is based on various factors, including the tumor's location and size. Cancer diagnostic tests and enzymatic probes, which are listed in the following section, can be used to diagnose cancer stages.

- *Stage 1:* Cancer has not spread to lymph nodes or other tissues, and it is limited to a restricted location.
- *Stage 2:* The tumor has started developing, but it has not spread.
- *Stage 3:* The tumor has increased in size and may have spread to lymph nodes or other tissues.
- *Stage 4:* The tumor has expanded to other organs or body parts. It is also known as advanced/metastatic cancer.

4.4.1 Diagnostic Test for Cancer

To detect cancer, the location of the tumor, its stage, and the individual's ability to take the therapy must all be recognized. A thorough examination is performed by three different tests: by using radiation, surgery, or other diagnostic methods.

i. Using Radiation
- Ultrasound (US) Imaging
- X-Ray Imaging
- Optical Imaging
- Noncontrast Optical Imaging of Tissue Surfaces
- Contrast-Enhanced Optical Imaging of Tissue Surfaces
- Magnetic Resonance Imaging (MRI)
- Hyperpolarization MRI
- Paramagnetic Chemical Exchange Saturation Transfer MRI
- Positron Emission Tomography (PET)
- PET Radiotracers
- Time-of-Flight PET
- Positron Emission Mammography (PEM)
- Single-Photon Emission Computerized Tomography (SPECT)
- Signal Amplification and Background Reduction

ii. Surgery

- Virtual Colonoscopy
- Cancer-Specific Targeting Ligands
- Image-Guided Chemotherapy
- Image-Guided Ion Beam Radiotherapy
- Image-Guided Surgery

iii. Others

- Stimuli-Responsive Systems for Nanotheranostics

4.4.2 Stimuli-Responsive Systems for Cancer Theranostics

This method usually focuses on a controlled response to physicochemical stimuli, including the composition of several components, with both external and internal stimuli.

Stimuli-responsive systems for nanotheranostics are categorized into two types, as in Table 4.6.

4.4.3 Enzymatic Probes for Cancer Diagnosis

Cancer development is inherently linked to abnormal enzymatic activity. It is crucial to locate these enzymes in live cancer cells and assess their amount of expression in order to detect malignancy

TABLE 4.6
Types of Stimuli-Responsive Systems in Nanotheranostics

Stimuli-Responsive Systems in Nanotheranostics

Biological Stimuli-Responsive	Physical Stimuli-Responsive Nanosystems
• pH-sensitive nanotheranostics	• Photo-triggered nanotheranostics
• Redox-dependent nanotheranostics	• Magneto-driven nanotheranostics
• Enzyme-based nanotheranostic	• Thermo-induced nanotheranostics

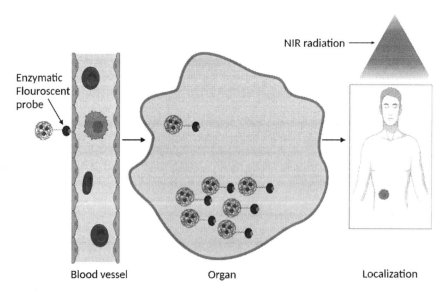

FIGURE 4.9 Cancer diagnosis via enzymatic fluorescent probes.

at an initial stage and assess the effectiveness of treatments. Because of their increased sensitivity, non-destructive quick analysis, and real-time detection capacities, small-molecule fluorescent probes are used to diagnose cancer and have become useful devices for detecting and imaging enzyme activity in biological systems (74).

A few enzymatic probes for cancer diagnosis are discussed in the following.

4.4.3.1 Probes for Transferases

- *Probes for γ-Glutamyltransferase (GGT):* It has been proposed that GGT stimulates cancer proliferation, invasion, and treatment resistance, presumably by modulating intracellular redox metabolism. As a result, GGT can be used as both a therapeutic and a diagnostic marker for cancer. Because GGT plays such a crucial role in pathological and physiological functions, it is essential to detect and visualize GGT function in biological samples. To detect and visualize GGT activity both in vivo and in vitro, Urano et al. designed a fluorescent probe (75).

4.4.3.2 Probes for Proteases

- *Probes for cathepsins:* Cathepsins play key functions in cell death, angiogenesis, and malignant cell invasion; thus they are associated with tumor growth. Acyloxymethyl ketone (AOMK)-based fluorescent probes with targets for cathepsin that appear fluorescent during activity-dependent covalent attachment have been formed by Bogyo and colleagues (76).
- *Probes for aminopeptidase:* Leucine aminopeptidase (LAP) is a catalytic enzyme that can hydrolyze leucine residues found at the N-terminus of peptides and proteins. LAP is involved in a variety of pathological and physiological processes, including drug resistance, tumor cell invading, and tumor growth. Furthermore, the report suggests that LAP is widely recognized as a diagnostic biomarker due to its excessive expression in the development of malignant growth. The attachment of leucine to a fluorophore has resulted in the development of many fluorescent LAP probes (77).
- *Probes for matrix metalloproteinase (MMP):* Invasion of cancer cells and angiogenesis associated with tumors are particularly linked to MMP-2 and MMP-9. A novel NIR probe with a fluorescence signal that self-quenched upon self-assembling into nanoparticles was designed by Liang et al. The fluorescence was induced upon the breakdown of nanoparticles by MMP-2.

4.4.3.3 Probes for Oxidoreductases

- *Probes for monoamine oxidase (MAO):* Tumor growth suppression and cancer proliferation are two of several ailments caused by an excess or lack of these enzymes. For the imaging and detection of MAOs in healthy tissues and cells, fluorescent probes that give a precise and reliable assessment of MAO activity are preferred. Zhau et al. developed a clorgyline (MAO-A inhibitor)-conjugating NIR fluorescence theranostic probe for tumor-specific MAO-A targeted treatment.
- *Probes for tyrosinase:* This enzyme is regarded to serve as a biomarker because of its high levels in malignant tumor cells. For the purpose of observing tyrosinase activity, Ma et al. designed a NIR fluorescent probe. In the presence of oxygen, the tyramine group on the probe potentially interacts with tyrosinase to form a quinone that might induce the fluorescence to be quenched via the PET process (78).
- *Probes for nitroreductase (NTR):* The amount of NTR in a solid tumor is proportional to the level of hypoxia. In order to assess the extent of tumor hypoxia, monitoring of the NTR expression may be significant. An NTR probe was formed by linking a p-nitrobenzyl moiety with a signaling moiety (naphthalimide derivative) via a carbamate group in order to recognize cellular hypoxia (79).

- *Probes for thioredoxin reductase (TrxR) and thioredoxin (Trx):* Cancer phenotypes are associated with the Trx system, which is also substantially overexpressed in malignant tumors. TrxR and Trx fluorescence probes have been formed owing to their essential function in physiological and pathological processes.

4.4.3.4 Probes for Hydrolases

- *Probes for phosphatases:* Many cellular signaling pathways are controlled by phosphatases. Phosphatase dysfunction leads to aberrant growth, invasion, and angiogenesis, all of which are related with the growth of many malignancies. Fluorescent probes with a significant turn-on fluorescence signal were formed by Kim et al. The probe was used in the efficient and accurate assessments of enzymatic activity at the single-cell level, including the real-time monitoring of alkaline phosphatase (ALP) activity in living cancer cells (80).
- *Probes for glycosidase:* In cancer cells, glycoprotein expression, metabolism, functions, stability, or localization are commonly altered as a result of glycosylation changes, contributing to the cancer cells' unchecked proliferation and aggressive behavior. Many fluorescent probes have been formed for visualizing β-gal activity in live cancer cells. These probes are mostly dependent on the process by which β-gal catalyzes the hydrolysis of β-D-galactosides from substrates, releasing D-galactoses, alcohols, or phenols (R-OH) (81).
- *Probes for carboxylic ester hydrolases:* For the imaging and detection of carboxylesterase in live cells, many fluorescent probes have been designed. According to Yang et al., a fluorescent probe relying on excited state proton transfer (ESIPT) may recognize CE2. They also designed two-photon and NIR fluorescent probes for CE2 action recognition and monitoring (82).
- *Probes for acyl-protein thioesterases (APTs):* Acyl-protein thioesterases (APTs) might potentially be found liable for cancer development. It has been suggested that APTs lack substrate selectivity and depalmitoylate intracellular membrane proteins across the board in a constitutive manner. Acyl-protein thioesterase 1 (APT1) and acyl-protein thioesterase 2 are thought to be cytosolic proteins in humans (APT2). Dickinson et al. formed numerous fluorescent APT probes, a novel class of molecular imaging probes that offer information on endogenous amounts of depalmitoylation activity in living cells (83).

4.5 CURRENT DEVELOPMENT AND FUTURE PROSPECTS OF ENZYME-RESPONSIVE CANCER THERANOSTICS

Enzyme-responsive nanocarriers have been shown to be a useful strategy because they provide a crucial catalytic function in several biological processes, and they have developed as a significant progressive means in cancer theranostics. Enzyme systems such as matrix metalloproteinase, phospholipase, cathepsin B, hyaluronidase, azoreductase, and others are described together with their associated nanosystems in the context of current development.

4.5.1 Advantages, Disadvantages, and Applications of Enzyme-Responsive Cancer Theranostics

Advantages

The following are some advantages of enzyme-responsive cancer theranostics:

- One advantage of enzyme-responsive systems is their selectivity.
- By adjusting the amounts of different enzymes, this method improves specificity.
- The system incorporates an internally stimulated mechanism that takes advantage of the diseased or physiological milieu, lowering the risk for cytotoxicity in healthy cells and tissues.

Disadvantages

The following are some of the disadvantages:

- Sometimes the drug is delivered before it reaches its targeted site; as a result, diminished or poor efficacy is observed. Hence, enzyme-responsive biomaterials have additional stimuli-responsive characteristics, such as pH, to preserve the delivery until it reaches its target.
- If the biomaterial is exposed to its enzyme trigger or a similarly comparable enzyme, the load may be released early.

4.5.2 Current Development of Enzyme-Responsive Cancer Theranostics

Enzyme-responsive nanocarriers have been demonstrated to be a useful approach, as they serve as vital enzymatic activity in several physiological systems and have developed a significant strategy of site-specific release. This technique increases affinity by modifying the levels of various enzymes. Some current developments are shown in Table 4.7.

4.5.3 Future Prospects of Enzyme-Responsive Cancer Theranostics

Enzyme-responsive systems have a higher potential for diagnostics, imaging, and therapies. Achieving successful clinical results is still challenging, despite discoveries that have already been noted in preclinical stages. However, further studies of sensitivity to varied stimuli are required,

TABLE 4.7
Enzyme-Responsive Cancer Theranostics Current Development

Enzyme	Action site	Drug	Nanosystem	Ref.
Matrix metalloproteinases (MMPs)	Colon cancerous cells (HT-29)	Doxorubicin	Mesoporous silica nanoparticle	(84)
MMP2/9	Solid tumors (HT1080) cells	Docetaxel	Micelles containing poloxamer oligopeptide (GK8)	(85)
MMP-7	Pancreatic cancer	Doxorubicin	Polymersomes	(86)
MMP-2	Tumor cells (A549 HT1080 cells)	Docetaxel	Liposomes containing polypeptide TMSP (trimethylsilyl3-propionic acid) with sequence PVGLIG at the outer surface, which is sensitive to MMP-2	(87)
Phospholipase A2(PLA2)	Tumor site	1-O-octadecyl-2-(5- fluorouracil)-N-acetyl-3- zidovudine phosphoryl glycerol (OFZG)	Amphiphilic prodrug nanoassembly	(88)
Cathepsin B	Colon carcinoma	Doxorubicin	Doxorubicin-conjugated magnetic nanoparticles modified by peptide (HPhe-Lys-OH)	(89)
Hyaluronidase	Breast cancer	Rhodamine B	MSNPs conjugated with a polysaccharide derived from hyaluronic acid	(90)
Prostate-specific antigen (PSA)	Prostate cancer	Thapsigargin	Prodrug analog developed by combining a PSA-sensitive peptide with thapsigargin	(91)
β-galactosidase	Colorectal cancerous cells	Doxorubicin	Galactose-DOX conjugated system	(92)

TABLE 4.8
List of Drugs under Clinical Trials

Drug	Enzyme	Cancer Type	Clinical Phase	Sponsor	Ref.
Niraparib+ dostarlimab	PARP	Pancreatic cancer	Phase II	Massachusetts General Hospital	(93, 94)
Abiraterone acetate+ enzalutamide	α-hydroxylase/17,20-lyase	Prostate cancer	Phase III	Alliance for Clinical Trials in Oncology	(95, 96)
Nivolumab	MAPK/PIK	Non-small cell lung cancer	Phase I	Bristol-Myers Squibb	(97, 98)
Ribociclib	(CDK) 4/6	Breast cancer	Phase II	Massachusetts General Hospital	(99, 100)
Adavosertib+carboplatin and paclitaxel	PARP	Lung cancer	Phase III	H. Lee Moffitt Cancer Centre and Research Institute	(101, 102)

which may aid in more complicated circumstances with more precise therapeutic, diagnostic, and imaging characteristics. It is expected that by using enzyme-responsive nanocarriers, novel approaches for localized delivery will be developed, minimizing existing barriers with greater consideration for biosafety and biocompatibility of systems. Table 4.8 shows a few drugs that are still under clinical trials.

PARP, poly adenine diphosphate (ADP)-ribose polymerase; CDK, cyclin D–cyclin-dependent kinase 4/6; MAPK, mitogen-activated protein kinase; PIK3, pathway/phosphoinositide 3-kinase.

REFERENCES

1. Siegel R, Naishadham D, Jemal A. Cancer statistics CA Cancer J. Clin. 2013; 63:11–30.
2. Poon SL, McPherson JR, P Tan, et al. Mutation signatures of carcinogen exposure: Genome-wide detection and new opportunities for cancer prevention. Genome Med. 2014; 6:24.
3. Cumberbatch MGK, Cox A, Teare D, et al. Contemporary occupational carcinogen exposure and bladder cancer: A systematic review and meta-analysis. JAMA Oncol. 2015; 1(9):1282–1290.
4. Parkin DM. The global health burden of infection-associated cancers in the year 2002. Int J Cancer. 2006; 118(12):3030–3044.
5. Seto M, Honma K, Nakagawa M. Diversity of genome profiles in malignant lymphoma. Cancer Sci. 2010; 10(3):573–578.
6. Shtivelman E, Lifshitz B, Gale R. et al. Fused transcript of ABL and BCR genes in chronic myelogenous leukaemia. Nature. 1985; 315:550–554.
7. Matlashewski P, Lamb D, Pim J.et al. Isolation and characterization of a human p53 cDNA clone: Expression of the human p53 gene. EMBO J. 1984; 3:3257.
8. Sreedhar A, Zhao Y. Dysregulated metabolic enzymes and metabolic reprogramming in cancer cells. Biomed Rep. 2018; 8(1):3–10.
9. Cross M. and Dexter TM. Growth factors in development, transformation, and tumorigenesis. Cell. 1991; 64:271–280.
10. Corrias m V, Croce M, De Ambrosis A, et al. Lack of HLA-class I antigens in human neuroblastoma cells: Analysis of its relationship to TAP and tapasin expression. Tissue Antigens. 2001; 57:110–117.
11. Dwm Oracki G, Kruk-Zagajewska A, et al. Tumor infiltrating lymphocytes in HLA + and HLA—laryngeal cancer—quantitative approach. Arch. Immunol. Ther. Exp. 1999; 47:161–168.
12. Pignatelli M, Vessey CJ. Adhesion molecules: Novel molecular tools in tumor pathology. Hum Pathol. 1994; 25(9):849–856.
13. Suminami Y, Kashii Y, Law JC, et al. Molecular analysis of the IL-2 receptor beta chain gene expressed in human tumor cells. Oncogene. 1998; 16(10):1309–1317.

14. Weidmann E, Sacchi M, Plaisance S, et al. Receptors for interleukin 2 on human squamous cell carcinoma cell lines and tumor in situ. Cancer Res. 1992; 52(21):5963–5970.
15. Kawakami K, Leland P, Puri RK. Structure, function, and targeting of interleukin 4 receptors on human head and neck cancer cells. Cancer Res. 2000; 60(11):2981–2987.
16. Janiszewska M, Primi MC, Izard T. Cell adhesion in cancer: Beyond the migration of single cells. J Biol Chem. 2020; 295(8):2495–2505.
17. Sikora J, Dwm Oracki G. and Z Eromski J. Expression of Fas and Fas ligand and apoptosis in tumor associated lymphocytes and in tumor cells from malignant effusions. Nat. Immun. 1998; 16:244–255.
18. Normanno N, de Luca A, Bianco C, et al. Epidermal growth factor receptor (EGFR) signaling in cancer. Gene. 2006; 366(1):2–16.
19. Mendelsohn J, Baselga J. The EGF receptor family as targets for cancer therapy. Oncogene. 2000; 19(56):6550–6565.
20. Wong AJ, Ruppert JM, Bigner SH, et al. Structural alterations of the epidermal growth factor receptor gene in human gliomas. Proc Natl Acad Sci USA. 1992; 89(7):2965–2969.
21. Resnicoff M., Abraham D., Yutanaviboonchai W., et al. The insulin-like growth factor I receptor protects tumor cells from apoptosis in vivo. Cancer Res. 1995; 55:2463–2469.
22. Basuli D, Tesfay L, Deng Z, et al. Iron addiction: A novel therapeutic target in ovarian cancer. Oncogene. 2017; 36(29):4089–4099.
23. Senyilmaz D, Virtue S, Xu X, et al. Regulation of mitochondrial morphology and function by stearoylation of TFR1. Nature. 2015; 525(7567):124–128.
24. Nawaz FZ, Kipreos ET. Emerging roles for folate receptor FOLR1 in signaling and cancer. Trends Endocrinol Metab. 2022; 33(3):159–174.
25. Schwartz AL, Fridovich SE, Lodish HF. Kinetics of internalization and recycling of the asialoglycoprotein receptor in a hepatoma cell. Lin J Biol Chem. 1982; 257:4230–4237.
26. Yousef S, Alsaab HO, Sau S, et al. Development of asialoglycoprotein receptor directed nanoparticles for selective delivery of curcumin derivative to hepatocellular carcinoma. Heliyon. 2018; 4(12):e01071.
27. Riese DJ 2nd, Stern DF. Specificity within the EGF family/ErbB receptor family signaling network. Bioessays. 1998; 20(1):41–48.
28. Van Der Geer P, Hunter T, Lindberg RA. Receptor protein-tyrosine kinases and their signal transduction pathways. Annu Rev Cell Biol. 1994; 10:251–337.
29. Liu, Y, Li Y, Wang Y, et al. Recent progress on vascular endothelial growth factor receptor inhibitors with dual targeting capabilities for tumor therapy. J Hematol Oncol. 2022; 15:89.
30. Zutter MM, Santoro SA, Staatz WD, et al. Re-expression of the alpha 2 beta 1 integrin abrogates the malignant phenotype of breast carcinoma cells. Proc Natl Acad Sci USA. 1999; 92(16):7411–1715.
31. Li M, Wang Y, Li M, et al. Integrins as attractive targets for cancer therapeutics. Acta Pharm Sin B. 2021; 11(9):2726–2737.
32. Jeelani S, Reddy RC, Maheswaran T, et al. Theranostics: A treasured tailor for tomorrow. J Pharm Bioallied Sci. 2014; 6:S6–S8.
33. Landais P, Méresse V, Ghislain JC. Evaluation and validation of diagnostic tests for guiding therapeutic decisions. Thérapie. 2009; 64:187–201.
34. Liu D, Yang F, Xiong F, et al. The smart drug delivery system and its clinical potential. Theranostics. 2016; 6(9):1306–1323.
35. Yan L, Li X. Biodegradable stimuli-responsive polymeric micelles for treatment of malignancy. Curr Pharm Biotechnol. 2016; 17(3):227–236.
36. Mitragotri S. Healing sound: The use of ultrasound in drug delivery and other therapeutic applications. Nat Rev Drug Discov. 2005; 4(3):255–260.
37. Kost J, Leong K, Langer R. Ultrasound-enhanced polymer degradation and release of incorporated substances. Proc Natl Acad Sci USA. 1989; 86(20):7663–7666.
38. Yoo D, Jeong H, Noh SH, et al. Magnetically triggered dual functional nanoparticles for resistance-free apoptotic hyperthermia. Angew Chem Int Ed. 2013; 52:13047–13051.
39. Mura S, Nicolas J, Couvreur P. Stimuli-responsive nanocarriers for drug delivery. Nature Materials. 2013; 12:991–1003.
40. Muhammad F, Wang A, Guo M, et al. pH Dictates the release of hydrophobic drug cocktail from mesoporous nanoarchitecture. ACS Appl. Mater. Interfaces. 2013; 5:11828.
41. Liu J, Huang Y, Kumar A, et al. PH-sensitive nano-systems for drug delivery in cancer therapy. Biotechnol Adv. 2014; 32(4):693–710.

42. Mi P. Stimuli-responsive nanocarriers for drug delivery, tumor imaging, therapy and theranostics. Theranostics. 2020; 10(10):4557–4588.
43. Wang J, Wang Z, Yu J, et al. Glucose-responsive insulin and delivery systems: Innovation and translation. Adv Mater. 2020; 32(13): e1902004.
44. Mura S, Nicolas J, Couvreur P. Stimuli-responsive nanocarriers for drug delivery. Nature Mater. 2013; 12:991–1003.
45. Du J, Lane LA, Nie S. Stimuli-responsive nanoparticles for targeting the tumor microenvironment. J Control Release. 2015; 219:205–214.
46. Sherwood RF, Melton RG, Alwan SM, et al. Purification and properties of carboxypeptidase g2 from pseudomonas sp. Strain rs-16. Use of a novel triazine dye affinity method. Eur J Biochem. 1985; 148(3):447–453.
47. Zhao Z, Ukidve A, Kim J, et al. Targeting strategies for tissue-specific drug delivery. Cell. 2020; 181(1):151–167.
48. Shin MC, Zhang J, Min KA, et al. PTD-Modified ATTEMPTS for enhanced toxin-based cancer therapy: An in vivo proof-of-concept study. Pharm Res. 2015; 32(8):2690–2703.
49. Narahari N. Palei, Bibhash C. Mohanta, Mohana L. Sabapathi, et al. Lipid-based nanoparticles for cancer diagnosis and therapy. Organic Materials as Smart Nanocarriers for Drug Delivery. 2018:415–470.
50. Chis AA, Arseniu AM, Morgovan C, et al. Biopolymeric prodrug systems as potential antineoplastic therapy. Pharmaceutics. 2022; 14:1773.
51. Torchilin VP. Multifunctional, stimuli-sensitive nanoparticulate systems for drug delivery. Nat Rev Drug Discov. 2014; 13(11):813–827.
52. Yu X., Gou XC, Wu P, et al. Activatable protein nanoparticles for targeted delivery of therapeutic peptides. Adv Mater. 2018; 30(7):1705383.
53. Zhang M, Xu C, Wen L, et al. A hyaluronidase-responsive nanoparticle-based drug delivery system for targeting colon cancer cells. Cancer Res. 2016; 76(24):7208–7218.
54. Han HJ, Valdeperez D, Jin Q, et al. Dual enzymatic reaction-assisted gemcitabine delivery systems for programmed pancreatic cancer therapy. ACS Nano. 2017; 11:1281–1291.
55. Ruan SB, Hu C, Tang X, et al. Increased gold nanoparticle retention in brain tumors by in situ enzyme-induced aggregation. ACS Nano. 2016; 10:10086–10098.
56. Fouladi F, Steffen KJ, Mallik S. Enzyme-responsive liposomes for the delivery of anticancer drugs. Bioconjug Chem. 2017; 28(4):857–868.
57. Changliang Liu, Zijian Zhao, Rui Gao, et al. Chen Matrix metalloproteinase-2-responsive surface-changeable liposomes decorated by multifunctional peptides to overcome the drug resistance of triple-negative breast cancer through enhanced targeting and penetrability. ACS Biomat Sci Eng. 2022; 8(7):2979–2994.
58. Mock JN, Costyn LJ, Wilding SL, et al. Evidence for distinct mechanisms of uptake and antitumor activity of secretory phospholipase A2 responsive liposome in prostate cancer. Integr Biol (Camb). 2013; 5(1):172–182.
59. Perner S, Hofer MD, Kim R, et al. Prostate-specific membrane antigen expression as a predictor of prostate cancer progression. Hum Pathol. 2007; 38(5):696–701.
60. Pak CC, Erukulla RK, Ahl PL, et al. Elastase activated liposomal delivery to nucleated cells. Biochim Biophys Acta. 1999; 1419(2):111–126.
61. Belfiore L, Saunders DN, Ranson M, et al. N-Alkylisatin-loaded liposomes target the urokinase plasminogen activator system in breast cancer. Pharmaceutics. 2020 Jul 7; 12(7):641.
62. Lee S, Song SJ, Lee J, et al. Cathepsin B-responsive liposomes for controlled anticancer drug delivery in hep G2 cells. Pharmaceutics. 2020; 12(9):876.
63. Barve A, Jain A, Liu H, et al. Enzyme-responsive polymeric micelles of cabazitaxel for prostate cancer targeted therapy. Acta Biomater. 2020; 113:501–511.
64. Lee PC, Chiou YC, Wong JM, et al. Targeting colorectal cancer cells with single-walled carbon nanotubes conjugated to anticancer agent SN-38 and EGFR antibody. Biomaterials. 2013; 34(34):8756–8765.
65. Li T, Murphy S, Kiselev B, et al. A New Interleukin-13 Amino-coated gadolinium metallofullerene nanoparticle for targeted MRI detection of glioblastoma tumor cells. J Am Chem Soc. 2015; 137:7881–7888.
66. Zhang C, Pan D, Li J, et al. Enzyme-responsive peptide dendrimer-gemcitabine conjugate as a controlled-release drug delivery vehicle with enhanced antitumor efficacy. Acta Biomater. 2017; 55:153–162.
67. Ding H, Zhang F, Zhao C, et al. Beyond a carrier: Graphene quantum dots as a probe for programmatically monitoring anti-cancer drug delivery, release, and response. ACS Appl Mater Interfaces. 2017; 9(33):27396–27401.

68. El-Far SW, Abo El-Enin HA, Abdou EM, et al. Targeting colorectal cancer cells with niosomes systems loaded with two anticancer drugs models; Comparative in vitro and anticancer studies. Pharmaceuticals (Basel). 2022; 15(7):816.
69. Wei H, Wang E. Nanomaterials with enzyme-like characteristics (nanozymes): Next-generation artificial enzymes. Chem Soc Rev. 2013; 42(14):6060–6093.
70. Gao S, Lin H, Zhang H, Yao H, Chen Y, Shi J. Nanocatalytic tumor therapy by biomimetic dual inorganic nanozyme-catalyzed cascade reaction. Adv Sci (Weinh). 2018; 6(3):1801733.
71. Li XS, Lovell JF, Yoon J, et al. Clinical development and potential of photothermal and photodynamic therapies for cancer. Nat Rev Clin Oncol. 2020; 17(11):657–674.
72. Tong X, Srivatsan A, Jacobson O, et al. Monitoring tumor hypoxia using 18F-FMISO PET and pharmacokinetics modeling after photodynamic therapy. Sci Rep. 2016; 6:31551.
73. Shahriari M, Zahiri M, Abnous K, Taghdisi SM, Ramezani M, Alibolandi M. Enzyme responsive drug delivery systems in cancer treatment. J Control Release. 2019; 308:172–189.
74. Li YX, Xie DT, Yang YX, et al. Development of small-molecule fluorescent probes targeting enzymes. Molecules. 2022; 27:4501.
75. Urano Y, Sakabe M, Kosaka N, et al. Rapid cancer detection by topically spraying a γ-glutamyltranspeptidase-activated fluorescent probe. Sci Transl Med. 2011; 3(110):110ra119.
76. Verdoes M, Oresic Bender K, Segal E, et al. Improved quenched fluorescent probe for imaging of cysteine cathepsin activity. J Am Chem Soc. 2013; 135(39):14726–14730.
77. Gong Q, Shi W, Li L, et al. Leucine aminopeptidase may contribute to the intrinsic resistance of cancer cells toward cisplatin as revealed by an ultrasensitive fluorescent probe. Chem Sci. 2016; 7(1):788–792.
78. Li X, Shi W, Chen S, et al. A near-infrared fluorescent probe for monitoring tyrosinase activity. Chem Commun (Camb). 2010; 46(15):2560–2562.
79. Cui L, Zhong Y, Zhu W, et al. A new prodrug-derived ratiometric fluorescent probe for hypoxia: High selectivity of nitro reductase and imaging in tumor cell. Org. Lett., 2011; 13:928–931.
80. Kim TI, Kim H, Choi Y, et al. A fluorescent turn-on probe for the detection of alkaline phosphatase activity in living cells. Chem Commun (Camb). 2011; 47(35):9825–9827.
81. Burke HM, Gunnlaugsson T, Scanlan EM. Chem. Commun. Recent advances in the development of synthetic chemical probes for glycosidase enzymes. Chem Commun. 2015; 51:10576–10588.
82. Feng L, Liu ZM, Hou J, et al. A highly selective fluorescent ESIPT probe for the detection of Human carboxylesterase 2 and its biological applications. Biosens Bioelectron. 2015; 65:9–15.
83. Dickinson BC, Kathayat RS, Elvira PD. A fluorescent probe for cysteine depalmitoylation reveals dynamic APT signaling. Nat Chem Biol. 2017; 13(2):150–152.
84. Xu JH, Gao, F.-P.; Li, L.-L.; et al. Gelatin-mesoporous silica nanoparticles as matrix metalloproteinases-degradable drug delivery systems in vivo. Micropor Mesopor Mater. 2013; 182:165–172.
85. Zhang X, Wang X, Zhong W, et al. Matrix metalloproteinases-2/9-sensitive peptide-conjugated polymer micelles for site-specific release of drugs and enhancing tumor accumulation: Preparation and in vitro and in vivo evaluation. Int J Nanomedicine. 2016; 11:1643–1661.
86. Anajafi T, Yu J, Sedigh A, et al. Nuclear localizing peptide-conjugated, redox-sensitive polymersomes for delivering curcumin and doxorubicin to pancreatic cancer microtumors. Mol Pharmaceutics. 2017; 14:1916–1928.
87. Ruan S, Yuan M, Zhang L, et al. Tumor microenvironment sensitive doxorubicin delivery and release to glioma using angiopep-2 decorated gold nanoparticles. Biomater. 2015; 37:425–435.
88. Jin Y, Yang F, Du L. Nanoassemblies containing a fluorouracil/zidovudine glyceryl prodrug with phospholipase A2-triggered drug release for cancer treatment. Colloids Surf B Biointerfaces. 2013; 112:421–428.
89. Yang Y, Aw J, Chen K, et al. Enzyme-responsive multifunctional magnetic nanoparticles for tumor intracellular drug delivery and imaging. Chem Asian J. 2011; 6(6):1381–1389.
90. Chen Z, Li Z, Lin Y, et al. Bioresponsive hyaluronic acid-capped mesoporous silica nanoparticles for targeted drug delivery. Chemistry. 2013; 19(5):1778–1783.
91. Denmeade SR, Jakobsen CM, Janssen S, et al. Prostate-specific antigen-activated thapsigargin prodrug as targeted therapy for prostate cancer. J Natl Cancer Inst. 2003; 95(13):990–1000.
92. Sharma A, Kim EJ, Shi H, et al. Development of a theranostic prodrug for colon cancer therapy by combining ligand-targeted delivery and enzyme-stimulated activation. Biomaterials. 2018; 155:145–151.
93. Parikh AR, Weekes CD, Blaszkowsky LS, et al. J Clin Oncol Suppl. 2022; 40(4):564–564.
94. https://clinicaltrials.gov/ct2/show/NCT04409002. Accessed on 5.11.2022.
95. Wright TC, Dunne VL, Alshehri AHD, et al. Abiraterone in vitro is superior to enzalutamide in response to ionizing radiation. Front Oncol. 2021; 11:700543.

96. https://clinicaltrials.gov/ct2/show/NCT01949337. Accessed on 5.11.2022.
97. Sundar R, Cho BC, Brahmer JR, et al. Nivolumab in NSCLC: Latest evidence and clinical potential. Ther Adv Med Oncol. 2015; 7(2):85–96.
98. www.clinicaltrials.gov/ct2/show/NCT04500535. Accessed on 5.11.2022.
99. Tripathy D, Bardia A, Sellers WR. Ribociclib (LEE011): Mechanism of action and clinical impact of this selective cyclin-dependent kinase 4/6 inhibitor in various solid tumors. Clin Cancer Res. 2017; 23(13):3251–3262.
100. www.clinicaltrials.gov/ct2/show/NCT03285412. Accessed on 5.11.2022.
101. Kato H, De Souza P, Kim SW, et al. Safety, pharmacokinetics, and clinical activity of adavosertib in combination with chemotherapy in Asian patients with advanced solid tumors: Phase Ib study. Target Oncol. 2020; 15(1):75–84.
102. www.clinicaltrials.gov/ct2/show/NCT02513563. Accessed on 5.11.2022.

5 Redox-Responsive Delivery Nanoplatforms in Cancer Theranostics

Pavan Walvekar, Nombeko Sikhosana, Samson A. Adeyemi, Philemon N. Ubanako and Yahya E. Choonara

CONTENTS

5.1 Introduction	77
5.2 Role of Redox States in TME	78
5.3 Glutathione-Responsive Nanoplatforms for Cancer Theranostics	78
5.3.1 Disulfide-Responsive Theranostic Nanoplatforms for Cancer Therapy	79
5.3.2 Diselenide-Responsive Theranostic Nanoplatforms for Cancer Therapy	82
5.3.3 Glutathione-Responsive Nanotheranostic Prodrug Delivery Against Cancer	85
5.3.4 Theranostic Nanoplatforms Undergoing Fenton/Fenton-Like Reactions	88
5.4 ROS Overview in Cancer	90
5.4.1 ROS-Responsive Nanotheranostics for Cancer Therapy	90
5.5 Summary and Prospects	93
References	94

5.1 INTRODUCTION

Despite significant advancements in modern medical technology, cancer continues to dominate current therapies and is regarded as one of the major reasons for premature mortality. According to a recent report by the World Health Organization (WHO), cancer accounted for 10 million deaths worldwide in 2020 [1], and if efforts are not made, cancer incidence and the death toll could increase to over 30.2 million and 16.3 million, respectively, by 2040 [2]. The failure of current cancer therapy can be attributed to drug resistance, poor drug pharmacokinetic profiles, poor tumor-selective drug delivery, inadequate drug uptake at cancer site, low therapeutic efficacy and unsustainable medical costs [3]. Further, the severe unwanted side effects associated with chemotherapy and inevitable damage caused to normal cells by radiation therapy have resulted in poor patient compliance [4, 5]. In this drug-resistant era, researchers have been devoting tremendous energy to developing or inventing novel chemotherapeutics. Discovering new antineoplastic agents and chemically modifying the existing ones might prove a solution to stop cancer cells from multiplying; however, there is no certainty that these therapeutics will be potent enough to eradicate tumors completely, and eventual development of resistance to these new drugs is also likely to occur [6]. In addition, clinical trials and complicated regulatory approval processes are likely to take many years. Therefore, there is a need to utilize the available therapeutics and technologies in a smarter way to treat cancer more efficiently.

The advantages and applications of nanotherapeutics in cancer treatment are well documented. Further, diagnosis is considered an extremely important parameter to track ongoing cancer treatment [7]. In recent decades, research has been focusing on combining both therapy and diagnosis in a single system to develop anticancer nanotheranostics. Since targeted delivery of nanosystems and the associated payloads (active drugs/non-drugs) to tumor sites is crucial to achieve theranostic outcomes, the microenvironment of cancer (redox species, acidic pH, hypoxia, over-expressed enzymes, etc.) has been widely used to device stimuli-responsive theranostic nanoplatforms. For example, Xiong et al. recently reported theranostic pH-responsive mRNA delivery and near infrared (NIR) imaging via dendrimer-based lipid nanoparticles prepared using pH-responsive PEGylated BODIPY-lipids [8]. In another study, Feng et al. reported chemotherapy and photodynamic therapy (PDT) by delivering banoxantrone (AQ4N), a hypoxia-activated prodrug, and chlorin e6 (Ce6), a photosensitizer via liposomes, and tracked them using positron emission tomography (PET), photoacoustic (PA) and fluorescence imaging [9]. Similarly, altered redox states (elevated levels of glutathione (GSH) and reactive oxygen species (ROS) such as hydrogen peroxide (H_2O_2)) in TME are important markers and have been explored extensively to synthesize a variety of theranostic nanoplatforms (organic and inorganic) for cancer therapy. In this chapter, we elucidate the role of redox differences for targeted delivery and highlight some relevant recent articles that report various redox-responsive theranostic nanoplatforms.

5.2 ROLE OF REDOX STATES IN TME

Physiological processes within organisms are maintained at a dynamic equilibrium (homeostasis) by precise and complex systems. Cellular redox reactions form part of these systems by ensuring a balance between oxidizing and reducing reactions within cells. Among various characteristic features of cancer environment, redox species/states are of utmost significance, as they are actively involved in tumorigenesis and apoptosis [10]. While maintaining redox balance plays a vital role in normal cell functioning and signaling, cancer cells or TME is characterized by redox differences. A dysregulation in redox homeostasis leads to ROS (H_2O_2) production, triggering oxidative stress [11]. Abnormal levels of ROS have deleterious effects on cell function, structure and genome stability; therefore, disruption of the redox steady state has been associated with various pathologies, such as cancer [11]. When there is an elevated concentration of ROS in TME, the mammalian cells tend to produce the most common antioxidant, GSH, as a defensive strategy to scavenge ROS, counteract the oxidative stress and maintain redox balance. However, elevated levels of GSH have been associated with chemotherapeutic resistance [12, 13]. In contrast, ROS can be pro-tumorigenic, but substantial levels of ROS results in extremely high oxidative stress that have the ability to produce cytotoxic effects and can inhibit cancer chemoresistance.

The adaptation of tumor cells to certain levels of oxidative stress promotes the upregulation of certain antioxidant factors to ameliorate non-beneficial levels of oxidative stress and optimize ROS promoted proliferation while avoiding ROS thresholds capable of eliciting senescence and apoptosis [11]. The tight regulation of the redox system in tumor cells contributes to their survival, progression and therapeutic resistance. Since there is an abnormal production of both oxidants and antioxidants in TME, understanding redox regulation in cancer development and targeting both ROS and GSH can provide strategies that might assist researchers to leverage the redox system in cancer treatment, particularly in overcoming drug resistance and facilitating tumor detection.

5.3 GLUTATHIONE-RESPONSIVE NANOPLATFORMS FOR CANCER THERANOSTICS

Increased ROS generation and oxidative stress are key features of cancer cells, which contribute to abnormal cell growth and tumor progression [14]. As a natural defensive approach, glutathione, a tripeptide antioxidant, is generated in abundance to neutralize the effects of ROS (mainly H_2O_2), thus intervening with oxidative stress and tumorigenesis. Considering the whole cancer environment, the intracellular GSH concentration (2–10 mM) is ~1000-fold higher compared to extracellular fluids (2–20 µM) [15]. Therefore, redox imbalance with high GSH concentration is recognized as a hotspot to develop redox-responsive drug delivery systems. Especially, nanoplatforms made up of disulfide- and diselenide-based materials have made tremendous progress in redox-responsive drug delivery [16, 17]. These nanosystems with disulfide and diselenide linkers are extremely sensitive to reductive environments of cancers, where they specifically get cleaved to release the encapsulated payloads. In addition to disulfide and diselenide linkers, inactive platinum (IV)–based prodrugs have been delivered to tumor tissues, where they get reduced to bioactive cytotoxic platinum (II) in the presence of GSH [18]. In another approach, intracellular glutathione has been targeted by Fenton/Fenton-like catalysts to promote GSH depletion and induce chemodynamic therapy [19]. Taking advantage of these biological processes and chemistries, various theranostic nanoplatforms have been developed as anticancer strategies; some of the relevant studies will be discussed in subsequent sections and are summarized in Table 5.1.

5.3.1 Disulfide-Responsive Theranostic Nanoplatforms for Cancer Therapy

Disulfide bonds have attracted considerable attention for stimuli-responsive anticancer drug delivery because of their high redox sensibility in TME. Nanoplatforms with disulfide bonds are easily cleaved in the reductive environment of cancer by thiol compounds such as glutathione to trigger release of the payloads [37]. Several anticancer drugs, biologics and photosensitizers have been loaded into various nanosystems for therapy and diagnosis. For instance, DOX, a commonly used chemotherapeutic, was loaded into disulfide containing PEGylated micelles. To provide bioimaging properties, an aggregation-induced emission (AIE) fluorophore, Tripp-COOH, was conjugated to the previously synthesized mPEG-SS-NH_2 via acid-amine coupling chemistry assisted by EDC/HOBt [20]. The DOX-loaded mPEG-SS-Tripp micelles with particle diameter and entrapment efficiency (EE) of ~120 nm and ~72%, respectively, showed enhancement of particle size to ~530 nm at 10 mM GSH environment (representative of the TME) with in 5 h of incubation and further released DOX rapidly compared to redox-insensitive micelles attributed to the cleavage of disulfide linker. In addition, the micelles were found to be rapidly internalized in the cytoplasm of the tested 4T1 cells, which were tracked by AIE bioimaging. When tested in mice, the DOX-loaded micelles were able to significantly reduce tumor volume and increase survival rates of mice compared to free DOX and redox-insensitive micelles.

Targeted gene delivery of chemotherapeutic DNA, mRNA, siRNA and others via non-viral vectors has provided promising outcomes in the treatment of various cancers [38]. Cheng et al. developed nanoplexes by electrostatically complexing cationic thiolated polyethyleneimine and anionic plasmid DNA (pDNA) [21]. These nanoplexes were further coated with GCS-PDP by grafting to form disulfide containing core-shell nanoparticles. As expected, in a high reducing environment of the tested prostate cancer mouse models, the fluorescent dye-labeled nanoparticles selectively released the pDNA in the cytosol to achieve transfection, which was confirmed by bioluminescence imaging.

Mesoporous silica nanoparticles (MSNs) are one of the important classes of inorganic nanosystems that have shown considerable potential in delivering therapeutics for cancer applications [39, 40]. Zhao et al. attempted to co-deliver doxorubicin and a gene (siRNA) to induce targeted and synergistic therapy using disulfide containing a silica-based nanocarrier [22]. MSNs prepared using tetraethylorthosilicate (TEOS) were further modified with 3-mercaptopropyltrimethoxysilane (MPTMS) and 2,2'-dithiodipyridine to provide disulfide bonds. While DOX was encapsulated in disulfide pyridyl mesopores (MSNs-SS-Py), the thiol-modified siRNA was conjugated to their surfaces to yield MSNs-SS-siRNA@DOX (Figure 5.1A). In vitro drug release and intracellular (MCF-7) studies suggested that the loaded therapeutics were rapidly released after MSNs-SS-siRNA@DOX reacted with 5 mM of glutathione (Figure 5.1B). Taking the advantage of doxorubicin's intrinsic fluorescence properties, MSNs-SS-siRNA@DOX displayed the highest fluorescence in tumor tissues in MCF-7 mice xenografts compared to other treatment controls, indicating selective targeting and drug accumulation (Figure 5.1C), resulting in significant reduction of tumor volume (Figure 5.1D).

TABLE 5.1
Summary of Various Glutathione-Responsive Theranostic Nanoplatforms for Cancer Therapy

Nanoplatform	Formulation Materials	Diagnostic Modality	Redox Responsiveness; Payload, Therapeutic Strategy	Cancer Cell Line; Targeted Cancer	Study
Reduction					
A. Disulfide-responsive					
Micelles	Methoxy polyethylene glycol-SS-Tripp	Aggregation-induced emission (AIE)	Cleavage of S-S bond; doxorubicin; chemotherapy	4T1 cells; breast cancer	[20]
Nanoparticles	Polyethyleneimine-SH and glycol chitosan-modified with succinimidyl 3-(2-pyridyldithio)propionate (GCS-PDP)	Bioluminescence imaging	Cleavage of S-S bond; pCMV-fLuc (therapeutic pDNA); gene therapy	PC3 cells; prostate cancer	[21]
Mesoporous silica nanoparticles	Cetyltrimethylammonium bromide (CTAB), tetraethylorthosilicate and 2,2′0-dithiodipyridine	Fluorescence imaging	Cleavage of S-S bond; doxorubicin and siRNA, chemotherapy and gene therapy	MCF-7 cells; breast cancer	[22]
Nanoassemblies	Human serum albumin (HAS)	Fluorescence, photoacoustic and magnetic resonance imaging	Cleavage of S-S bond; Ce6; Photodynamic therapy	4T1 cells; Breast cancer	[23]
B. Diselenide-responsive					
Micelles	Hyaluronic acid-SeSe-Ce6	Fluorescence imaging	Cleavage of Se-Se bond; Ce6; photodynamic therapy	4T1 cells; breast cancer	[24]
Molybdenum nanoparticles	Molybdenum diselenide (MoSe$_2$) and polyvinylpyrrolidone (PVP)	Fluorescence imaging	Cleavage of Se-Se bond; Ce6 and PD-1 monoclonal antibody; immunotherapy, photothermal therapy and photodynamic therapy	HT29 cells; colorectal adenocarcinoma	[25]
Bismuth selenide nanoplates	Bismuth selenide	X-ray computed tomography (CT) imaging	Cleavage of Se-Se bond; photothermal therapy	H22 cells; hepatocarcinoma	[26]
Nanocomposites	Polycaprolactone and polyethylene glycol	Fluorescence imaging	Cleavage of Se-Se bond; indocyanine green (ICG); photodynamic therapy	4T1 cells; breast cancer	[27]
Mesoporous silica nanoparticles	Bis [3-(triethoxysilyl) propyl] diselenide (BTESePD) and tetraethyl orthosilicate	Fluorescence imaging	Cleavage of Se-Se bond; doxorubicin, methylene blue and anti-PD-1 antibody; chemotherapy, photodynamic therapy and immunotherapy	4T1 and MCF-7 cells; breast cancer	[28]

C. Prodrug delivery					
Nanoparticles	IR820-SS-camptothecin	Fluorescence and photoacoustic imaging	Cleavage of S-S bond; camptothecin, chemotherapy and photothermal therapy	4T1 cells; breast cancer	[29]
Lipid polymer hybrid nanoparticles	Camptothecin-SS-croconaine-SS-camptothecin PLGA, phosphatidylcholine and DSPE-PEG-folic acid	Fluorescence and photoacoustic imaging	Cleavage of S-S bond; camptothecin, chemotherapy and photothermal therapy	MCF-7 cells; breast cancer	[30]
Nanoparticles	Doxorubicin-SS-capreomycin and DSPE-PEG	Fluorescence and photoacoustic imaging	Cleavage of S-S bond; doxorubicin and ICG; chemotherapy and photothermal therapy	SCLC cells; small-cell lung cancer	[31]
Nanoparticles	Paclitaxel-Se, mPEG$_{5k}$-pPhe and uPA-mPEG$_{5k}$-pPhe	Fluorescence imaging	Chemotherapy	MDA-MB-231 cells; triple negative breast cancer (TNBC)	[32]
Nanoparticles	Polydopamine, β-cyclodextrin and 1-adamantyl bromomethyl ketone	Fluorescence and photoacoustic imaging	Reduction of Pt (IV) to Pt (II); cisplatin; chemotherapy and photothermal therapy	143B cells; osteosarcoma	[33]
D. Fenton/Fenton-like reactions					
Nanoparticles	Zinc phthalocyanine, dioleoylphosphatidic acid (DOPA) and 1,2-dioleoyl-sn-glycero-3-phosphocholine (DOPC)	Fluorescence imaging	Redox reaction with Fe; ferric pyrophosphate; photodynamic therapy and chemodynamic therapy	4T1 cells; breast cancer	[34]
Nanoparticles	$Zn_{0.2}Fe_{2.8}O_4$, polydopamine and KMnO$_4$	Magnetic resonance imaging	Fenton-like reaction caused by Mn^{2+}; MnO_2; photothermal therapy	4T1 cells; breast cancer	[35]
Nanoparticles	Calcium chloride, hydrogen peroxide, sodium hydroxide, tetraethyl orthosilicate; KMnO$_4$ and 4T1 cell membrane	Magnetic resonance imaging	Fenton-like reaction caused by Mn^{2+}; MnO_2 and CaO_2; chemotherapy and chemodynamic therapy	4T1 cells; breast cancer	[36]

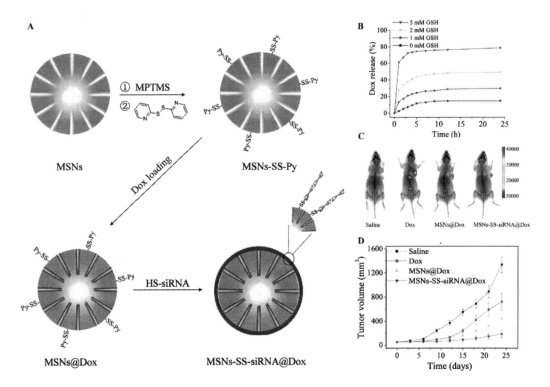

FIGURE 5.1 (A) Formulation method of MSNs-SS-siRNA@DOX. (B) Rapid release of DOX from MSNs-SS-siRNA@DOX due to cleavage of SS bonds in 5 mM GSH environment. (C) Fluorescent images showing selective accumulation of DOX at tumor site. (D) Reduced tumor volume following administration of MSNs-SS-siRNA@DOX compared to other treatment controls. Reproduced with permission from [22]. Copyright (2017) Royal Society of Chemistry.

While chemotherapy has been the mainstay of cancer treatment for several decades, PDT has garnered immense popularity in recent years due to precise targeting, no or less invasiveness, lower side effects and short-term treatment [41]. Hu et al. reported thermal modulated theranostic redox-responsive HSA nanoassemblies loaded with a photosensitizer, Ce6 (HSA-Ce6 NAs), to promote PDT at tumor sites upon GSH exposure [23]. The prepared nanoassemblies were further chelated with Mn^{2+} to facilitate magnetic resonance imaging (MRI). Since hypoxia at tumor tissues does not encourage PDT and retards it because of lack of oxygen supply, the temperature of TME was mildly elevated from 37 °C to 43 °C to subsequently increase oxygen saturation and accelerate photosensitization reaction rate (promote PDT). The increase in body temperature of 4T1 tumor-bearing mice to 43 °C resulted in improved blood circulation and simultaneously increased oxygen supply (saturation increased from 52% to 79%) in tumor blood vessels, which was visualized by photoacoustic tomography scanner. At reduced GSH environments in tumor-bearing mice, the nanoassemblies were disassembled because of the cleavage of disulfide bonds to release the photosensitizer. High accumulation of NAs and Ce6 was confirmed by triple modal imaging (fluorescence, PA and MR), where each of them showed significantly higher intensities for Ce6/NAs compared to free counterparts.

5.3.2 Diselenide-Responsive Theranostic Nanoplatforms for Cancer Therapy

Selenium's duality, as an antioxidant and a prooxidant, provides the foundation for its application in cancer therapies and prevention. The common mechanism of action by seleno-compounds involves the reduction of free radicals, activation of pro-apoptotic pathways, inhibition of angiogenesis and efflux pumps in chemotherapeutic-resistant cancer cell lines and induction of cytotoxic activity in cancer cells [42]. Thoroughly investigated antitumor and chemo-preventive organic selenocompounds include diselenides, slelenocynates, selenoesters, selenide and selenium nanoparticles, among others. Diselenides in particular have received enormous attention, as they are key elements in redox recycling. In consonance with this, the naturally occurring diselenide selenocysteine displays considerable antioxidant properties and sensitivity to $^1O_2$ and 600 nm or higher wavelength light in PDT drug delivery [43]. A self-assembling micelle nanosystem (HA-SeSe-Ce6) constructed from the conjugation of hyaluronic acid, selenocystamine dihydrochloride and the PDT related photosensitizers, Ce6 was shown to drive a redox responsive drug release for enhanced PDT and possess cancer targeting capabilities via the overexpressed CD44 receptors [24]. Redox sensitivity of the nanosystem is endowed by the presence of the diselenide bond within the IIA-ScSc-Cc6 micelles, which is cleaved in redox environments and promotes a reduction in particle size under oxidizing and reducing environments. The incorporation of the diselenide bond in conjunction with hyaluronic acid within the nanosystem therefore aided targeted delivery of the nanosystem, accumulation of Ce6 at the tumor site and redox-responsive release of the photosensitizer rather than being cleared by the kidneys as seen in the Ce6 only control when intravenously administered to 4T1 tumor-bearing mice. Apart from the HA-SeSe-Ce6 micelle-treated mice showing improved PDT efficacy compared to Ce6 control (increased anticancer activity), the nanosystem promoted apoptosis of cancer cells and inhibition of metastasis.

Thus far, the use of tumor phototherapy (particularly PDT) in conjunction with photosensitizer-loaded nanoplatforms as novel alternatives approaches to cancer therapy has been reviewed. The combinatorial approach of phototherapy involving PDT and photothermal therapy (PTT), which requires the delivery of photothermal agents to tumors, where they drive the conversion of near-infrared laser energy into heat resulting in selective tumor cell death, has been explored [44]. Zhao et al. designed Ce6-loaded polyvinylpyrrolidone (PVP)-molybdenum diselenide $MoSe_2$ nanoparticles ($MoSe_2$-PVP) for combined PTT, PDT and immunotherapy-based cancer therapy [25]. $MoSe_2$, particularly in the nanoscale, shows superior NIR absorption and therefore was utilized as a photothermal agent in the nanosystem. The addition of the immune checkpoint inhibitor PD-1 monoclonal antibody (mAb) to the $MoSe_2$-PVP nanosystem was conducted to facilitate cancer therapy by activating autologous immune cells that recognize and eradicate tumor cells [45, 46]. The anticancer potential of the prepared theranostic nanosystem was investigated by monitoring the tumor inhibition in HT29 tumor-bearing mice. A PTT/PDT combinatorial therapy approach showed excellent suppression of tumor growth; the tumor was eliminated by day 5 post treatment, no recurrence was observed within 28 d and even in vitro treatment of HT29 cells with the nanosystem plus phototherapy showed significant tumor cell survival rate credited to the catalase-impersonation ability of $MoSe_2$-PVP/Ce6 nanoparticles. On the contrary, PTT- and PDT-treated mice with HT29 tumors showed some tumor suppression, with the volume increasing ~4.5- and 6-fold, respectively, compared to the initial tumor volume. Since selenium is essential in maintaining immune function, importantly in inhibition of cancer growth via the activation of T cells and natural killer cells, a tumor immune therapy analysis was conducted. This was carried out to assess the potential of $MoSe_2$PVP NP degradation products ($MoSeO_3$) to activate dendritic cells (DCs) and enhance the apoptotic effects of CD8+ T lymphocytes. Additionally, the nanosystem was combined with PD-1 mAb to assess its effect on the tumor environment. $MoSe_2$-PVP NPs were shown to efficiently promote CD8+ T cell-mediated immune responses to inhibit tumor growth. Interestingly, the combination of immunotherapy and PTT in HT29 tumor-bearing mice displayed significant cancer inhibitory effect, suggesting once again a combinatorial approach to cancer therapy is much more effective.

The impressive advances made in theranostic nanoplatforms encompass various materials having imaging properties, such as the topological insulator bismuth selenide (Bi_2Se_3) [47]. Bi_2Se_3 has attractive photoelectric and optical properties necessary for the development of devices involving photonic and quantum computing [48]. Li et al. therefore designed Bi_2Se_3 nanoplates and investigated their potential role in photothermal conversion of NIR laser and tumor ablation in vivo [26]. Subsequently, Bi_2Se_3-based in vivo CT imaging was analyzed. When tested on mice carrying H22 cancer tumors, Bi_2Se_3 nanoplates were able to induce photothermal ablation upon laser irradiation, which led to reduction in tumor volume. The diagnostic capability of the Bi_2Se_3 nanoplates was analyzed by comparing the X-ray absorption of the nanosystem with that of the clinically approved CT contrast agent iopamidol in vitro and in vivo. Bi_2Se_3 provided a higher contrast at lower dose compared to iodopamidol under both conditions; this was attributed to Bi_2Se_3's higher X-ray attenuation coefficient, thus making it a promising contrast agent for theranostic purposes.

The relative sensitivity of diselenide-containing drug delivery systems to redox state, temperature and light limits their application for therapeutic delivery [49]. To bypass these limitations, supercritical carbon dioxide (SC-CO_2) technology was adopted for the fabrication of diselenide-based nanosystems loaded with indocyanine green (ICG) NIR dye for PDT therapy of breast cancer [27]. Briefly, the initiator, 11-bromo-1-undecanol, was reacted with selenium to synthesize diselenide precursor; a subsequent reaction with ε-caprolactone (PCL) in the presence of stannous octoate yielded PCL-SeSe-PCL. Thereafter, the hydroxyl and carboxyl groups of PCL-SeSe-PCL and PEG-COOH, respectively, were esterified using carbodiimide chemistry resulting in the formation of a diselenide containing block polymer poly(ethylene glycol)-poly(ε-caprolactone)-poly(ethylene glycol) (PSe) (Figure 5.2A). Further, ICG was loaded into the block copolymer via interaction of PSe and ICG facilitated by SCF technology. In 10 mM GSH and 100 μM H_2O_2 environments, ICG was rapidly released, revealing that the diselenide linker was sensitive to both reductive and

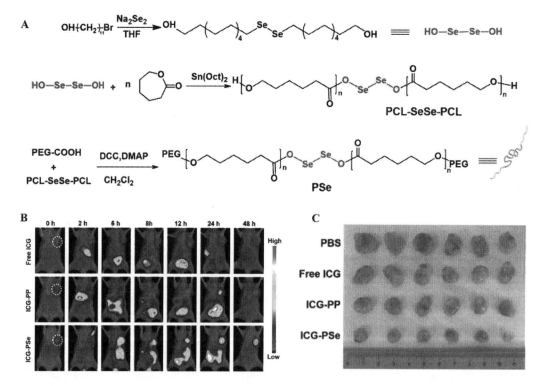

FIGURE 5.2 (A) Reaction steps involved in the synthesis of PSe. (B) In vivo NIR fluorescence imaging of 4T1 tumor-bearing mice intravenously injected with free ICG, ICG-PP nanoparticles and ICG-PSe nanoparticles. (C) Photos of harvested tumors at the end of various treatments. Reproduced with permission from [27]. Copyright (2021) Elsevier.

oxidative conditions. The efficacy of the nanosystem in tumor growth inhibition and ablation was assessed by intravenous administration in mice with 4T1 tumors followed by laser irradiation. Attributed to the diselenide cleavage and selective release of ICG at the tumor site, ICG-PSe nanoparticles displayed prolonged retention until 48 h; however, the same was not observed for free ICG and ICG-PP nanoparticles (without Se-Se linkers), as confirmed by fluorescence imaging (Figure 5.2B). The ICG-PSe nanoparticle-treated group displayed a significant inhibition of tumor growth (Figure 5.2C), and histological data showed no substantial abnormality in the organs of treated mice compared to the untreated group.

Immune checkpoint blockade therapies aid in the elimination of some tumors and guard against metastasis by enhancing the antitumor immune response; these include monoclonal antibodies specific for programmed cell death protein (ligand) 1 (PD-1/L1) and cytotoxic T lymphocyte antigen 4 [50]. However, despite initial encouraging findings, these therapies do not significantly assist the majority of patients with certain aggressive cancers due to low response rates that limit their clinical relevance [51, 52]. This might be because some tumors frequently display a tumor microenvironment regarded as immunologically "cold" due to the restricted expression of activation markers and decreased tumor infiltration [53, 54]. Therefore, the promotion of intra-tumoral T cell infiltration/activation when combined with other treatment modalities like PDT represents a more promising approach for cancer treatment. In view of this, the effectiveness of combining PD-1 mAb and redox-responsive nanosystems to treat breast cancer has been assessed [28]. The sol-gel method was adopted to synthesize red light-responsive MSNs with a ROS-cleavable diselenide-bond to enable the degradation of the organosilica matrix and the delivery of the photosensitizer, methylene blue, and the chemotherapeutic, doxorubicin. The combination of immunotherapy, chemotherapy and PDT elicited a sustained anti-neoplastic behavior for both primary and metastatic tumors as well as preventing the recurrence of 4T1 tumors in vitro and in vivo.

5.3.3 GLUTATHIONE-RESPONSIVE NANOTHERANOSTIC PRODRUG DELIVERY AGAINST CANCER

A prodrug approach has been used widely to overcome some drug-associated limitations such as toxicity, solubility, low permeability and low bioavailability [55]. For example, platinum (II)-based drugs such as cisplatin, oxaliplatin and carboplatin, highly potent antineoplastic drugs, show severe unwanted side effects and are prone to acquire resistance. Therefore, these square-planar Pt (II) drugs are often converted to inactive octahedral Pt (IV) prodrug complexes, which, upon exposure to in GSH in TME, are reduced to bioactive Pt (II) drugs [18, 56, 57]. Similarly, the functional groups of some anticancer drugs can be conjugated to various polymers or lipids containing disulfide or diselenide linkages. Utilizing a prodrug strategy, various GSH-responsive theranostic nanoplatforms have been reported.

Ao et al. recently constructed prodrug-based nanoparticles comprising of camptothecin (CPT) and a photothermal agent, IR820, to produce chemotherapy and photothermal therapy [29]. Camptothecin was conjugated to IR820 via a disulfide linker, which self-assembled in an aqueous environment to form nanoparticles of ~72 nm. Attributed to the disulfide linker present in the amphiphile, these nanoparticles disassembled rapidly in the presence of 10 mM GSH to release ~85% CPT and IR820. In contrast, only about 4% of the drug was released in the same time period of 32 h. When tested in 4T1 tumor-bearing mice, the nanoparticles displayed strong NIR fluorescence at the tumor site compared to free IR820, highlighting the ability of the nanosystem to achieve passive targeting. According to the authors, the enhanced accumulation of nanoparticles was due to the enhanced permeability and retention (EPR) effect. Furthermore, while CPT showed chemotherapeutic effects, NIR laser irradiation rapidly elevated the temperature to ~50 °C within 2 min (confirmed by photoacoustic signals) to produce PTT. Chemotherapy and PTT, when combined, showed better anticancer effects with respect to reduction of tumor volume and body weight compared to free CPT, free IR820 and nanoparticles without NIR irradiation.

With a similar aim to produce chemophotothermal therapy, CPT (two molecules) was conjugated to croconaine (CR) dye via two-SS-linkers [30]. The synthesized CPT-SS-CR was further loaded into folate-modified lipid polymer hybrid nanoparticles (LPHNs) made up of 1,2-distearoyl-sn-glycero-3-phosphoethanolamine-N-[folate (polyethylene glycol)-2000] (DSPE-PEG-FA), soybean phosphatidylcholine (SPC) and PLGA. The chemical structure of CPT-SS-CR and theranostic application of LPHNs is illustrated in Figure 5.3A. Having released CPT and CR rapidly following exposure to 10mM GSH, folic acids (FAs) present on the surfaces of LPHNs facilitated enhanced cell internalization by active targeting folate receptors of MCF-7 cells. In tumor-bearing

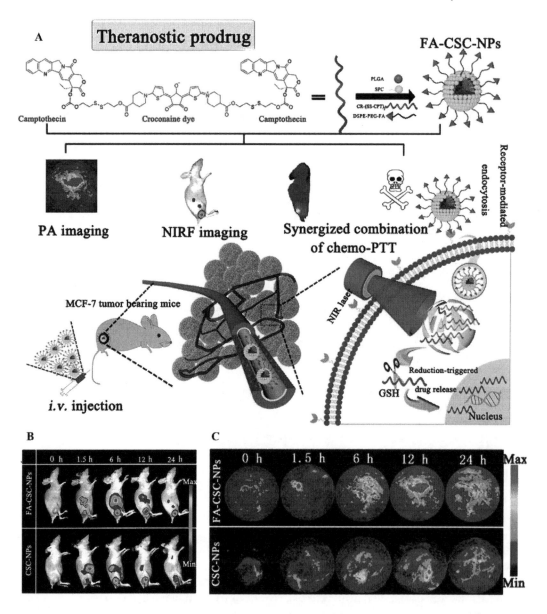

FIGURE 5.3 (A) Chemical structure of CPT-SS-CR prodrug and theranostic application of LPHNs. (B) Fluorescence images following treatment in MCF-7 tumor-bearing mice. (C) Photoacoustic images following treatment in MCF-7 tumor-bearing mice. Reproduced with permission from [30]. Copyright (2018) Royal Society of Chemistry.

mice, LPHNs exhibited 2-fold enhanced accumulation in tumor tissues compared to folate absent nanoparticles, as confirmed by fluorescence (Figure 5.3B) and PA imaging (Figure 5.3C). Upon irradiation with NIR laser, CPT-SS-CR-loaded LPHNs increased the temperature of TME to 54.6 °C to induce PTT, while the CPT showed chemotherapeutic effects to show complete tumor inhibition. Although nanoparticles without folic acid were also able to inhibit tumor growth, the anticancer effects were inferior compared to FA-modified LPHNs. The combination of CPT and CR-mediated PTT was able to produce synergistic effects compared to other treatment controls crediting to FA-based targeting.

Dual-responsive systems that respond to the GSH and acidic pH of TME have also been investigated. Zhu et al. recently synthesized an amphiphilic prodrug made up of DOX, dithiodiacetic acid and capreomycin (used as a CD56 targeting ligand) to treat small-cell lung cancer (SCLC) [31]. The two-step synthetic procedure involved amidation reactions between carboxylic acids of dithiodiacetic acid and amines of DOX and capreomycin using carbodiimide chemistry. The synthesized DOX-SS-Cm could self-assemble into nano-sized particles of ~170 nm to encapsulate photosensitizer ICG. These particles were further coated with DSPE-mPEG to improve in vivo long circulation ability. The prepared ICG@Dox-SS-Cm nanoparticles (~200 nm) displayed rapid release of payload (~83% in 48 h) in glutathione and acidic environments of cancer. The authors attributed the pH-dependent release to the protonation of amine groups of DOX-SS molecule in acidic medium. When tested in an H446 tumor mice model with SCLC overexpressed with CD56 receptors, the nanoparticles showed efficient accumulation in tumor sites within one hour post-injection, as confirmed by NIRF and PA imaging; however, this was not observed for capreomycin free particles, thus indicating the important role of the ligand in targeting. After laser irradiation, the temperature of the tumor site in the ICG@Dox-SS-Cm nanoparticle-treated group increased to 46.2 °C, displaying PTT. Having confirmed in vitro cytotoxicity, the mice treated with ICG@Dox-SS-Cm nanoparticles + laser could inhibit tumor growth significantly compared to other treatments displaying synergism.

As paclitaxel (PTX) is hydrophobic and therefore tends to aggregate during formulation, compromising drug loading efficiency, He et al. synthesized a PTX-selenide dimeric prodrug loaded onto uPA peptide functionalized amphiphilic nanoparticles (uPA-PTXD NPs) targeting the urokinase-type plasminogen activator receptor, which is overexpressed in TNBCs [32]. To investigate the redox responsiveness of the nanosystem (uPA-PTXD NPs), in vitro PTX release under varied GSH concentrations mirroring the physiological and tumor intracellular redox microenvironments respectively was investigated. In the presence of 10 mM GSH, the Se-Se bond in the dimeric PTX-Se prodrug nanoparticles (uPA-PTXD nanoparticles) was selectively cleaved therefore releasing PTX. Subsequently, the anti-cancer effect of the nanosystem was assessed by comparing it to the commercially available PTX drug Taxol. MDA-MB-231 cells were incubated with Taxol, uPA-PTXD nanoparticles and PTXD-Se nanoparticles, respectively. After 72 h, uPA-PTXD NP-treated cells showed lower IC50 value (461.3 ng/mL) compared to PTXD-Se treated cells (585.3 ng/mL); however, both displayed higher IC50 values than Taxol (258.5 ng/mL). According to the authors, the difference in IC50 values could be due to relative burst release of the free Taxol compared to nanoformulations. Finally, the nanosystems were investigated in vivo by intravenously injecting them into orthotopic MDA-MB-231 tumor-bearing mice. BODIPY was loaded onto the nanoparticles to facilitate the tracking of nanoparticle biodistribution and visualize the TNBC tumorigenic tissue. The uPA-tagged nanosystem showed targeted accumulation of the anti-cancer drug and significantly lower toxicity compared to the PTXD-Se nanoparticles. Although free Taxol had lower IC50 values in vitro, uPA-PTXD nanoparticles treated mice exhibited a stronger anti-tumor effect compared to free Taxol and PTXD-Se nanoparticle-treated mice by significantly reducing tumor volume.

Polydopamine (PDA) has been documented to induce strong photothermal therapy [58]. In line to offer chemotherapy and photothermal therapy, Du et al. reported a supramolecular theranostic nanosystem by combining cisplatin (IV)-conjugated β cyclodextrin (βCD) and adamantane-functionalized PDA nanoparticles [33]. A marked binding affinity between hydrophobic adamantly groups present on PDA nanoparticles and hydrophobic cavities of βCD resulted in strong host-guest recognition, therefore achieving high encapsulation for cisplatin (IV). In 10 mM GSH-reducing environment, an accelerated conversion of inactive cisplatin (IV) to bioactive cisplatin (II) was observed facilitating its rapid release; however, <10% of the drug was released in PBS without GSH. In tumor-bearing mice, both PDA and a combinatorial system induced PTT after irradiating with laser as confirmed by photoacoustic tomography signal intensity; meanwhile, the latter was more promising in reducing tumor volume and inhibiting tumor growth, which can be attributed to PTT and chemotherapy produced by PDA and redox-responsive release of cisplatin (II), respectively.

5.3.4 Theranostic Nanoplatforms Undergoing Fenton/Fenton-Like Reactions

As is known, oxidative stress produced by ROS such as H_2O_2, hydroxyl radical (·OH) and singlet oxygen ($^1O_2$) can actively drive clearance of cancer cells via apoptosis. Various antineoplastic drugs have been known to trigger tumor apoptosis by promoting the endogenous generation of these cytotoxic ROS [59]. Furthermore, emerging strategies such as chemodynamic therapy are absolutely dependent on ROS-mediated apoptosis [60]. The glutathione suppressed H_2O_2 in tumor cells is at a low concentration and often too weak to induce cell death; therefore there is a need to covert the existing H_2O_2 into highly reactive ·OH to induce programmed cell death, and this can be achieved through Fenton/Fenton-like reactions [61].

Fenton/Fenton-like reactions have been regarded as important architects in inducing oxidative stress in tumor cells. Generally, transition metal ions such as Fe^{2+}, Mn^{2+}, Cu^{2+} and Co^{2+} participate in catalyzing H_2O_2 to generate ·OH [62]. Examples of typical reactions involving Fe^{2+} and Mn^{2+} are shown in Equations (1) and (2). Several Fenton and Fenton-like catalysts such as Fe_3O_4, MnO_2 and Cu_2Se are often used as nanomaterials to build various nanoplatforms to achieve chemodynamic therapy (CDT) [62]. Apart from taking part in catalyzing H_2O_2 to produce ·OH, these catalysts have been able to interact with abundantly available GSH to cause its depletion and improve the efficacies of ROS-based therapies [63]. Moreover, some of the metal ions, mainly Fe^{2+} and Mn^{2+}, can act as MRI contrast agents to assist cancer diagnosis [64, 65]. Considering these advantages, a number of potential nanotheranostic strategies have been developed to fight various cancers.

$$Fe^{2+} + H_2O_2 \rightarrow Fe^{3+} + \cdot OH + OH^- \qquad (5.1)$$

$$Mn^{2+} + H_2O_2 \rightarrow Mn^{3+} + \cdot OH + OH^- \qquad (5.2)$$

With an aim to induce PDT and chemodynamic therapies, Huang et al. prepared multifunctional ferric pyrophosphate and zinc phthalocyanine-based FeP-ZnPc nanoparticles [34]. The authors hypothesized that FeP-ZnPc nanoparticles in TME would release Fe^{3+} ions which would undergo redox reaction with glutathione to cause its depletion and production of Fe^{2+} ions. Further, following laser irradiation, ZnPc (photosensitizer) would generate ROS such as singlet oxygen ($^1O_2$) and H_2O_2 to facilitate PDT. Meanwhile, the generated H_2O_2 would interact with Fe^{2+} ions via Fenton reaction to produce highly toxic ROS such as ·OH, thus resulting in chemodynamic therapy (Figure 5.4A). When incubated with MCF-7 cells, ferric pyrophosphate present in the nanoparticles released Fe^{3+}, which reacted with intracellular GSH to deplete its concentration (Figure 5.4B) and subsequently reduced to Fe^{2+} ions (Fenton reaction catalyst). Thereafter the Fe^{2+} ions formed in this way underwent Fenton reaction with the elevated H_2O_2 to generate highly reactive ·OH in the tested cancer cells. After irradiation in an 4T1 xenograft tumor mice model, FeP-ZnPc nanoparticles could induce both PDT and chemodynamic therapies and showed better results in reducing tumor volume and weight compared to saline, FeP-DOPC, FeP-ZnPc (without irradiation), calcium phosphate (CaP)-ZnPc, CaP-ZnPc (without irradiation) and CaP-ZnPc (with irradiation) (Figure 5.C-D). It was an interesting study where FeP-ZnPc nanoparticles were able to induce photo/chemodynamic therapy while depleting intracellular GSH.

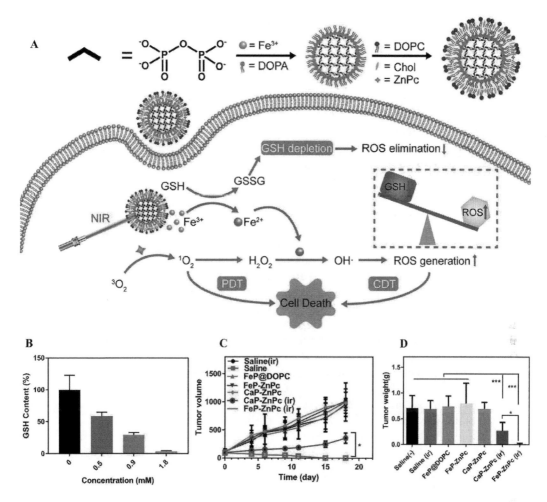

FIGURE 5.4 (A) Formulation method and cancer cell apoptosis mechanism (GSH depletion, PDT and chemodynamic therapy) of FeP-ZnPc nanoparticles. (B) Intracellular depletion of GSH following treatment with various concentrations of FeP@DOPC. (C) Tumor volume after various treatments. (D) Tumor weight after various treatments. Reproduced with permission from [34]. Copyright (2020) Elsevier.

Manganese ions (Mn^{2+}) are known to initiate Fenton-like reactions to catalyze H_2O_2 into highly reactive ·OH [66]. In addition, Mn^{2+}-complexes can be used as MRI contrast agents that can assist tumor imaging. Taking these advantages into account, Yao et al. constructed theranostic MnO_2-coated $Zn_{0.2}Fe_{2.8}O_4$-PDA nanoparticles to induce photothermal therapy [35]. Upon laser irradiation, MnO_2-coated $Zn_{0.2}Fe_{2.8}O_4$-PDA nanoparticles elevated the temperature to ~52 °C to show photothermal effects. Further, these particles were able to react with GSH to release Mn^{2+} and cause GSH depletion. From MTT assay performed on 4T1 cells, it was found that MnO_2-coated $Zn_{0.2}Fe_{2.8}O_4$-PDA nanoparticles did not induce cytotoxicity initially, with >80% cells being viable; however, combination with laser irradiation resulted in significant cell death, which can be attributed to photothermal effects and generation of toxic ·OH. These results were in complete agreement with in vivo studies, where tumor growth was significantly reduced compared to saline and MnO_2-coated $Zn_{0.2}Fe_{2.8}O_4$-PDA nanoparticles (without laser irradiation). Apart from playing a vital role in inducing photothermal effects, Mn^{2+} and $Zn_{0.2}Fe_{2.8}O_4$ assisted in T1/T2-weighted magnetic resonance imaging.

Calcium overloading in tumor cells is characterized by dysfunction of mitochondria followed by amplification of intracellular oxidative stress [67]. With an aim to self-generate H_2O_2, Liu et al. developed a CaO_2-MnO_2 theranostic nanoreactor loaded with DOX to induce chemo-chemodynamic therapy [36]. The prepared CaO_2-MnO_2 nanoreactors were further coated with 4T1 cancer cell membrane (CM) to provide homotypic targeting. According to the authors, following uptake of CaO_2-MnO_2-CM nanoreactors by tumor cells, GSH that is present in high concentrations would react with CaO_2-MnO_2 to release Mn^{2+}, DOX and CaO_2. The released CaO_2 would react with water (H_2O) present in the cytosol to substantially generate and self-supply H_2O_2. Mn^{2+} that acts as a Fenton-like catalyst would further react with H_2O_2 to produce highly toxic ·OH to induce chemodynamic therapy, while DOX would produce chemotherapy. From flow cytometry results, it was also found that the excessive calcium ions produced inside the cells could damage mitochondria to amplify oxidative stress and enhance CDT. Having confirmed the generation of highly reactive ·OH and cytotoxicity potential in vitro, CaO_2-MnO_2-CM nanoreactors were tested in 4T1 tumor-bearing mice. As confirmed by T1-weighted MR images, CaO_2-MnO_2-CM nanoreactors were specifically accumulated at the tumor site, which can be attributed to homotypic targeting provided by the 4T1 cell membrane. The nanosystem showed enhanced anticancer activity compared to individual counterpart treatments by significantly reducing tumor volume.

5.4 ROS OVERVIEW IN CANCER

Cellular compartments such as the endoplasmic reticulum, phagosomes, cell membranes and mitochondria are key ROS producers. Through ROS production, mitochondria have emerged as an integral part in the control of cell signaling under both physiological and pathological conditions. Highly reactive molecules such as O_2^-, ·OH, $^1O_2$ and H_2O_2 resulting from electron leakage and/or incomplete electron transfer through the electron transport chain migrate to the cytosol and modify/damage DNA, lipids and proteins, resulting in the activation of oncogenes, inhibitions of tumor suppressors and support of migratory signaling [68]. Additionally, cyclooxygenases, NADPH oxidases (NOXs), xanthine oxidases, lipoxygenases and the iron-catalyzed Fenton reactions all continuously produce ROS. External factors such as exposure to physical agents (UV rays and heat), chemotherapy and radiotherapy in cancer are also key ROS producers. The less reactive but versatile ROS, H_2O_2, unlike other oxygen species, can reach different cellular compartments. It acts as a second messenger in pathways involving extracellular signal transduction and gene expression regulation and contributes to redox signaling [59]. In fact, several studies have shown an interest in the typical unrestricted growth pattern of tumor cells in response to ROS build-up [69, 70]. While ROS has been recognized to be mitogenic in tumor cells, elevated ROS can also induce cancer cell apoptosis or necroptosis if the antioxidant system is not efficient enough [68]. This demonstrates that cancer cells must attain a delicate redox balance during cellular development and proliferation in order to sustain survival. Thus, attaining redox balance is essential for tumor growth, progression and proliferation. Synonymous with normal cells, imbalances in redox homeostasis are damaging to tumors. Hence, the effect of ROS on cancer cells is dependent on ROS concentration, type of oxidant and site of action. Low to moderate concentrations of oxidants have beneficial effects on tumor cells, as they can trigger signaling cues involved in cancer cell growth, proliferation and metabolism [71]. Thus, a plethora of ROS responsive nanosystems have drawn much attention for the treatment and diagnosis of cancer. For instance, a dual responsive (ROS-pH) smart nanoplatform capable of achieving desired drug distribution, bioavailability via EPR effect and intra-tumoral penetrating efficiency, thereby bypassing tumor-associated fibroblasts along with the associated dense extracellular matrix [72], was constructed.

5.4.1 ROS-Responsive Nanotheranostics for Cancer Therapy

The tumor microenvironment is characterized by a plethora of factors, one of which is the elevated levels of H_2O_2 (50–100 μM) produced by superoxide dismutase in the mitochondria through the oxidation of O_2^- [73, 74]. The strong oxidative capacity of H_2O_2 has been leveraged in cancer treatment and diagnoses as a key stimulus [75]. Accumulation of H_2O_2 is thus used to generate O_2 to reduce hypoxia and improve O_2-mediated treatments. Hypoxia, playing a prominent role in the recurrence, invasion and metastasis of tumor cells, remains a major barrier in cancer theranostics and reduces the therapeutic effects of chemotherapy and O_2-mediated therapies such as radiotherapy and PDT [76]. Nanoplatforms designed for the catalysis of H_2O_2 to O_2, particularly cytotoxic singlet oxygen ($^1O_2$), with the assistance of catalase and PDT (dependent on the presence of O_2) for the inhibition of tumor growth have been investigated. Recently, Phua et al. constructed β cyclodextrin functionalized, catalase-conjugated hyaluronic acid nanoparticles loaded with adamantane modified chlorin e6 to yield HA-CAT@aCe6 nanosystem [77]. In this work, catalase was used to decompose H_2O_2 to O2 and help relieve hypoxia. The nanosystem selectively targeted cancer cells overexpressing CD44 receptors via hyaluronic acid binding; in vivo experiments also demonstrated tumor accumulation of the nanosystem in mice bearing the MDA-MB-231 tumor. Catalase decomposed endogenous H_2O_2, supplying oxygen and thus improving PDT efficacy following light irradiation. Following intravenous administration of the nanosystem under light irradiation, substantial tumor regression was observed compared to the catalase-free control nanosystem.

To take advantage of the tumor microenvironment and modulate tumor hypoxia for efficient combinatorial therapy, MnO_2-based theranostic with redox activity was developed with focus on preventing premature photosensitizer leakage from nanocarriers during delivery, thus increasing bioactive accumulation at the tumor site and achieving in vivo magnetic resonance imaging under oxidative stress [78]. PDT was enhanced by synthesizing a silicon dioxide-methylene blue (SiO_2-MB) core encapsulating the photosensitizer and methylene blue and introducing a layer of MnO2 to act as a "gatekeeper" for pH/redox dependent payload release. The SiO_2-MB core was synthesized by the condensation of tetraethoxysilane and MB in the presence of $NH_3 \cdot H_2O$ to obtain a high MB loading efficiency. Subsequently, the SiO_2-MB core was coated with a MnO_2 shell through of $KMnO_4$ in situ reduction by hydroxylated polyethylene glycol (PEG) forming SiO_2-MB@MnO_2. This was conducted to prevent the premature release of the MB photosensitizer. In vivo studies using mice with established uterine cervical cancer (U14 cell line) showed a reduction of the MnO_2 shell to Mn^{2+} upon endocytosis by tumor cells in response to tumor microenvironment (acidic pH and elevated H_2O_2). Subsequently, PDT efficacy was enhanced through the production of $^1O_2$ generation. This was accompanied by tumor cell death and inhibition of tumor growth compared to the control which had a 6-to-9-fold increase in tumor size compared to the initial volumes. Additionally, a reduction in the expression of the hypoxia inducible factor (HIF)-1α in the SiO_2-MB@MnO_2 treatment group was observed compared to the control group, which experienced high HIF-1α expression levels, indicating that the developed theranostic nanoplatforms can reduce hypoxia within tumor tissues, thus eliminating PDT efficacy limitations. Importantly, the reduced Mn^{2+} ions facilitated in the tracking, imaging and detection of the tumor in vivo by MRI, demonstrating the possible novel routes that can be employed to develop TME-sensitive theranostic nanoplatforms.

The chemotherapeutic pro-drug capecitabine (cap), used for malignant tumors, is converted to its active form 5-fluorouracil in vivo following oral administration [79]. The severe side effects, including inadequate targeting capabilities and short-elimination half-life of the drug, significantly impact its efficacy [80]. Therefore, the use of nano drug delivery systems helps improve their effectiveness by extending the blood circulation of the drug, minimizing toxicity and ensuring targeted delivery via EPR effect. Ma et al. designed a dual pH and ROS-responsive triblock prodrug polymer micelle system designated as $PMMTA_b$-Cap [81]. The micelle system was synthesized using the versatile polymer poly (2-azepane ethyl methacrylate) (PAEMA), which gets protonated and undergoes hydrophobic to hydrophilic transition to become electropositive at pH 6.8, thus enhancing endocytosis. In addition to PAEMA and other polymers, the hydrophilic poly(2methacryloyloxyethyl phosphorylcholine) (PMPC) was also used and conjugated to the micelle core via a pH-sensitive benzoyl imide bond, improving system biocompatibility and biostability during circulation. Later, cap was conjugated via a ROS-sensitive boron ester bond, which facilitated directed release of the pro-drug cap, and a two-photon fluorophore (TP) was added to produce a strong fluorescence emission for tumor bioimaging. The pH- and ROS-responsive mechanisms are illustrated in Figure 5.5A. The PMMTA$_b$-Cap micelle system was intravenously administered to tumor-bearing mice. The biodistribution and accumulation of the theranostic nanoplatform was observed by a distinguishable fluorescence signal in the tumor, kidney and liver across different time points (Figure 5.5B). It was found that PMMTA$_b$-Cap micelles showed highest fluorescence signals in tumor tissues after 48 h, thus confirming their accumulation. Subsequently, significant anti-tumor activity and tumor growth inhibition facilitated by the micelle system was demonstrated compared to the control (Figure 5.5C), therefore demonstrating an exceptional effect in cancer therapy.

Recognizing the ability of PDA to induce PTT via NIR irradiation, a dual PTT/PDT-PA/MR imaging-guided theranostic nanoplatform was synthesized for the detection and destruction of cancer cells. Accordingly, polydopamine was used to fabricate the nanoparticles via a self-polymerization method; subsequently, PEG was utilized to modify the nanosystem, and IR820 and iron ions (Fe^{3+}) were loaded into the system resulting in the formation of PDA/IR820/Fe^{3+} (PPIF)

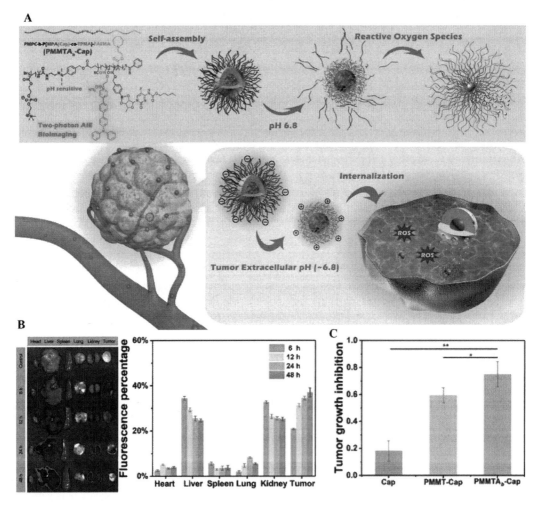

FIGURE 5.5 (A) pH- and ROS-triggered drug release from PMMTA$_b$-Cap micelles. (B) Accumulation of PMMTA$_b$-Cap micelles in various organs at different time points *ex vivo* fluorescence imaging. (C) PMMTA$_b$-Cap micelles exhibiting enhanced tumor growth inhibition. Reproduced with permission from [81]. Copyright (2019) American Chemical Society.

nanoparticles [82]. Flow cytometry was used to detect cellular uptake and intracellular localization of the nanosystem by HeLa cells. An increased photoacoustic and magnetic resonance signal was correlated to the cumulative nanosystem concentration within the cells. Subsequently, the PTT/PDT therapeutic impact of the nanosystem was analyzed using the CCK-8 method. Under NIR light irradiation, ROS was produced to facilitate PDT/PTT photo-responsive therapeutic properties. Additionally, the use of IR820 and Fe^{3+} ions pegylated polydopamine nanosystem yielded discrete PA and T1-weighted MR signals to facilitate cancer detection.

Molecular dynamic therapy (MDT) has shown considerable potential in the improvement of PDT efficacy and in reducing the requirement for high doses of photosensitizers through the exploitation of endogenous H_2O_2 [83]. Zhang et al. prepared biodegradable zinc peroxide nanoparticles loaded with the photosensitizer sinoporphyrin sodium (SPS) with an aim to induce ROS-mediated mitochondrial disruption and simultaneous apoptosis [84]. In vitro treatment evaluation of SPS@ZnO_2 NPs was conducted by treating 4T1 cells with the nanosystem. Upon SPS@ZnO_2 NPs cellular uptake, acidic degradation of the ZnO_2 NPs resulted in H_2O_2 release followed by SPS-mediated $^1O_2$ generation to induce PDT. Notably, the treatment of 4T1 cells with the nanosystem and the adoption of an MDT/PDT combinatorial therapeutic approach resulted in an increase in ROS production and higher cancer cell death rate (>81%) in comparison to MDT (54.1%) or PDT (25%) alone. Subsequently, in vivo studies were conducted on mice-bearing 4T1 tumors, to which TME-responsive SPS@ZnO_2 NP was intravenously administered. SPS@ZnO_2 NP enabled the production of ROS (facilitating tumor cell death via MDT) and intracellular depletion of GSH to enhance PDT effects. Therefore, combination therapy (MDT/PDT) once again displayed almost complete tumor inhibition and significantly promoted therapeutic efficacy compared monotherapy. Additionally, the nanosystem enabled the in vivo visualization of its distribution, tumor location and possibly effective visual guidance for PDT via by fluorescence imaging. This theranostic nanoplatform therefore provides an interesting venue for TME/redox-responsive therapeutic strategies with interactive enhancement of anticancer activity.

5.5 SUMMARY AND PROSPECTS

The redox state of cancer that encompasses both glutathione and ROS is a potential target for delivery of a variety of drugs and non-drugs. The nanoplatforms discussed in this chapter have integrated functions as both treatment and diagnostic imaging for effective cancer therapy. Interesting strategies utilizing polymers, lipids, silica and metals have been reported to prepare both organic- and inorganic-based redox-responsive theranostic nanoplatforms. Various nanoplatforms such as self-assembled micelles, core-shell nanoparticles, nanocomposites, LPHNs and MSNs have been developed as theranostic nanoplatforms against cancer. These nanosystems were mainly loaded with therapeutic drugs, photosensitizers, dyes and so on, which facilitated inducing chemo/chemodynamic/photodynamic/photothermal therapies. Apart from loading of therapeutic drugs and photosensitizers, these nanosystems have used encapsulation of contrast agents, thus providing diagnostic properties. Prodrug therapy employed combinations of therapeutic drugs, photothermal agents or dyes to provide multimodal therapies. Furthermore, inactive platinum (IV) drugs have been selectively reduced to bioactive platinum (II) drugs in the GSH environment, which displayed enhanced apoptotic effects and subsequently decreased off-target toxicity. While disulfide and diselenide linkers were cleaved in the presence of high GSH concentrations to specifically deliver therapeutic payloads in TME, Fenton and Fenton-like agents reacted with intracellular GSH to cause its depletion in order to facilitate ROS-mediated chemodynamic therapy.

The intrinsic fluorescence property of doxorubicin has been widely used to study accumulation of drug or nanoplatform in tumor sites. Similarly, the utilization of polydopamine has facilitated nanoparticle construction and surface functionalization and also provided photothermal conversion effects to induce photothermal therapy. Some of the nanosystems have incorporated ligands such as hyaluronic acid, folic acid and uPA peptide to facilitate enhanced cancer targeting.

Although nanotheranostics is a promising approach for cancer therapy and diagnosis, there are a few issues that need to be addressed. In order to provide multimodal therapies, responses and imaging in a single system, the nanostructures have been constructed using multiple excipients, dyes, linkers, photosensitizers and so on that might prove toxic. Therefore, research must focus on providing multiple properties using less complex systems. Besides, a long-term in vivo toxicity evaluation is recommended. In line, clearance of such multicomponent systems from the body is another issue, which warrants a detailed pharmacokinetic investigation. Furthermore, the mechanisms relating to intracellular metabolism and synergistic effects of these systems remain unclear; future research must investigate these mechanisms in detail to provide better understanding. Although redox-responsive theranostic nanomedicines are still far from real diagnostic and therapeutic applications, overall, they have shown great promise to be clinically translated in the near future.

REFERENCES

[1] Cancer (2022). www.who.int/news-room/fact-sheets/detail/cancer.
[2] A.-I. Zaromytidou, Cancer research that matters, Nat Cancer, 2 (2021) 1268–1270.
[3] H. Maeda, M. Khatami, Analyses of repeated failures in cancer therapy for solid tumors: Poor tumor-selective drug delivery, low therapeutic efficacy and unsustainable costs, Clin Transl Med, 7 (2018) 1–20.
[4] U. Anand, A. Dey, A.K.S. Chandel, R. Sanyal, A. Mishra, D.K. Pandey, V. De Falco, A. Upadhyay, R. Kandimalla, A. Chaudhary, Cancer chemotherapy and beyond: Current status, drug candidates, associated risks and progress in targeted therapeutics, Genes Dis, (2022).
[5] D. De Ruysscher, G. Niedermann, N.G. Burnet, S. Siva, A.W. Lee, F. Hegi-Johnson, Radiotherapy toxicity, Nat Rev Dis Primers, 5 (2019) 1–20.
[6] R. Nussinov, C.-J. Tsai, H. Jang, Anticancer drug resistance: An update and perspective, Drug Resist Updat, (2021) 100796.
[7] I. Brigger, C. Dubernet, P. Couvreur, Nanoparticles in cancer therapy and diagnosis, Adv Drug Del Rev, 64 (2012) 24–36.
[8] H. Xiong, S. Liu, T. Wei, Q. Cheng, D.J. Siegwart, Theranostic dendrimer-based lipid nanoparticles containing PEGylated BODIPY dyes for tumor imaging and systemic mRNA delivery in vivo, J Control Release, 325 (2020) 198–205.
[9] L. Feng, L. Cheng, Z. Dong, D. Tao, T.E. Barnhart, W. Cai, M. Chen, Z. Liu, Theranostic liposomes with hypoxia-activated prodrug to effectively destruct hypoxic tumors post-photodynamic therapy, ACS Nano, 11 (2017) 927–937.
[10] L. Kennedy, J.K. Sandhu, M.-E. Harper, M. Cuperlovic-Culf, Role of glutathione in cancer: From mechanisms to therapies, Biomolecules, 10 (2020) 1429.
[11] J.D. Hayes, A.T. Dinkova-Kostova, K.D. Tew, Oxidative stress in cancer, Cancer Cell, 38 (2020) 167–197.
[12] G.K. Balendiran, R. Dabur, D. Fraser, The role of glutathione in cancer, Cell Biochemistry and Function: Cellular biochemistry and its modulation by active agents or disease, Cell Biochem Funct, 22 (2004) 343–352.
[13] H.H. Chen, M.T. Kuo, Role of glutathione in the regulation of Cisplatin resistance in cancer chemotherapy, Met-Based Drugs, 2010 (2010).
[14] J.N. Moloney, T.G. Cotter, ROS signalling in the biology of cancer, Semin Cell Dev Biol, Elsevier (2018) 50–64.
[15] R. Cheng, F. Feng, F. Meng, C. Deng, J. Feijen, Z. Zhong, Glutathione-responsive nano-vehicles as a promising platform for targeted intracellular drug and gene delivery, J Control Release, 152 (2011) 2–12.
[16] Z. Deng, J. Hu, S. Liu, Disulfide-based self-immolative linkers and functional bioconjugates for biological applications, Macromol Rapid Commun, 41 (2020) 1900531.
[17] W. Sun, Y. Yang, Recent advances in redox-responsive nanoparticles for combined cancer therapy, Nanoscale Adv (2022); 4:3504-3516.
[18] C. Zhang, C. Xu, X. Gao, Q. Yao, Platinum-based drugs for cancer therapy and anti-tumor strategies, Theranostics, 12 (2022) 2115.
[19] C. Cao, X. Wang, N. Yang, X. Song, X. Dong, Recent advances of cancer chemodynamic therapy based on Fenton/Fenton-like chemistry, Chem Sci, 13 (2022) 863–889.

[20] C. Sun, J. Lu, J. Wang, P. Hao, C. Li, L. Qi, L. Yang, B. He, Z. Zhong, N. Hao, Redox-sensitive polymeric micelles with aggregation-induced emission for bioimaging and delivery of anticancer drugs, J. Nanobiotechnol, 19 (2021) 1–15.

[21] B. Cheng, H.-H. Ahn, H. Nam, Z. Jiang, F.J. Gao, I. Minn, M.G. Pomper, A Unique core—shell structured, glycol chitosan-based nanoparticle achieves cancer-selective gene delivery with reduced off-target effects, Pharmaceutics, 14 (2022) 373.

[22] S. Zhao, M. Xu, C. Cao, Q. Yu, Y. Zhou, J. Liu, A redox-responsive strategy using mesoporous silica nanoparticles for co-delivery of siRNA and doxorubicin, J Mater Chem B, 5 (2017) 6908–6919.

[23] D. Hu, Z. Sheng, G. Gao, F. Siu, C. Liu, Q. Wan, P. Gong, H. Zheng, Y. Ma, L. Cai, Activatable albumin-photosensitizer nanoassemblies for triple-modal imaging and thermal-modulated photodynamic therapy of cancer, Biomaterials, 93 (2016) 10–19.

[24] C. Feng, D. Zhu, L. Chen, Y. Lu, J. Liu, N.Y. Kim, S. Liang, X. Zhang, Y. Lin, Y. Ma, Targeted delivery of chlorin e6 via redox sensitive diselenide-containing micelles for improved photodynamic therapy in cluster of differentiation 44-overexpressing breast cancer, Front Pharmacol, 10 (2019) 369.

[25] J. Zhao, Y. Zhang, J. Zhang, H. Wu, J. Li, Y. Zhao, L. Zhang, D. Zou, Z. Li, S. Wang, Synthetic and biodegradable molybdenum (IV) Diselenide triggers the cascade photo-and immunotherapy of tumor, Adv Healthc Mater, 11 (2022) 2200524.

[26] J. Li, F. Jiang, B. Yang, X.-R. Song, Y. Liu, H.-H. Yang, D.-R. Cao, W.-R. Shi, G.-N. Chen, Topological insulator bismuth selenide as a theranostic platform for simultaneous cancer imaging and therapy, Sci Rep, 3 (2013) 1998.

[27] C. Fu, R. Wei, P. Xu, S. Luo, C. Zhang, R.K. Kankala, S. Wang, X. Jiang, X. Wei, L. Zhang, Supercritical fluid-assisted fabrication of diselenide-bridged polymeric composites for improved indocyanine green-guided photodynamic therapy, Chem Eng J, 407 (2021) 127108.

[28] Y. Yang, F. Chen, N. Xu, Q. Yao, R. Wang, X. Xie, F. Zhang, Y. He, D. Shao, W.-f. Dong, Red-light-triggered self-destructive mesoporous silica nanoparticles for cascade-amplifying chemo-photodynamic therapy favoring antitumor immune responses, Biomaterials, 281 (2022) 121368.

[29] M. Ao, F. Yu, Y. Li, M. Zhong, Y. Tang, H. Yang, X. Wu, Y. Zhuang, H. Wang, X. Sun, Carrier-free nanoparticles of camptothecin prodrug for chemo-photothermal therapy: The making, in vitro and in vivo testing, J Nanobiotechnology, 19 (2021) 1–15.

[30] F. Yu, F. Zhang, L. Tang, J. Ma, D. Ling, X. Chen, X. Sun, Redox-responsive dual chemophotothermal therapeutic nanomedicine for imaging-guided combinational therapy, J Mater Chem B, 6 (2018) 5362–5367.

[31] Q. Zhu, Z. Fan, W. Zuo, Y. Chen, Z. Hou, X. Zhu, Self-distinguishing and stimulus-responsive carrier-free theranostic nanoagents for imaging-guided chemo-photothermal therapy in small-cell lung cancer, ACS Appl. Mater. Interfaces, 12 (2020) 51314–51328.

[32] X. He, J. Zhang, C. Li, Y. Zhang, Y. Lu, Y. Zhang, L. Liu, C. Ruan, Q. Chen, X. Chen, Enhanced bioreduction-responsive diselenide-based dimeric prodrug nanoparticles for triple negative breast cancer therapy, Theranostics, 8 (2018) 4884.

[33] X.-F. Du, Y. Li, J. Long, W. Zhang, D. Wang, C.-R. Li, M.-X. Zhao, Y. Lai, Fabrication of cisplatin-loaded polydopamine nanoparticles via supramolecular self-assembly for photoacoustic imaging guided chemo-photothermal cancer therapy, Appl. Mater. Today, 23 (2021) 101019.

[34] Y. Huang, Y. Jiang, Z. Xiao, Y. Shen, L. Huang, X. Xu, G. Wei, C. Xu, C. Zhao, Three birds with one stone: A ferric pyrophosphate based nanoagent for synergetic NIR-triggered photo/chemodynamic therapy with glutathione depletion, Chem Eng J, 380 (2020) 122369.

[35] J. Yao, F. Zheng, F. Yang, C. Yao, J. Xing, Z. Li, S. Sun, J. Chen, X. Xu, Y. Cao, An intelligent tumor microenvironment responsive nanotheranostic agent for T 1/T 2 dual-modal magnetic resonance imaging-guided and self-augmented photothermal therapy, Biomater Sci, 9 (2021) 7591–7602.

[36] Y. Liu, S. Chi, Y. Cao, Z. Liu, Glutathione-responsive biodegradable core—shell nanoparticles that self-generate H2O2 and deliver doxorubicin for chemo—chemodynamic therapy, ACS Appl. Nano Mater, 5 (2022) 2592–2602.

[37] D. Yang, W. Chen, J. Hu, Design of controlled drug delivery system based on disulfide cleavage trigger, J Phys Chem B, 118 (2014) 12311–12317.

[38] R. Mohammadinejad, A. Dehshahri, V.S. Madamsetty, M. Zahmatkeshan, S. Tavakol, P. Makvandi, D. Khorsandi, A. Pardakhty, M. Ashrafizadeh, E.G. Afshar, In vivo gene delivery mediated by non-viral vectors for cancer therapy, J Control Release, 325 (2020) 249–275.

[39] A.F. Moreira, D.R. Dias, I.J. Correia, Stimuli-responsive mesoporous silica nanoparticles for cancer therapy: A review, Microporous Mesoporous Mater, 236 (2016) 141–157.

[40] M. Gisbert-Garzarán, M. Vallet-Regí, Redox-responsive mesoporous silica nanoparticles for cancer treatment: Recent updates, Nanomaterials, 11 (2021) 2222.

[41] G.M.F. Calixto, J. Bernegossi, L.M. De Freitas, C.R. Fontana, M. Chorilli, Nanotechnology-based drug delivery systems for photodynamic therapy of cancer: A review, Molecules, 21 (2016) 342.

[42] M. Álvarez-Pérez, W. Ali, M.A. Marć, J. Handzlik, E. Domínguez-Álvarez, Selenides and diselenides: A review of their anticancer and chemopreventive activity, Molecules, 23 (2018) 628.

[43] C. Sun, S. Ji, F. Li, H. Xu, Diselenide-containing hyperbranched polymer with light-induced cytotoxicity, ACS Appl. Nano Mater, 9 (2017) 12924–12929.

[44] X. Li, J.F. Lovell, J. Yoon, X. Chen, Clinical development and potential of photothermal and photodynamic therapies for cancer, Nat. Rev. Clin. Oncol, 17 (2020) 657–674.

[45] J. Park, M. Kwon, E.-C. Shin, Immune checkpoint inhibitors for cancer treatment, Arch Pharm Res, 39 (2016) 1577–1587.

[46] X. Jiang, J. Wang, X. Deng, F. Xiong, J. Ge, B. Xiang, X. Wu, J. Ma, M. Zhou, X. Li, Role of the tumor microenvironment in PD-L1/PD-1-mediated tumor immune escape, Mol Cancer, 18 (2019) 1–17.

[47] M.-A. Shahbazi, L. Faghfouri, M.P. Ferreira, P. Figueiredo, H. Maleki, F. Sefat, J. Hirvonen, H.A. Santos, The versatile biomedical applications of bismuth-based nanoparticles and composites: Therapeutic, diagnostic, biosensing, and regenerative properties, Chem Soc Rev, 49 (2020) 1253–1321.

[48] D. Kong, Y. Cui, Opportunities in chemistry and materials science for topological insulators and their nanostructures, Nat Chem, 3 (2011) 845–849.

[49] J. Wang, D. Li, W. Tao, Y. Lu, X. Yang, J. Wang, Synthesis of an oxidation-sensitive polyphosphoester bearing thioether group for triggered drug release, Biomacromolecules, 20 (2019) 1740–1747.

[50] L. Chen, W. Xue, J. Cao, S. Zhang, Y. Zeng, L. Ma, X. Qian, Q. Wen, Y. Hong, Z. Shi, TiSe2-mediated sonodynamic and checkpoint blockade combined immunotherapy in hypoxic pancreatic cancer, J Nanobiotechnology, 20 (2022) 1–14.

[51] A.H. Morrison, K.T. Byrne, R.H. Vonderheide, Immunotherapy and prevention of pancreatic cancer, Trends in Cancer, 4 (2018) 418–428.

[52] R. Winograd, D.M. Simeone, D. Bar-Sagi, A novel target for combination immunotherapy in pancreatic cancer: IL-1β mediates immunosuppression in the tumour microenvironment, Br J Cancer, 124 (2021) 1754–1756.

[53] J. Leinwand, G. Miller, Regulation and modulation of antitumor immunity in pancreatic cancer, Nat Immunol, 21 (2020) 1152–1159.

[54] C. Falcomatà, S. Bärthel, S.A. Widholz, C. Schneeweis, J.J. Montero, A. Toska, J. Mir, T. Kaltenbacher, J. Heetmeyer, J.J. Swietlik, Selective multi-kinase inhibition sensitizes mesenchymal pancreatic cancer to immune checkpoint blockade by remodeling the tumor microenvironment, Nat Cancer, 3 (2022) 318–336.

[55] V. Abet, F. Filace, J. Recio, J. Alvarez-Builla, C. Burgos, Prodrug approach: An overview of recent cases, Eur J Med Chem, 127 (2017) 810–827.

[56] S. van Zutphen, J. Reedijk, Targeting platinum anti-tumour drugs: Overview of strategies employed to reduce systemic toxicity, Coord Chem Rev, 249 (2005) 2845–2853.

[57] P. Xie, Y. Wang, D. Wei, L. Zhang, B. Zhang, H. Xiao, H. Song, X. Mao, Nanoparticle-based drug delivery systems with platinum drugs for overcoming cancer drug resistance, J Mater Chem B, 9 (2021) 5173–5194.

[58] S. Fan, W. Lin, Y. Huang, J. Xia, J.-F. Xu, J. Zhang, J. Pi, Advances and potentials of polydopamine nanosystem in photothermal-based antibacterial infection therapies, Front Pharmacol, 13 (2022) 193.

[59] B. Perillo, M. Di Donato, A. Pezone, E. Di Zazzo, P. Giovannelli, G. Galasso, G. Castoria, A. Migliaccio, ROS in cancer therapy: The bright side of the moon, Exp Mol Med, 52 (2020) 192–203.

[60] W. Zhang, J. Liu, X. Li, Y. Zheng, L. Chen, D. Wang, M.F. Foda, Z. Ma, Y. Zhao, H. Han, Precise chemodynamic therapy of cancer by trifunctional bacterium-based nanozymes, ACS nano, 15 (2021) 19321–19333.

[61] Z. Tang, Y. Liu, M. He, W. Bu, Chemodynamic therapy: Tumour microenvironment-mediated Fenton and Fenton-like reactions, Angew Chem, 131 (2019) 958–968.

[62] X. Wang, X. Zhong, Z. Liu, L. Cheng, Recent progress of chemodynamic therapy-induced combination cancer therapy, Nano Today, 35 (2020) 100946.

[63] Y. Xiong, C. Xiao, Z. Li, X. Yang, Engineering nanomedicine for glutathione depletion-augmented cancer therapy, Chem Soc Rev, 50 (2021) 6013–6041.

[64] O. Dietrich, J. Levin, S.-A. Ahmadi, A. Plate, M.F. Reiser, K. Bötzel, A. Giese, B. Ertl-Wagner, MR imaging differentiation of Fe 2+ and Fe 3+ based on relaxation and magnetic susceptibility properties, Neuroradiology, 59 (2017) 403–409.

[65] M. Devreux, C. Henoumont, F. Dioury, S. Boutry, O. Vacher, L.V. Elst, M. Port, R.N. Muller, O. Sandre, S. Laurent, Mn2+ Complexes with Pyclen-based derivatives as contrast agents for magnetic resonance imaging: Synthesis and relaxometry characterization, Inorg Chem, 60 (2021) 3604–3619.

[66] S. Hussain, E. Aneggi, D. Goi, Catalytic activity of metals in heterogeneous Fenton-like oxidation of wastewater contaminants: A review, Environ Chem Lett, 19 (2021) 2405–2424.

[67] Z. Dong, L. Feng, Y. Hao, Q. Li, M. Chen, Z. Yang, H. Zhao, Z. Liu, Synthesis of CaCO3-based nanomedicine for enhanced sonodynamic therapy via amplification of tumor oxidative stress, Chem, 6 (2020) 1391–1407.

[68] M.d.P.S. Idelchik, U. Begley, T.J. Begley, J.A. Melendez, Mitochondrial ROS control of cancer, Semin Cancer Biol, Elsevier (2017) 57–66.

[69] S.S. Sabharwal, P.T. Schumacker, Mitochondrial ROS in cancer: Initiators, amplifiers or an Achilles' heel? Nat Rev Cancer, 14 (2014) 709–721.

[70] C.R. Reczek, N.S. Chandel, The two faces of reactive oxygen species in cancer, Annu Rev Cancer Biol, 1 (2017) 79–98.

[71] F. Xing, Q. Hu, Y. Qin, J. Xu, B. Zhang, X. Yu, W. Wang, The relationship of redox with hallmarks of cancer: The importance of homeostasis and context, Front Oncol, 12 (2022).

[72] F. Yuan, M. Dellian, D. Fukumura, M. Leunig, D.A. Berk, V.P. Torchilin, R.K. Jain, Vascular permeability in a human tumor xenograft: Molecular size dependence and cutoff size, Cancer Res, 55 (1995) 3752–3756.

[73] L.H. Fu, Y. Wan, C. Qi, J. He, C. Li, C. Yang, H. Xu, J. Lin, P. Huang, Nanocatalytic theranostics with glutathione depletion and enhanced reactive oxygen species generation for efficient cancer therapy, Adv Mater, 33 (2021) 2006892.

[74] S. Gao, X. Lu, P. Zhu, H. Lin, L. Yu, H. Yao, C. Wei, Y. Chen, J. Shi, Self-evolved hydrogen peroxide boosts photothermal-promoted tumor-specific nanocatalytic therapy, J Mater Chem B, 7 (2019) 3599–3609.

[75] Z. Yu, P. Zhou, W. Pan, N. Li, B. Tang, A biomimetic nanoreactor for synergistic chemiexcited photodynamic therapy and starvation therapy against tumor metastasis, Nat. Commun, 9 (2018) 5044.

[76] A. Sahu, I. Kwon, G. Tae, Improving cancer therapy through the nanomaterials-assisted alleviation of hypoxia, Biomaterials, 228 (2020) 119578.

[77] S.Z.F. Phua, G. Yang, W.Q. Lim, A. Verma, H. Chen, T. Thanabalu, Y. Zhao, Catalase-integrated hyaluronic acid as nanocarriers for enhanced photodynamic therapy in solid tumor, ACS Nano, 13 (2019) 4742–4751.

[78] Z. Ma, X. Jia, J. Bai, Y. Ruan, C. Wang, J. Li, M. Zhang, X. Jiang, MnO2 gatekeeper: An intelligent and O2-evolving shell for preventing premature release of high cargo payload core, overcoming tumor hypoxia, and acidic H2O2-sensitive MRI, Adv Funct Mater, 27 (2017) 1604258.

[79] S. Chen, L. Blaney, P. Chen, S. Deng, M. Hopanna, Y. Bao, G. Yu, Ozonation of the 5-fluorouracil anticancer drug and its prodrug capecitabine: Reaction kinetics, oxidation mechanisms, and residual toxicity, Front Environ Sci Eng, 13 (2019) 1–14.

[80] G. Aprile, M. Mazzer, S. Moroso, F. Puglisi, Pharmacology and therapeutic efficacy of capecitabine: Focus on breast and colorectal cancer, Anti-Cancer Drugs, 20 (2009) 217–229.

[81] B. Ma, W. Zhuang, H. Xu, G. Li, Y. Wang, Hierarchical responsive nanoplatform with two-photon aggregation-induced emission imaging for efficient cancer theranostics, ACS Appl Nano Mater, 11 (2019) 47259–47269.

[82] J. Wang, Y. Guo, J. Hu, W. Li, Y. Kang, Y. Cao, H. Liu, Development of multifunctional polydopamine nanoparticles as a theranostic nanoplatform against cancer cells, Langmuir, 34 (2018) 9516–9524.

[83] Z.m. Tang, Y.y. Liu, D.l. Ni, J.j. Zhou, M. Zhang, P.r. Zhao, B. Lv, H. Wang, D.y. Jin, W.b. Bu, Biodegradable nanoprodrugs:"Delivering" ROS to cancer cells for molecular dynamic therapy, Adv Mater, 32 (2020) 1904011.

[84] D.-Y. Zhang, F. Huang, Y. Ma, G. Liang, Z. Peng, S. Guan, J. Zhai, Tumor microenvironment-responsive theranostic nanoplatform for guided molecular dynamic/photodynamic synergistic therapy, ACS Appl Nano Mater, 13 (2021) 17392–17403.

6 ROS-Responsive Delivery Nanoplatforms in Cancer Theranostics

Naina Rajak, Praveen Kumar, Namdev L. Dhas, Yashasvi Singh and Neha Garg

CONTENTS

6.1 Introduction 98
6.2 ROS-Responsive Materials 100
6.3 ROS-Responsive Nanoplatforms for Cancer Drug Delivery 100
    6.3.1 ROS-Responsive Nanoparticles and Their Applications in Cancer Theranostic 101
6.4 Types of ROS-Responsive Nanoplatforms to Prevent Cancer 104
    6.4.1 ROS-Responsive Cyclo-Dextrin Nanoplatforms 104
6.5 ROS and Glutathione (GSH)-Based Dual-Sensitive Nanocarriers for Combined Cancer Treatment 105
6.6 Synergistic Combined Cancer Theranostic Based on ROS-Responsive Polymers 106
    6.6.1 Combination Therapy with Chemotherapy and Photodynamic Therapy 107
    6.6.2 Combination Therapy with Immunotherapy 108
6.7 Conclusion and Future Outlook 109
Acknowledgements 109
References 109

6.1 INTRODUCTION

In 1954, Commoner et al. discovered free radicals in many freeze-dried biological materials [1]. The partial reduction of molecular oxygen produces reactive oxygen species (ROS) [2]. ROS contain free radicals such as superoxide anion ($O_2\bullet-$), hydrogen peroxide (H_2O_2), singlet oxygen $^1O_2$, hypochlorous acid (HOCL), and hydroxyl radicals ($\bullet OH$) [3]. In the human biological system, ROS can be derived from exogenous and endogenous metabolism. Exogenous ROS can be produced from exposure to ultraviolet radiation, air pollutants, tobacco, and metals [4, 5]. Endogenous ROS can be produced through two main sources: a mitochondrial respiratory chain (which forms ROS as a byproduct) and activated NADPH oxidases (NOXs), whose main function is ROS production. Endoplasmic reticulum and peroxisomes have also been reported as cellular sites of ROS formation, as shown in Figure 6.1 [6]. Typically, a balance of ROS levels is maintained in living beings, which prevents cellular damage. However, some exogenous exposure can lead to increased ROS levels, which is called oxidative stress [7]. At normal concentrations, ROS is essential for cellular metabolism such as cell growth, migration, apoptosis, and fighting against foreign pathogens [8]. However, the overproduction of ROS can lead to several pathological conditions, including cancer, chronic inflammation, diabetes, rheumatoid arthritis, cardiovascular disease, and neurodegenerative diseases in human beings [9]. The involvement of ROS in the pathogenesis of a disease state is not limited to macromolecular damage. There is an increasing body of experimental evidence that ROS signalling is influenced the action of chemotherapeutic agents and ionizing radiation [10].

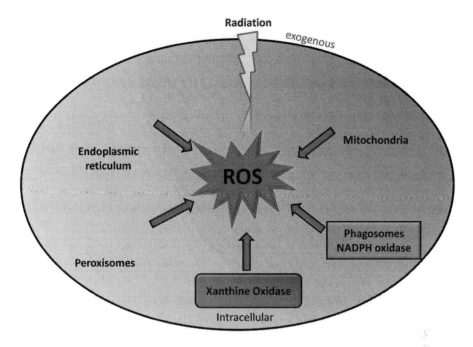

FIGURE 6.1 Various sources of reactive oxygen species in the cells.

Biochemical, physiological, or chemical changes in ROS-responsive materials are prompted by stimuli from the microenvironment such as enzymes, light, pH, ionic strength, and temperature [11–13]. A wide range of investigations has been carried out to characterize such distinctive properties of ROS-sensitive materials, which is vital for their application in biotechnology and biomedical sectors, which includes drug delivery systems for cancer targeting and novel platforms for cell therapy and inflammation-related diseases [14].

Elevated levels of ROS are detected in various cancers, which are produced through the partial depletion of molecular oxygen and are categorized as free radicals, ions, or molecules. The free radicals play crucial roles in the activation of pro-tumourigenic signalling, cell proliferation, metastasis, angiogenesis, resistance to apoptosis, and genetic instability [15]. ROS formation is implicated in arbitrating cancer therapy via modulation of various cell prosurvival or apoptotic signalling cascades [16]. These findings argued that ROS modulators can be used for preventing cancer or amplifying therapeutic effects [17]. Recently, ROS-sensitive nanocarriers have been intensively investigated in drug delivery systems and pro-drug systems due to their potential prospect for cancer-specific diagnoses and effective therapy. Several ROS-sensitive functional moieties have been investigated on a variety of carrier systems for effective ROS-mediated drug delivery [18]. The idea of creating nanoparticles based on ROS-responsive materials has garnered a lot of interest because it has the potential to boost drug availability at the site of action, improve therapeutic efficacy, and lessen unwanted side effects. [19]. Various strategies have been reported to treat cancer such as surgery, gene therapy, chemotherapy, radiotherapy, and immunotherapy, but cancer remains a life-threatening disease worldwide [20]. The development of various types of ROS-responsive nanocarriers/nanoparticles can be combined with chemotherapeutic drugs, such as chemo-photodynamic therapy, chemo-radiotherapy, and chemo-immunotherapy, which offers a promising strategy to target numerous cancers effectively.

In recent years, a special category of multifunctional nanosystems (including nanoparticles (NPs)) has been developed which could serve both diagnostic and therapeutic purposes and is referred to as 'theranostics nanoparticles'. By merging diagnostic and therapeutic characteristics into a single biocompatible and biodegradable nanoparticle, theranostic NPs are well-suited for more targeted and individualized therapy. Theranostic NPs should be safe to consume and also be able to (1) quickly accumulate in a target-specific manner, (2) allow reporting of various biochemical and morphological parameters of the target site, (3) support sufficient and effective delivery of drugs without additional side effects, and (4) easily clear the body or biodegrade into nontoxic byproducts. So far, no NPs have been developed which satisfy all the conditions of ideal theranostic NPs despite the development of numerous types in the last decade for cancer treatment. Nanoparticles targeting cancer markers such as folate and β-integrin receptors are some examples of theranostic nanoparticles used in cancer research [21].

6.2 ROS-RESPONSIVE MATERIALS

ROS are produced naturally as by-products of normal cellular metabolism and play a crucial part in modulating numerous cellular functions. Mitochondria are the main source of ROS production because of their prime roles in oxidative ATP production, wherein H_2O in the electron transport chain is reduced to O_2. ROS is a significant characteristic that differentiates diseased tissues from normal tissues, and high levels of ROS cause harmful effects on cell homeostasis [22]. ROS (H_2O_2, $O_2^{\bullet-}$, $^1O_2$, HOCL, •OH) are very reactive and enhance oxidative stress, leading to serious series of diseases such as neurodegenerative diseases and cancer. ROS are over-produced in tumour cells, and this feature has been utilized to synthesize ROS-sensitive materials for targeted therapy. ROS-responsive material and its medical applications are attracting significant attention to contribute greatly to the cellular redox state and promising enhanced therapeutic efficacy over conventional drugs. ROS-responsive materials are synthesized materials of a new generation (including nanoparticles, scaffolds, and hydrogels), which show potential results in preclinical testing. Generally, ROS-responsive materials are polymerics that allow targeted delivery of the drug, control drug release, and induce therapeutic responses while diminishing toxicity and unwanted effects on normal cells (Figure 6.2) [11]. The two major categories of ROS-responsive materials are ROS-induced solubility switch [poly(propylene sulfide), selenium-containing polymers, poly(thioetherketal)] and ROS-induced degradation [arylboronic acid/ester-containing polymers, aryl oxalate esters, mesoporous silicon microparticles, oligoproline peptide-crosslinked polymer and poly(thioketal)] [23]. These ROS-responsive materials/polymers lead to a hydrophobic-hydrophilic transformation that permits the solubilization of water and also induces the breakdown of the chemical bonds described in Table 6.1 [24].

6.3 ROS-RESPONSIVE NANOPLATFORMS FOR CANCER DRUG DELIVERY

Several approaches have been invented over the years to cure cancer, including chemotherapy and radiotherapy. The main goal of cancer chemotherapy is to enhance the therapeutic efficacy by preventing abnormal cell proliferation and tumour multiplication and diminishing the toxicity and side effects. But systemic administration of chemotherapy can cause several side effects, including damage to healthy cells, hair loss, and nausea [32]. To reduce these side effects, nanotechnology-based platforms are playing a potential role in the diagnosis and treatment of cancer. The development of nanotechnology products can play a crucial role to enhance the delivery of less water-soluble drugs, delivery of drugs to cell/tissue-specific target sites, delivery of a drug to cross epithelial and endothelial barriers, and delivery of macromolecular drugs at disease sites [33]. The growth and clinical trials of biodegradable nanocarriers are greatly obstructed by several factors, such as low drug targeting, weak drug delivery efficiency, and poor drug penetration at the disease site. Various polymeric nanoparticles/nanocarriers designed by spontaneous self-assembly of hydrophobic-hydrophilic transition copolymers have been broadly distinguished as proficient carriers for the delivery of a broad array of therapeutic agents [34]. Numerous stimuli-based nanocarrier/nanoparticle

ROS-Responsive Delivery Nanoplatforms in Cancer Theranostics

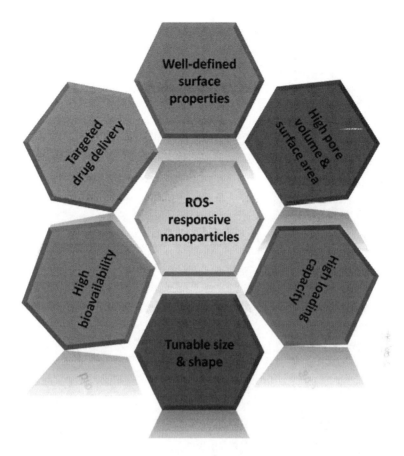

FIGURE 6.2 Advantages of ROS-responsive nanoparticles in cancer treatment.

drug-delivery systems have been reported, and among these, ROS-responsive nanoparticles have drawn significant attention for improved therapeutic effects [35].

6.3.1 ROS-Responsive Nanoparticles and Their Applications in Cancer Theranostic

Elevated levels of ROS, (in particular H_2O_2) are a key characteristic of the cancer tissue. Higher levels of H_2O_2 in tumour tissue retain a robust oxidative capacity to create an appropriate stimulus for responsive nanoparticles [36]. Numerous representative nanoplatforms based on thioketal-, aryl boronic ester-, and amino acrylate moieties have been reported [37].

6.3.1.1 Thioketal-Based ROS-Responsive Nanoparticles

Thioketal is a sulfur-containing analogue of ketal, which is synthesized by the condensation of thiols with ketones. Thioketal reacts with ROS, comprising H_2O_2, $^1O_2$, ˙OH, and $O_2^{˙-}$ [38], which demonstrates that the thioketal bond can be utilized as a functional group in ROS-sensitive polymers for delivery of drugs in tumour site [39]. Like disulfide bonds, thioketal is integrated into polymer chains to develop polymer-based prodrugs or can be used to conjugate hydrophobic-hydrophilic prodrugs to synthesize amphipathic nanosystems/nanocarriers [40].

Liu et al. synthesized a new fluorescent photosensitive prodrug TPP-L-GEM by conjugating gemcitabine with a meso-tetraphenylporphyrin (TPP) with the help of a ROS-responsive thioketal linker. Upon the irradiation of red light, gemcitabine release occurs from TPP-L-GEM through thioketal linker breakdown in the presence of $^1O_2$ produced by TPP. Thus, with

TABLE 6.1
ROS-Responsive Polymer Structure, Mechanism of Action and Application in Cancer Theranostics

Polymers	Structure Oxidation	Application	References
Thioether Selenium Tellurium		Use in combined chemotherapy and photodynamic therapy	[25]
Structural cleavage Thioketal		Use in combined photothermal/photodynamic and chemotherapy	[26, 27]
Aryl boronic acid/esters containing polymers		Use in radiotherapy, photodynamic therapy, chemotherapy	[11, 28]
Aryl oxalate ester		Use in radiotherapy or chemotherapy	[24, 29]
Proline oligomer		Use in photodynamic and chemotherapy	[30]
Ferrocene		Use in chemotherapy, immunotherapy, and photodynamic therapy	[18, 31]

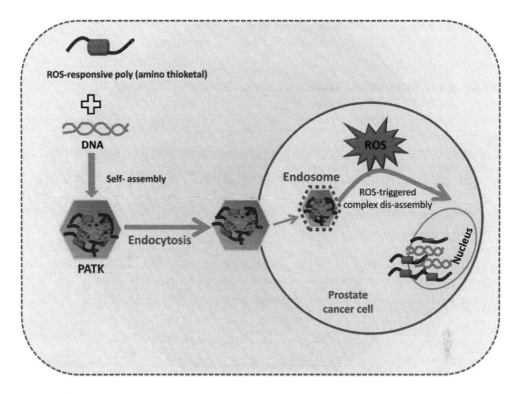

FIGURE 6.3 Mechanism of cancer cell killing by thioketal-based ROS-responsive nanoparticles.

combined photo-chemodynamic therapy, the prodrug TPP-L-GEM inhibits tumour growth in vivo and in vitro [41].

Xia et al. synthesized a ROS-sensitive cationic polymer poly (amino thioketal) (PATK), which is soluble in water, for safer targeted genetic material delivery for gene therapy in prostate cancer. PATK builds stable complexes with DNA to allow plasmid DNA entry into cancer cells. When complexes enter the cancer cells, the thioketal bonds are broken in PATK because of elevated levels of ROS (H_2O_2) in the tumour microenvironment, leading to nanoparticle breakdown, as well as releasing of plasmid DNA in prostate cancer cells shown (Figure 6.3). This research shows that the higher concentrations of ROS in the tumour microenvironment are uniquely capable of stimulating ROS-sensitive materials and could be used for targeted delivery of DNA in tumour sites to reduce toxicity and avoid random accumulation in healthy cells [42].

6.3.1.2 Selenium-Containing ROS-Responsive Nanoparticles/Polymers

Selenium, an essential micronutrient, can regulate the redox balance and increase the production of ROS, which enhances mitochondrial dysfunction and leads to induced cellular apoptosis [43]. Using this feature, the integration of selenium in the polymer backbone can develop a hydrophobic/hydrophilic transition, which can break drug-loaded nanocarriers in the presence of elevated concentrations of ROS. Selenium-containing polymers can also be oxidized under ROS regulation via photosensitizers and exogenous light.

Chen et al. designed a ROS-sensitive amphipathic selenium-containing copolymer (Se-polymer). The Se-polymer has been utilized to co-co-enclose ICG as well as the chemodrug DOX to develop I/D-Se-NPs. Drug-loaded Se-polymer nanoparticles are speedily separated in a few minutes by ROS-dependent oxidation of selenium oxidation at near-infrared laser wavelengths, and invariant separation of I/D-Se-nanoparticles through lower irradiation promotes sustained drug release. As a result, within 5 minutes post-irradiation, DOX is preferentially distributed into the nucleus, obtaining synergistic tumour ablation without tumour regrowth [44].

6.3.1.3 Aryl Boronic Ester-Based ROS-Sensitive Nanoparticles

Aryl boronic ester is a ROS-sensitive moiety, which can be used to synthesize stimuli-responsive nanomaterials. ROS (H_2O_2) reacts with the boron centre leading to the release of phenol and boric acid upon hydrolysis. Under the exposure of H_2O_2, the nanoplatform-installed aryl boronic ester group breaks and releases cargo [45].

Cui et al. synthesized ROS and light dual-responsive nanocarriers, which consist of a cross-linkage between phenylboronic acid and polyethylene glycol-polycaprolactone, indocyanine green (ICG), and pyridine endoperoxide (PE) (ICG/PE-TSPBA). ICG/PE-TSPBA can act as a photosensitizer and photothermal species to produce three types of ROS and hyperthermia. Elevated levels of ROS in the tumour microenvironment degrade the phenylboronic acid to increase the release of ICG and PE. Hyperthermia enhances the release of O_2 from PE, which can reduce tumour hypoxia and promote the synthesis of ROS and photodynamic therapy (PDT), and thereafter stimulate fast depolymerization of the nanocarrier, encourage penetration, and ultimately stimulate dual programmed cell death by multistage reactions. These nanocarriers with multistep collegial effects have exhibited wonderful effects to induce tumour double apoptosis and may also maximize the therapeutic effects of PTT/PDT [46].

6.3.1.4 Aminoacrylate-Based ROS-Responsive Nanoparticles

Aminoacrylate-based ROS-sensitive moiety was demonstrated to react with mostly $^1O_2$. The amino acrylate group can be broken to release the parent prodrug that undergoes fast oxidative degradation by singlet oxygen, called "photo-unclick chemistry". Moses Bio and co-workers synthesized a far-red-light-activated prodrug by the conjugation of combretastatin A-4(CA4), dithiaporphyrin (CMP), a photosensitizer, and L (amino acrylate linker) (CMP-L-CA4). Under irradiation of the laser, the amino acrylate linker of the CMP-L-CA4 (prodrug) degrades, leading to a fast release of CA4 in $CDCL_3$. In an in vitro study, it was demonstrated that the CMP-L-CA4 has a superior anticancer effect, and it could be a new low-energy light-sensitive prodrug [47].

6.4 TYPES OF ROS-RESPONSIVE NANOPLATFORMS TO PREVENT CANCER

6.4.1 ROS-Responsive Cyclo-Dextrin Nanoplatforms

Cyclodextrins (CDs) are nanopolymeric, biocompatible oligosaccharides obtained from biodegradable starch that utilize the glucanotransferase enzyme [48]. Cyclodextrin contains external hydrophilic and internal hydrophobic properties that help in the delivery of chemotherapeutic drugs [49]. Cyclodextrin-based nanoparticles can carry drugs in solid or liquid form and enhance the solubility of less water-soluble molecules, diminishing the toxicity and adverse effects of the drugs [50]. Nowadays, various stimulus-responsive nanomedicines are developing through using cyclodextrin, including pH, temperature, ROS, and redox (Figure 6.4) [51, 52].

Applications of Cyclodextrin-Based Nanoparticles [53]

1. Increased drug loading
2. Enhanced stability
3. Site- or target-specific drug delivery
4. Improved solubility profile
5. Improved permeability, absorption, and bioavailability
6. Improved bio adhesion
7. Controlled and sustained release

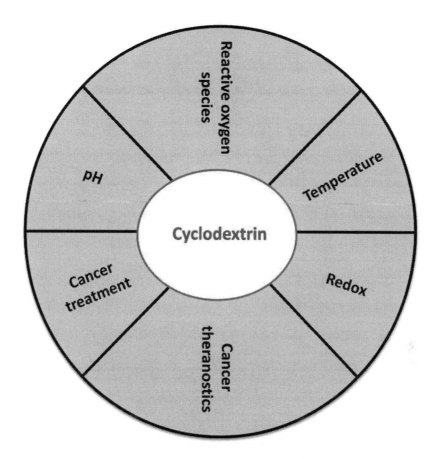

FIGURE 6.4 Cyclodextrin-based stimuli responsive nanoplatforms. Cyclodextrin-based nanocarriers can respond to various stimuli, and such nanoplatforms can be utilized for cancer treatment and theranostics.

ROS-responsive cyclodextrin-based micelles have been reported to deliver anti-tumour drugs on their target sites. One of the ROS-responsive cyclodextrins, MPEG-CD, was created through the polymerization of mono (6-amino-6-deoxy)-β- cyclodextrin and hydrophilic polymPEG-NHS. The MPEG-CD reacts with the compound PHB-CDI that has a borate bond to produce ROS-responsive MPEG-CD-PHB (PCP) to wrap doxorubicin and purpurin-18. The tumour microenvironment contains a higher concentration of H_2O_2. When the nanoparticles enter tumour cells by the enhanced permeability and retention (EPR) effect, the loaded doxorubicin and purpurin-18 are released under a higher concentration of ROS in the tumour microenvironment (TME). This produces cytotoxic ROS under the near-infrared laser to promote cell apoptosis and thus achieve successful anti-tumour therapy. These ROS-responsive cyclodextrin-based micelles provide excellent stability, higher biocompatibility, fast phagocytosis, and biological safety effects [54].

6.5 ROS AND GLUTATHIONE (GSH)-BASED DUAL-SENSITIVE NANOCARRIERS FOR COMBINED CANCER TREATMENT

ROS and GSH dual-responsive nanoparticles have attracted increasing interest to enhance the therapeutic effects and reduce the adverse effects of chemotherapeutic drugs in cancer theranostics [55]. The difference between healthy and diseased cells is due to the heterogeneous co-existence of

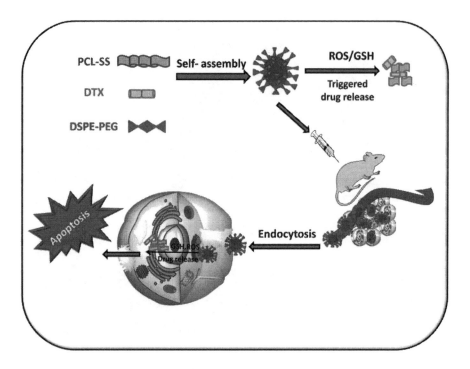

FIGURE 6.5 Mechanism of ROS-GSH–based nanoparticles in targeting cancer cells.

higher concentrations of ROS and overproduced GSH, which is found in cancer cells. Thus, it could be used as a stimulus to trigger the release of drugs from nanocarriers.

One innovative ROS and GSH dual responsive polymer is amphipathic diblock copolymer prodrug (BCP) (GR-BCP). GR-BCP includes side chains of poly (ethylene glycol) (PEG) and camptothecin (CPT)-conjugated poly(methacrylate) through single thioether bonds. Wei Yin et al. compared the three different types of redox-responsive BCP-based nanocarriers for antitumor drug releases, such as GSH single-responsive BCP nanoparticles (GBCPs), ROS single-responsive BCP nanoparticles (R-BCPs), and dual-responsive BCP nanoparticles (GR-BCPs). They found that the release of free drugs at the target site of cancer cells for the GR-BCP–treated group is considerably quicker than single-responsive nanocarriers. Due to the heterogeneity in the tumour micro-niche, elevated levels of ROS and intracellular GSH cleave the thioether linker of GR-BCPs. In this study, researchers found that GR-BCPs release the drug more effectively at a higher level of ROS and overproduced GSH intracellularly in comparison to GBCPs or R-BCPs. GR-BCPs are an effective tumour growth suppressor with fewer side effects [56, 57].

Other ROS and GSH dual-responsive nanoparticles, such as DTX-loaded PCL-SS nanocarriers (PCL-SS@DTX nanoparticles), can be used to enhance prostate cancer therapy in mice. The PCL-SS@DTX nanoparticles can rapidly and selectively release DTX in tumour sites, increase cell apoptosis, and prevent cell migration and invasion, as shown in Figure 6.5 [58].

6.6 SYNERGISTIC COMBINED CANCER THERANOSTIC BASED ON ROS-RESPONSIVE POLYMERS

ROS regulation can be achieved through numerous methods, including photodynamic therapy, radiotherapy, sonodynamic therapy, and chemo-dynamic therapy, as shown in Figure 6.6 [59]. The combinations of these therapies with ROS-sensitive nanoparticles can directly kill cancerous cells and trigger the breakdown of these ROS-sensitive nanoparticles, leading to the fast release of loaded

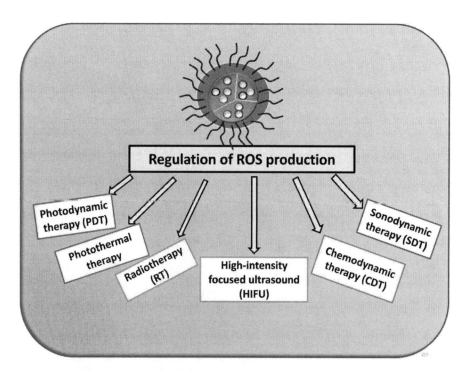

FIGURE 6.6 Regulation of ROS production using various approaches.

drugs in tumour regions. Nowadays, many combined synergistic treatment perspectives based on ROS-sensitive nanoparticles have been synthesized to overcome cancer-related drug resistance [60].

6.6.1 Combination Therapy with Chemotherapy and Photodynamic Therapy

Chemotherapy has become one of the most routinely used treatment options in some stages of certain cancer types. However, in numerous cases, it has been shown that cancer cells develop resistance against chemotherapeutic drugs, while in photodynamic therapy, lasers, photosensitizers, and molecular oxygen generate ROS, which stimulates apoptosis [61]. Combination therapy, by utilizing photodynamic therapy and chemotherapy, can promote synergistic therapeutic effects against cancer and diminish events of resistance to chemotherapy. This combination therapy could be an efficient strategy to deal with tumour cell heterogeneity [62, 63].

To use this characteristic, Hohyeon Lee et al. synthesized doxorubicin enclosing an HSA (human serum albumin) nanocarrier/chlorine e6 encapsulating microbubble (DOX-NPs/Ce6-MBs) complex system to promote the co-delivery of both chemo-drugs and photosensitizing agents. Systematic loading of chemo-drugs and photosensitizers into nanoparticles and microbubbles also solves the problem of steric hindrance. Additionally, microbubbles serve as a drug carrier and functional structure for sonoporation (increasing the cell membrane permeability and extravasations through utilizing ultrasound waves and microbubbles) effects. The DOX-NPs/Ce6-MBs were synthesized by the combination of NPs on the surface of a microbubble. The synthesized DOX-NP/Ce6-MB nanocarrier complex can be confirmed in vivo and in vitro to transport drugs efficiently with greater penetrability into tumour cells via (1) nanoparticle release because of the cavitation of DOX-NPs/Ce6-MBs, (2) the increased extravasation of nanoparticles because of the sonoporation phenomenon, and (3) the induced penetration of extravasated nanocarriers into the deeper cancer tissues because of the mechanical energy of ultrasound (US). This DOX-NP/Ce6-MB nanocarrier with US

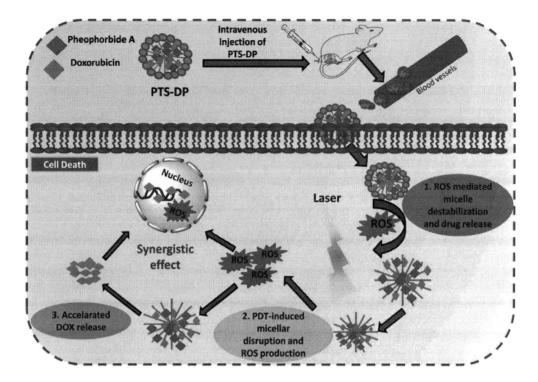

FIGURE 6.7 Schematic illustration of ROS-responsive drug release of PTS-DP by the combination of chemotherapy and photodynamic therapy.

irradiation demonstrated enhanced therapeutic efficacy and can be an alternative way to overcome the limitations of drug delivery in combined therapy for cancer theranostics [64].

To initiate combined chemo-photodynamic therapy, Park and colleagues synthesized photoactivatable nanomicelles such as PTS-DP. Researchers have used thioketal bonds to construct poly (ethylene glycol) (PEG) and stearine (C18) to achieve a ROS-responsive polymer PEG-TK-stearine (PTS) co-loaded with doxorubicin (DOX) and photosensitizer PhA (pheophorbide A). In the tumour microenvironment, PTS nanocarriers encapsulating DOX and PhA (PTS-DP) exhibited ROS response–mediated release of the DOX and PhA loaded inside. PTS-DP is triggered through intracellular ROS within the tumour microenvironment, leading to the release of the DOX and PhA. Furthermore, upon LASER irradiation on the tumour microenvironment, increased $^1O_2$ is produced by PhA released initially in tumour cells and kills tumour cells directly, achieving combined chemo-photodynamic therapy (shown in Figure 6.7) [65, 66].

6.6.2 Combination Therapy with Immunotherapy

Immunotherapy has achieved the status of care for the treatment of many cancers [67]. The synthesis of immune checkpoint inhibitors (ICIs) has attracted significant attention in immunotherapy to treat metastatic diseases. However, the efficacy of immunotherapeutic agents is curbed by their poor response rates, which limits their scope of applications in biomedical fields [68]. Partial cancer treatment includes some chemotherapeutic agents (doxorubicin, oxaliplatin, paclitaxel, etc.), PDT, radiotherapy that can stimulate death of tumour tissue in an immune response dependent manner, called immunogenic cell death (ICD) [69–71]. Immunotherapy is a generic therapeutic perspective to use antibodies that can target the immune checkpoints, programmed death-1 (PD-1), programmed death-1 ligand (PD-1L), and cytotoxic T lymphocyte antigen-4 (CTLA4) to interrupt regulatory checkpoints on cancer and immune cells [72]. Several research studies have reported that ROS enhance immunogenicity of cancer antigens and infiltration of T cells via cancer immunogenic death of tumour cells [73]. Black phosphorus (BP) is an ideal biodegradable, biocompatible semiconductor which possess characteristic optical features in the NIR and demonstrates exquisite potential when developed as nanomedicine for theranostic applications. [74]. BP quantum dots (BPQDs) and their nanocarrier-based drug delivery system develop a multi-modal synergistic anticancer nanoplatform combined with photothermal therapy, photodynamic therapy, immunotherapy, and radiotherapy to further improve the efficacy of cancer therapy and reduce its side effects [75]. BP quantum dots can produce an immense amount of ROS because of high carrier mobility and demonstrate a direct band gap which is layer dependent tunable. BPQDs are highly acceptable PDT agents with a strong potential to stimulate apoptosis in tumour cells [76].

Zhi Li and coworkers developed vesicles of ROS-sensitive black phosphorus quantum dot immunoadjuvant carriers for cancer-targeted photodynamic immunotherapy. ROS-responsive BPQD vesicles are designed by self-assembly of amphiphilic BPQDs implanted with polyethylene glycol and ROS-responsive poly (propylene sulfide) (PPS). BPNVs demonstrate enhanced photo-absorption in the NIR region and have a high loading efficiency of immunoadjuvant CpG oligodeoxynucleotides (CpG ODNs) in the cavity of the BPNVs. On NIR laser, an enormous amount of ROS is produced by BPNVs to trigger the hydrophobic to hydrophilic transition of PPs polymers, which stimulates the controlled release of free BPQDs and CpG at the cancer region to inhibit cancer cell proliferation and invasion to achieve efficacious photodynamic immunotherapy in vivo [77].

6.7 CONCLUSION AND FUTURE OUTLOOK

The recently redefined role of ROS in cancer treatment is gaining an ever-increasing spotlight. In light of this, developing nanoparticle-based approaches to deliver materials that would help kill cancer is one of the recent and novel methods. In the current chapter, we have reviewed various approaches in which ROS-responsive materials can be successfully used to target cancer cells. ROS-responsive NP-based technology can be combined with other cancer therapies such as chemo-, radio-, or immunotherapy to maximize the efficacy of the treatment. However, more preclinical followed by clinical investigations are required in this direction.

ACKNOWLEDGEMENTS

NR is supported by a Junior Research Fellowship through CSIR, India. PK acknowledges a Malviya Post-Doctoral Fellowship from the Institute of Eminence Scheme, Banaras Hindu University, for financial support. NG acknowledges the seed grant under the Institute of Eminence Scheme, Banaras Hindu University, for financial support.

REFERENCES

[1] B. Commoner, J. Townsend, G.E. Pake, Free radicals in biological materials, Nature. 174 (1954) 689–691. https://doi.org/10.1038/174689a0.

[2] D. Trachootham, J. Alexandre, P. Huang, Targeting cancer cells by ROS-mediated mechanisms: A radical therapeutic approach? Nat. Rev. Drug Discov. 8 (2009) 579–591. https://doi.org/10.1038/nrd2803.

[3] J. Liu, Y. Li, S. Chen, Y. Lin, H. Lai, B. Chen, T. Chen, Biomedical application of reactive oxygen species–responsive nanocarriers in cancer, inflammation, and neurodegenerative diseases, Front. Chem. 8 (2020). https://doi.org/10.3389/fchem.2020.00838.

[4] R.S. Balaban, S. Nemoto, T. Finkel, Mitochondria, oxidants, and aging, Cell. 120 (2005) 483–495. https://doi.org/10.1016/j.cell.2005.02.001.

[5] J.L. Bolton, M.A. Trush, T.M. Penning, G. Dryhurst, T.J. Monks, Role of quinones in toxicology, Chem. Res. Toxicol. 13 (2000) 135–160. https://doi.org/10.1021/tx9902082.

[6] L.A. del Río, E. López-Huertas, ROS generation in peroxisomes and its role in cell signaling, Plant Cell Physiol. (2016) pcw076. https://doi.org/10.1093/pcp/pcw076.

[7] G. Pizzino, N. Irrera, M. Cucinotta, G. Pallio, F. Mannino, V. Arcoraci, F. Squadrito, D. Altavilla, A. Bitto, Oxidative stress: Harms and benefits for human health, Oxid. Med. Cell. Longev. 2017 (2017) 1–13. https://doi.org/10.1155/2017/8416763.

[8] J. Checa, J.M. Aran, Reactive oxygen species: Drivers of physiological and pathological processes, J. Inflamm. Res. 13 (2020) 1057–1073. https://doi.org/10.2147/JIR.S275595.

[9] B. Uttara, A. Singh, P. Zamboni, R. Mahajan, Oxidative stress and neurodegenerative diseases: A review of upstream and downstream antioxidant therapeutic options, Curr. Neuropharmacol. 7 (2009) 65–74. https://doi.org/10.2174/157015909787602823.

[10] U.S. Srinivas, B.W.Q. Tan, B.A. Vellayappan, A.D. Jeyasekharan, ROS and the DNA damage response in cancer, Redox Biol. 25 (2019) 101084. https://doi.org/10.1016/j.redox.2018.101084.

[11] F. Gao, Z. Xiong, Reactive oxygen species responsive polymers for drug delivery systems, Front. Chem. 9 (2021). https://doi.org/10.3389/fchem.2021.649048.

[12] S. Bertoni, A. Machness, M. Tiboni, R. Bártolo, H.A. Santos, Reactive oxygen species responsive nanoplatforms as smart drug delivery systems for gastrointestinal tract targeting, Biopolymers. 111 (2020). https://doi.org/10.1002/bip.23336.

[13] D. Lee, N. Rejinold, S. Jeong, Y.-C. Kim, Stimuli-responsive polypeptides for biomedical applications, Polymers (Basel). 10 (2018) 830. https://doi.org/10.3390/polym10080830.

[14] Q. Xu, C. He, C. Xiao, X. Chen, Reactive oxygen species (ROS) responsive polymers for biomedical applications, Macromol. Biosci. 16 (2016) 635–646. https://doi.org/10.1002/mabi.201500440.

[15] J.N. Moloney, T.G. Cotter, ROS signalling in the biology of cancer, Semin. Cell Dev. Biol. 80 (2018) 50–64. https://doi.org/10.1016/j.semcdb.2017.05.023.

[16] V. Aggarwal, H. Tuli, A. Varol, F. Thakral, M. Yerer, K. Sak, M. Varol, A. Jain, M. Khan, G. Sethi, Role of reactive oxygen species in cancer progression: Molecular mechanisms and recent advancements, Biomolecules. 9 (2019) 735. https://doi.org/10.3390/biom9110735.

[17] Y. Ma, J. Chapman, M. Levine, K. Polireddy, J. Drisko, Q. Chen, High-dose parenteral ascorbate enhanced chemosensitivity of ovarian cancer and reduced toxicity of chemotherapy, Sci. Transl. Med. 6 (2014). https://doi.org/10.1126/scitranslmed.3007154.

[18] G. Saravanakumar, J. Kim, W.J. Kim, Reactive-oxygen-species-responsive drug delivery systems: Promises and challenges, Adv. Sci. 4 (2017) 1600124. https://doi.org/10.1002/advs.201600124.

[19] S. Bertoni, A. Machness, M. Tiboni, R. Bártolo, H.A. Santos, Reactive oxygen species responsive nanoplatforms as smart drug delivery systems for gastrointestinal tract targeting, Biopolymers. 111 (2020). https://doi.org/10.1002/bip.23336.

[20] R. Baskar, K.A. Lee, R. Yeo, K.-W. Yeoh, Cancer and radiation therapy: Current advances and future directions, Int. J. Med. Sci. 9 (2012) 193–199. https://doi.org/10.7150/ijms.3635.

[21] F. Chen, E.B. Ehlerding, W. Cai, Theranostic nanoparticles, J. Nucl. Med. 55 (2014) 1919–1922. https://doi.org/10.2967/jnumed.114.146019.

[22] A. V. Snezhkina, A. V. Kudryavtseva, O.L. Kardymon, M. V. Savvateeva, N. V. Melnikova, G.S. Krasnov, A.A. Dmitriev, ROS generation and antioxidant defense systems in normal and malignant cells, Oxid. Med. Cell. Longev. 2019 (2019) 1–17. https://doi.org/10.1155/2019/6175804.

[23] S.H. Lee, M.K. Gupta, J.B. Bang, H. Bae, H. Sung, Current progress in reactive oxygen species (ROS)-responsive materials for biomedical applications, Adv. Healthc. Mater. 2 (2013) 908–915. https://doi.org/10.1002/adhm.201200423.

[24] H. Ye, Y. Zhou, X. Liu, Y. Chen, S. Duan, R. Zhu, Y. Liu, L. Yin, Recent advances on reactive oxygen species-responsive delivery and diagnosis system, Biomacromolecules. 20 (2019) 2441–2463. https://doi.org/10.1021/ACS.BIOMAC.9B00628.

[25] J. Liang, B. Liu, ROS-responsive drug delivery systems, Bioeng. Transl. Med. 1 (2016) 239–251. https://doi.org/10.1002/btm2.10014.

[26] B. Chen, Y. Zhang, R. Ran, B. Wang, F. Qin, T. Zhang, G. Wan, H. Chen, Y. Wang, Reactive oxygen species-responsive nanoparticles based on a thioketal-containing poly(β-amino ester) for combining photothermal/photodynamic therapy and chemotherapy, Polym. Chem. 10 (2019) 4746–4757. https://doi.org/10.1039/C9PY00575G.

[27] A. Rinaldi, R. Caraffi, M.V. Grazioli, N. Oddone, L. Giardino, G. Tosi, M.A. Vandelli, L. Calzà, B. Ruozi, J.T. Duskey, Applications of the ROS-responsive thioketal linker for the production of smart nanomedicines, Polymers (Basel). 14 (2022) 687. https://doi.org/10.3390/polym14040687.

[28] Z. Wang, F.-J. Xu, B. Yu, Smart polymeric delivery system for antitumor and antimicrobial photodynamic therapy, Front. Bioeng. Biotechnol. 9 (2021). https://doi.org/10.3389/fbioe.2021.783354.

[29] C.-C. Song, F.-S. Du, Z.-C. Li, Oxidation-responsive polymers for biomedical applications, J. Mater. Chem. B. 2 (2014) 3413–3426. https://doi.org/10.1039/C3TB21725F.
[30] B. Hu, Z. Lian, Z. Zhou, L. Shi, Z. Yu, Reactive oxygen species-responsive adaptable self-assembly of peptides toward advanced biomaterials, ACS Appl. Bio Mater. 3 (2020) 5529–5551. https://doi.org/10.1021/acsabm.0c00758.
[31] D. Li, R. Zhang, G. Liu, Y. Kang, J. Wu, Redox-responsive self-assembled nanoparticles for cancer therapy, Adv. Healthc. Mater. 9 (2020) 2000605. https://doi.org/10.1002/adhm.202000605.
[32] A. Saini, M. Kumar, S. Bhatt, V. Saini, A. Malik, Cancer causes and treatments, Int. J. Pharm. Sci. Res. 11 (2020) 3121. https://doi.org/10.13040/IJPSR.0975-8232.11(7).3121-34.
[33] O.C. Farokhzad, R. Langer, Impact of nanotechnology on drug delivery, ACS Nano. 3 (2009) 16–20. https://doi.org/10.1021/nn900002m.
[34] K. Kuperkar, D. Patel, L.I. Atanase, P. Bahadur, Amphiphilic block copolymers: Their structures, and self-assembly to polymeric micelles and polymersomes as drug delivery vehicles, Polymers (Basel). 14 (2022) 4702. https://doi.org/10.3390/polym14214702.
[35] S.H. Pham, Y. Choi, J. Choi, Stimuli-responsive nanomaterials for application in antitumor therapy and drug delivery, Pharmaceutics. 12 (2020) 630. https://doi.org/10.3390/pharmaceutics12070630.
[36] R. Weinstain, E.N. Savariar, C.N. Felsen, R.Y. Tsien, In vivo targeting of hydrogen peroxide by activatable cell-penetrating peptides, J. Am. Chem. Soc. 136 (2014) 874–877. https://doi.org/10.1021/ja411547j.
[37] D. Zhu, W. Chen, W. Lin, Y. Li, X. Liu, Reactive oxygen species-responsive nanoplatforms for nucleic acid-based gene therapy of cancer and inflammatory diseases, Biomed. Mater. 16 (2021) 042015. https://doi.org/10.1088/1748-605X/ac0a8f.
[38] X. Xu, P.E. Saw, W. Tao, Y. Li, X. Ji, S. Bhasin, Y. Liu, D. Ayyash, J. Rasmussen, M. Huo, J. Shi, O.C. Farokhzad, ROS-responsive polyprodrug nanoparticles for triggered drug delivery and effective cancer therapy, Adv. Mater. 29 (2017) 1700141. https://doi.org/10.1002/adma.201700141.
[39] Y. Zhang, Q. Guo, S. An, Y. Lu, J. Li, X. He, L. Liu, Y. Zhang, T. Sun, C. Jiang, ROS-switchable polymeric nanoplatform with stimuli-responsive release for active targeted drug delivery to breast cancer, ACS Appl. Mater. Interfaces. 9 (2017) 12227–12240. https://doi.org/10.1021/acsami.6b16815.
[40] Y. Yuan, J. Liu, B. Liu, Conjugated-polyelectrolyte-based polyprodrug: Targeted and image-guided photodynamic and chemotherapy with on-demand drug release upon irradiation with a single light source, Angew. Chemie Int. Ed. 53 (2014) 7163–7168. https://doi.org/10.1002/anie.201402189.
[41] L.-H. Liu, W.-X. Qiu, B. Li, C. Zhang, L.-F. Sun, S.-S. Wan, L. Rong, X.-Z. Zhang, A red light activatable multifunctional prodrug for image-guided photodynamic therapy and cascaded chemotherapy, Adv. Funct. Mater. 26 (2016) 6257–6269. https://doi.org/10.1002/adfm.201602541.
[42] M.S. Shim, Y. Xia, A reactive oxygen species (ROS)-responsive polymer for safe, efficient, and targeted gene delivery in cancer cells, Angew. Chemie. 125 (2013) 7064–7067. https://doi.org/10.1002/ange.201209633.
[43] T. Li, F. Li, W. Xiang, Y. Yi, Y. Chen, L. Cheng, Z. Liu, H. Xu, Selenium-containing amphiphiles reduced and stabilized gold nanoparticles: Kill cancer cells via reactive oxygen species, ACS Appl. Mater. Interfaces. 8 (2016) 22106–22112. https://doi.org/10.1021/acsami.6b08282.
[44] Y. Wang, Y. Deng, H. Luo, A. Zhu, H. Ke, H. Yang, H. Chen, Light-responsive nanoparticles for highly efficient cytoplasmic delivery of anticancer agents, ACS Nano. 11 (2017) 12134–12144. https://doi.org/10.1021/acsnano.7b05214.
[45] C.-C. Song, R. Ji, F.-S. Du, Z.-C. Li, Oxidation-responsive poly(amino ester)s containing arylboronic ester and self-immolative motif: Synthesis and degradation study, Macromolecules. 46 (2013) 8416–8425. https://doi.org/10.1021/ma401656t.
[46] H. Luo, C. Huang, J. Chen, H. Yu, Z. Cai, H. Xu, C. Li, L. Deng, G. Chen, W. Cui, Biological homeostasis-inspired light-excited multistage nanocarriers induce dual apoptosis in tumors, Biomaterials. 279 (2021) 121194. https://doi.org/10.1016/j.biomaterials.2021.121194.
[47] M. Bio, P. Rajaputra, G. Nkepang, S.G. Awuah, A.M.L. Hossion, Y. You, Site-specific and far-red-light-activatable prodrug of combretastatin A-4 using photo-unclick chemistry, J. Med. Chem. 56 (2013) 3936–3942. https://doi.org/10.1021/jm400139w.
[48] Z. Xiao, Y. Zhang, Y. Niu, Q. Ke, X. Kou, Cyclodextrins as carriers for volatile aroma compounds: A review, Carbohydr. Polym. 269 (2021) 118292. https://doi.org/10.1016/j.carbpol.2021.118292.
[49] B. Tian, S. Hua, J. Liu, Cyclodextrin-based delivery systems for chemotherapeutic anticancer drugs: A review, Carbohydr. Polym. 232 (2020) 115805. https://doi.org/10.1016/j.carbpol.2019.115805.
[50] R. Cavalli, F. Trotta, W. Tumiatti, Cyclodextrin-based nanosponges for drug delivery, J. Incl. Phenom. Macrocycl. Chem. 56 (2006) 209–213. https://doi.org/10.1007/s10847-006-9085-2.

[51] X. Song, Y. Wen, J. Zhu, F. Zhao, Z.-X. Zhang, J. Li, Thermoresponsive delivery of paclitaxel by β-cyclodextrin-based poly(N-isopropylacrylamide) star polymer via inclusion complexation, Biomacromolecules. 17 (2016) 3957–3963. https://doi.org/10.1021/acs.biomac.6b01344.

[52] X. Yao, J. Mu, L. Zeng, J. Lin, Z. Nie, X. Jiang, P. Huang, Stimuli-responsive cyclodextrin-based nanoplatforms for cancer treatment and theranostics, Mater. Horizons. 6 (2019) 846–870. https://doi.org/10.1039/C9MH00166B.

[53] D.D. Gadade, S.S. Pekamwar, Cyclodextrin based nanoparticles for drug delivery and theranostics, Adv. Pharm. Bull. 10 (2020) 166–183. https://doi.org/10.34172/apb.2020.022.

[54] D. Jia, X. Ma, Y. Lu, X. Li, S. Hou, Y. Gao, P. Xue, Y. Kang, Z. Xu, ROS-responsive cyclodextrin nanoplatform for combined photodynamic therapy and chemotherapy of cancer, Chinese Chem. Lett. 32 (2021) 162–167. https://doi.org/10.1016/j.cclet.2020.11.052.

[55] D. Chen, G. Zhang, R. Li, M. Guan, X. Wang, T. Zou, Y. Zhang, C. Wang, C. Shu, H. Hong, L.-J. Wan, Biodegradable, hydrogen peroxide, and glutathione dual responsive nanoparticles for potential programmable paclitaxel release, J. Am. Chem. Soc. 140 (2018) 7373–7376. https://doi.org/10.1021/jacs.7b12025.

[56] M. Criado-Gonzalez, D. Mecerreyes, Thioether-based ROS responsive polymers for biomedical applications, J. Mater. Chem. B. 10 (2022) 7206–7221. https://doi.org/10.1039/D2TB00615D.

[57] W. Yin, W. Ke, N. Lu, Y. Wang, A.A.-W.M.M. Japir, F. Mohammed, Y. Wang, Y. Pan, Z. Ge, Glutathione and reactive oxygen species dual-responsive block copolymer prodrugs for boosting tumor site-specific drug release and enhanced antitumor efficacy, Biomacromolecules. 21 (2020) 921–929. https://doi.org/10.1021/acs.biomac.9b01578.

[58] L. Zhang, S. Zhang, M. Li, Y. Li, H. Xiong, D. Jiang, L. Li, H. Huang, Y. Kang, J. Pang, Reactive oxygen species and glutathione dual responsive nanoparticles for enhanced prostate cancer therapy, Mater. Sci. Eng. C. 123 (2021) 111956. https://doi.org/10.1016/j.msec.2021.111956.

[59] Y. Li, J. Yang, X. Sun, Reactive oxygen species-based nanomaterials for cancer therapy, Front. Chem. 9 (2021). https://doi.org/10.3389/fchem.2021.650587.

[60] Y. Yao, Y. Zhou, L. Liu, Y. Xu, Q. Chen, Y. Wang, S. Wu, Y. Deng, J. Zhang, A. Shao, nanoparticle-based drug delivery in cancer therapy and its role in overcoming drug resistance, Front. Mol. Biosci. 7 (2020). https://doi.org/10.3389/fmolb.2020.00193.

[61] A. El-Hussein, S.L. Manoto, S. Ombinda-Lemboumba, Z.A. Alrowaili, P. Mthunzi-Kufa, A review of chemotherapy and photodynamic therapy for lung cancer treatment, Anticancer. Agents Med. Chem. 21 (2020) 149–161. https://doi.org/10.2174/1871520620666200403144945.

[62] A. Khdair, Di Chen, Y. Patil, L. Ma, Q.P. Dou, M.P.V. Shekhar, J. Panyam, Nanoparticle-mediated combination chemotherapy and photodynamic therapy overcomes tumor drug resistance, J. Control. Release. 141 (2010) 137–144. https://doi.org/10.1016/j.jconrel.2009.09.004.

[63] H. Broxterman, N. Georgopapadakou, Anticancer therapeutics: "Addictive" targets, multi-targeted drugs, new drug combinations, Drug Resist. Updat. 8 (2005) 183–197. https://doi.org/10.1016/j.drup.2005.07.002.

[64] H. Lee, J. Han, H. Shin, H. Han, K. Na, H. Kim, Combination of chemotherapy and photodynamic therapy for cancer treatment with sonoporation effects, J. Control. Release. 283 (2018) 190–199. https://doi.org/10.1016/j.jconrel.2018.06.008.

[65] Z. Cao, D. Li, J. Wang, X. Yang, Reactive oxygen species-sensitive polymeric nanocarriers for synergistic cancer therapy, Acta Biomater. 130 (2021) 17–31. https://doi.org/10.1016/j.actbio.2021.05.023.

[66] S. Uthaman, S. Pillarisetti, A.P. Mathew, Y. Kim, W.K. Bae, K.M. Huh, I.-K. Park, Long circulating photoactivable nanomicelles with tumor localized activation and ROS triggered self-accelerating drug release for enhanced locoregional chemo-photodynamic therapy, Biomaterials. 232 (2020) 119702. https://doi.org/10.1016/j.biomaterials.2019.119702.

[67] D. Banik, S. Moufarrij, A. Villagra, Immunoepigenetics combination therapies: An overview of the role of HDACs in cancer immunotherapy, Int. J. Mol. Sci. 20 (2019) 2241. https://doi.org/10.3390/ijms20092241.

[68] S. Lantuejoul, M. Sound-Tsao, W.A. Cooper, N. Girard, F.R. Hirsch, A.C. Roden, F. Lopez-Rios, D. Jain, T.-Y. Chou, N. Motoi, K.M. Kerr, Y. Yatabe, E. Brambilla, J. Longshore, M. Papotti, L.M. Sholl, E. Thunnissen, N. Rekhtman, A. Borczuk, L. Bubendorf, Y. Minami, M.B. Beasley, J. Botling, G. Chen, J.-H. Chung, S. Dacic, D. Hwang, D. Lin, A. Moreira, A.G. Nicholson, M. Noguchi, G. Pelosi, C. Poleri, W. Travis, A. Yoshida, J.B. Daigneault, I.I. Wistuba, M. Mino-Kenudson, PD-L1 Testing for lung cancer in 2019: Perspective from the IASLC pathology committee, J. Thorac. Oncol. 15 (2020) 499–519. https://doi.org/10.1016/j.jtho.2019.12.107.

[69] S. Spranger, T.F. Gajewski, Impact of oncogenic pathways on evasion of antitumour immune responses, Nat. Rev. Cancer. 18 (2018) 139–147. https://doi.org/10.1038/nrc.2017.117.

[70] T. Chen, L. Su, X. Ge, W. Zhang, Q. Li, X. Zhang, J. Ye, L. Lin, J. Song, H. Yang, Dual activated NIR-II fluorescence and photoacoustic imaging-guided cancer chemo-radiotherapy using hybrid plasmonic-fluorescent assemblies, Nano Res. 13 (2020) 3268–3277. https://doi.org/10.1007/s12274-020-3000-9.

[71] T. Wang, D. Wang, H. Yu, B. Feng, F. Zhou, H. Zhang, L. Zhou, S. Jiao, Y. Li, A cancer vaccine-mediated postoperative immunotherapy for recurrent and metastatic tumors, Nat. Commun. 9 (2018) 1532. https://doi.org/10.1038/s41467-018-03915-4.

[72] W. Huang, J.-J. Chen, R. Xing, Y.-C. Zeng, Combination therapy: Future directions of immunotherapy in small cell lung cancer, Transl. Oncol. 14 (2021) 100889. https://doi.org/10.1016/j.tranon.2020.100889.

[73] D. Ti, X. Yan, J. Wei, Z. Wu, Y. Wang, W. Han, Inducing immunogenic cell death in immuno-oncological therapies, Chinese J. Cancer Res. 34 (2022) 1–10. https://doi.org/10.21147/j.issn.1000-9604.2022.01.01.

[74] W. Tao, X. Ji, X. Zhu, L. Li, J. Wang, Y. Zhang, P.E. Saw, W. Li, N. Kong, M.A. Islam, T. Gan, X. Zeng, H. Zhang, M. Mahmoudi, G.J. Tearney, O.C. Farokhzad, Two-dimensional antimonene-based photonic nanomedicine for cancer theranostics, Adv. Mater. 30 (2018) 1802061. https://doi.org/10.1002/adma.201802061.

[75] N. Gao, L. Mei, Black phosphorus-based nano-drug delivery systems for cancer treatment: Opportunities and challenges, Asian J. Pharm. Sci. 16 (2021) 1–3. https://doi.org/10.1016/j.ajps.2020.03.004.

[76] H. Wang, S. Jiang, W. Shao, X. Zhang, S. Chen, X. Sun, Q. Zhang, Y. Luo, Y. Xie, Optically switchable photocatalysis in ultrathin black phosphorus nanosheets, J. Am. Chem. Soc. 140 (2018) 3474–3480. https://doi.org/10.1021/jacs.8b00719.

[77] Z. Li, Y. Hu, Q. Fu, Y. Liu, J. Wang, J. Song, H. Yang, NIR/ROS-responsive black phosphorus QD vesicles as immunoadjuvant carrier for specific cancer photodynamic immunotherapy, Adv. Funct. Mater. 30 (2020) 1905758. https://doi.org/10.1002/adfm.201905758.

7 Hypoxia-Responsive Delivery Nanoplatforms in Cancer Theranostics

Priyanka Mishra, Yamini B Tripathi, Namdev L. Dhas and Neha Garg

CONTENTS

7.1 Introduction	114
7.2 Hypoxia in the Pathogenesis of Tumours	115
7.3 Role of Hypoxia-Responsive Nanotechnology as Therapeutics	116
7.4 Role of Hypoxia-Responsive Nanotechnology as Diagnostics	117
7.5 Conclusion	118
Acknowledgements	118
References	118

7.1 INTRODUCTION

Cancer is a broad term that refers to diseases that can affect any region of the body. Malignant tumours and neoplasms are other common terminology. One distinguishing aspect of cancer is metastasis, where there is rapid formation of aberrant cells that can grow beyond their normal borders and can then infect neighbouring sections of the body and migrate to other organs. The leading cause of cancer-related mortality is widespread metastasis. Cancer is the world's largest cause of death per the World Health Organization (WHO), accounting for roughly 10 million fatalities in 2021 (Sung et al., 2021) The number of cases and cancer-related mortality are daunting in their diversity and represented in Figure 7.1, suggesting a rough sketch of most common types of cancer, such as breast cancer, and common causes of cancer-related death, such as lung cancer.

The pathogenesis of tumour formation can be summarized through the multistep evolution of the eight cardinal biological abilities also known as cancer's hallmarks (Lazebnik, 2010). These are avoiding growth inhibitors, accumulating skills for maintaining proliferative signalling, resisting cell death, allowing for reproducible immortality, accessing vasculature, activating metastasis and invasion, and altering cellular metabolism while preventing immune system deterioration. These characteristics serve as an organizing foundation for understanding the intricacies of neoplastic disease. Among them, one of the distinguishing essential features is cancer hypoxia, which influences gene expression, metabolism, and, ultimately, tumour biology-related activities (Sebestyén et al., 2021). Inadequate vascularization and systemic hypoxia in the patient are the fundamental causes of cancer hypoxia, which leads to a unique sort of genetic reprogramming via hypoxia-induced transcription factors (HIF). Hypoxia could be either acute, chronic, toxic, or systemic (Brown & Wilson, 2004). The toxic form of hypoxia has no effect on cancer. Acute hypoxia is described as perfusion hypoxia, whereas chronic hypoxia is attributed to diffusion hypoxia, and both follow different pathological routes. Cancer patients frequently experience systemic hypoxia. Tumour hypoxia results from the amalgamation of acute, chronic, and systemic hypoxia, which leads to tumour progression and is also a major cause of resistance to many therapy methods.

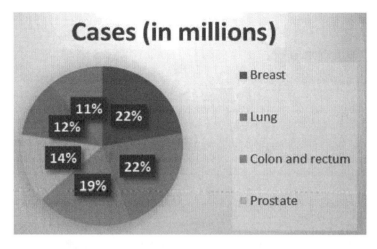
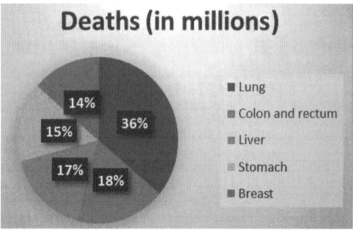

FIGURE 7.1 Representation of the most common types of cancer and the most common causes of cancer-related deaths.

7.2 HYPOXIA IN THE PATHOGENESIS OF TUMOURS

Aggressively proliferating tumours have insufficient oxygen concentrations because of the disparity between the rapid pace of tumour development and blood supply, creating a profoundly hypoxic intra-tumor microenvironment. More specifically, tumour cells' rapid rates of proliferating and metastasizing increase oxygen consumption, which results in hypoxia, while a lack of oxygen contributes to poor angiogenesis in solid tumours, resulting in intra-tumoral hypoxia. Hypoxia-inducible factor-1 alpha (HIF-1) can be elevated as a tumour grows, promoting the synthesis of vascular endothelial growth factor (VEGF) via the HIF-1 signalling pathway, which drives blood vessel expansion (Axelson et al., 2005). However, due to the inadequate efficiency of enhanced blood circulation, the deformed neovascular tissue is unable to rescue the oxygen supply. The molecular oxygen (O_2) level in normal tissue ranges from 2% to 9% (40 mm Hg). In a tumour microenvironment, "hypoxia" and "anoxia" are generally described as O_2 levels ranging from 0.02% to 2% (2.5 mmHg pO_2) (Gnaiger, 2003). In tumour cells, abnormally active metabolism and cell development can deplete intracellular oxygen. Second, there is an inadequate supply of oxygen due to the disorganized vascular system in the tumour tissue. Finally, the requirement of tumour cells further away from the blood arteries

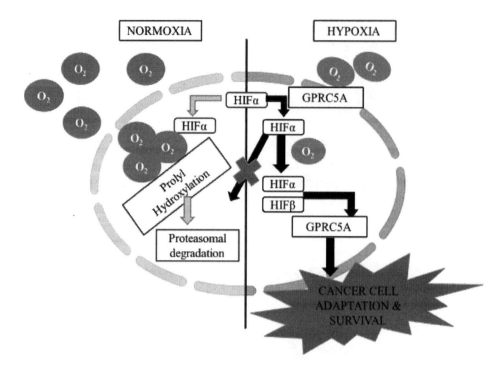

FIGURE 7.2 Hypoxia signalling in the tumour microenvironment. Under normoxia, with the help of prolyl hydroxylase, it follows the normal path of proteasomal degradation. However, under hypoxia, the enzymes are suppressed due to lower oxygen concentration, and HIFα binds HIF1β and translocates itself to a hypoxia-responsive element and upregulates, leading to cancer cell adaptation and survival.

cannot be satisfied by the short O_2 diffusion distance (less than 200 um). Furthermore, tumour cells survive better under hypoxia due to an increased HIF-1 pathway, which can change the glucose metabolism pathway (Figure 7.2). On the other hand, hypoxia is linked to poor overall survival because it maintains cancer stem-like cells that are not cycling, which leads to the development of chemoresistance and radio resistance (Oei et al., 2017). Because of its impact on the control of the cell cycle, stem cell maintenance and quiescence, apoptosis evasion, and hypoxia are critical factors to be considered for the selection of treatment in cancer relapse.

Hypoxia is a defining feature of malignant tumours as it is frequently associated with increased metastasis and poor prognosis. For the management of tumour growth and dissemination, early diagnosis and effective killing of hypoxic cancer cells are therefore essential. Currently, interdisciplinary research is targeted at generating functional compounds and nanomaterials that can be utilized to screen and treat hypoxic tumours noninvasively and efficiently. These include nanoparticles that are hypoxia-active, anti-hypoxia drugs, and treatments targeting tumour hypoxia biomarkers (Y. Wang et al., 2019).

By producing O_2, activating nanoparticles or medications in the hypoxic tumour microenvironment, and focusing on tumour hypoxia biomarkers, the aim is to reverse hypoxia and increase the effectiveness of therapy.

7.3 ROLE OF HYPOXIA-RESPONSIVE NANOTECHNOLOGY AS THERAPEUTICS

Hypoxia is recognized as the primary target for cancer diagnosis and treatment due to its particular role in the growth of tumours. Because of their miniature size and customizable physicochemical features, nanocarriers are a rapidly developing class of materials that find extensive

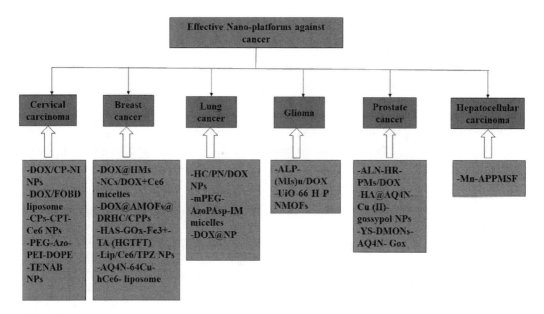

FIGURE 7.3 Flow chart: nano-platform effective against cancer utilizing hypoxia-responsive strategy.

usage in biological applications (Thambi et al., 2016). Materials of particular interest are stimuli-responsive nanocarriers, which release their payloads, particularly in the tumour microenvironment. Transporting anticancer drugs to hypoxic regions is challenging since they are remote from the blood arteries and tumours have aberrant vascular properties. Hypoxia also enhances the expression of genes that contribute to drug resistance, including P-glycoprotein. To address the problems associated with hypoxia, nanocarriers must be developed capable of releasing imaging agents or anticancer medicines in hypoxic environments.

Tumour tissue hypoxia is generally thought to signify a bad prognosis for tumour treatment, but specific molecular properties can make it a specific target for cancer therapy. Tumour cells prefer aerobic glycolysis to the conventional oxidative phosphorylation route due to the Warburg effect (Zheng, 2012). As a result, several enzymes involved in electron donation or reduction response, such as nitro reductase, methionine synthase reductase, azo-reductase, inducible nitric synthase, and DT-diaphorase (DTD), are overproduced in hypoxic tumour cells (Zhang et al., 2020). The aforementioned findings have prompted considerable effort to be put into developing hypoxia-responsive nanoplatforms that can be activated by these enzymes for improved tumour treatment. In the creation of hypoxia-responsive nanoparticles, azo groups, quinone, nitro groups, and N-oxide compounds are also utilized as chemical linkages that are hypoxia responsive. By gaining or shedding electrons, they can alter their structure and physicochemical characteristics including hydrophobicity and electron affinity. These hypoxia-responsive nanoparticles have proven effective at delivering drugs, as expected (Figure 7.3). They offer a lot of potential for tumour treatment since they feature longer blood circulation time, hypoxia-responsive cargo release, and improved tumour penetration and accumulation.

7.4 ROLE OF HYPOXIA-RESPONSIVE NANOTECHNOLOGY AS DIAGNOSTICS

Acute hypoxia is emerging as a major target in developing diagnostics and therapies because there is a well-known distinctive feature of disordered vasculature of growing solid tumour physiology that is rarely observed in healthy tissue. Physical barriers are produced by hypoxia, a key microenvironment of solid tumours. It is a crucial factor in tumour metastasis, angiogenesis, and increased treatment resistance. Most anticancer medications will struggle to penetrate this barrier and reach the tumour site, which reduces their ability to have a therapeutic effect (Kumari et al., 2020). Following the evaluation of the impact of the hypoxic microenvironment on tumour treatment, two approaches—overcoming hypoxia and dodging hypoxia—are suggested. The former refers to raising the oxygen content of tumour tissue, whilst the latter refers to the utilization of water or gas molecules in tumour tissue to strengthen oxygen reliance and improve the therapeutic effect of hypoxic tumours. In addition to those two approaches, the tactic of "using hypoxia" is recognized. The pathogenic hypoxic situation causes an elevation of several bioreductive enzymes, which activates prodrugs. Different quinone, nitroaromatic, and azobenzene derivatives that comprise hypoxia-sensitive groups have been exploited in the design and development of prodrugs or molecular imaging diagnostic agents based on intracellular hypoxia (Liu et al., 2021). Through one- or two-electron oxidoreductase routes, they are quickly broken down to produce transitory radical anion intermediates that are back-oxidized effectively by oxygen present in normal cells but further reduced to amino aromatics in hypoxic malignancies. As a result, various hypoxia-sensitive prodrugs, including RSU-1069, TH-302, and PR-104, have been created for the treatment of cancer and are inspired by the special property of bioreductive hypoxic TME (Wilson et al., 2014).

7.5 CONCLUSION

Here, we have described the unique properties of the microenvironment in hypoxic tumour tissues. Cancer is a leading cause of death worldwide, and effective theranostics (diagnosis and treatment) are urgently needed. Nanotechnology offers great promise in this area, due to its ability to specifically target cancer cells and deliver therapeutics. Nanotechnology can be used as a tool to target cancer cells and deliver therapies effectively. Nanoparticles (NPs) have unique properties that make them ideal for this purpose: they are small enough to penetrate tissues in which blood flow has been blocked, and they can carry many different types of cargo, including drugs, radioisotopes, and DNA constructs for gene therapy. Nanoparticles have been used successfully to deliver cancer treatments in preclinical and clinical trials, including chemotherapy drugs, radiotherapies, and gene therapies.

Based on these features, we have presented the principles in the design of hypoxia-triggered nano-theranostic agents and summarized the recent progress in this area. As to the versatility of nanomaterials, these principles are often used on their own and in combination. Although various strategies to treat hypoxic tumours have already been developed and applied, there remain several challenges to overcome with additional future research.

ACKNOWLEDGEMENTS

NG acknowledges the seed grant under the Institute of Eminence Scheme, Banaras Hindu University, for financial support.

REFERENCES

Axelson, H., Fredlund, E., Ovenberger, M., Landberg, G., & Påhlman, S. (2005). Hypoxia-induced dedifferentiation of tumor cells—A mechanism behind heterogeneity and aggressiveness of solid tumors. *Seminars in Cell & Developmental Biology*, *16*(4–5), 554–563. https://doi.org/10.1016/J.SEMCDB.2005.03.007.

Brown, J. M., & Wilson, W. R. (2004). Exploiting tumour hypoxia in cancer treatment. *Nature Reviews Cancer*, *4*(6), 437–447. https://doi.org/10.1038/nrc1367.

Feng, L., Cheng, L., Dong, Z., Tao, D., Barnhart, T. E., Cai, W., Chen, M., & Liu, Z. (2017). Theranostic liposomes with hypoxia-activated prodrug to effectively destruct hypoxic tumors post-photodynamic therapy. *ACS Nano*, *11*(1), 927–937. https://doi.org/10.1021/ACSNANO.6B07525

Gnaiger, E. (2003). Oxygen conformance of cellular respiration: A perspective of mitochondrial physiology. *Advances in Experimental Medicine and Biology*, *543*, 39–55. https://doi.org/10.1007/978-1-4419-8997-0_4/COVER.

Kumari, R., Sunil, D., & Ningthoujam, R. S. (2020). Hypoxia-responsive nanoparticle based drug delivery systems in cancer therapy: An up-to-date review. *Journal of Controlled Release*, *319*, 135–156. https://doi.org/10.1016/J.JCONREL.2019.12.041.

Lazebnik, Y. (2010). What are the hallmarks of cancer? *Nature Reviews Cancer*, *10*(4), 232–233. https://doi.org/10.1038/nrc2827.

Liu, X., Wu, Z., Guo, C., Guo, H., Su, Y., Chen, Q., Sun, C., Liu, Q., Chen, D., & Mu, H. (2021). Hypoxia responsive nano-drug delivery system based on angelica polysaccharide for liver cancer therapy, *29*(1), 138–148. https://doi.org/10.1080/10717544.2021.2021324.

Oei, A. L., Vriend, L. E. M., Krawczyk, P. M., Horsman, M. R., Franken, N. A. P., & Crezee, J. (2017). Targeting therapy-resistant cancer stem cells by hyperthermia, *33*(4), 419–427. https://doi.org/10.1080/02656736.2017.1279757.

Sebestyén, A., Kopper, L., Dankó, T., & Tímár, J. (2021). Hypoxia signaling in cancer: From basics to clinical practice. *Pathology and Oncology Research*, *27*, 100. https://doi.org/10.3389/PORE.2021.1609802/BIBTEX.

Sung, H., Ferlay, J., Siegel, R. L., Laversanne, M., Soerjomataram, I., Jemal, A., & Bray, F. (2021). Global cancer statistics 2020: GLOBOCAN estimates of incidence and mortality worldwide for 36 cancers in 185 countries. *CA: A Cancer Journal for Clinicians*, *71*(3), 209–249. https://doi.org/10.3322/CAAC.21660.

Thambi, T., Park, J. H., & Lee, D. S. (2016). Hypoxia-responsive nanocarriers for cancer imaging and therapy: Recent approaches and future perspectives. *Chemical Communications*, *52*(55), 8492–8500. https://doi.org/10.1039/C6CC02972H.

Wang, Y., Shang, W., Niu, M., Tian, J., & Xu, K. (2019). Hypoxia-active nanoparticles used in tumor theranostic. *International Journal of Nanomedicine*, *14*, 3705. https://doi.org/10.2147/IJN.S196959.

Wilson, W. R., Hicks, K. O., Wang, J., & Pruijn, F. B. (2014). Prodrug strategies for targeting tumour hypoxia, 283–328. https://doi.org/10.1007/978-1-4614-9167-5_13.

Zhang, K., Meng, X., Yang, Z., Dong, H., & Zhang, X. (2020). Enhanced cancer therapy by hypoxia-responsive copper metal-organic frameworks nanosystem. *Biomaterials*, *258*, 120278. https://doi.org/10.1016/J.BIOMATERIALS.2020.120278.

Zheng, J. (2012). Energy metabolism of cancer: Glycolysis versus oxidative phosphorylation (review). *Oncology Letters*, *4*(6), 1151–1157. https://doi.org/10.3892/OL.2012.928/HTML.

8 Dual/Multi-Responsive Delivery Nanoplatforms in Cancer Theranostics

Kunj Vyas and Mayur M Patel

ABBREVIATIONS¹

CONTENTS

8.1 Introduction 120
8.2 Dual Stimuli–Responsive DDSs for Imagining and Diagnosis of Tumors 122
8.3 Dual-Stimuli Response DDSs for Cancer Therapy and Treatment 124
    8.3.1 pH and Temperature 124
    8.3.2 pH and Redox 127
    8.3.3 Temperature and Redox 128
    8.3.4 pH and Light Responsive 129
    8.3.5 Double pH 130
    8.3.6 Other Dual Stimuli–Responsive Systems 131
8.4 Multi-Stimuli–Response DDSs for Cancer Therapy and Treatment 132
8.5 Multi-Stimuli–Responsive System for Tumor Immunotherapy 134
8.6 Conclusion 135
References 136

8.1 INTRODUCTION

Cancer is a serious public health concern worldwide, accounting for the second largest incidence of death in the United States and the fifth greatest cause of mortality in India and for 5.7% of all deaths [1, 2]. Globally, 9.6 million people died from cancer in 2018, according to GLOBOCAN [1]. Recent coronavirus disease 2019 (COVID-19) pandemic effects negatively impacted cancer diagnosis and treatment. Despite this, there are more cancer survivors in the United States than ever before, largely due to the combined effects of an aging and growing population as well as higher cancer survival rates brought on by advancements in early detection and treatment [3]. The expected increases in cancer incidence and mortality by 2040 are 29.5 million and 16.3 million, respectively [1]. The primary cancer treatments currently in clinics, such as chemotherapy, surgery and radiotherapy, face a number of difficulties, including a lack of tumor specificity, a limited ability to penetrate the extremely dense extracellular matrix, a high level of systemic toxicity, insufficient clearance and multi-drug resistance. Furthermore, there are a number of cutting-edge treatment options available, including immunotherapy and gene therapy, both of which have recently entered the market and are known to be successful [4–6]. Despite this, there is a high risk of therapeutic failure or tumor recurrence, as well as metastasis, if a timely diagnosis is not obtained. To achieve satisfactory clinical results, effective and specialized treatment plans that incorporate both diagnosis and therapeutic approaches are required.

Theranostics is the term used to describe the combination of diagnosis and treatment that enables monitoring of drug distribution, the simultaneous detection of targets and evaluation of therapeutic responses to achieve personalized medicine [6–8]. Numerous more recent delivery platforms have demonstrated significantly improved theranostic effects, particularly through nanotechnology. However, conventional nanoplatforms failed to achieve the anticipated therapeutic effect due to a variety of barriers, including scavenging effect by the mononuclear phagocytic system (MPS), inconsistent enhanced permeability and retention (EPR) effect in different tumors, inadequate therapeutics and toxicity and complex tumor microenvironment (TME). Further problems of nanoplatforms include inadequate tumor cell uptake. Most solid tumors in humans do not exhibit the EPR effect because of the presence of a dense tumor matrix or for other reasons. Another major issue is the stability of the circulation and the targeted release of nanoparticles (NPs). As a result, more sophisticated systems are used, and stimuli-responsive drug delivery systems (DDSs), also referred to as smart DDSs, have helped to partially overcome delivery barriers for nanotechnology. Smart DDS characteristics will drastically alter in response to various stimuli, leading to changes in size, charge, ligand emergence or disintegration for better tumor targeting, internalization and permeability at an incredibly precise rate [9]. They can enhance the effectiveness of targeted drugs to successfully reach distinct tumor cells and tissues and prevent drug leakage [10, 11]. Stimuli-responsive nanomaterials are gaining popularity due to their use in cancer treatment, drug delivery and fluorescence imaging (FLI).

The idea of a stimuli-responsive delivery system was first put forth in the 1970s; since then, there have been numerous related studies and applications [12–14]. Rapid drug delivery aims to accomplish prolonged and regulated drug release by creating a stimuli-responsive delivery system that can recognize the multiphase nature of the body's biochemical processes and changes in the microenvironment. This mimics the response of living organs. They are easily capable of changing their structural makeup through chemical processes like polymerization, hydrolysis, isomerization and others that are based on physiochemical stimulations both inside and outside of the cell. This is done to hasten the release of active ingredients in particular physiological settings [15]. They enable on-demand or sustained drug release, sensing of specific pathological factors/molecules, promoted tumor accumulation, signaling in specific positions, drug activation, tumor-specific diagnosis and theranostics. Additionally, it was reported that the stimuli-responsive nanocarriers could circumvent multidrug resistance in the treatment of cancer [16]. Besides that, the stimuli could be used to activate specific prodrug-formulated nanocarriers biologically in diseased tissues or cells for targeted therapy. They can also draw attention to biological functions of nanocarriers. For instance, an external magnetic field may promote the buildup of magnetic nanocarriers in tumors.

In reality, there are three primary types of reactions: endogenous trigger reactions, exogenous trigger reactions and multi stimuli–triggered reactions. These reactions are further divided into categories based on their interactions. There are four broad categories of responses for drug delivery. The main one is extracorporeal stimuli, like light, magnetic fields, electric field, temperature and ultrasound. The second is based on TME, which can easily reduce in size or change in charge to improve tumor permeation or cellular uptake [17]. It is thought that ideal DDS can regulate the release of drugs by reacting to external physical stimuli and signals present in the TME. The third is internal stimuli responsive systems, which may require a wider distinction between health and illness circumstances since they require a significant amount of drug to be delivered to the target spot. For instance, pH-responsive systems need a high pH gradient to dissolve in an acidic environment. The final one relies on the overexpression of particular enzymes, which operates when it comes into contact with a high concentration of that enzyme and is also an internal stimulus. Some other types of internal responses are adenosine triphosphate (ATP), H_2O_2, redox-potential and hypoxia [18, 19]. Multi stimuli responses are composed of two or more stimuli-responsive factors. Furthermore, stimuli-responsive DDSs can deliver a variety of bioactive substances, including anti-neoplastic agents, peptides, oligonucleotides, vaccines and others, with great potential for clinical use.

They have a significant advantage and can effectively overcome biological barriers, but there are some issues that can impede effective drug delivery. Also, because of the off-target responses caused by typical cells or tissues with an identical triggering environment, stimulus responsive DDSs are not sufficiently accurate [20, 21]. Internal stimuli DDSs are primarily based on abnormal environments in cancer tissues, and in order to carry out effective treatment, they must first identify the target. Photo stimuli-responsive DDSs, for example, are frequently affected by the penetration of light from various in situ carcinomas [15]. A stimulus might also not be powerful enough to cause a reaction; for instance, the extrinsic pH in the TME might be slightly different from that of healthy tissues, causing the drug carrier to react slowly or not at all [22].

In this chapter, various stimuli-sensitive strategies and their applications in drug delivery, therapy, tumor imaging and theranostics are presented, along with a summary of recent advancements and successes in DDSs that are responsive to external or internal stimuli. This review also includes the design strategies of multiple stimuli responsive DDS with varying stimulation thresholds and critically discusses their benefits and drawbacks, which may provide new insights into tumor treatment potential. The clinical application of stimuli-responsive DDSs is illustrated in the sections that follow, and then conclusions are drawn.

8.2 DUAL STIMULI–RESPONSIVE DDSS FOR IMAGINING AND DIAGNOSIS OF TUMORS

Stimuli-responsive systems can be used to treat tumors precisely as well as to offer a comprehensive, patient-centered solution. To ensure a more precise diagnosis, multimodal imaging can incorporate structural or functional data to reduce the impact of high amounts of background signals, complicated biological contexts and probe concentration [23]. For instance, magnetic fields and light are crucial for imaging and diagnosis in the context of magnetic resonance imagining (MRI) and photoacoustic imaging (PAI). Quenching or disintegrating agents can be used to make materials visible by causing an external trigger to emit light or to reduce signal transmission from the carriers themselves. These techniques can be used for multimodal imaging, theranostics and real-time in vivo agent monitoring.

Recently it was noticed that melanin could be used for PAI and PTT because of its thermal conversion and potent absorption in the near-infrared region (NIR) [24]. Researchers tested this by creating pH/NIR-responsive magnetic nanoparticles (MNPs) coated with melanin that would deliver the drug obatoclax (OBX), an inhibitor of wingless type (Wnt)/β-catenin signaling. It was created by hydrophobic interactions and π–π stacking, enabling dual-mode MRI/PAI for more precise diagnosis and guided chemotherapy of tumors, which was strengthened by mild hyperthermia [25]. Compared to a single-mode platform, this imaging platform with dual mode can offer more precise diagnosis and imaging. NIR light's limited absorption allows for deep penetration into biological tissues, making it perfect for the beginning of light-activated chemotherapy [26]. There has been a lot of interest in the up-conversion approach, which transforms light into photoactivated chemotherapy at the tumor site [27]. The researchers used specialized up-converting NPs (UCNPs), a unique class of optical nanomaterials doped with lanthanide ions. UCNPs made it possible to combine dual-mode computed tomography (CT) and photoactivated luminescence chemotherapy. Researchers labelled ^{64}Cu on nanosurfaces using a straightforward chelating technique in order to use single-photon emission CT to track the response, accumulation and distribution of these materials in tumor-bearing mice in real time [28, 29]. Another study prepared a theranostic nanoprobe for accurate diagnosis. It could be used as a drug carrier to deliver drug directly to the tumor site, allowing for both rapid tumor detection and direct tumor treatment [30].

For the purpose of diagnosis, numerous imaging techniques have recently been developed and used. PAI is a modern and effective biomedical imaging approach that combines ultrasonic imaging's deep tissue penetration with optical imaging's great specificity to offer molecular information with high sensitivity [31, 32]. Another such method is positron emission tomography (PET), a popular clinical imaging method with extreme sensitivity and quantitative precision [33]. Consequently, combining PAI, MRI and PET would provide trimodal imaging with good selectivity, sensitivity and resolution. It is common knowledge that MRI has poor sensitivity. Moreover, tracking and monitoring the specifics of intracellular drug release are also crucial for the development of multi-stimuli DDSs. As a result, a new method was created using gadolinium (Gd)-chelators, which are contrast agents for MR but have a low relaxivity and a high risk of toxicity related to released Gd ions. Here, a core-satellite nanotheranostic agent with a ^{64}Cu-labeled polydopamine (PDA)-gadolinium-metallofullerene structure loaded with doxorubicin (DOX) was employed. Additionally, pH/NIR-triggered drug release was carried out to improve delivery and effectiveness, which resulted in an efficient chemo-photothermal combination therapy for PDA [34]. In a similar vein, a second pH/light dual-stimuli responsive agent was created for photothermally enhanced chemotherapy and dual targeting of the U87MG tumor. RGD (Arg-Gly-Asp) peptide, melanin-coated MNPs, DOX and indocyanine green (ICG), collectively known as RMDI, make up the compound. As demonstrated by in vivo trimodal PAI/MRI/FLI, the tumor accumulation of RMDI was simultaneously improved through physical magnetic targeting by an external magnetic field at tumor tissues and biological active targeting by RGD [35].

Stimulus-responsive FLI has been found to hold a lot of promise for tumor detection. It offers benefits like high sensitivity, ease of use and non-invasiveness. However, certain non-cancerous regions, such as inflammation and nodules, may also display an aspect of response to stimulus, leading to the issue of general responsiveness and ultimately leading to false positive results for tumor recognition. A new study was conducted in which acidic pH and hypoxia, two typical attributes strongly linked to cancer progression, invasion and metastasis in the TME, were selected as dual stimuli. They were fabricated into fluorescent nanoprobes for extremely precise and specific imaging of tumor cells. The nanocarrier used was mesoporous silica-coated gold nanorods (AuNRs) to load the pH-sensitive fluorescent reporter (Rho-TP). Azobenzene (azo) was chosen to act as the efficient gatekeeper for AuNR@mSiO2 by forming a complex with the polymer cyclodextrin through host-guest interaction (azo/CDP). This is due to the fact that under hypoxic conditions, highly expressed azoreductase can convert them to amines. It has been claimed that sensitive and precise imaging of tumor cells can be done by properly merging the pH-responsive fluorescent signal reporter and the hypoxia-responsive gatekeeper into a single nanoprobe. When this test was further performed in vivo on tumor-bearing mice, it suggested a potential early cancer diagnosis and cancer treatment application [36].

Clinical tumor resection faces a number of challenges, especially tumor-specific imaging. Numerous triggerable probes have been created for use in imaging of tumor to solve this issue. If a single factor stimulation is done, there are also problems with false signals that are irreversible. Therefore, a new dual stimuli agent based on light, enzyme and pH probes was created for tumor imaging by fluorescence and photoacoustic ratiometry. This was backed up by the fact that the ATP and H+ content in the TME is considerably higher than that in healthy tissues. The ATP-pH Förster resonance energy transfer-based probe was designed with silicon rhodamine as the donor, CS dye as the recipient and H+/ATP recognition units that triggered only when both ATP and H+ were coupled to the acceptor. Also, it was discovered that reversible stimulation of the ATP-pH probe by both H+ and ATP significantly reduced the cumulative response of the probe in circulation after intravenous injection. As an NIR ratiometric probe, it also not only detects deep tissue but also employs self-tuning to lower the number of false signals brought on by the buildup of the probe in tumor regions and other environmental factors. Furthermore, the shortcomings of single imaging could be compensated for by dual-mode imaging that uses both the photoacoustic effect and fluorescence. These characteristics enabled the probe to be used successfully for accurate ratiometric photoacoustic and fluorescence mapping of malignancies in mice [37]. In another study, a dual-responsive probe called βgal-BP-PMB (β-galactosyl-3,3'-dihydroxy-2,2'-bipyridyl-p-methoxybenzyl), whose photoluminescence can be activated by β-galactosidase (β-gal), as well as acidic conditions, was recently developed for the detection of ovarian cancer cells (OCs). Since senescent cells can cause false positive responses from drugs that only respond to one stimulus, pH- and enzyme-based dual responsive stimuli were used instead. This is because probe βgal-BP-PMB can be specifically activated in OCs but silenced in senescent cells due to the increased lysosomal pH in senescent cells. The researchers discovered from the outcomes that the probe could tell senescent cells from cancer cells with ease. Additionally, they anticipated that such dual-responsive small molecules would realize their enormous potential in the field of bioimaging and provide fresh viewpoints for a deeper comprehension of challenging physiological processes [38].

8.3 DUAL-STIMULI RESPONSE DDSS FOR CANCER THERAPY AND TREATMENT

Multiple trigger coexistence may enable more accurate and flexible control over the spatial location of agents [39]. For the precise treatment of tumors, multiple stimuli-responsive DDSs were developed. These systems are based on a combination of internal and external stimuli. They can overcome a variety of ongoing physiological and pathological challenges. DDSs that are responsive to multiple stimuli can deliver drugs to specific sites over several stages, ultimately realizing the greatest therapeutic efficacy. Multiple stimuli-responsive DDSs can dismantle barriers and offer a promising platform for the fusion of cytotoxic nanomedicine, tumor theranostics and immunotherapy to realize more effective personalized therapy [40]. Although it has many advantages, there are always some drawbacks. In this case, one of these is the design complexity of multiple stimuli responsive-DDSs, which presents a challenge because it makes it difficult to accurately control the spatiotemporal distribution of agents. The multi-stimuli responsive DDS has two or more stimuli. Combining pH and temperature; light and redox; light and reactive oxygen species (ROS); pH and ROS; dual pH; light; PH and magnet; pH, ROS and light; and many other types of responses are examples of possible stimuli. Some of the studies are given in Table 8.1. The stimuli-triggered release mechanism is shown in Figure 8.1.

8.3.1 pH and Temperature

pH and temperature-responsive DDSs are among the most-studied dual-sensitive delivery systems. The majority of this type of delivery system is covered by nanosystems. For instance, pH-sensitive NPs have been created and developed to release drugs in the tumor site and/or endo/lysosomal compartments, taking advantage of the slightly acidic environments in cancerous tissues (pH 6.5–7.2), endosomes (pH 5.0–6.5) and lysosomes (pH 4.5–5.0), as opposed to the physiological pH of 7.4 in the blood and normal tissues [48]. The majority of pH-responsive systems use chemical bonds that can be coated or grafted onto the surface of the NPs, which are pH labile [49].

On the other hand, the temperature-responsive system is one system where drug release is controlled by changes in the environment around the tumor's temperature. The development of temperature-responsive DDSs depends on the use of gatekeeping molecules known as thermosensitive polymers that go through a temperature-dependent phase change process. This is especially useful for tumor tissue with elevated inflammatory markers that exhibit a considerable temperature variation and which can be used to good effect by thermosensitive DDSs [50]. The thermo-responsive carriers hold onto the drug load when the body temperature is around 37 °C. However, the drug is released when the surrounding tumor environment reaches the local temperature (40–42 °C) [51].

Furthermore, incorporating pH-sensitive elements like weak acids into thermosensitive polymer such as poly(N-isopropylacrylamide) (PNIPAM) allows for the outline and preparation of numerous pH and temperature triggered DDS. These dual-sensitive nanosystems are among the most-researched ones. Incorporating pH-sensitive elements like weak acids into thermosensitive PNIPAM allows for the design and preparation of numerous pH and temperature-responsive polymers. This optimization of phase transformation by subtle pH change has allowed for the development of tumor pH-sensitive drug release systems [48]. Combining multiple monomers that are responsive to various stimuli depending on their applicability results in polymers that are responsive to multiple

TABLE 8.1
Dual-Stimuli DDS for Theranostics

Triggers	Platform	Method	Functions	Methods	Ref
NIR/GSH	NPs	POEGMA-b-P(CPT-CyOH) (PCC) prodrug loaded with camptothecin	The breakage of the PCC disulfide bonds was mediated by GSH, resulting in the release of the chemotherapeutic agent CPT. NIR was used for fluorescence in vivo imaging.	Chemotherapy, fluorescence imaging	[41]
pH/GSH	NPs	TME-sensitive nanocarriers based on DOX, NIR-emitting carbon dots (C-dots), hollow mesoporous silica nanoparticles (HMSN) and anionic polymer citraconic anhydride-modified polylysine (PLL(cit)). The NIR-emitting C-dots are conjugated on HMSN via disulfide bonds	The NIR emitting C-dots is reduced by using GSH for DOX release. The drug was released using PLL(cit), which prolongs blood circulation time and improves cellular internalization. A promising NIR fluorescent probe called DOX@HMSN-SS-C-dots-PLL(cit) can be used for specialized imaging of tumor tissue.	Chemotherapy, fluorescence imaging	[42]
pH/GSH	Nanoprobe	A new intelligent theranostic probe based on NIR called Cy-1 which can react to GSH and low intrinsic pH at the same time	This probe has been shown to be capable of intermolecularly undergoing a biologically compatible CBT-Cys condensation process to selectively create large nanoaggregates in the TME, resulting in improved tumor formation and retention.	Photoacoustic imaging and PTT	[43]
pH/NIR	NPs	PDA functionalized copper ferrite nanospheres (PDA@CFNs)	PDA@CFNs' great relaxivity allows them to be a viable MRI contrast agent. PTT and pH/NIR-triggered chemotherapy results in highly successful tumor ablation in a mouse model.	Chemotherapy, MRI and PTT	[44]
pH/ROS	NPs	A dextran-drug conjugate (Nap-Dex) and blended Nap-Dex with an acid-sensitive acetal-dextran polymer (Ac-DEX) was used. The PBA-modified anti-inflammatory drug naproxen was loaded	Dual stimuli-responsive nanoparticles reduced the levels of proinflammatory cytokines IL-6 and TNFα by 120 and 6 times, respectively.	Anti-inflammatory	[45]
pH/Thermal	NPs	The NP was created by conjugation of a thiol end-capped HS-PNIPAAm-b-PAA onto Fe3O4@AuNPs which is then loaded with DOX	When stimulated by pH and temperature, theranostic nanomedicine showed superior in vitro drug release behavior. Also, the magnetite property of the manufactured theranostic NPs was discovered to be 19.4 emu per g, qualifying them as an appropriate MRI contrast agent.	Chemotherapy, MRI	[46]
Theram/pH	Nanomicelle	A micellar system loaded with DOX and IR780 was created using a pH- and thermal-responsive polymer (mPEG-PAAV) with an upper critical solution temperature (UCST)	Under NIR laser irradiation, high local temperatures could be produced at the tumor site as a result of the photothermal conversion of the mPEG-PAAV micelle/IR780+DOX (200 nm, 3.82 mV). In addition to speeding up tumor necrosis after NIR laser irradiation, this hyperthermia also caused micelles in the lower pH environment to break down, causing rapid DOX release.	Chemotherapy, photoacoustic imaging	
pH/MMP-13	Nanoprobe	For osteoarthritis (OA) imaging and treatment, a unique class of biocompatible ferritin nanocages (CMFn) that target cartilage and are MMP-13/pH responsive were incorporated with the anti-inflammatory drug hydroxychloroquine (HCQ)	Under acidic pH conditions, CMFn@HCQ nanocages could specifically target the cartilage and deliver HCQ in the OA joint in a sustained manner, prolonging the drug retention time to 14 days to significantly lessen synovial inflammation in the OA joints. They also found that when MMP-13 is overexpressed in the OA microenvironment, CMFn can be triggered for OA imaging.	Imaging-guided precision therapy	[47]

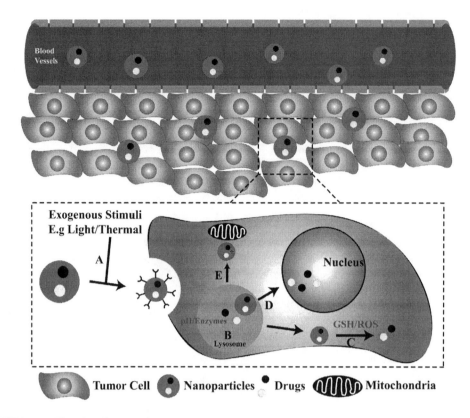

FIGURE 8.1 The stimuli responsive system can promote cellular delivery by (A) charge reversal or ligand emergence in response to external stimuli; (B) disintegration of NPs; (C) release of NPs by breakdown; (D) nuclear release by localization; (E) NP charge reversal and effectively targeting mitochondria [17].

stimuli. Temperature induces a reversible hydrophilic/hydrophobic phase change in PNIPAM-based polymers [52]. In one study, biodegradable and biocompatible chitosan (CS) was combined with a pH responsive polymer and a temperature-responsive co-polymer PNIPAM to create a material that is both temperature and pH responsive. In a preliminary study, the impact of chitosan and N,N-methylenebisacrylamide (MBA)-cross-linking agent concentrations on the response time of CS-g-PNIPAM co-polymer was examined. Based on the findings, the TME responded well to the CS-g-PNIPAM co-polymer. The targeted delivery of the anti-cancer drug oxaliplatin (OXA) at TME temperature and pH was then assessed using the co-polymer that had been loaded with the drug [53, 54].

Hydrogels that are dual responsive to pH and temperature have also recently attracted a lot of attention. A system that is simple to inject, undergoes an in-situ sol-to-gel transition in response to body temperature and promotes drug release at a particular tumoral pH is of great interest. For instance, a dual thermo-pH-responsive plasmonic nanogel (AuNP@Ng) that functions as a DDS was created by cross-linking PNIPAM to CS in the vicinity of a chemical crosslinker. Gold nanoparticles (AuNPs) were additionally added to the nanogel to provide combined photothermal therapy (PTT) and drug delivery. Curcumin was used as an anticancer agent that was loaded into the nanogel to increase the anticancer efficiency. When tested under various temperature and pH conditions, it was discovered that nanogel had a hydrodynamic size of about 167 nm and could release curcumin for up to 72 hours with dual thermo-pH responsive drug release behavior. Studies on NP uptake revealed that cancer cells internalized more NP than non-tumorigenic cells. Curcumin loaded AuNP@Ng/Cur showed dose- and time-dependent drug delivery in in vitro cytotoxicity studies. Additionally, in vitro exposure to a near-infrared (808 nm) laser improved the chemotherapy efficacy of the developed NPs. This research showed that synthetic curcumin-loaded plasmonic nanogels (AuNP@Ng/Cur) can function as stimuli-responsive nanocarriers and have the potential to deliver hydrophobic drugs and deliver photothermal therapy [55]. PNIPAM has been widely used in thermosensible delivery systems and copolymerization to improve tumor targeting. A PNIPAM-based polymer with active targeting and dual responsiveness was created to more effectively accumulate controlled payloads inside of the tumor through cellular internalization. The lactoferrin conjugated PNIPAM-acrylic acid (LF-PNIPAM-co-AA) copolymer was synthesized with Honokiol (HK) as an anticancer drug. The developed HK-loaded LF-PNIPAM-co-AA nanohydrogels had a excellent drug loading capacity of 18.65 wt% and great physical and serum stability. In addition to inducing apoptosis by increasing the expression level of active caspase-3 in breast cancer-bearing mice, in vivo experiments with HK-loaded nanohydrogels showed suppression of the levels of Ki-67 and VEGF-1 expression. Overall, the nanoparticles designed with temperature and pH responsiveness offer a promising anticancer treatment [56].

8.3.2 pH and Redox

The most alluring stimuli are pH and redox, which are both naturally present in some pathological locations and in all tumor cells. Redox-sensitive systems respond only to intrinsic intracellular signals unique to the TME [50]. Redox and pH double-sensitive NPs have been prepared to facilitate NP synthesis in aqueous settings through pH change, to promote in vivo stability of NPs using disulfide crosslinking, to release drug in the endo/lysosomal compartments, to trigger drug release or to boost tumor cell uptake through inversion of surface charges at tumor pH and/or to accomplish quick drug release in the cytoplasm and nucleus. For example, a brand-new, highly selective nanoscale coordination polymer that is pH and redox responsive was created. DOX was added as an anticancer agent, and then a coating of -tocopheryl succinate-polyethylene glycol polymer was applied. The prepared platform showed a high capacity for DOX loading and release that was responsive to pH and redox. Compared to non-targeted drugs, the drug with the AS1411 aptamer showed increased DOX internalization and toxicity toward 4T1 and MCF7 breast cancer cells [57]. Dendrimers are well known for DDSs that are pH and redox sensitive. In a recent study, the redox- and pH-responsive DDS was made by incorporating redox-sensitive disulfide linkage between hydrophilic Heparin (Hep) and poly (amidoamine) (PAMAM) dendrimers G3.5 (P) (P-SS-Hep). The P-SS-Hep system was loaded with letrozole, which could enhance target focus and lessen systemic phenomena. Here, the disulfide bond has a high potential for facilitating redox-sensitive release in the subcellular region as well as strengthening the survivability during blood circulation. Indeed, the P-SS-Hep particle size that could be controlled at around 11 nm had a significant potential for drug release as well as a relatively high effective drug loading carrier. The preliminary findings highlight the potential for chemotherapy using pH and redox-responsive nanocarriers [58].

Hydrogels may experience sol-to-gel transitions in response to environmental stimuli. These transitions may enable controlled drug release and gelation at particular sites. Non-sensitive hydrogels, on the other hand, swell as a result of water absorption rather than reacting to changes in the environment. Hydrogel was used alongside a magnetic property which created a magnetic redox and pH triggered DDS. This DDS was based on tragacanth gum (TG). A copolymer was also prepared containing acrylic acid (AA) monomer which was grafted onto maleic anhydride-functionalized TG macromonomer. The magnetic stimuli-responsive hydrogel for targeted cancer therapy was made using modified-MNPs, folic acid (FA) and simultaneous crosslinking of TG-g-PAA copolymer by using cystamine (Cys) moiety. The release of DOX under pH- and redox-triggered conditions was studied after it was loaded into the manufactured magnetic hydrogel (MH). It was discovered that the developed DDS had higher anti-cancer activity (24%) as a result of its slow drug release behavior, pH and redox responsiveness and synergistic effects of hyperthermia therapy [59]. Another study developed the polypeptide crosslinked methoxy poly(ethylene glycol)-g-poly(aspartic acid)-g-tyrosine

(CPPT) for pH and redox-responsive systems. The dual-sensitive polypeptide-based organic-inorganic hybrid NPs loaded DOX on the hydrophilic segment of micelles as well as the CaP shell and encapsulated curcumin into the hydrophobic core of micelles. At physiological pH, the increased structural integrity effectively prevented the premature leakage of drugs from the NPs, but in acidic and hypoxic environments, the rapid dissolution of the CaP shell and the breakdown of the disulfide crosslinked network promoted the release of both drugs, making it easier to control the release of drugs in response to stimuli. It also demonstrated greater toxicity and improved anti-cancer effectiveness. The dual-sensitive drug co-delivery system based on self-assembling polypeptides may offer a promising platform for effective chemotherapy [60]. The most frequently studied thermoresponsive polymers today are those that exhibit a reversible phase transition with temperature. The polymers with adjustable phase transition temperatures are particularly preferred for practical applications. Pegylated polypeptides is one such polymer that is easily synthesized. The polypeptoids have reversible thermoresponsive qualities and are easily soluble in aqueous solution. They are easily tunable from 25 °C to 60 °C by varying the chemical composition and degree of polymerization. Moreover, the CPs are significantly affected by the oxidation and reduction of thioether groups, offering a second stimulus for fine-tuning phase transition behaviors. The dual-responsive polypeptides have potential for use in a variety of biomedical applications because of their biodegradability, biocompatibility and dual responsiveness [61].

8.3.3 Temperature and Redox

There is a huge difference in the redox states of healthy tissues, and the TME is significant; redox-responsive PNMs are promising candidates for DDSs selective to the TME. This can help to promote redox-based treatment options. Furthermore, when temperature and redox are combined, we get better controlled release and tumor accumulation. This was demonstrated in a study involving N-isopropylacrylamide (NIPAM) hydrogel. NIPAM is temperature sensitive but has low strength, so 2,2,6,6-tetramethylpiperidine-1-oxyl (TEMPO) oxidized cellulose nanofiber (CNF) is used to address this issue. The strength nearly tripled to 360%. The hydrogels are crosslinked with N, N'-bis(acryloyl) cystamine (BACy) to give them redox-responsive properties. DOX and Berberine (BBR) found to have maximum cumulative release rates of 39.56% and 99.50%, respectively, after 60 hours. Also, at temperatures between 30 °C and 40 °C, hydrogel swelling and transparency underwent significant changes. The hydrogel had almost no cytotoxicity, according to cytotoxicity tests. These findings showed the potential of redox and temperature responsive systems as effective therapeutics [62].

When considering thermo/redox sensitive delivery systems, hydrogels have been essential. The versatility of hydrogels as a delivery system for stimuli-responsive drugs is reflected in their capacity to alter their conformation in response to environmental changes. This type of stimuli responsive system also makes use of PNIPAM. In a recent study, a hydrogel based on a dual crosslinked matrix is developed. The primary chemical crosslinking copolymer was made up of thermoresponsive PNIPAM and poly (ionic liquid), while the secondary physical crosslinking component was created by the ionic coordination of iron ions and carboxyl groups in the poly (ionic liquid). The hydrogel was used as an excellent thermo-responsive switch. Besides that, the hydrogels have redox responsiveness, which allows them to be chemically oxidized and reduced reversibly thanks to the trapped iron ions in the network. The ability of iron ions in the gel matrix to crosslink could also be used to adjust the mechanical strength of hydrogels. These stimuli-responsive hydrogels may be used as intelligent materials for stimuli-responsive systems [63]. In another study, injectable hydrogels prepared from diselenide linked methoxy poly (ethylene glycol)-block-poly (ε-caprolactone-co-p-dioxanone) (Bi(PPCD)-Se2) were synthesized for dual responsive delivery. The amphiphilic behavior of the Bi(PPCD)-Se2 solution allowed it to remain in a free-flowing state at low temperatures. However, it spontaneously transformed into a semisolid hydrogel at physiologic temperatures, which is crucial for long-lasting drug release. It was observed that the amount of solvent in the copolymer affects the temperature at which the phase change occurs. Most importantly, the diselenide linkages enabled the thermosensitive hydrogels to resist degradation brought on by oxidation and reduction. Bi(PPCD)-Se2 hydrogel degradation was consequently significantly accelerated. When the system was loaded with DOX, it was discovered to be effective in treating tumors while causing no toxicity [64].

8.3.4 pH and Light Responsive

Combining external and internal stimuli has always been regarded as best due to multiple physiology involved in the process. Light has a distinct advantage over other external stimuli because it can be precisely focused, enable spatiotemporal control and deliver therapeutic agents on demand [65]. Furthermore, it is possible to control a number of aspects of light, including wavelength, duration and intensity. For light-responsive delivery, three different types of light are used: ultraviolet (UV; 10–400 nm), visible (400–750 nm) or near-infrared (750–900 nm) light [66]. The most widely used light is UV because it is more attractive to photoactive chromophores. However, UV has a number of disadvantages, including poor depth penetration and high phototoxicity that harms healthy cells, both of which are noteworthy when considering bioapplication [67]. NIR light, on the other hand, is less potent but penetrates body tissue deeper and is more congruent with cells [68]. UCNPs are used to optically convert NIR to UV into tissues because the NIR light outside body tissue is not easily scattered, while the visible light inside the tissue can effectively control the switch of the ion channel [69]. For instance, in a recent study, transformable poly-o-nitrobenzyl shell and lanthanide-doped UCNP were prepared for NIR/pH responsiveness DDS. The system was loaded with DOX to have an anticancer effect. For the delivery of stimuli-responsive light, the UCNP easily converted NIR to UV visible light. Moreover, the release of drugs at low pH can be triggered by the pH responsiveness brought on by the hydrogen bond and charge interactions between DOX and nanocapsules. In contrast to the synergistic combination of NIR radiation and acidic conditions, visible light irradiation and neutral (pH 7.4) conditions resulted in a cumulative release of DOX that was only 8.35% after 300 minutes [70].

There are three different kinds of light responsive DDS release mechanisms. The first is the morphological change in the nanocarrier brought on by photoisomerization, while the drug component can also be released from the nanocarrier's degradation brought on by light reaction. The rupture of nanocarriers caused by photothermal stimulation also facilitates in the delivery of drugs. For instance, mesoporous PDA (mPDA) shells encapsulating AuNR were used for NIR responsive delivery. The release was facilitated by using the degradation mechanism of NIR. AuNRs@mPDA can produce heat when exposed to a NIR laser, which breaks down the hydrogen bonds and intermolecular forces between the PDA shell and the drug, resulting in a photothermally enhanced drug release [71]. The use of polymer is necessary for pH-responsive delivery systems. The polymer's ability to hold the drug and shield it from physiological action as well as its propensity to degrade at a particular pH aid its use in pH-sensitive delivery systems. Consequently, a coating of polymer is required in addition to the photothermal effect when combining pH/light-responsive drugs. For example, in a study to deliver the hydrophobic anticancer drug bortezomib (BTZ), it was loaded onto mPDAs through acid-sensitive ester bonding. The surface of mPDA was conjugated with the pH-responsive carboxymethyl chitosan (CMCS) polymer, which assisted in the co-loading of DOX and demonstrated modified release behavior in an acidic TME. They provided a controlled drug release when combined with NIR laser irradiation, and the hyperthermia produced by mPDA efficiently annihilated cancer cells [72]. Furthermore, in the AuNRs@mPDA study, polyethyleneimine (PEI) polymer was employed to secure the pore channels of mPDA and for pH responsive delivery. PEI was used effectively as a gating switch to regulate drug release. The anticancer drug DOX, which is cytotoxic to HeLa cells, was carried and released by AuNRs@mPDA@PEI nanocomposite. The combination of phototherapy with NIR radiation and chemotherapy is lethal to HeLa cells. It has enhanced therapeutic outcome and provided a superb pH/NIR response [71]. It has been largely established that light-actuated mesoporous delivery systems can hold onto cargo molecules while they are being stored and transported, but once they reach the intended target, the delivery of their payload is precisely controlled. In light of this, a recent study uses N-doped mesoporous carbon (NMCS) as the core and again coated it with polyethylene glycol (PEG)-PEI polymers for dual pH/NIR delivery. Gemcitabine is used as an active drug for controlled drug release by the system via dual stimuli. The NMCS core converts NIR light to UV, which is absorbed by a photosensitive molecular gate, causing it to cleave and release the drug. In vitro tests demonstrated that when the NMCS-linker-PEG-PEI-GEM hybrid particle is exposed to NIR light, it can produce synergistic therapeutic effects in FADU cells [73]. Despite the benefits, there are some drawbacks to using pH/light systems, such as their mechanical strength, the possibility of chromophore leakage and compatibility issues. Certainly, resolving the issues will result in this being one of the best stimuli responsive delivery systems.

8.3.5 Double pH

Numerous studies have been conducted on pH-responsive polymer-drug conjugates that attach an acid-cleavable linker to the polymer backbone and drug molecule. These polymer-drug conjugates have the ability to self-assemble into 3D structured polymeric NP, which can be induced to break down in the internal pH environment and release the cargo accurately as intended, hence increasing the efficacy of drug delivery. This was demonstrated in a study involving the development of dual-pH responsive charge polymeric NP 2,3-dimethylmaleic-anhydride-poly(ethyleneglycol)–poly-L-lysine-DOX/lapatinib (DMMA-P-DOX/LAP) based on PEG and EPLYS. Polymeric NPs have physicochemical properties that promote sustained circulation in physiological settings, but they ultrasuspectively switch from having a negative to a positive surface charge in a mildly acidic TME, encouraging internalization of cells and profound tumor penetration Then, in reaction to the intracellular environment's significantly increased acidity, LAP and DOX are released simultaneously into the cytoplasm. In vivo results also indicated that combination therapy assisted in the contraction and eradication of MCF-7 tumor models. This pH/pH-responsive DDS offers great potentials for secure and efficient cancer therapy [74]. Currently, the main barrier to treating any disease is multidrug resistance (MDR). The key determinant in overcoming it is efficient intracellular drug delivery and accumulation. In one study, pH/pH-responsive chitosan NPs with DOX (DCCA/DOX-NPs) were created to treat MDR in tumors. The particles were primarily responsive to the tumor's external pH, which was 6.5. This sensitivity resulted in the reversal of surface charge (−6.32 →11.45 mV) caused by the cleavage of the β-carboxylic amide, which significantly improved cellular uptake efficiency. Lower intracellular pH 5.0 prompted DCCA/DOX-NPs to respond further, causing DOX to be released by the cleavage of Schiff base. In vivo antitumor testing confirmed that the particles prompted an increase in apoptosis, with an inhibition rate of 84.94%. Furthermore, the chitosan NPs demonstrated higher biosafety and significantly reduced the adverse effect of DOX [75].

As stated earlier, the pH-responsive system is easily capable of changing surface charge while transitioning from a physiological microenvironment to an extracellular tumor. This property may also contribute to improved cellular uptake and intracellular release. Polymeric micelles are also effective delivery systems for pH-responsive systems. In a recent study, Poly(DEA)-block-Poly(PgMA) (PDPP) block copolymers were created to improve in vitro cellular uptake and anticancer efficacy. mPEG-N3 and a DOX-based prodrug (DOX-hyd-N3) were co-conjugated onto PDPP. It was simple to self-assemble the pH responsive system (mPEG-g-PDPP-g-hyd-DOX) into core-shell polymeric micelles. The surface charge of the prodrug-based micelles could easily change from negative to positive (−6.64→5.35 mV) after being incubated in extracellular microenvironment in tumor under physiologically similar conditions. This was made possible by lowering the pH value from 7.4 in blood circulation to 6.5 in extracellular tumor tissues. Furthermore, it was found that the charge-reversal of micelles significantly increased anticancer efficacy and cellular uptake when cells were incubated with them at pH 6.5 [76]. Thus, it is clear that the dual pH-responsive micelle with charge-reversal platform is a viable system and an effective drug delivery strategy.

8.3.6 OTHER DUAL STIMULI–RESPONSIVE SYSTEMS

There are several other stimuli responsive systems which are at the growing stage but have proven their worth. Temperature and pH in tumor tissues rise as a result of the malignant cells' rapid rate of growth. Intracellular drug release took advantage of the higher redox potential of the cytosol and cell nuclei, which have up to a thousand times more glutathione (GSH) tri-peptide than that of the blood and external environment (0.5e10 versus 2e20 mM GSH). Combining GSH with NIR for therapy can significantly increase drug effectiveness and allow for more accurate targeting. In a study, oxidized glycogen was generated using disulfide linkage for GSH and NIR therapy, and it was enclosed in polypyrrol NP and coated with functional phospholipids. The DOX model drug used in the experiment was Gly-ss-DOX@ppy@Lipid-RGD. Disulfide linkers are cleaved by GSH, which allows Gly-ss-DOX@ppy@Lipid-RGD to release drugs that cause cell nucleus damage. The polypyrrole NPs, meanwhile, can radiate heat energy from tumors while also absorbing near-infrared light. Heat has other benefits than facilitating drug release, such as photothermal tumor treatment. Gly-ss-DOX@ppy@Lipid-RGD is a unique candidate for synergistic chemo photothermal therapy with strong anticancer therapeutic activity and low systematic toxicity, effectively reducing tumor growth, as demonstrated by in vitro and in vivo investigations [77]. When compared to normal cells, cancer cells exhibit a heterogeneous coexistence of abundant intracellular GSH and a significant amount of ROS. Owing to this, they are frequently employed in the treatment of cancer. Researchers created an amphiphilic diblock copolymer prodrug (BCP) (GR-BCP) that is dual responsive to GSH and ROS and has side chains of PEG and poly(methacrylate) linked to camptothecin (CPT). These systems were also contrasted with single response systems made up of GSH (G-BCP) and redox (R-BCP). The result suggested that with a half-inhibitory concentration (IC50) of 6.3 M, which is substantially lower than the values of 17.8 M for G-BCP and 28.9 M for R-BCP, GR-BCP demonstrated the maximum cytotoxicity against HeLa cells. Moreover, after intravenous injection, tumor formation and blood circulation efficiency for in vivo antitumor performance were similar for G-BCP, R-BCP and GR-BCP. The most effective tumor suppression was, however, achieved by GR-BCP with minimal adverse effects. The results imply that both intracellular GSH and ROS dual-responsive BCPs showed a more potent responsive drug release inside tumor cells for increasing the antitumor efficacy than ROS or GSH single-responsive BCPs [78].

Furthermore, a variety of enzymes, such as carboxylesterases, glucuronidase or proteases, that are differentially expressed by healthy and malignant cells can potentially be used as biochemical triggers. For instance, a system was developed utilizing lysosomal enzyme mannose-6-phosphate glycopolypeptide (^{M6P}GP). By combining alkyne end-functionalized M6PGP15 with pH-responsive, biocompatible azide end-functionalized acetal PPO and azide end-functionalized branching poly (ε-caprolactone) PCL, respectively, two amphiphilic M6P block copolymers were created. They were discovered to deconstruct at acidic pH or when esterase was present, but they were stable at physiological pH. These M6PGP-based micellar NPs have the potential to be used for receptor-mediated lysosomal cargo delivery because they can specifically target lysosomes in breast cancer cells like MCF-7 and MDA-MB-231 [79]. Matrix metalloproteinase (MMP) enzymes are frequently targeted for the treatment of diseases. They are simple to target and can be used for stimuli responsive delivery systems. A recently developed MMP-cleavable linker poly(l-methionine-block-l-lysine)-PLGLAG-PEG (MLMP) was used to create a protease-activatable cell-penetrating peptide that had a ROS-responsive methionine and a cell permeable lysine chain, and a DOX was used as the core drug and loaded into nanomicelles. The MLMP displayed MMP-sensitive cleavage and DOX release caused by ROS. Also, researchers verified effective DOX transport into cancer cells and in vitro development of the apoptotic potential. Dual stimuli-MLMP has excellent potential as a drug delivery platform for anticancer medications because it also shown amazing tumor inhibitory capability while being less toxic than free DOX [80]. Thus, it is obvious that dual stimuli–responsive systems are still in their infancy yet have already made a lot of good progress. They can be employed for further human clinical studies and undoubtedly boost efficiency and targeting and are non-toxic.

8.4 MULTI-STIMULI–RESPONSE DDSS FOR CANCER THERAPY AND TREATMENT

There are only two stimuli-based reactions used for drug administration in a dual responsive system. Despite having conquered two hurdles, a dual stimuli-responsive drug delivery vehicle's therapeutic impact is still constrained by numerous challenges [81]. For example, the therapeutic efficacy of a pH-responsive nanocarrier is constrained because the extracellular pH level is marginally out of the ordinary, which could result in a sluggish reaction [82]. By combining many stimuli into a single nanocarrier, a series of physiological challenges could be solved one by one, thereby facilitating the best therapeutic results. Responses in a multi-responsive system either show up concurrently in the same place or in a pattern across different environment and/or compartments. These systems in nano form offer extraordinary drug release control, leading to better anti-cancer potency in vitro and/or in vivo. The creation of carriers that can react to several stimuli at once can compensate for a single external and internal response's deficit, delivering comprehensive and superior illness imaging and treatment. The drug can be delivered considerably more precisely and with better stability when we combine the triple responsive system of pH, temperature and redox. This was demonstrated in a study in which polymer nanogels P(NIPAM-AA) were synthesized for drug delivery by utilizing mesoporous NPs and P (N-isopropylacrylamide-acrylic acid) hybrid nanogel for thermal/pH/redox-triple sensitive drug delivery. The cross-linker was N, N'-bis(acryloyl)cystamine (BAC), and the active compound was DOX. The resulting system was found to have exceptional stability and exhibit improved drug delivery in conditions that simulated solid tumors and endolysosomal compartments while displaying strong anticancer bioactivity. Moreover, the system was able to disintegrate into tiny drug-loaded mesoporous NPs in the decreasing microenvironment of tumor tissue, which are more readily absorbed by tumor cells, enabling multistage anticancer drug delivery [83]. In another study, thermosensitive NIPAM, pH-responsive AA and N,N'-dimethylamino ethyl methacrylate (DMAMEA) were used to create triple-responsive poly(N-isopropylacrylamide-N,N'-dimethylaminoethyl methacrylate-acrylic acid) (PNDA) nanogels. The triple-responsive PNDA nanogels outperformed PND nanogels in terms of cumulative release efficiency, thermosensitivity and drug encapsulation efficiency (encapsulation efficiency: 89%). In vitro tests revealed that the PNDA nanogels with DOX were highly cytotoxic to A549 cancer cells while having few adverse effects [84]. This study was furthered by synthesizing PND-BAC and PND-MBA nanogels with pH/redox/thermal responsive characteristics. In contrast to PND-MBA nanogels, they demonstrated a better encapsulation efficiency and non-redox reactivity. DOX-loaded PND-BAC nanogel demonstrated a good TME triple-responsive DOX release property in vitro. PND-BAC might be a good candidate for targeted delivery of drugs [85]. The target-specific delivery system can be provided by including enzymes in the pH- and temperature-responsive system. This was the subject of a study using triple-responsive protamine/PAA-b-PNIPAAm nanogels to deliver drugs and photosensitizers. Loading of DOX and rose bengal, a photosensitizer was found to be effective into nanogels. The pH, temperature and enzyme (trypsin) responses of the nanogel systems were controlled by changes in helix conformation, disaggregation of PAA-b-PNIPAAm and enzymatic hydrolysis of protamine. The cationic feature of the nanogels allowed for easy passage through the cell membrane and improved accumulation in tumor MCF-7/ADR cells. Due to nanogel swelling and surface charge conversion, the acidic tumor microenvironment may facilitate the intracellular release of drug (from negative to positive) [86].

NIR can also be combined with pH and redox for an improved drug effect. A drug delivery platform was developed comprising both organosilica and copper sulfide NPs (DOX-CuS@PMO) that were crosslinked by thiol bonds. In both glioblastoma cell lines and mouse model, mild laser irradiation caused moderate hyperthermia in DOX-CuS@PMO, which improved its cellular internalization. DOX release was then induced by higher GSH levels as well as acidic pH in tumor cells. The multifaceted tristimuli DDS was dependent on the TME's acidic pH and increased GSH levels, among other factors. Furthermore, it used external laser irradiation to control the release of DOX more precisely, which helped to maximize tumor selectivity and reduce systemic toxicity [87]. Another study created a multifunctional theranostic nanoplatform that worked with magnetic and thermal imaging to deliver photothermal-chemotherapy combo therapy. After being coated with hyaluronic acid (HA), Fe3O4@polydopamine NPs containing redox-sensitive disulfide linkers were then placed with the chemotherapeutic drug DOX and it was named FPCH-DOX NPs. GSH concentration, UV exposure and pH all have the potential to promote drug release. The findings suggested that compared to CD44-negative NIH 3T3 normal cell lines, CD44-positive HeLa cell lines exhibit better biocompatibility and good cellular uptake of these nanoplatforms. In vitro tests show that the combined therapy of FPCH-DOX NPs reduces cell viability to 16.2%, less than that of separate chemotherapy (55.3%) or PTT (52.1%). The accumulation of FPCH-DOX NPs in tumors appears to enhance MRI contrast, according to in vivo MRI, and in vivo thermal imaging has confirmed their regional photothermal transformation action in tumor tissues. Importantly, photothermal-chemotherapy combination therapy with FPCH-DOX NPs exhibits significant anti-tumor activity [88].

Another type of multiple stimuli is light, pH and temperature responsive. Temperature and light sensitivity make a good team for a powerful domino trigger. The NPs typically convert light to thermal and form temperature-responsive polymers. The NPs first transform light energy into heat energy that is then focused by NIR light. The temperature-sensitive polymers then take up heat and go through volume conversion or decomposition to release drugs [89]. A hydrophobic and hydrophilic product can be delivered and released selectively using a temperature, pH and UV light tri-responsive nanogel. These nanogels were made of an o-nitrobenzyl (ONB) linkage that was UV light cleavable and thermally and pH-responsive poly DMAEMA (PDMAEMA). These dual-loaded nanogels showed redox-sensitive release behavior for the hydrophilic cargoes, while the hydrophobic cargoes showed pH, redox, UV light, temperature and sensitive released behavior. The dual-loaded nanogels technique has many potential uses in the treatment of diseases. The multi stimulus-responsive nanogels' diversity and complexity offer a useful method for precisely controlling the release of numerous payloads under varied circumstances [90].

There are studies that use more than three stimuli to determine how well a drug works. Both their development and their behavior are more complicated. An amphiphilic diblock copolymer called poly(2-methacryloyloxyethyl ferrocenecarboxylate)-(5-propargylether-2-nitrobenzyl bromoisobutyrate) PDMAEMA (PMAEFc-ONB-PDMAEMA) was self-assembled in water to create shell crosslinked quintuple-stimuli-responsive nanocontainers that could react to light, pH, temperature and redox species. The nanocontainer was composed of the following components: a crosslinked PDMAEMA shell that might improve the stability of the nanocarriers in harsh settings and cause their dissolution in intracellular reductive circumstances. When the pH is acidic, it can also swell and shrink over the lower critical solution temperature. A photocleavage linker that links the hydrophilic shell with hydrophobic core using o-nitrobenzyl methyl esters was used primarily because it was thought that the light stimulus could be easily and precisely controlled. This system's ferrocenyl-based hydrophobic core provided redox activity and chemical stability. Multiple stimuli have a higher chance of successfully initiating the release of a hydrophobic drug model than a single stimulus. More crucially, the crosslinked shell's disulfide bond enabled an easy regulated release of visitor molecules in addition to providing NPs with a reduction stimulus. Moreover, it was discovered that in harsh environments, the shell crosslinked structure can effectively prevent the drug's burst release. There was also a difference in light-stimulated behavior between non-crosslinked and crosslinked micelles [91].

8.5 MULTI-STIMULI–RESPONSIVE SYSTEM FOR TUMOR IMMUNOTHERAPY

Tumor immunotherapy is quickly becoming a viable cancer therapeutic technique because of the anti-tumor immune response (ACIR), which reflects immune cell detection, activation and proliferation, effector T cell migration and penetration into tumor tissue, ultimately targeting, combating and finally eradicating cancer cells without causing the death of healthy cells in tumor tissue [92]. T-helper cells (CD4+) and cytotoxic T cells (CD8+) recognize antigenic epitopes that promote the growth of cancers to begin immunologic effects [93]. The human immune system may be resilient, but the TME will still manage to elude it through a variety of forms of reprogramming. For instance, it may activate immune checkpoint proteins, polarize macrophages to become tumor-genic macrophages, highly express Indoleamine 2.3-Dioxygenase (IDO), neutralize cytotoxic CD8+ T cells, reduce T cell activity and other mechanisms [94]. The primary categories of immunotherapy techniques are blocking tumor immune escape and promoting ACIR by enhancing tumor immunogenicity or administering immune-stimuli drugs to directly stimulate immune system activity, and so on. Immunotherapeutic drugs can be delivered on the multi-stimuli–responsive DDS platform to accomplish targeted delivery with few negative autoreactive immune responses. When a foreign material stimulates and activates the human body, the immune system responds. However, immune-checkpoint proteins, including CTLA-4, LAG3 and programmed cell death-1 (PD-1), can create immunosuppressive cytokines, preventing the immune system from becoming overly active. Because T cells and B cells express the PD-1 protein, tumor cellular ligands can bind to it and prevent immune cells from recognizing and attacking them. This process is known as immunological escape [94]. Dual redox and pH-responsive NPs provide the release of chemotherapeutic and immunoadjuvant drug in the GSH-rich and low-pH TME through two different methods. This mechanism causes immunoadjuvant to be more specifically localized in the tumor tissue, improving cellular immunity. This has been demonstrated in a study that used the GSH- and pH-responsive nanostructure (ASPN) to release OXA at lower pH and the reduction-responsive NP (NLG919) to inactivate IDO-1, an enzyme that is essential for the development of the immunosuppressive TME. Their findings showed a 2.5-fold increase in CRT positive cells when compared to free OXA. Moreover, ASPN treated at pH 6.5 showed 2.1-fold more CRT exposure than at pH 7.4. IDO-1 was deactivated by the GSH-responsive release of NLG919, which also helped to reverse the immunosuppressive TME [95]. In a related study, DOX-loaded metal phenolic manganese dioxide NPs (Fe 3+/MnO2 NPs) were used to induce immunogenic cell death (ICD) in B16F10 cells. To release DOX into the tumor microenvironment, phenolic NPs (MDP NPs) degrade MnO_2 NPs based on pH and GSH. The ICD induction behavior of MDP NPs was studied in B16F10 tumor-bearing mice. CRT expression on the tumor tissue was investigated when mice carrying B16F10 tumors were intravenously injected with MDP NPs and PD-1. The release of DOX from this nanosystem in response to pH and GSH strongly activates ICD, according to the enhanced expression of CRT in groups that had received PD-1 treatment in addition to MDP NPs [96]. To provide dual stimuli triggered release of chemotherapeutic or immunoadjuvant drug for improved immunomodulation, another pH- and enzyme-responsive NP has been designed. A new hyaluronic acid and DOX linked system was created using a dual stimuli responsive system, and it was further modified with 3-diethylaminopropyl isothiocyanate and MMP-2 to create formulations for NPs. Celecoxib and DOX are released from the NP, designated DHPDB, using a enzymatic cleavage and pH-dependent chemistry. Their findings suggested that the dual MMP-2 and pH responsive nano system enhanced the ICD because DHPDB nanoparticles generated the greatest exposure of calreticulin and release of HMGB1 when compared to free DOX and celecoxib [97]. Another study developed docetaxel (DTX) releasing nanoconstructs that are pH/cathepsin B responsive (HRN). Lysosomal cathepsin and the extracellular matrix's acidity hasten DTX's cleavage, which heightens the anticancer effect. By stimulating cytotoxic T lymphocytes (CD8+), the nanosystem proved very successful in suppressing tumors that expressed the B16 ovalbumin (B16OVA) [98]. A DOX and palladium nanoparticles (Pd NPs) were combined with trigycerol monostearate in a different study (TGM) Due to TGM and Pd NPs, respectively, the synthesized NP (Pd-DOX@TGM) possesses affinity for photothermal treatment and MMP-2. A NIR laser (808 nm, 0.5 W/cm2, 5 min) was used after the Pd-DOX@TGM was incubated with the CT26 cell line. The enhanced expression of CRT, as well as the release of extracellular ATP, were likewise suggested by the immunohistochemical labelling of tumor sections treated with Pd-DOX@TGM + NIR laser. They came to the conclusion that the NPs dual stimuli responsiveness strongly promotes ICD in colorectal cancer, producing the strongest anticancer effects [99].

The importance of multi-stimuli triggered NPs for strong cellular immunity has been emphasized in order to elicit greater ICD and create synergistic effects. For instance, a system made of paclitaxel (PTX), indoximod (IND) and chlorine 6 was designed to be cell membrane-masking NPs (I-P@NPs@M) (Ce6). The created I-P@NPs@M was capable of responding to NIR as well as GSH and ROS in order to release PTX and IND. In 4T1 tumor-bearing mice, the created I-P@NPs@M NPs was injected intratumorally and then laser-irradiated. The researchers discovered that the photothermal therapy with the release of PTX and IND via GSH and ROS was particularly effective at causing the breast cancer to regress, with I-P@NPs@M + laser having the highest ratio of cytotoxic T lymphocytes to regulatory T cells [100].

8.6 CONCLUSION

Dual and multi-responsive systems with planned targeted drug delivery have undergone several years of rapid advancement. These systems are infinite in their design and chemical make-up. These technologies undoubtedly addressed the issues with present nanosystems, such as intracellular drug administration, system shielding and deshielding, in vivo stability, unparalleled self-regulated controlled release and many more. The ability to change drug release through the use of intelligent polymers could enable the transition from passive to active stimuli-controlled drug delivery. The therapeutic success is finally improved through targeted site-specific drug administration, which also lowers the toxic and unfavorable effects.

Despite numerous developments, it is still difficult to adapt stimuli-responsive drug delivery devices for therapeutic use. Furthermore, multi-stimuli–responsive DDSs were created for theranostics and multimodal imaging, which resolves many of the problems with conventional or individual imaging models, such as inadequate resolution or sensitivity [40]. Regardless of the fact that there have been numerous successes with smart DDSs so far, proof-of-concept studies are still mostly what are published due to the relative youth of nanotechnologies. ThermoDox (temperature-responsive), AuroShell (temperature-sensitive), Opaxio (enzyme-responsive) and many other are still in the clinical testing stage, while Visudyne (UV light-responsive) was the only smart DDS to receive FDA approval [101]. Therefore, there are still a lot of problems that need to be handled when transitioning from lab to bedside applications for the associated scientific, inventive and technical concerns, such as low pH, changes in the expression and distribution of stimuli in TME, elevated reductant concentration, or up-regulation of enzymes in healthy cells or tissues may generate off-target effects, variances in patient illness and cellular heterogeneity of tumor cells, among many others. Compared to single-stimulus responsive NPs, multi stimuli-DDSs can significantly lessen the effects of individual variation and tumor heterogeneity with very few off-target tissues, but their use is constrained by their complex designs, which make quality control and industrial production challenging on a massive scale. Designing logical multi-stimuli DDSs with straightforward preparation techniques is urgent. To better the accuracy of agent release, lessen the effects of individual differences and adapt to complicated clinical applications, more precise stimuli and innovative stimulus combinations should be tested. Numerous multi-responsive polymer systems act merely as proof for in vitro concept research. Therefore, a thorough evaluation of the in vivo performance of multi-stimuli responsive devices is required. We anticipate that novel responsive systems will continue to be developed, with highly developed and adaptable drug carriers being made available in the future.

NOTE

1 Enhanced permeability and retention (EPR), mononuclear phagocytic system (MPS), tumor microenvironment (TME), nanoparticles (NPs), drug delivery system (DDS), fluorescence imaging (FLI), magnetic resonance imagining (MRI) and photoacoustic imaging (PAI), near infrared region (NIR), magnetic nanoparticles (MNPs), obatoclax (OBX), up-converting NPs (UCNP), positron emission tomography (PET), polydopamine (PDA), gadolinium (Gd), doxorubicin (DOX), indocyanine green (ICG), poly(N-isopropylacrylamide) (PNIPAM), N,N-methylenebisacrylamide (MBA), chitosan (CS), gold nanoparticles (AuNP), photothermal therapy (PTT), Heparin (Hep), poly (amidoamine) (PAMAM), acrylic acid (AA), poly(ethylene glycol)-g-poly(aspartic acid)-g-tyrosine (CPPT), gold nanorods (AuNRs), near-infrared (NIR), mesoporous PDA (mPDA), N-doped mesoporous carbon (NMCS), polyethylene glycol (PEG), polyethyleneimine (PEI), glutathione (GSH), osteoarthritis (OA), matrix metalloproteinases (MMPs), reactive oxygen species (ROS), oxaliplatin (OXA), immunogenic cell death (ICD), anti-tumor immune response (ACIR).

REFERENCES

[1] V. Kulothungan, K. Sathishkumar, S. Leburu, T. Ramamoorthy, S. Stephen, D. Basavarajappa, N. Tomy, R. Mohan, G.R. Menon, P. Mathur, Burden of cancers in India—estimates of cancer crude incidence, YLLs, YLDs and DALYs for 2021 and 2025 based on national cancer registry program, BMC Cancer. 22 (2022) 527. https://doi.org/10.1186/s12885-022-09578-1.

[2] R.L. Siegel, K.D. Miller, H.E. Fuchs, A. Jemal, Cancer statistics, CA. Cancer J. Clin. 72 (2022) 7–33. https://doi.org/10.3322/caac.21708.

[3] K.D. Miller, L. Nogueira, T. Devasia, A.B. Mariotto, K.R. Yabroff, A. Jemal, J. Kramer, R.L. Siegel, Cancer treatment and survivorship statistics, CA. Cancer J. Clin. 72 (2022) 409–436. https://doi.org/10.3322/caac.21731.

[4] Z. Yang, J. Shi, J. Xie, Y. Wang, J. Sun, T. Liu, Y. Zhao, X. Zhao, X. Wang, Y. Ma, V. Malkoc, C. Chiang, W. Deng, Y. Chen, Y. Fu, K.J. Kwak, Y. Fan, C. Kang, C. Yin, J. Rhee, P. Bertani, J. Otero, W. Lu, K. Yun, A.S. Lee, W. Jiang, L. Teng, B.Y.S. Kim, L.J. Lee, Large-scale generation of functional mRNA-encapsulating exosomes via cellular nanoporation, Nat. Biomed. Eng. 4 (2020) 69–83. https://doi.org/10.1038/s41551-019-0485-1.

[5] A.S. Semkina, M.A. Abakumov, A.S. Skorikov, T.O. Abakumova, P.A. Melnikov, N.F. Grinenko, S.A. Cherepanov, D.A. Vishnevskiy, V.A. Naumenko, K.P. Ionova, A.G. Majouga, V.P. Chekhonin, Multimodal doxorubicin loaded magnetic nanoparticles for VEGF targeted theranostics of breast cancer, Nanomedicine Nanotechnology, Biol. Med. 14 (2018) 1733–1742. https://doi.org/10.1016/j.nano.2018.04.019.

[6] V.J. Yao, S. D'Angelo, K.S. Butler, C. Theron, T.L. Smith, S. Marchiò, J.G. Gelovani, R.L. Sidman, A.S. Dobroff, C.J. Brinker, A.R.M. Bradbury, W. Arap, R. Pasqualini, Ligand-targeted theranostic nanomedicines against cancer, J. Control. Release. 240 (2016) 267–286. https://doi.org/10.1016/j.jconrel.2016.01.002.

[7] G. Feng, B. Liu, Multifunctional AIEgens for future theranostics, Small. 12 (2016) 6528–6535. https://doi.org/10.1002/smll.201601637.

[8] S. Li, Q. Zou, R. Xing, T. Govindaraju, R. Fakhrullin, X. Yan, Peptide-modulated self-assembly as a versatile strategy for tumor supramolecular nanotheranostics, Theranostics. 9 (2019) 3249–3261. https://doi.org/10.7150/thno.31814.

[9] L. Li, J. Wang, H. Kong, Y. Zeng, G. Liu, Functional biomimetic nanoparticles for drug delivery and theranostic applications in cancer treatment, Sci. Technol. Adv. Mater. 19 (2018) 771–790. https://doi.org/10.1080/14686996.2018.1528850.

[10] W. Zhang, M. Liu, A. Liu, G. Zhai, Advances in functionalized mesoporous silica nanoparticles for tumor targeted drug delivery and theranostics, Curr. Pharm. Des. 23 (2017). https://doi.org/10.2174/1381612822666161025153619.

[11] M. Zhou, K. Wen, Y. Bi, H. Lu, J. Chen, Y. Hu, Z. Chai, The application of stimuli-responsive nanocarriers for targeted drug delivery, Curr. Top. Med. Chem. 17 (2017). https://doi.org/10.2174/1568026617666170224121008.

[12] C. Alvarez-Lorenzo, A. Concheiro, Smart drug delivery systems: From fundamentals to the clinic, Chem. Commun. 50 (2014) 7743–7765. https://doi.org/10.1039/C4CC01429D.

[13] H. Mekaru, J. Lu, F. Tamanoi, Development of mesoporous silica-based nanoparticles with controlled release capability for cancer therapy, Adv. Drug Deliv. Rev. 95 (2015) 40–49. https://doi.org/10.1016/j.addr.2015.09.009.
[14] V. Biju, Chemical modifications and bioconjugate reactions of nanomaterials for sensing, imaging, drug delivery and therapy, Chem. Soc. Rev. 43 (2014) 744–764. https://doi.org/10.1039/C3CS60273G.
[15] S. Mura, J. Nicolas, P. Couvreur, Stimuli-responsive nanocarriers for drug delivery, Nat. Mater. 12 (2013) 991–1003. https://doi.org/10.1038/nmat3776.
[16] P. Huang, G. Wang, Y. Su, Y. Zhou, W. Huang, R. Zhang, D. Yan, Stimuli-responsive nanodrug self-assembled from amphiphilic drug-inhibitor conjugate for overcoming multidrug resistance in cancer treatment, Theranostics. 9 (2019) 5755–5768. https://doi.org/10.7150/thno.36163.
[17] R. Jia, L. Teng, L. Gao, T. Su, L. Fu, Z. Qiu, Y. Bi, Advances in multiple stimuli-responsive drug-delivery systems for cancer therapy, Int. J. Nanomedicine. 16 (2021) 1525–1551. https://doi.org/10.2147/IJN.S293427.
[18] M. Liu, H. Du, W. Zhang, G. Zhai, Internal stimuli-responsive nanocarriers for drug delivery: Design strategies and applications, Mater. Sci. Eng. C. 71 (2017) 1267–1280. https://doi.org/10.1016/j.msec.2016.11.030.
[19] H. He, L. Sun, J. Ye, E. Liu, S. Chen, Q. Liang, M.C. Shin, V.C. Yang, Enzyme-triggered, cell penetrating peptide-mediated delivery of anti-tumor agents, J. Control. Release. 240 (2016) 67–76. https://doi.org/10.1016/j.jconrel.2015.10.040.
[20] K.T. Vo, K.K. Matthay, S.G. DuBois, Targeted antiangiogenic agents in combination with cytotoxic chemotherapy in preclinical and clinical studies in sarcoma, Clin. Sarcoma Res. 6 (2016) 9. https://doi.org/10.1186/s13569-016-0049-z.
[21] A. Jhaveri, P. Deshpande, V. Torchilin, Stimuli-sensitive nanopreparations for combination cancer therapy, J. Control. Release. 190 (2014) 352–370. https://doi.org/10.1016/j.jconrel.2014.05.002.
[22] B. Chen, W. Dai, B. He, H. Zhang, X. Wang, Y. Wang, Q. Zhang, Current multistage drug delivery systems based on the tumor microenvironment, Theranostics. 7 (2017) 538–558. https://doi.org/10.7150/thno.16684.
[23] Y. Zhao, C. Shi, X. Yang, B. Shen, Y. Sun, Y. Chen, X. Xu, H. Sun, K. Yu, B. Yang, Q. Lin, pH- and Temperature-sensitive hydrogel nanoparticles with dual photoluminescence for bioprobes, ACS Nano. 10 (2016) 5856–5863. https://doi.org/10.1021/acsnano.6b00770.
[24] L. Zhang, D. Sheng, D. Wang, Y. Yao, K. Yang, Z. Wang, L. Deng, Y. Chen, bioinspired multifunctional melanin-based nanoliposome for photoacoustic/magnetic resonance imaging-guided efficient photothermal ablation of cancer, Theranostics. 8 (2018) 1591–1606. https://doi.org/10.7150/thno.22430.
[25] T. Feng, L. Zhou, Z. Wang, C. Li, Y. Zhang, J. Lin, D. Lu, P. Huang, Dual-stimuli responsive nanotheranostics for mild hyperthermia enhanced inhibition of Wnt/β-catenin signaling, Biomaterials. 232 (2020) 119709. https://doi.org/10.1016/j.biomaterials.2019.119709.
[26] S. Wu, H.-J. Butt, Near-infrared-sensitive materials based on upconverting nanoparticles, Adv. Mater. 28 (2016) 1208–1226. https://doi.org/10.1002/adma.201502843.
[27] H. Zhang, X. Jiao, Q. Chen, Y. Ji, X. Zhang, X. Zhu, Z. Zhang, A multi-functional nanoplatform for tumor synergistic phototherapy, Nanotechnology. 27 (2016) 085104. https://doi.org/10.1088/0957-4484/27/8/085104.
[28] X. Dong, W. Yin, X. Zhang, S. Zhu, X. He, J. Yu, J. Xie, Z. Guo, L. Yan, X. Liu, Q. Wang, Z. Gu, Y. Zhao, Intelligent MoS 2 nanotheranostic for targeted and enzyme-/pH-/NIR-responsive drug delivery to overcome cancer chemotherapy resistance guided by PET imaging, ACS Appl. Mater. Interfaces. 10 (2018) 4271–4284. https://doi.org/10.1021/acsami.7b17506.
[29] J. Yu, W. Yin, T. Peng, Y. Chang, Y. Zu, J. Li, X. He, X. Ma, Z. Gu, Y. Zhao, Biodistribution, excretion, and toxicity of polyethyleneimine modified NaYF 4 :Yb,Er upconversion nanoparticles in mice via different administration routes, Nanoscale. 9 (2017) 4497–4507. https://doi.org/10.1039/C7NR00078B.
[30] X. Zhao, C.-X. Yang, L.-G. Chen, X.-P. Yan, Dual-stimuli responsive and reversibly activatable theranostic nanoprobe for precision tumor-targeting and fluorescence-guided photothermal therapy, Nat. Commun. 8 (2017) 14998. https://doi.org/10.1038/ncomms14998.
[31] Y. Liu, S. Wang, Y. Ma, J. Lin, H.-Y. Wang, Y. Gu, X. Chen, P. Huang, Ratiometric photoacoustic molecular imaging for methylmercury detection in living subjects, Adv. Mater. 29 (2017). https://doi.org/10.1002/adma.201606129.
[32] S. Wang, J. Lin, T. Wang, X. Chen, P. Huang, Recent advances in photoacoustic imaging for deep-tissue biomedical applications, Theranostics. 6 (2016) 2394–2413. https://doi.org/10.7150/thno.16715.

[33] L. Cheng, A. Kamkaew, H. Sun, D. Jiang, H.F. Valdovinos, H. Gong, C.G. England, S. Goel, T.E. Barnhart, W. Cai, Dual-modality positron emission tomography/optical image-guided photodynamic cancer therapy with chlorin e6-containing nanomicelles, ACS Nano. 10 (2016) 7721–7730. https://doi.org/10.1021/acsnano.6b03074.

[34] S. Wang, J. Lin, Z. Wang, Z. Zhou, R. Bai, N. Lu, Y. Liu, X. Fu, O. Jacobson, W. Fan, J. Qu, S. Chen, T. Wang, P. Huang, X. Chen, Core—satellite polydopamine—gadolinium-metallofullerene nanotheranostics for multimodal imaging guided combination cancer therapy, Adv. Mater. 29 (2017) 1701013. https://doi.org/10.1002/adma.201701013.

[35] T. He, J. He, M.R. Younis, N.T. Blum, S. Lei, Y. Zhang, P. Huang, J. Lin, Dual-stimuli-responsive nanotheranostics for dual-targeting photothermal-enhanced chemotherapy of tumor, ACS Appl. Mater. Interfaces. 13 (2021) 22204–22212. https://doi.org/10.1021/acsami.1c03211.

[36] S. Chen, M. Chen, J. Yang, X. Zeng, Y. Zhou, S. Yang, R. Yang, Q. Yuan, J. Zheng, Design and engineering of hypoxia and acidic pH dual-stimuli-responsive intelligent fluorescent nanoprobe for precise tumor imaging, Small. 17 (2021) 2100243. https://doi.org/10.1002/smll.202100243.

[37] X. Liu, X. Gong, J. Yuan, X. Fan, X. Zhang, T. Ren, S. Yang, R. Yang, L. Yuan, X.-B. Zhang, Dual-stimulus responsive near-infrared reversible ratiometric fluorescent and photoacoustic probe for in vivo tumor imaging, Anal. Chem. 93 (2021) 5420–5429. https://doi.org/10.1021/acs.analchem.0c04804.

[38] W. Huo, K. Miki, D. Tokunaga, H. Mu, M. Oe, H. Harada, K. Ohe, Dual-stimuli-responsive probes for detection of ovarian cancer cells and quantification of both pH and enzyme activity, Bull. Chem. Soc. Jpn. 94 (2021) 2068–2075. https://doi.org/10.1246/bcsj.20210168.

[39] A. Raza, T. Rasheed, F. Nabeel, U. Hayat, M. Bilal, H. Iqbal, Endogenous and exogenous stimuli-responsive drug delivery systems for programmed site-specific release, Molecules. 24 (2019) 1117. https://doi.org/10.3390/molecules24061117.

[40] H.S. El-Sawy, A.M. Al-Abd, T.A. Ahmed, K.M. El-Say, V.P. Torchilin, Stimuli-responsive nano-architecture drug-delivery systems to solid tumor micromilieu: Past, present, and future perspectives, ACS Nano. 12 (2018) 10636–10664. https://doi.org/10.1021/acsnano.8b06104.

[41] X. Zhang, D. Jia, Y. Wang, F. Wen, X. Zhang, Engineering glutathione-responsive near-infrared polymeric prodrug system for fluorescence imaging in tumor therapy, Colloids Surfaces B Biointerfaces. 206 (2021) 111966. https://doi.org/10.1016/j.colsurfb.2021.111966.

[42] Z. Chen, T. Liao, L. Wan, Y. Kuang, C. Liu, J. Duan, X. Xu, Z. Xu, B. Jiang, C. Li, Dual-stimuli responsive near-infrared emissive carbon dots/hollow mesoporous silica-based integrated theranostics platform for real-time visualized drug delivery, Nano Res. 14 (2021) 4264–4273. https://doi.org/10.1007/s12274-021-3624-4.

[43] A. Wang, Q. Mao, M. Zhao, S. Ye, J. Fang, C. Cui, Y. Zhao, Y. Zhang, Y. Zhang, F. Zhou, H. Shi, pH/reduction dual stimuli-triggered self-assembly of NIR theranostic probes for enhanced dual-modal imaging and photothermal therapy of tumors, Anal. Chem. 92 (2020) 16113–16121. https://doi.org/10.1021/acs.analchem.0c03800.

[44] Y. Du, D. Wang, S. Wang, W. Li, J. Suo, A new pH/NIR responsive theranostic agent for magnetic resonance imaging guided synergistic therapy, RSC Adv. 11 (2021) 6472–6476. https://doi.org/10.1039/D0RA09538A.

[45] S. Lee, A. Stubelius, N. Hamelmann, V. Tran, A. Almutairi, Inflammation-responsive drug-conjugated dextran nanoparticles enhance anti-inflammatory drug efficacy, ACS Appl. Mater. Interfaces. 10 (2018) 40378–40387. https://doi.org/10.1021/acsami.8b08254.

[46] B. Massoumi, A. Farnudiyan-Habibi, H. Derakhshankhah, H. Samadian, R. Jahanban-Esfahlan, M. Jaymand, A novel multi-stimuli-responsive theranostic nanomedicine based on Fe_3O_4@Au nanoparticles against cancer, Drug Dev. Ind. Pharm. 46 (2020) 1832–1843. https://doi.org/10.1080/03639045.2020.1821052.

[47] H. Chen, Z. Qin, J. Zhao, Y. He, E. Ren, Y. Zhu, G. Liu, C. Mao, L. Zheng, Cartilage-targeting and dual MMP-13/pH responsive theranostic nanoprobes for osteoarthritis imaging and precision therapy, Biomaterials. 225 (2019) 119520. https://doi.org/10.1016/j.biomaterials.2019.119520.

[48] R. Cheng, F. Meng, C. Deng, H.-A. Klok, Z. Zhong, Dual and multi-stimuli responsive polymeric nanoparticles for programmed site-specific drug delivery, Biomaterials. 34 (2013) 3647–3657. https://doi.org/10.1016/j.biomaterials.2013.01.084.

[49] T. Chen, W. Wu, H. Xiao, Y. Chen, M. Chen, J. Li, Intelligent drug delivery system based on mesoporous silica nanoparticles coated with an ultra-pH-sensitive gatekeeper and poly(ethylene glycol), ACS Macro Lett. 5 (2016) 55–58. https://doi.org/10.1021/acsmacrolett.5b00765.

[50] T. Moodley, M. Singh, Current stimuli-responsive mesoporous silica nanoparticles for cancer therapy, Pharmaceutics. 13 (2021) 71. https://doi.org/10.3390/pharmaceutics13010071.

[51] M.A. Rahim, N. Jan, S. Khan, H. Shah, A. Madni, A. Khan, A. Jabar, S. Khan, A. Elhissi, Z. Hussain, H.C. Aziz, M. Sohail, M. Khan, H.E. Thu, Recent advancements in stimuli responsive drug delivery platforms for active and passive cancer targeting, Cancers (Basel). 13 (2021) 670. https://doi.org/10.3390/cancers13040670.

[52] Y. Hiruta, Poly(N-isopropylacrylamide)-based temperature- and pH-responsive polymer materials for application in biomedical fields, Polym. J. (2022). https://doi.org/10.1038/s41428-022-00687-z.

[53] A.S. Patil, A.P. Gadad, R.D. Hiremath, P.M. Dandagi, Exploration of the effect of chitosan and crosslinking agent concentration on the properties of dual responsive chitosan-g-poly (N-Isopropylacrylamide) Co-polymeric particles, J. Polym. Environ. 26 (2018) 596–606. https://doi.org/10.1007/s10924-017-0971-z.

[54] A.S. Patil, A.P. Gadad, R.D. Hiremath, S.D. Joshi, Biocompatible tumor micro-environment responsive CS-g-PNIPAAm co-polymeric nanoparticles for targeted oxaliplatin delivery, J. Polym. Res. 25 (2018) 77. https://doi.org/10.1007/s10965-018-1453-2.

[55] F. Howaili, E. Özliseli, B. Küçüktürkmen, S M Razavi, M. Sadeghizadeh, J.M. Rosenholm, Stimuli-responsive, plasmonic nanogel for dual delivery of curcumin and photothermal therapy for cancer treatment, Front. Chem. 8 (2021). https://doi.org/10.3389/fchem.2020.602941.

[56] O.R.M. Metawea, M.A. Abdelmoneem, N.S. Haiba, H.H. Khalil, M. Teleb, A.O. Elzoghby, A.F. Khafaga, A.E. Noreldin, F. Albericio, S.N. Khattab, A novel 'smart' PNIPAM-based copolymer for breast cancer targeted therapy: Synthesis, and characterization of dual pH/temperature-responsive lactoferrin-targeted PNIPAM-co-AA, Colloids Surfaces B Biointerfaces. 202 (2021) 111694. https://doi.org/10.1016/j.colsurfb.2021.111694.

[57] M. Falsafi, N. Hassanzadeh Goji, A. Sh. Saljooghi, K. Abnous, S.M. Taghdisi, S. Nekooei, M. Ramezani, M. Alibolandi, Synthesis of a targeted, dual pH and redox-responsive nanoscale coordination polymer theranostic against metastatic breast cancer in vitro and in vivo, Expert Opin. Drug Deliv. 19 (2022) 743–754. https://doi.org/10.1080/17425247.2022.2083602.

[58] T.L. Nguyen, T.H. Nguyen, C.K. Nguyen, D.H. Nguyen, Redox and pH responsive poly (amidoamine) dendrimer-heparin conjugates via disulfide linkages for letrozole delivery, Biomed Res. Int. 2017 (2017) 1–7. https://doi.org/10.1155/2017/8589212.

[59] R. Jahanban-Esfahlan, K. Soleimani, H. Derakhshankhah, B. Haghshenas, A. Rezaei, B. Massoumi, A. Farnudiyan-Habibi, H. Samadian, M. Jaymand, Multi-stimuli-responsive magnetic hydrogel based on Tragacanth gum as a de novo nanosystem for targeted chemo/hyperthermia treatment of cancer, J. Mater. Res. 36 (2021) 858–869. https://doi.org/10.1557/s43578-021-00137-1.

[60] Q. Li, D. Fu, J. Zhang, H. Yan, H. Wang, B. Niu, R. Guo, Y. Liu, Dual stimuli-responsive polypeptide-calcium phosphate hybrid nanoparticles for co-delivery of multiple drugs in cancer therapy, Colloids Surfaces B Biointerfaces. 200 (2021) 111586. https://doi.org/10.1016/j.colsurfb.2021.111586.

[61] X. Fu, J. Tian, Z. Li, J. Sun, Z. Li, Dual-responsive pegylated polypeptoids with tunable cloud point temperatures, Biopolymers. 110 (2019) e23243. https://doi.org/10.1002/bip.23243.

[62] S. Zong, H. Wen, H. Lv, T. Li, R. Tang, L. Liu, J. Jiang, S. Wang, J. Duan, Intelligent hydrogel with both redox and thermo-response based on cellulose nanofiber for controlled drug delivery, Carbohydr. Polym. 278 (2022) 118943. https://doi.org/10.1016/j.carbpol.2021.118943.

[63] N. Sun, P. Sun, A. Wu, X. Qiao, F. Lu, L. Zheng, Facile fabrication of thermo/redox responsive hydrogels based on a dual crosslinked matrix for a smart on—off switch, Soft Matter. 14 (2018) 4327–4334. https://doi.org/10.1039/C8SM00504D.

[64] H.T. Gebrie, K.D. Addisu, H.F. Darge, T.W. Mekonnen, D.T. Kottackal, H.-C. Tsai, Development of thermo/redox-responsive diselenide linked methoxy poly (ethylene glycol)-block-poly(ε-caprolactone-co-p-dioxanone) hydrogel for localized control drug release, J. Polym. Res. 28 (2021) 448. https://doi.org/10.1007/s10965-021-02776-8.

[65] H.J. Cho, M. Chung, M.S. Shim, Engineered photo-responsive materials for near-infrared-triggered drug delivery, J. Ind. Eng. Chem. 31 (2015) 15–25. https://doi.org/10.1016/j.jiec.2015.07.016.

[66] S. Zhou, C. Ding, C. Wang, J. Fu, UV-light cross-linked and pH de-cross-linked coumarin-decorated cationic copolymer grafted mesoporous silica nanoparticles for drug and gene co-delivery in vitro, Mater. Sci. Eng. C. 108 (2020) 110469. https://doi.org/10.1016/j.msec.2019.110469.

[67] Y. Tang, G. Wang, NIR light-responsive nanocarriers for controlled release, J. Photochem. Photobiol. C Photochem. Rev. 47 (2021) 100420. https://doi.org/10.1016/j.jphotochemrev.2021.100420.

[68] T. Cheng, R. Marin, A. Skripka, F. Vetrone, Small and bright lithium-based upconverting nanoparticles, J. Am. Chem. Soc. 140 (2018) 12890–12899. https://doi.org/10.1021/jacs.8b07086.

[69] C. Yu, L. Li, P. Hu, Y. Yang, W. Wei, X. Deng, L. Wang, F.R. Tay, J. Ma, Recent advances in stimulus-responsive nanocarriers for gene therapy, Adv. Sci. 8 (2021) 2100540. https://doi.org/10.1002/advs.202100540.

[70] X. Wang, Y. Yang, C. Liu, H. Guo, Z. Chen, J. Xia, Y. Liao, C.-Y. Tang, W.-C. Law, Photo- and pH-responsive drug delivery nanocomposite based on o-nitrobenzyl functionalized upconversion nanoparticles, Polymer (Guildf). 229 (2021) 123961. https://doi.org/10.1016/j.polymer.2021.123961.

[71] Z. Han, M. Gao, Z. Wang, L. Peng, Y. Zhao, L. Sun, pH/NIR-responsive nanocarriers based on mesoporous polydopamine encapsulated gold nanorods for drug delivery and thermo-chemotherapy, J. Drug Deliv. Sci. Technol. 75 (2022) 103610. https://doi.org/10.1016/j.jddst.2022.103610.

[72] S. Li, Y. Gan, C. Lin, K. Lin, P. Hu, L. Liu, S. Yu, S. Zhao, J. Shi, NIR-/pH-responsive nanocarriers based on mesoporous hollow polydopamine for codelivery of hydrophilic/hydrophobic drugs and photothermal synergetic therapy, ACS Appl. Bio Mater. 4 (2021) 1605–1615. https://doi.org/10.1021/acsabm.0c01451.

[73] S. Panda, C.S. Bhol, S.K. Bhutia, S. Mohapatra, PEG—PEI-modified gated N-doped mesoporous carbon nanospheres for pH/NIR light-triggered drug release and cancer phototherapy, J. Mater. Chem. B. 9 (2021) 3666–3676. https://doi.org/10.1039/D1TB00362C.

[74] Z. Guo, J. Sui, M. Ma, J. Hu, Y. Sun, L. Yang, Y. Fan, X. Zhang, pH-Responsive charge switchable PEGylated ε-poly-l-lysine polymeric nanoparticles-assisted combination therapy for improving breast cancer treatment, J. Control. Release. 326 (2020) 350–364. https://doi.org/10.1016/j.jconrel.2020.07.030.

[75] Q. Chen, C. Jia, Y. Xu, Z. Jiang, T. Hu, C. Li, X. Cheng, Dual-pH responsive chitosan nanoparticles for improving in vivo drugs delivery and chemoresistance in breast cancer, Carbohydr. Polym. 290 (2022) 119518. https://doi.org/10.1016/j.carbpol.2022.119518.

[76] J. Liao, H. Peng, C. Liu, D. Li, Y. Yin, B. Lu, H. Zheng, Q. Wang, Dual pH-responsive-charge-reversal micelle platform for enhanced anticancer therapy, Mater. Sci. Eng. C. 118 (2021) 111527. https://doi.org/10.1016/j.msec.2020.111527.

[77] J. Zhou, Y. Han, Y. Yang, L. Zhang, H. Wang, Y. Shen, J. Lai, J. Chen, phospholipid-decorated glycogen nanoparticles for stimuli-responsive drug release and synergetic chemophotothermal therapy of hepatocellular carcinoma, ACS Appl. Mater. Interfaces. 12 (2020) 23311–23322. https://doi.org/10.1021/acsami.0c02785.

[78] W. Yin, W. Ke, N. Lu, Y. Wang, A.A.-W.M.M. Japir, F. Mohammed, Y. Wang, Y. Pan, Z. Ge, Glutathione and reactive oxygen species dual-responsive block copolymer prodrugs for boosting tumor site-specific drug release and enhanced antitumor efficacy, Biomacromolecules. 21 (2020) 921–929. https://doi.org/10.1021/acs.biomac.9b01578.

[79] B. Mondal, B. Pandey, N. Parekh, S. Panda, T. Dutta, A. Padhy, S. Sen Gupta, Amphiphilic mannose-6-phosphate glycopolypeptide-based bioactive and responsive self-assembled nanostructures for controlled and targeted lysosomal cargo delivery, Biomater. Sci. 8 (2020) 6322–6336. https://doi.org/10.1039/D0BM01469A.

[80] J. Yoo, N. Sanoj Rejinold, D. Lee, S. Jon, Y.-C. Kim, Protease-activatable cell-penetrating peptide possessing ROS-triggered phase transition for enhanced cancer therapy, J. Control. Release. 264 (2017) 89–101. https://doi.org/10.1016/j.jconrel.2017.08.026.

[81] C. Li, J. Wang, Y. Wang, H. Gao, G. Wei, Y. Huang, H. Yu, Y. Gan, Y. Wang, L. Mei, H. Chen, H. Hu, Z. Zhang, Y. Jin, Recent progress in drug delivery, Acta Pharm. Sin. B. 9 (2019) 1145–1162. https://doi.org/10.1016/j.apsb.2019.08.003.

[82] R. Zhang, R. Liu, C. Liu, L. Pan, Y. Qi, J. Cheng, J. Guo, Y. Jia, J. Ding, J. Zhang, H. Hu, A pH/ROS dual-responsive and targeting nanotherapy for vascular inflammatory diseases, Biomaterials. 230 (2020) 119605. https://doi.org/10.1016/j.biomaterials.2019.119605.

[83] Y. Zhan, H. Wang, M. Su, Z. Sun, Y. Zhang, P. He, Mesoporous silica and polymer hybrid nanogels for multistage delivery of an anticancer drug, J. Mater. Sci. 56 (2021) 4830–4842. https://doi.org/10.1007/s10853-020-05576-5.

[84] L. Duan, Y. Wang, Y. Zhang, Z. Wang, Y. Li, P. He, pH/redox/thermo-stimulative nanogels with enhanced thermosensitivity via incorporation of cationic and anionic components for anticancer drug delivery, Int. J. Polym. Mater. Polym. Biomater. 67 (2018) 288–296. https://doi.org/10.1080/00914037.2017.1323215.

[85] D. Gao, L. Duan, M. Wu, X. Wang, Z. Sun, Y. Zhang, Y. Li, P. He, Preparation of thermo/redox/pH-stimulative poly(N-isopropylacrylamide- co - N, N′-dimethylaminoethyl methacrylate) nanogels and their DOX release behaviors, J. Biomed. Mater. Res. Part A. 107 (2019) 1195–1203. https://doi.org/10.1002/jbm.a.36611.

[86] T.-M. Don, K.-Y. Lu, L.-J. Lin, C.-H. Hsu, J.-Y. Wu, F.-L. Mi, Temperature/pH/Enzyme triple-responsive cationic protein/PAA-b-PNIPAAm nanogels for controlled anticancer drug and photosensitizer delivery against multidrug resistant breast cancer cells, Mol. Pharm. 14 (2017) 4648–4660. https://doi.org/10.1021/acs.molpharmaceut.7b00737.

[87] N. Lu, P. Huang, W. Fan, Z. Wang, Y. Liu, S. Wang, G. Zhang, J. Hu, W. Liu, G. Niu, R.D. Leapman, G. Lu, X. Chen, Tri-stimuli-responsive biodegradable theranostics for mild hyperthermia enhanced chemotherapy, Biomaterials. 126 (2017) 39–48. https://doi.org/10.1016/j.biomaterials.2017.02.025.

[88] X. Lin, X. Song, Y. Zhang, Y. Cao, Y. Xue, F. Wu, F. Yu, M. Wu, X. Zhu, Multifunctional theranostic nanosystems enabling photothermal-chemo combination therapy of triple-stimuli-responsive drug release with magnetic resonance imaging, Biomater. Sci. 8 (2020) 1875–1884. https://doi.org/10.1039/C9BM01482A.

[89] Y. Cao, Y. Cheng, G. Zhao, Near-Infrared Light-, Magneto-, and pH-Responsive GO—Fe 3 O 4/Poly(N-isopropylacrylamide)/alginate Nanocomposite hydrogel microcapsules for controlled drug release, Langmuir. 37 (2021) 5522–5530. https://doi.org/10.1021/acs.langmuir.1c00207.

[90] Z. Cao, X. Zhou, G. Wang, Selective release of hydrophobic and hydrophilic cargos from multi-stimuli-responsive nanogels, ACS Appl. Mater. Interfaces. 8 (2016) 28888–28896. https://doi.org/10.1021/acsami.6b10360.

[91] K. Zhang, J. Liu, Y. Guo, Y. Li, X. Ma, Z. Lei, Synthesis of temperature, pH, light and dual-redox quintuple-stimuli-responsive shell-crosslinked polymeric nanoparticles for controlled release, Mater. Sci. Eng. C. 87 (2018) 1–9. https://doi.org/10.1016/j.msec.2018.02.005.

[92] J. Wang, Y. Chang, H. Luo, W. Jiang, L. Xu, T. Chen, X. Zhu, Designing immunogenic nanotherapeutics for photothermal-triggered immunotherapy involving reprogramming immunosuppression and activating systemic antitumor responses, Biomaterials. 255 (2020) 120153. https://doi.org/10.1016/j.biomaterials.2020.120153.

[93] R.E. Tay, E.K. Richardson, H.C. Toh, Revisiting the role of CD4+ T cells in cancer immunotherapy—new insights into old paradigms, Cancer Gene Ther. 28 (2021) 5–17. https://doi.org/10.1038/s41417-020-0183-x.

[94] S. Sau, H.O. Alsaab, K. Bhise, R. Alzhrani, G. Nabil, A.K. Iyer, Multifunctional nanoparticles for cancer immunotherapy: A groundbreaking approach for reprogramming malfunctioned tumor environment, J. Control. Release. 274 (2018) 24–34. https://doi.org/10.1016/j.jconrel.2018.01.028.

[95] B. Feng, F. Zhou, B. Hou, D. Wang, T. Wang, Y. Fu, Y. Ma, H. Yu, Y. Li, Binary cooperative prodrug nanoparticles improve immunotherapy by synergistically modulating immune tumor microenvironment, Adv. Mater. 30 (2018) 1803001. https://doi.org/10.1002/adma.201803001.

[96] L. Xie, G. Wang, W. Sang, J. Li, Z. Zhang, W. Li, J. Yan, Q. Zhao, Y. Dai, Phenolic immunogenic cell death nanoinducer for sensitizing tumor to PD-1 checkpoint blockade immunotherapy, Biomaterials. 269 (2021) 120638. https://doi.org/10.1016/j.biomaterials.2020.120638.

[97] L. Wang, K. Ding, C. Zheng, H. Xiao, X. Liu, L. Sun, R. Omer, Q. Feng, Z. Zhang, detachable nanoparticle-enhanced chemoimmunotherapy based on precise killing of tumor seeds and normalizing the growing soil strategy, Nano Lett. 20 (2020) 6272–6280. https://doi.org/10.1021/acs.nanolett.0c01415.

[98] H. Du, S. Zhao, Y. Wang, Z. Wang, B. Chen, Y. Yan, Q. Yin, D. Liu, F. Wan, Q. Zhang, Y. Wang, pH/Cathepsin B hierarchical-responsive nanoconjugates for enhanced tumor penetration and chemo-immunotherapy, Adv. Funct. Mater. 30 (2020) 2003757. https://doi.org/10.1002/adfm.202003757.

[99] Y. Wen, X. Chen, X. Zhu, Y. Gong, G. Yuan, X. Qin, J. Liu, Photothermal-chemotherapy integrated nanoparticles with tumor microenvironment response enhanced the induction of immunogenic cell death for colorectal cancer efficient treatment, ACS Appl. Mater. Interfaces. 11 (2019) 43393–43408. https://doi.org/10.1021/acsami.9b17137.

[100] R. Liu, Y. An, W. Jia, Y. Wang, Y. Wu, Y. Zhen, J. Cao, H. Gao, Macrophage-mimic shape changeable nanomedicine retained in tumor for multimodal therapy of breast cancer, J. Control. Release. 321 (2020) 589–601. https://doi.org/10.1016/j.jconrel.2020.02.043.

[101] D. Liu, F. Yang, F. Xiong, N. Gu, The smart drug delivery system and its clinical potential, Theranostics. 6 (2016) 1306–1323. https://doi.org/10.7150/thno.14858.

9 Temperature-Responsive Delivery Nanoplatforms in Cancer Theranostics

Mduduzi N. Sithole and Yahya E. Choonara

CONTENTS

9.1 Introduction 142
9.2 Nanoplatforms for Cancer Therapy 144
9.3 Imaging (Detection) as a Diagnostic Tool for Cancer 146
9.4 Nanoplatforms for Cancer Theranostic 146
9.5 Temperature-Responsive Nanoplatform-Facilitated Diagnosis and Therapy of Cancer 148
9.6 Conclusion 150
References 150

9.1 INTRODUCTION

Despite excellent global efforts over the years, cancer is still the deadliest disease and leading cause of death worldwide, accounting for more than 9 million deaths per year. Cancer cases have been reported to increase annually (1–4). Different solutions (e.g., nanoplatforms) such as drug delivery systems have been developed in an attempt to achieve enhanced therapeutic efficacy and reduced side effects (5). Chemotherapy, tumour surgery resection, and radiotherapy are the approved anti-cancer treatments, which have been developed over time; however, aforementioned therapies do not achieve a complete cure of the disease and responsible for comprised lifestyle of patience (6). For an example, surgery is likely to cause the development of infections and recurrence (7, 8), while chemotherapy is limited by multidrug resistance and inability of tumour-targeting drug delivery systems (9, 10), whereas radiotherapy is limited by hypoxic conditions in tumours which make them resistant to radiotherapy (11–13). However, chemotherapy is the dominant treatment option. Furthermore, the development of a multi-functional nanoplatform that merges stimuli-responsive moieties, imaging, and therapy will be of great advantage in cancer treatment (14).

Intelligent novel nanoplatform technologies that can effectively treat cancer disease tissue with limited damage to the healthy tissue of the patient are of great necessity, since they can reduce the chance of cancer recurrence, having an immediate impact in clinic studies (15). When designing nanoplatform technologies, it is critical to consider its functionalities, such as specific delivery to the cancer tissue only, lack of an immune response, concomitant treatment and non-invasive monitoring, evasion of normal tissue and accumulation in the organs, and adequate circulation time for successful drug delivery to occur. The use of triggered treatment merged with imaging modalities such as magnetic resonance imaging, positron emission tomography, computed tomography, and ultrasound could address many of the current issues in cancer treatment, possibly leading to significant clinical outcomes (16–18).

The term "theranostic" was introduced by Funkhouser in 2002 for treatment systems (platforms) merging both therapy and diagnosis (19). An applicable nano-system as an anti-cancer treatment needs to comprehend the micro-environment idiosyncrasies in the tumour and the cellular phenotype. Hence, theranostics promotes individualised drug therapy and avoid unwanted distribution, since it offers an all-in-one package (20). Cancer heterogeneity necessitates robust development of imaging procedures for disease management (21). Imaging assists with disclosing the status and drug response of cancer in a non-invasive manner at the cellular and molecular levels (22–24). Furthermore, imaging procedures can be used to examine extensively the retention of cancer nano-systems in tumours or other organs and the degradation route plus their clearance from the body (25).

The ever-growing demand in medical therapy and the rapid development in materials science, mostly for cancer therapy, has directed research towards the construction of efficient anticancer platforms (26). These advancements in the development of novel nanoplatforms have inspired researchers to strive to design efficient and safe drug delivery systems for cancer therapy. Stimuli-responsive nanoplatforms are an alternative for designing controllable cancer delivery systems on account of their spatiotemporally controllable properties. External stimuli such as temperature are developed/reinforce the cancer theranostic nanoplatform application for personalised systems because of their unique characteristics (27). A great number of well-tailored cancer nanoplatform carriers have been developed as result of advancements in nanotechnology and insight into the pathology of cancer at the cellular and molecular level; Figure 9.1 shows an example of a variety of developed cancer nanoplatform carriers such as dendrimer; liposomes; polymer nanoparticles; and inorganic nanoparticles made of quantum dots, iron oxide, gold, or other metal frameworks (28).

Stimuli-responsive cancer theranostics are typically made of at least two components, which are therapeutic and imaging agent moieties, with one of the components being responsive to external stimuli or to the micro-environment of the tumour. Stimuli-responsive theranostics are classified according to the stimuli that trigger the drug release at the tumour site. The frequency of the trigger results in a simultaneous alteration of the image and therapy components. The release of the drug from theranostics systems can be divided into two categories: a) a remotely activated stimuli-responsive cancer theranostic therapy, whereby the external signal is used as a therapeutic response trigger, such as light or heat, or b) an environmentally activated stimuli-responsive cancer theranostic therapy, whereby an endogenous signal is used as a drug release trigger, such as aberrant hypoxia in tumours (30). Heat was traditionally applied regionally, systematically, or locally either through local insertion of a needle or using external machines transmitting high-energy waves. Furthermore, magnetic nanoparticles which contain iron are mostly utilised as temperature-responsive nanoparticles. The usage in the form of nanofluids with external application of a magnetic field is predicted to generate enough heat and is postulated to produce adequate sensitivity and specificity to achieve clinical diagnostics and treatments effectively (31). Therefore,

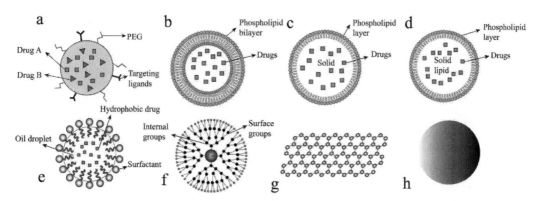

FIGURE 9.1 A variety of nanoplatforms employed in cancer therapy: **a** nanoparticles, **b** liposomes, **c** solid lipid nanoparticles, **d** nanostructure lipid carriers, **e** nanoemulsion, **f** dendrimers, **g** graphene, **h** metallic nanoparticles (29).

researchers are predominantly interested in integrating diagnosis, treatment, or multimodal therapies as one platform for personalised cancer delivery systems (26). This chapter summarises in general the nanosystems utilised in cancer therapy, images used as diagnostic tools, and nanoplatforms for cancer theranostics. Finally, the chapter will discuss the progress in the inclusion of a stimuli-responsive moieties, specifically focusing on temperature-responsive nanoplatforms for cancer theranostics.

9.2 NANOPLATFORMS FOR CANCER THERAPY

Over the years, the use of nanoplatform systems as drug delivery systems has been investigated to address the limitations associated with conventional therapeutic dosage limitations, predominantly cancer chemotherapy, such as nonspecific biodistribution, low therapeutic indices, poor oral bioavailability, limited aqueous solubility, the lack of targeting cancer tissue, and multidrug resistance (32, 33). Nanoplatforms refers to a technology that performs at the nanoscale. In this chapter, nanoplatform refers to ultrafine particles of a size ranging from 1 to 1000 nm. Nanoplatforms have been developed specifically for intervention at the molecular level for diagnosis, prevention, and treatment of diseases (34). Nanoplatforms are advantageous due to advanced improvement for therapeutic effects and controllable drug delivery with minimal adverse reactions. Currently many materials are used to develop nanoplatforms as anti-cancer drug delivery systems, which include organic materials (e.g., micelles, liposomes, and dendrimers) and inorganic materials (e.g., mesoporous silica materials, gold nanoparticles), amongst others; see Table 9.1 for few examples of such nanoplatforms (35, 36).

There has been much great development made toward engineering nanoplatform systems specifically for cancer treatment with sufficient efficacy and sensitivity. Furthermore, nanoplatforms can be functionalised with specific ligands to target specific cancer cells in a predictable manner and deliver anti-cancer drugs effectively. Moreover, nanoplatform systems can be designed to improve the drug half-life in the body; control release; and increase drug loading and selective distribution by modifying the composition, size, morphology, and surface chemistry (37). Nevertheless, a very small number of the developed therapeutic nanosystems are involved in clinical use (29). The different nanosystems have their own limitations in application, such as dendrimers, liposomes, and micelles mostly have low loading capacity, and most inorganic porous materials suffer from undesirable toxicity and unacceptable degradation (38). Nanoplatforms that are ideal for drug loading have these requirements: carriers should be nanoscale to facilitate the release of drugs by intravenous administration, carriers must be biocompatible—they should be nontoxic or have low toxicity and should be easily degraded by the body's metabolism, and carriers should have a large loading capacity. Also, drug release from the nanoplatform delivery system is expected to be controlled effectively. Consequently, there is an extensive effort dedicated to investigating different controlled drug release nanosystems for cancer chemotherapy. Scientists are mostly focusing on developing a multimodal treatment systems to satisfy an ever-growing demand for efficient therapeutic strategies (39, 40).

Nanoplatforms in oncology (the study and treatment of tumours) can lead to targeted drug delivery systems, drug development, diagnosis, and efficient anticancer therapies (41). For example, gold nanoparticles have extensively been employed as a drug delivery vehicle for breast and prostate cancers (29, 42). It is apparent that overcoming cancer is a battle against complexities and intricacies due to mutations, the metastatic nature of cancer cells, lack of early detection techniques, and the inability of the scientific community to solve many cancer-related problems. However, the recent developments in nanoplatforms can be a key to unlock the secrets of cancer diagnosis and treatment strategies (32, 37, 43). Therefore, currently there is an exponential growth in developing nanosystems that contain both imaging techniques and therapeutic treatments; consequently the following sections will expand on the benefit of nanosystems that contain both (imaging and treatment).

TABLE 9.1
A Range of Particles on the Nanometre Scale Utilised for Cancer Treatment

Organic	Inorganic
Dendrimer	Carbon nanotube 
Micelle	Quantum dots
Solid lipid nanoparticle	Metal oxide nanoparticle
Liposome	Upconversion nanoparticle
Polymeric nanoparticle	Metal nanoparticle
Protein nanoparticle	Mesoporous silica nanoparticle
Viral particle	

9.3 IMAGING (DETECTION) AS A DIAGNOSTIC TOOL FOR CANCER

Imaging is currently one of the techniques used in cancer diagnostic by making pictures inside the body. Imaging is valuable in detecting changes inside the body and finding tumours. It indicates how much tumour is present and significantly indicates if the treatment used is effective. Imaging is also useful during surgical procedures such as biopsies (44). The following are imaging techniques available to clinicians who diagnose, stage, and treat human cancer:

  i) Ultrasound (US) (sonography)—It uses high-energy sound waves and a computer to make images of tissues, blood vessels, and organs. However, US is not useful in the chest, because the ribs block the sound waves. US is used to see the kidneys, liver, and tumours in the belly (abdomen). It can show blood flow through the vessels and look at how well organs are currently functioning.
  ii) X-rays—They use a low dose of radiation to make images of tissue, organs, and bones. They can be taken in any part of the body in the search for tumours.
  iii) Computed tomography (CT) scan—It uses X-ray scan and a computer to make 3D images (often called slices) of the body. A CT scan can be used to show body parts such as bones, organs, and muscle, and it is more detailed compared to X-ray scan only.
  iv) Magnetic resonance imaging (MRI)—It uses a computer, radio, and strong magnet to make detailed images of organs and other structures inside the body. An MRI shows even small changes in tissues and hence can detect tumours and diagnose many forms of cancer, assess blood flow, evaluate infections, and assess injuries to bones and joints.
  v) Single-photon emission computed tomography (SPECT)—It uses a radioactive substance and a specialised camera to make 3D pictures; it analyses tissue, bones, and body organs.
  vi) Optical imaging—It uses a specialised photon and light properties to gain more detailed images of tissues, organs, molecules, and cells. This technique looks inside the body through a non-invasive method.
  vii) Nuclear medicine scan—This imaging method, also referred to as a radionuclide scan, employs a radioactive tracer that is injected into the bloodstream to produce informative images. The scan captures the locations where the tracer has accumulated and moved throughout the body. The specific type of scanner used depends on the area of the body being examined. Some commonly used nuclear scans include scans of the thyroid, gallium scans, multigated acquisition scans, PET scans, and PET scans with metal.
  viii) Mammogram—It is primarily utilised for X-ray examinations of the breast and is effective in identifying and diagnosing breast issues in women who have symptoms such as pain, discharge from the nipples, or lumps. Additionally, screening mammograms are used to detect breast diseases before any symptoms appear. However, it is important to note that while a mammogram can indicate the presence of a possible abnormality, it cannot confirm that it is cancer; further testing is required for that.

It is crucial to understand the historical development and advancements of these imaging techniques, as they have their roots in chemistry and physics rather than being specifically designed for oncologists' needs (45). Engineered nanoplatforms play a significant role in the advancement of treatment and diagnosis of diseases (46). Consequently, the use of nanoplatform systems has been proposed as a method for creating systems that combine both therapeutic and imaging capabilities to achieve the goal of cancer theranostics (47). The next section will provide an overview of cancer theranostics using nanoplatforms.

9.4 NANOPLATFORMS FOR CANCER THERANOSTIC

A variety of nanomedicine platforms have been created; however, at present, the emphasis in the design and development of nanoplatform drug delivery systems is on those that incorporate imaging nanoagents (48). Theranostics is a term used to describe a platform that can be utilised for both diagnostic and therapeutic purposes (19). Having a platform/system that can grasp the unique features of the tumour microenvironment and the specific characteristics of the cancer cells is crucial. Theranostics presents a comprehensive solution due to its clear advantages; it prevents unnecessary distribution and promotes personalised drug therapy (20). In recent decades, interest in theranostic nanoplatforms has grown significantly due to their ability to combine various diagnostic and therapeutic components into a single unit (20, 49). Theranostics is capable of characterising, monitoring molecular events, and imaging while simultaneously delivering therapy (20, 50). As a result, theranostics often integrates one or more therapeutic modalities, such as photodynamic therapy (PDT), immunotherapy, photothermal therapy (PTT), magnetic hyperthermia (MH), chemotherapy, and radiotherapy (RT), with imaging techniques like PET, MRI, US, and SPECT (20, 51, 52). These nanoplatforms have been proven effective in treating a wide range of cancers, as they enable more personalised delivery of cancer drugs to the site of the disease (53–57). The integration of therapy and monitoring/diagnosis offers benefits that surpass using treatment or standard imaging alone. Well-designed theranostic agents have the potential to aid in the early diagnosis and treatment of cancer, provide insight into drug distribution at target sites, and allow for monitoring of therapy response (58–62).

There is a pressing need to enhance the development and production of nanoplatform systems that can enhance the effectiveness of cancer therapy for various conditions, to benefit patients and provide lasting remission and success. To meet the needs in cancer therapy, researchers are combining various forms of treatment to create synergies that have the potential to minimise therapy-related complications (63, 64). These nanoplatforms act as "catalysts", providing multiple functions, such as therapy and imaging, and have been developed in different designs and fabrication methods (65, 66). It is accepted that theranostic nanoplatforms are a promising approach for early detection and treatment of cancer. In the design and development of these platforms, several key considerations must be taken into account, including the ability to accurately detect cancer and the ability to control the delivery of cancer therapeutics (66). Figure 9.2 shows an example design of a cancer nanoplatform theranostic agent. Such nanoplatform cancer theranostics can be obtained from polymeric nanoparticles (NPs) (67), dendrimers (68), liposomes (69), carbon-based nanomaterials (70), metal or inorganic nanocarriers (71), and systems which integrate both categories, such as

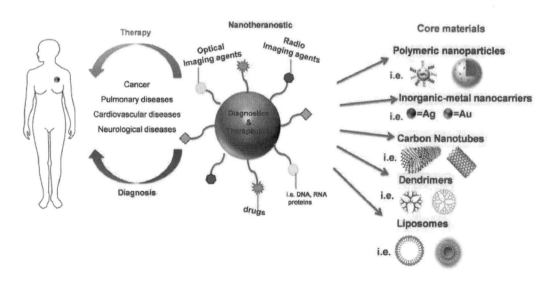

FIGURE 9.2 Use of nanotheranostics for concurrent release and imaging. Nanotheranostics can be equipped with various functional components such as drugs, DNA, RNA, or imaging agents (80).

polymeric coated nanocarriers (72, 73). Carbon-based nanomaterials, such as carbon nanotubes either alone or decorated with other materials (74, 75), graphene oxide (76–78), fullerenes, carbon quantum dots exhibit good potential as a theranostic platform (79).

The use of theranostics in cancer therapy is highly desirable; however, incorporating a stimuli-responsive component in a nanoplatform would make it even more effective for targeted delivery. The inclusion of nanosized theranostics with multiple capabilities such as targeted delivery, sustained release, improved transport efficiency via endocytosis, and responsive systems, as well as the combination of multiple diagnostic and therapeutic approaches, is a major advancement in cancer therapy (81). Additionally, the following section specifically delves into the use of temperature-responsive nanoplatforms for cancer theranostics therapy.

9.5 TEMPERATURE-RESPONSIVE NANOPLATFORM-FACILITATED DIAGNOSIS AND THERAPY OF CANCER

Achieving minimal harm to normal tissues while effectively treating tumours has long been a difficult task in the field of cancer therapy (82). Nanoplatform theranostics that respond to specific stimuli, such as temperature, are designed to release or activate therapeutic agents once they reach the target tissue. Two commonly used methods for inducing a response in cancer theranostics are by exploiting differences in the pathological conditions of the target site compared to healthy tissue or by utilising hypoxia conditions (83, 84), and/or it can be locally induced through an external stimuli (such as irradiation with light) (30).

The aim of targeted therapies for cancer treatment is to deliver chemotherapy drugs specifically to the cancer cells, minimise damage to healthy cells, and reduce side effects. This is often achieved by administering drugs intravenously or orally for systemic distribution; however, these methods often lack specificity (32). The use of laser as an external source can cause a change in temperature above the lower critical solution temperature, resulting in a change in the phase (phase transition) of the nanomaterial, releasing its contents. This process, known as phase transition, releases the active drug by converting the gel phase to a crystalline liquid phase (85), Furthermore, the release of active drugs from nanoplatform systems can be achieved through modifications of the interaction between hydrophobic and hydrophilic regions, as well as changes in temperature that affect the material's structure. These materials are typically stable at normal body pH but can quickly release their cargo in areas of increased temperature. This allows for targeted delivery of drugs to specific regions while reducing overall toxicity (86).

One of the most-researched physical signals in oncology is temperature, which has the ability to alter the cancerous condition. As previously mentioned, the release of drugs can be controlled by the difference in temperature between tumour cells and normal cells. The temperature of the tumour microenvironment (TME) is reported to be higher than that of normal cells, with TME temperatures reported to be around 40–42 degrees Celsius and normal cells around 37 degrees Celsius. This difference in temperature can be utilised to target mild hyperthermia-based drug delivery systems to cancer cells while leaving normal cells mostly unaffected (87). Temperature-responsive systems are considered one of the most effective and convenient methods for controlling drug release in cancer therapy. However, another strategy for improving drug release within the tumour microenvironment is the application of external triggers, such as ultrasound or magnetic fields, to heat the site of the tumour. Figure 9.1 illustrates the general concept of temperature-stimulated drug release. The current challenge in developing thermo-responsive nanoplatforms is to ensure their safety while maintaining their sensitivity to slight temperature changes (88).

A wide variety of nanoplatform therapeutics have been created that incorporate various types of molecules for cancer diagnosis, treatment, and responsiveness to stimuli (89–92). Temperature-responsive theranostic agents typically exhibit changes in material physical conformation, such as swelling, in response to temperature increases, which subsequently leads to drug release. Some commonly used materials for these systems include lipids, core-shell nanoparticles, self-assembling

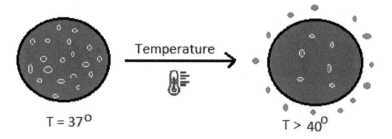

FIGURE 9.3 Illustrative representation of how a drug release is triggered by diffusion-based stimuli.

amphiphilic micelles, and polymers. As a result, a wide range of nanoplatform therapeutics systems have been developed that incorporate various molecules for cancer imaging, treatment, and stimulus responsiveness. Among the various polymeric systems, poly(hydroxypropyl methacrylamide-lactate) (p(HPMAm-Lacn)), pluronics, and poly(N-isopropylacrylamide) (pNIPAAm) are often utilised in clinical trials (93). One way to develop temperature-responsive nanoplatforms for cancer therapy is by modifying the ratio of hydrophobic and hydrophilic groups in polymeric systems. This can make the systems more responsive to changes in temperature, which can be used to trigger the release of therapeutic agents. This section will explore various examples of temperature-responsive nanoplatforms that have been developed for cancer theranostics.

A study by Xi et al. (2017) created a temperature-responsive nanoplatform delivery system made of poly (lactic-co-glycolic acid) (PLGA) that was loaded with doxorubicin (DOX) and modified with polydopamine (PDA) and manganese (II) (Mn^{2+}) ions on the surface. This nanoplatform was designed to be both a chemotherapy agent and a photothermal (PTT) photosensitiser for MRI due to the presence of PDA on the surface and DOX in the core. The release of DOX was analysed using UV-vis spectroscopy and inductively coupled plasma atomic emission spectroscopy (ICP-AES). It was found that drug release was triggered by the absorption of near-infrared (NIR) light by PDA. Magnetic resonance (MR) images also showed an improvement after 24 hours of injection with the Mn^{2+}-PDA@PLGA nanoplatform due to the presence of Mn^{2+} ions on the surface. The efficacy of this nanoplatform was confirmed in both in vitro and in vivo assays, where a synergistic effect between chemotherapy and PTT was observed in mice treated with DOX-loaded nanoparticles and exposed to an 808-nm laser (92).

Tang et al. (2018) designed and fabricated a multifunctional, temperature-responsive, and targeted cancer theranostic nanoplatform using five key components: perfluorohexane, the anticancer drug doxorubicin, poly (lactic-co-glycolic acid), iron (II,III) oxide, and folic acid. This nanoplatform was created for efficient ultrasound/magnetic resonance dual-modality imaging-guided high-intensity focused ultrasound/chemotherapy treatment. The folic acid was conjugated to the surface of the nanoplatform to actively target hepatoma cells and allow for accumulation at the tumour site. The iron oxide nanoparticles also provided a contrast enhancement in T_2-weighted magnetic resonance imaging. Overall, the developed nanoplatform was considered a promising candidate for future cancer therapy studies due to its temperature-responsive phase transition, synergistic therapeutic efficiency, responsive drug delivery, and efficient targeting (94).

To improve the safety and effectiveness of temperature-responsive nanoplatforms for cancer diagnosis and treatment, it is necessary to carefully design, develop, and optimise these systems to minimise any potential negative side effects (95, 96). From this perspective, many cancer-targeting agents based on nanomaterials have been developed by utilising the properties of the tumour microenvironment. By combining radiotherapy with heat treatments, the efficiency of cancer treatment has been enhanced, particularly in randomised trials for high-risk soft-tissue sarcomas (97). There is a significant demand for advanced drug delivery systems that can effectively target and control the delivery of drugs while also providing sensitive imaging capabilities for early detection of disease symptoms and affected areas. To address this need, we are working to create a system that combines both therapeutic and diagnostic capabilities (theranostics) plus a temperature-responsive component (98).

9.6 CONCLUSION

Thermoregulation is a fundamental aspect of maintaining bodily equilibrium in all living organisms and can be harnessed for therapeutic purposes. Hyperthermia, which involves treating the body at temperatures above its normal range (37 degrees Celsius), is one such application. This chapter highlighted that the integration of different modes of functionalities for cancer treatment, such as the merging of imaging techniques, chemotherapy and/or temperature-responsive nanoplatform into one nanoplatform system, can personalise medication. These nanoplatforms can be triggered to release the drug in a targeted manner, enhance the synergistic effects of drugs and inducing cell death, and enhance drug uptake. However, further research is needed to fully understand the impact of the tumour microenvironment on nanotherapy for improved palliative care, diagnosis, and treatment.

REFERENCES

1. Bagherifar R, Kiaie SH, Hatami Z, et al. Nanoparticle-mediated synergistic chemoimmunotherapy for tailoring cancer therapy: Recent advances and perspectives. *J Nanobiotechnology*. 2021; 19(1):1–18.
2. Journal Article: Sung H, Ferlay J, Siegel RL, et al. Global cancer statistics 2020: GLOBOCAN estimates of incidence and mortality worldwide for 36 cancers in 185 countries. *CA Cancer J Clin*. 2021; 71(3):209–249.
3. Journal Article: Xie P, Wang Y, Wei D, et al. Nanoparticle-based drug delivery systems with platinum drugs for overcoming cancer drug resistance. *J Mater Chem B*. 2021; 9(26):5173–5194.
4. Journal Article: Yang D, Wang T, Su Z, Xue L, Mo R, Zhang C. Reversing cancer multidrug resistance in xenograft models via orchestrating multiple actions of functional mesoporous silica nanoparticles. *ACS Appl Mater Interfaces*. 2016; 8(34):22431–22441.
5. Journal Article: Doane TL, Burda C. The unique role of nanoparticles in nanomedicine: Imaging, drug delivery and therapy. *Chem Soc Rev*. 2012; 41(7):2885–2911.
6. Journal Article: Singh R, Sharma A, Saji J, Umapathi A, Kumar S, Daima HK. Smart nanomaterials for cancer diagnosis and treatment. *Nano Converg*. 2022; 9(1):1–39.
7. Journal Article: Hu Q, Li H, Archibong E, et al. Inhibition of post-surgery tumour recurrence via a hydrogel releasing CAR-T cells and anti-PDL1-conjugated platelets. *Nat Biomed Eng*. 2021; 5(9):1038–1047.
8. Journal Article: Wang C, Fan W, Zhang Z, Wen Y, Xiong L, Chen X. Advanced nanotechnology leading the way to multimodal imaging-guided precision surgical therapy. *Adv Mater*. 2019; 31(49):1904329.
9. Journal Article: Cao J, Huang D, Peppas NA. Advanced engineered nanoparticulate platforms to address key biological barriers for delivering chemotherapeutic agents to target sites. *Adv Drug Deliv Rev*. 2020; 167:170–188.
10. Journal Article: Wei G, Wang Y, Yang G, Wang Y, Ju R. Recent progress in nanomedicine for enhanced cancer chemotherapy. *Theranostics*. 2021; 11(13):6370.
11. Journal Article: Peng S, Song R, Lin Q, et al. A Robust oxygen microbubble radiosensitizer for iodine-125 brachytherapy. *Adv Sci*. 2021; 8(7):2002567.
12. Journal Article: Zhong D, Li W, Hua S, et al. Calcium phosphate engineered photosynthetic microalgae to combat hypoxic-tumor by in-situ modulating hypoxia and cascade radio-phototherapy. *Theranostics*. 2021; 11(8):3580.
13. Journal Article: Zou MZ, Liu WL, Li CX, et al. A multifunctional biomimetic nanoplatform for relieving hypoxia to enhance chemotherapy and inhibit the PD-1/PD-L1 axis. *Small*. 2018; 14(28):1801120.
14. Journal Article: Shao L, Hu T, Fan X, et al. Intelligent nanoplatform with multi therapeutic modalities for synergistic cancer therapy. *ACS Appl Mater Interfaces*. 2022; 14(11):13122–13135.
15. Sneider A, VanDyke D, Paliwal S, Rai P. Remotely triggered nano-theranostics for cancer applications. *Nanotheranostics*. 2017; 1(1):1.
16. Journal Article: Giacomantonio CA, Temple WJ. Quality of cancer surgery: Challenges and controversies. *Surg Oncol Clin N Am*. 2000; 9(1):51–60.

17. Journal Article: Ikeda T, Mitsuyama S. New challenges in the field of breast cancer therapy-do we need surgery for the patients with breast cancer? *Breast Cancer.* 2007; (1):37–38.
18. Journal Article: Meurisse M, Defechereux T, Meurisse N, Bataille Y, Hamoir E. New challenges in the treatment of early breast cancer or surgery for early breast cancer . . . can less be more? *Acta Chir Belg.* 2002; 102(2):97–109.
19. Journal Article: Funkhouser J. Reinventing pharma: The theranostic revolution. *Curr Drug Discov.* 2002; 2:17–19.
20. Journal Article: Kelkar SS, Reineke TM. Theranostics: Combining imaging and therapy. *Bioconjug Chem.* 2011; 22(10):1879–1903.
21. Fass L. Imaging and cancer: A review. *Mol Oncol.* 2008; 2(2):115–152.
22. Journal Article: Choi JH, Lee YJ, Kim D. Image-guided nanomedicine for cancer. *J Pharm Investig.* 2017; 47(1):51–64.
23. Journal Article: Weissleder R. Molecular imaging in cancer. *Science.* 2006; 312(5777):1168–1171.
24. Journal Article: Willmann JK, Van Bruggen N, Dinkelborg LM, Gambhir SS. Molecular imaging in drug development. *Nat Rev Drug Discov.* 2008; 7(7):591–607.
25. Journal Article: Poudel K, Gautam M, Jin SG, Choi HG, Yong CS, Kim JO. Copper sulfide: An emerging adaptable nanoplatform in cancer theranostics. *Int J Pharm.* 2019; 562:135–150.
26. Journal Article: Wu MX, Gao J, Wang F, et al. Multistimuli responsive core—shell nanoplatform constructed from $Fe_3O_4@$ MOF equipped with pillar [6] arene nanovalves. *Small.* 2018; 14(17):1704440.
27. Journal Article: Yao J, Feng J, Chen J. External-stimuli responsive systems for cancer theranostic. *Asian J Pharm Sci.* 2016; 11(5):585–595.
28. Journal Article: Hu Q, Katti PS, Gu Z. Enzyme-responsive nanomaterials for controlled drug delivery. *Nanoscale.* 2014; 6(21):12273–12286.
29. Journal Article: Cheng Z, Li M, Dey R, Chen Y. Nanomaterials for cancer therapy: Current progress and perspectives. *J Hematol OncolJ Hematol Oncol.* 2021; 14(1):1–27.
30. Journal Article: Brito B, Price TW, Gallo J, Bañobre-López M, Stasiuk GJ. Smart magnetic resonance imaging-based theranostics for cancer. *Theranostics.* 2021; 11(18):8706.
31. Journal Article: Liu XL, Fan HM. Innovative magnetic nanoparticle platform for magnetic resonance imaging and magnetic fluid hyperthermia applications. *Curr Opin Chem Eng.* 2014; 4:38–46.
32. Journal Article: Chidambaram M, Manavalan R, Kathiresan K. Nanotherapeutics to overcome conventional cancer chemotherapy limitations. *J Pharm Pharm Sci.* 2011; 14(1):67–77.
33. Journal Article: Gelperina S, Kisich K, Iseman MD, Heifets L. The potential advantages of nanoparticle drug delivery systems in chemotherapy of tuberculosis. *Am J Respir Crit Care Med.* 2005; 172(12):1487–1490.
34. Journal Article: Park K. Nanotechnology: What it can do for drug delivery. *J Control Release Off J Control Release Soc.* 2007; 120(1–2):1.
35. Journal Article: Wen T, Quan G, Niu B, et al. Versatile nanoscale metal—organic frameworks (nMOFs): An emerging 3D nanoplatform for drug delivery and therapeutic applications. *Small.* 2021; 17(8):2005064.
36. Journal Article: Zhou Q, Zhang L, Yang T, Wu H. Stimuli-responsive polymeric micelles for drug delivery and cancer therapy. *Int J Nanomedicine.* 2018; 13:2921.
37. Journal Article: Navya PN, Kaphle A, Srinivas SP, Bhargava SK, Rotello VM, Daima HK. Current trends and challenges in cancer management and therapy using designer nanomaterials. *Nano Converg.* 2019; 6(1):1–30.
38. Journal Article: Tang S, Huang X, Chen X, Zheng N. Hollow mesoporous zirconia nanocapsules for drug delivery. *Adv Funct Mater.* 2010; 20(15):2442–2447.
39. Journal Article: Gaumet M, Vargas A, Gurny R, Delie F. Nanoparticles for drug delivery: The need for precision in reporting particle size parameters. *Eur J Pharm Biopharm.* 2008; 69(1):1–9.
40. Journal Article: Wu MX, Yang YW. Metal—organic framework (MOF)-based drug/cargo delivery and cancer therapy. *Adv Mater.* 2017; 29(23):1606134.
41. Journal Article: Sharma A, Goyal AK, Rath G. Recent advances in metal nanoparticles in cancer therapy. *J Drug Target.* 2018; 26(8):617–632.
42. Journal Article: Cherian AM, Nair SV, Lakshmanan VK. The role of nanotechnology in prostate cancer theranostic applications. *J Nanosci Nanotechnol.* 2014; 14(1):841–852.
43. Journal Article: Umapathi A, Kumawat M, Daima HK. Engineered nanomaterials for biomedical applications and their toxicity: A review. *Environ Chem Lett.* 2022; 20(1):445–468.
44. Journal Article: Zhang Y, Li M, Gao X, Chen Y, Liu T. Nanotechnology in cancer diagnosis: Progress, challenges and opportunities. *J Hematol OncolJ Hematol Oncol.* 2019; 12(1):1–13.

45. Journal Article: Frangioni JV. New technologies for human cancer imaging. *J Clin Oncol*. 2008; 26(24):4012–4021.
46. Journal Article: Mitchell MJ, Billingsley MM, Haley RM, Wechsler ME, Peppas NA, Langer R. Engineering precision nanoparticles for drug delivery. *Nat Rev Drug Discov*. 2021; 20(2):101–124.
47. Journal Article: Liu N, Chen X, Sun X, Sun X, Shi J. Persistent luminescence nanoparticles for cancer theranostics application. *J Nanobiotechnology*. 2021; 19(1):1–24.
48. Journal Article: Chen X, Wang T, Le W, et al. Smart sorting of tumor phenotype with versatile fluorescent Ag nanoclusters by sensing specific reactive oxygen species. *Theranostics*. 2020; 10(8):3430.
49. Journal Article: Zhang L, Forgham H, Huang X, et al. All-in-one inorganic nanoagents for near-infrared-II photothermal-based cancer theranostics. *Mater Today Adv*. 2022; 14:100226.
50. Journal Article: Jeong Y, Hwang HS, Na K. Theranostics and contrast agents for magnetic resonance imaging. *Biomater Res*. 2018; 22(1):1–13.
51. Journal Article: Gong N, Zhang Y, Teng X, et al. Proton-driven transformable nanovaccine for cancer immunotherapy. *Nat Nanotechnol*. 2020; 15(12):1053–1064.
52. Journal Article: Shen Z, Liu T, Li Y, et al. Fenton-reaction-acceleratable magnetic nanoparticles for ferroptosis therapy of orthotopic brain tumors. *ACS Nano*. 2018; 12(11):11355–11365.
53. Journal Article: Cattaneo M, Froio A, Gallino A. Cardiovascular imaging and theranostics in cardiovascular pharmacotherapy. *Eur Cardiol Rev*. 2019; 14(1):62.
54. Journal Article: Ramanathan S, Archunan G, Sivakumar M, et al. Theranostic applications of nanoparticles in neurodegenerative disorders. *Int J Nanomedicine*. 2018; 13:5561.
55. Journal Article: Shanbhag PP, Jog SV, Chogale MM, Gaikwad SS. Theranostics for cancer therapy. *Curr Drug Deliv*. 2013; 10(3):357–362.
56. Journal Article: Sharma R, Mody N, Agrawal U, Vyas SP. Theranostic nanomedicine; a next generation platform for cancer diagnosis and therapy. *Mini Rev Med Chem*. 2017; 17(18):1746–1757.
57. Journal Article: Tang J, Lobatto ME, Read JC, Mieszawska AJ, Fayad ZA, Mulder WJ. Nanomedical theranostics in cardiovascular disease. *Curr Cardiovasc Imaging Rep*. 2012; 5(1):19–25.
58. Journal Article: Gao Z, Mu W, Tian Y, et al. Self-assembly of paramagnetic amphiphilic copolymers for synergistic therapy. *J Mater Chem B*. 2020; 8(31):6866–6876.
59. Journal Article: Guo Y, Jiang K, Shen Z, et al. A small molecule nanodrug by self-assembly of dual anticancer drugs and photosensitizer for synergistic near-infrared cancer theranostics. *ACS Appl Mater Interfaces*. 2017; 9(50):43508–43519.
60. Journal Article: Li M, Liang Y, He J, Zhang H, Guo B. Two-pronged strategy of biomechanically active and biochemically multifunctional hydrogel wound dressing to accelerate wound closure and wound healing. *Chem Mater*. 2020; 32(23):9937–9953.
61. Journal Article: Muthu MS, Leong DT, Mei L, Feng SS. Nanotheranostics- application and further development of nanomedicine strategies for advanced theranostics. *Theranostics*. 2014; 4(6):660.
62. Journal Article: Shi C, Zhou Z, Lin H, Gao J. Imaging beyond seeing: Early prognosis of cancer treatment. *Small Methods*. 2021; 5(3):2001025.
63. Wang Y, Yang T, Ke H, et al. Smart albumin-biomineralized nanocomposites for multimodal imaging and photothermal tumor ablation. *Adv Mater*. 2015; 27(26):3874–3882.
64. Journal Article: Zheng M, Liu S, Li J, et al. Integrating oxaliplatin with highly luminescent carbon dots: An unprecedented theranostic agent for personalized medicine. *Adv Mater*. 2014; 26(21): 3554–3560.
65. Journal Article: Cabral H, Miyata K, Kishimura A. Nanodevices for studying nano-pathophysiology. *Adv Drug Deliv Rev*. 2014; 74:35–52.
66. Journal Article: Ray S, Li Z, Hsu CH, et al. Dendrimer-and copolymer-based nanoparticles for magnetic resonance cancer theranostics. *Theranostics*. 2018; 8(22):6322.
67. Journal Article: Shao L, Li Q, Zhao C, et al. Auto-fluorescent polymer nanotheranostics for self-monitoring of cancer therapy via triple-collaborative strategy. *Biomaterials*. 2019; 194:105–116.
68. Journal Article: Fu F, Wu Y, Zhu J, Wen S, Shen M, Shi X. Multifunctional lactobionic acid-modified dendrimers for targeted drug delivery to liver cancer cells: Investigating the role played by PEG spacer. *ACS Appl Mater Interfaces*. 2014; 6(18):16416–16425.
69. Journal Article: Wang JT, Hodgins NO, Wafa'T Al-Jamal JM, Sosabowski JK, Al-Jamal KT. Organ biodistribution of radiolabelled γδ T cells following liposomal alendronate administration in different mouse tumour models. *Nanotheranostics*. 2020; 4(2):71.
70. Journal Article: Costa PM, Wang JTW, Morfin JF, et al. Functionalised carbon nanotubes enhance brain delivery of amyloid-targeting Pittsburgh compound B (PiB)-derived ligands. *Nanotheranostics*. 2018; 2(2):168.

71. Journal Article: Mardhian DF, Vrynas A, Storm G, Bansal R, Prakash J. FGF2 engineered SPIONs attenuate tumor stroma and potentiate the effect of chemotherapy in 3D heterospheroidal model of pancreatic tumor. *Nanotheranostics*. 2020; 4(1):26.
72. Journal Article: Guo F, Li G, Ma S, Zhou H, Chen X. Multi-responsive nanocarriers based on β-CD-PNIPAM star polymer coated MSN-SS-Fc composite particles. *Polymers*. 2019; 11(10):1716.
73. Journal Article: Li G, Pei M, Liu P. pH/Reduction dual-responsive comet-shaped PEGylated CQD-DOX conjugate prodrug: Synthesis and self-assembly as tumor nanotheranostics. *Mater Sci Eng C*. 2020; 110:110653.
74. Journal Article: Govindasamy M, Manavalan S, Chen SM, et al. Determination of neurotransmitter in biological and drug samples using gold nanorods decorated f-MWCNTs modified electrode. *J Electrochem Soc*. 2018; 165(9):B370.
75. Journal Article: Mani V, Chen TW, Selvaraj S. Determination of Non-Steroidal Anti-Inflammatory Drug (NSAID) azathioprine in human blood serum and tablet samples using Multi-Walled Carbon Nanotubes (MWCNTs) decorated manganese oxide microcubes composite film modified electrode. *Int J Electrochem Sci*. 2017; 12:7446–7456.
76. Journal Article: Govindasamy M, Mani V, Chen SM, et al. Highly sensitive determination of non-steroidal anti-inflammatory drug nimesulide using electrochemically reduced graphene oxide nanoribbons. *RSC Adv*. 2017; 7(52):33043–33051.
77. Journal Article: Karthik R, Govindasamy M, Chen SM, et al. A facile graphene oxide based sensor for electrochemical detection of prostate anti-cancer (anti-testosterone) drug flutamide in biological samples. *Rsc Adv*. 2017; 7(41):25702–25709.
78. Journal Article: Keerthi M, Akilarasan M, Chen SM, et al. One-pot biosynthesis of reduced graphene oxide/prussian blue microcubes composite and its sensitive detection of prophylactic drug dimetridazole. *J Electrochem Soc*. 2018; 165(2):B27.
79. Journal Article: Sung SY, Su YL, Cheng W, et al. Graphene quantum dots-mediated theranostic penetrative delivery of drug and photolytics in deep tumors by targeted biomimetic nanosponges. *Nano Lett*. 2018; 19(1):69–81.
80. Journal Article: Siafaka PI, Okur NÜ, Karantas ID, Okur ME, Gündoğdu EA. Current update on nanoplatforms as therapeutic and diagnostic tools: A review for the materials used as nanotheranostics and imaging modalities. *Asian J Pharm Sci*. 2021; 16(1):24–46.
81. Journal Article: Qi B, Crawford AJ, Wojtynek NE, et al. Tuned near infrared fluorescent hyaluronic acid conjugates for delivery to pancreatic cancer for intraoperative imaging. *Theranostics*. 2020; 10(8):3413.
82. Journal Article: An D, Wu X, Gong Y, et al. Manganese-functionalized MXene theranostic nanoplatform for MRI-guided synergetic photothermal/chemodynamic therapy of cancer. *Nanophotonics*. 2022; 11(22):5177–5188.
83. Journal Article: Caldorera-Moore ME, Liechty WB, Peppas NA. Responsive theranostic systems: Integration of diagnostic imaging agents and responsive controlled release drug delivery carriers. *Acc Chem Res*. 2011; 44(10):1061–1070.
84. Journal Article: Penet MF, Krishnamachary B, Chen Z, Jin J, Bhujwalla ZM. Molecular imaging of the tumor microenvironment for precision medicine and theranostics. *Adv Cancer Res*. 2014; 124:235–256.
85. Journal Article: Ta T, Porter TM. Thermosensitive liposomes for localized delivery and triggered release of chemotherapy. *J Controlled Release*. 2013; 169(1–2):112–125.
86. Journal Article: Pereira Gomes I, Aparecida Duarte J, Chaves Maia AL, et al. Thermosensitive nanosystems associated with hyperthermia for cancer treatment. *Pharmaceuticals*. 2019; 12(4):171.
87. Journal Article: Tagami T, Foltz WD, Ernsting MJ, et al. MRI monitoring of intratumoral drug delivery and prediction of the therapeutic effect with a multifunctional thermosensitive liposome. *Biomaterials*. 2011; 32(27):6570–6578.
88. Journal Article: Liu D, Yang F, Xiong F, Gu N. The smart drug delivery system and its clinical potential. *Theranostics*. 2016; 6(9):1306–1323.
89. Journal Article: de Moura CL, Gallo J, García-Hevia L, Pessoa OD, Ricardo NM, Bañobre-López M. Magnetic hybrid wax nanocomposites as externally controlled theranostic vehicles: High MRI enhancement and synergistic magnetically assisted thermo/chemo therapy. *Chem Eur J*. 2020; 26(20):4531–4538.
90. Journal Article: Hiruta Y, Nemoto R, Kanazawa H. Design and synthesis of temperature-responsive polymer/silica hybrid nanoparticles and application to thermally controlled cellular uptake. *Colloids Surf B Biointerfaces*. 2017; 153:2–9.
91. Journal Article: Qiao ZY, Zhang R, Du FS, Liang DH, Li ZC. Multi-responsive nanogels containing motifs of ortho ester, oligo (ethylene glycol) and disulfide linkage as carriers of hydrophobic anti-cancer drugs. *J Controlled Release*. 2011; 152(1):57–66.

92. Journal Article: Xi J, Da L, Yang C, et al. Mn^{2+}-coordinated PDA@ DOX/PLGA nanoparticles as a smart theranostic agent for synergistic chemo-photothermal tumor therapy. *Int J Nanomedicine.* 2017; 12:3331.
93. Journal Article: Rao S, Song Y, Peddie F, Evans AM. Particle size reduction to the nanometer range: A promising approach to improve buccal absorption of poorly water-soluble drugs. *Int J Nanomedicine.* 2011; 6:1245.
94. Journal Article: Tang H, Guo Y, Peng L, et al. In vivo targeted, responsive, and synergistic cancer nanotheranostics by magnetic resonance imaging-guided synergistic high-intensity focused ultrasound ablation and chemotherapy. *ACS Appl Mater Interfaces.* 2018; 10(18):15428–15441.
95. Journal Article: Morales-Cruz M, Delgado Y, Castillo B, et al. Smart targeting to improve cancer therapeutics. *Drug Des Devel Ther.* 2019; 13:3753.
96. Journal Article: Pucci C, Martinelli C, Ciofani G. Innovative approaches for cancer treatment: Current perspectives and new challenges. *Ecancermedicalscience.* 2019; 13.
97. Journal Article: Issels RD, Lindner LH, Verweij J, et al. Neo-adjuvant chemotherapy alone or with regional hyperthermia for localised high-risk soft-tissue sarcoma: A randomised phase 3 multicentre study. *Lancet Oncol.* 2010; 11(6):561–570.
98. Journal Article: Baek S, Singh RK, Kim TH, et al. Triple hit with drug carriers: pH-and temperature-responsive theranostics for multimodal chemo-and photothermal therapy and diagnostic applications. *ACS Appl Mater Interfaces.* 2016; 8(14):8967–8979.

10 Light-Responsive Delivery Nanoplatforms in Cancer Theranostics

Atul Garkal, Anam Sami, Jahanvi Patel, Lajja Patel, Namdev L. Dhas and Tejal Mehta

CONTENTS

10.1 Introduction 155
10.2 Classification of Light-Responsive Drug Delivery Systems 156
    10.2.1 Photochemical Reactions 156
    10.2.2 Photoisomerization 159
10.3 Near-Infrared Light-Responsive Drug Delivery Systems 160
    10.3.1 Design Strategies and Mechanisms of Drug Release 160
    10.3.2 Two-Photon Absorption-Based NIRDDSs 163
    10.3.3 Up-Converting Nanoparticles 163
10.4 Applications of Photosensitive Nanoparticles in the Treatment of Cancer 164
    10.4.1 Photosensitive Nanosystems for Cancer Chemotherapy 164
    10.4.2 Cancer Therapies Based on Phototriggered Photosensitizer 164
    10.4.3 Combined Application of Photodynamic or Photothermal Treatment with Chemotherapy 165
10.5 Concluding Remarks and Future Perspectives 166
Acknowledgments 166
References 166

10.1 INTRODUCTION

In the twenty-first century, cancer is a key barrier to raising life expectancy globally. According to the International Agency for Research on Cancer, by 2030, there would be 13.2 million cancer-related deaths and 21.4 million new cases worldwide. 2 The most successful treatments for tumors at the moment are still surgery and chemotherapy, although, during the past several years, gene therapy and immunotherapy have been used more and more in cancer treatment plans. However, due to non-specificity, biodistribution, cycle stability, or systemic toxicity, anti-tumor medicines have not reached their full clinical therapeutic potential [1, 2].

The development of therapeutic and diagnostic approaches targeted at treating cancer has received significant attention. One of the most extensively used therapies is chemotherapy, which includes the medication doxorubicin (DOX). However, due to its toxicity, DOX delivered by conventional drug delivery systems (DDSs) can harm healthy tissues. Therefore, although it has strong anticancer effects, DOX can significantly reduce patient survival because of unfavorable side effects [3].

Innovative intelligent DDSs have been created as a result of unprecedented advancements in nanotechnology. By shielding healthy tissue from poisonous drugs, focusing on the location of the disease in the body, and displaying stimuli responsiveness for controlled drug release, these intelligent

DDSs successfully remove the negative side effects of conventional DDSs. Toxic medications have been shielded from healthy tissues using a variety of nanotechnologies. Numerous formulations are included in these nanotechnologies, such as polymer nanogels; dendrimers; micelles; liposomes; nanovesicles; and metal, magnetic, and inorganic nanoparticles (NPs) [4].

With the ongoing advancement of nanotechnology, drug delivery systems have seen a sharp increase in their use for tumor therapy. Nano-carriers were extensively used in oncotherapy due to their properties as tumor targets, drug-controlled release, or blood long circulation. However, due to several barriers, including inconsistent enhanced permeability and retention effects in various tumor types, rummaging effects by the mononuclear phagocytic system, the premature release of agents from NPs in the blood circulation, and complex tumor microenvironments (TMEs), conventional nanoplatforms did not produce desired therapeutic effects. The TME is made up of a wide variety of cell types, including immune cells, tumor cells, peripheral blood arteries, tumor stem cells, and cytokines extracellular matrix [5].

Endogenous triggers include unchecked cell proliferation, an odd nutritional environment, and metabolic pattern changes in tumor tissues that lead to an acidic pH, raised temperature, overexpressed proteins and enzymes, and high levels of the oxidation-deoxidation environment (GSH/ROS). Single stimulus-responsive drug-delivery system designs also employ exogenous stimuli. To make it easier for patients to adjust to complicated therapeutic applications, a manually regulated external stimulus can help to lessen the variances between people and different therapy phases. There are four main categories into which the trigger levels can be categorized. Exogenous stimuli, such as light, magnetic fields, and ultrasound, fall within the first category. Based on metal materials, heat-triggered DDSs could effectively convert absorbed magnetic field or light energy (often near-infrared (NIR)) into thermal energy. The second type of NP includes those that can alter a specific tumor microenvironment's size or charge to improve tumor penetration or cellular absorption. To facilitate endosome elopement or a "burst" release of "drug payloads", the pH of endosomes and lysosomes is lowered in the third category. The last group relies on overexerted proteins and enzymes, a high level of adenosine triphosphate, and chemical reduction in the cytoplasm to hasten the release of the drug payload or to reach nuclear or mitochondrial targets [6–8].

10.2 CLASSIFICATION OF LIGHT-RESPONSIVE DRUG DELIVERY SYSTEMS

10.2.1 Photochemical Reactions

Active singlet oxygen interacts with organic compounds to create derivatives of hydrobioxides, epoxides, and epoxides that play a significant role in rupturing the DNA of cells like different tumors. In photochemistry, photooxidation reactions of organic compounds are crucial. There is hope that deadly diseases can be treated with this new understanding. Chemicals absorb photons and shift from their ground states to their excited states, then to their triplet states in a photochemical reaction, in contrast to the majority of chemical reactions in organic chemistry, which take place between molecules in a stable electronic state (ground state). In photooxidation reactions, the oxygen molecule transforms into a more active oxygen in the presence of light and a catalyst ($^1O_2$). S1, S2, S3 . . . . . . If the electron's half-life is brief, it loses the energy it gained by producing fluorescence-related photon emissions and returns to the ground state. However, if the electron half-life is quite lengthy, electron transit takes place in the triplet state (T1, T2, . . .), where the electron spin is altered and turns parallel. The energy is then partially relaxed, and if the half-life period is brief (nanoseconds), it loses energy in the form of phosphorescence light. The half-life is raised to its active singlet state if it is a little bit longer (microseconds–seconds) [9].

10.2.1.1 Photooxidation

This process, often referred to as photosensitization-induced oxidation, produces radical oxygen species (ROS), including singlet oxygen (1O2), hydroxyl radicals (•OH), superoxide anions (•O2), and hydrogen peroxide (H2O2). Each of these ROS has a unique reactivity towards biological targets. The oxidation of certain nanocarrier components, such as unsaturated lipids, cholesterol, or other photooxidizable lipids, can be used to induce the release of cargo from nanoparticles by mechanically disrupting the structure of the nanocarrier. The production of 1O2, a potent oxidizing agent, when the photosensitizer molecule is exposed to light with the proper wavelength is noteworthy. Photodynamic treatment (PDT) uses this unpaired electron pair on 1O2 to damage cancer cells' membranes and trigger apoptosis since it confers high responsiveness to most cellular or molecular systems. There is evidence of these physicochemical changes resulting from photooxidation, which has been reviewed and validated by numerous studies. According to tests, different encapsulated dyes displayed different levels of permeability to photosensitive liposomes containing unsaturated lipids, with the smaller molecular weight colors releasing at a slower rate [10].

In one work, reactive unsaturated lipid 1,2-dio-leoyl-sn-glycero-3-phosphocholine (DOPC) and a saturated lipid 2-distearoyl-sn-glycero-3-phosphocholine (DSPC) were combined to create photosensitive liposomes. The inclusion of DOPC led to higher encapsulated doxorubicin release rates following permeabilization brought on by exposure to light at 665 nm. When compared to liposomes made only from DSPC, the addition of 2 mol% DOPC enhanced the rate of doxorubicin release by 11.6 times, and the addition of 5 mol% DOPC raised the released rate by 16.6 times. However, concentrations of DOPC greater than 5 mol% produced unstable liposomes with large rates of passive leakage, making them unsuitable for use in actual applications.

10.2.1.2 Photocleavage

Photocleavage strategy refers to the approach where molecules of interest in photolabile protecting groups (PPGs), which can be irreversibly removed to reinstate the molecules' biological activity upon photoirradiation. Several PPGs, such as coumarin, pyrenylene, ruthenium (II) complex, o-nitrobenzyl and its derivatives, and o-nitrobenzyl and derivatives, have been developed as phototriggers with varying chemical reactivity and photosensitivity. Small-molecular anticancer medicines, small interfering RNA (siRNA), enzymes, genes, and other therapeutics have been caged and photo-released at the region of interest by nanoagents such as micelles, magnetic nanoparticles, gold nanoparticles, UCNPs, and others.

The PPGs primarily carry out two crucial tasks: first, they act as linkers between the therapeutics and nanoagents, and second, they offer exquisite photosensitive control over therapeutics release. The PPGs undergo chemical bond breaking when exposed to light, releasing the medicines contained in the nanoagents. A high degree of photocontrollability in drug release and improved therapeutic efficacy with greater resistance to the biological environment before reaching the target locations are benefits of such an approach. A straightforward photocleavable DNA-drug nanostructure for the delivery of nucleic acids and camptothecin was described by Zhang and colleagues (CPT). The amphiphilic DNA-drug conjugates, which included a hydrophilic DNA strand and three hydrophobic CPT molecules joined by a phenol-based self-immolative linker and an o-nitrobenzyl moiety, allowed the nanoparticles to self-assemble. Upon being exposed to 365 nm radiation for 5 minutes, this conjugate was broken into free DNA and CPT. The targeted and advantageous light-controlled cytotoxicity of the photolabile DNA-drug nanostructures against cancer cells was evident.

The usage of 3 (bromomethyl)perylene and coumarin-4-ylmethyl derivatives, which absorb visible light, is a result of attempts to build protherapeutic nanoagents that respond to light of longer wavelengths. By encasing chlorambucil with 3-(bromomethyl)perylene, Singh and colleagues reported the formation of an organic perylene-3-ylmethanol nanoparticle. The absorption and emission peaks of the nanoparticles were localized at 350 to 550 nm and 625 nm, respectively. As a result, it is possible to photocontrollably monitor the medication release process. After being exposed to visible light (>410 nm) for 20 minutes, the polymer broke down and the nanoparticles separated, releasing chlorambucil. The observation of a considerable increase in cytotoxicity towards cancer cells was attributed to nanoparticles' greater intracellular drug absorption than free drug. The observation of a considerable increase in cytotoxicity towards cancer cells was attributed to nanoparticles' greater intracellular drug absorption than free drug. The same tactic was used by Zhao et al., who included the medication in perylene-3,4,9,10-tetrayltetramethanol. Two-photon excitation and upconversion processes are the typical methods for extending the irradiation wavelength to the biological optical window. Li and colleagues loaded the chlorambucil prodrug into yolkshell–shaped nanocages to show the first NIR-regulated prodrug release in a real animal model. The UCNPs (NaYF4:Yb,Tm@NaLuF4) covered with a mesoporous silica shell made up the yolkshell-shaped nanocage. Two octyl groups were added to the "phototrigger" amino-coumarin, and it was then conjugated with chlorambucil to create the hydrophobic prodrugs. To provide photo-controllable drug release, the mesoporous silica shell not only provided space for loading the prodrugs but also prevented interaction between the absorbed prodrug and endogenous enzymes. The amino-coumarin was broken by the high UV emissions at 345 and 360 nm upon 980 nm stimulation, releasing the chlorambucil that was then diffused from the nanoparticles and exerted cytotoxicity on tumor cells [11].

10.2.1.3 Photopolymerization

In a process called photopolymerization, unsaturated bonds between molecules in the drug carrier are polymerized in response to light, causing the medication to release. When photosensitive polymerizable moieties were exposed to the proper wavelength, the ensuing crosslinking results in the structure "shrinking" as a result. Regions within the lipid bilayer contract as a result of the crosslinking that takes place in the hydrophobic domain of the lipid bilayer in the case of carrier drug delivery systems like liposomes. Hence, a structural shift for improved drug release results.

The photon energy, photoinitiators, and photoreactive precursor are typically three key factors that control the photopolymerization process [4, 5]. When the integrated photoinitiators inside the composite absorb the incoming photons from the light source, photopolymerization begins, while the photoreactive precursor is subjected to radiation from a photon source of the right wavelength. In this active state, the stimulated monomers in the structure crosslink with one another to form a polymeric network. The phototriggered use of DC8,9PC in a drug delivery system showed that when exposed to light at 254 and 514 nm for 30 minutes, liposomes containing DC8,9PC, DPPC, stabilized with 1,2-distearoyl-sn-glycero-3-phosphoethanolamine-N-[amino(polyethylene glycol)-2000] (DSPE-PEG2000), were able to release over 40% of the anticancer drug. Recently, the scientists improved the formulation and quickened the distribution of a PDT medication called 2-(1-hexyloxyethyl)-2-devinyl pyropheophorbide-alpha (HPPH). After 5 minutes of exposure to radiation at 660 nm, the novel formulation showed a close to 40% release of the medication that was trapped inside, demonstrating quick photopolymerization pore development in the liposome. Although thorough biophysical studies are being carried out to fully understand the molecular interactions between HPPH and DC8,9PC within the lipid bilayer, it is believed that the preferential intercalation of HPPH into pockets of DC8,9PC is the cause of the significant reduction in time for phototriggered lipid destabilization and drug release.

Isolated coumarin monomers can polymerize to produce dimers through cycloaddition, which eliminates the need for additional monomers or photoinitiators while still enabling the reversal of the process through photocleavage. When coumarin moieties are exposed to UV light at wavelengths greater than 310 nm, they undergo [2s + 2s] photocycloaddition, which results in the formation of cyclobutane derivatives within the polymerization reaction. Two new Csingle bondC bonds are formed as a result of a photochemical interaction between photoexcited antibonding orbitals of one double bond and antibonding orbital of other double bond, is followed by spin inversion and ring closure to produce the cyclobutane ring adduct. With increasing levels of dimerization, these semi-stable linked polymers rigidify and can be used as polymeric networks to enclose cargo in photosensitive drug delivery systems.

10.2.2 Photoisomerization

The term "photoisomerization", sometimes known as "photoswitch", describes molecules that can reversibly alter their structural composition in response to UV or visible light exposure. As molecular photoswitches, azobenzene (azo) and its derivatives, spiropyrans (SP), and diarylethenes have all found extensive use. The stable trans isomer and the metastable cis isomer are the two reversible isomers for azo. When exposed to UV and visible light, two phenyl groups connected by an N-N connection in the azo isomerize into the trans and cis conformations, respectively. On the other hand, when exposed to UV light, SP can go through ring-opening isomerization, changing from a hydrophobic closed-ring structure to a hydrophilic zwitterionic merocyanine structure. Visible light can cause its trans-to-cis isomerization, just like it can with azo. The disruption of the nanoagents due to the reversible trans-cis structural alterations of the photoswitches causes on-demand medication release to occur in a photocontrollable way [11].

10.2.2.1 Photoisomerization Based on Azobenzene

When exposed to light irradiation reversibly 105–106 times, azobenzene can undergo isomerization on a timescale of microseconds or perhaps sub-nanoseconds before being fatigued. The rapid, reversible, and risk-free cis-trans geometrical isomerization of azobenzene moieties allows for the induction of large intra- and intermolecular modifications in polymeric materials. Azobenzene-containing polymeric materials can thus be used in a variety of systems, including drug delivery, photomechanical responsiveness, reversibly wettable surfaces, and biological and bio-inspired systems [12].

It has been researched to distribute DOX using micelles with azo inserted into their hydrophobic tails. Upon exposure to UV light, the photoisomerizable linkers caused the conformational change and broke apart to cause drug release. Also investigated were nanofibers for photocontrolled medication release. Electrospinning and a subsequent coupling with sodium azide were used to create the nanofibers. By a "click-reaction", the photoactive 4-progargyloxyazobenzene was attached to the nanofibers' surface. As a prodrug, 5-fluoroucil with -cyclodextrin was employed. By host–guest interactions between the cyclodextrin cavity and the trans-azo isomer, prodrug loading was made possible by the surfaces of nanofibers with transazo isomers. When exposed to UV light, the cis-azo isomer formed and released the prodrug.

Ju and colleagues described a UCNP-fueled DNA-azo nanopump for effective and regulated anticancer drug release, making use of UCNPs' capacity as light transducers. The azo-functionalized DNA chains were assembled on the NaYF4:Tm,Yb@ NaYF4 UCNPs to create the nanopump. DOX was specifically intercalated in the DNA helix that was created. In order to prevent drug exportation into the cytoplasm and reduce multidrug resistance, nuclear localization peptide (HIV-1 TAT), acting as the nucleus targeting moiety, was initially attached to the DNAazo nanopump (Figure 5a). Hyaluronic acid (HA) was further coated on the nanopump to target TAT by HA-mediated endocytosis, which can be degraded by hyaluronidase (HAase) in the tumor microenvironment. The UV and visible emissions from UCNPs under 980 nm light irradiation caused the rotation-inversion movement of azo, which started DNA hybridization and dehybridization to release DOX.

An outstanding in vivo therapeutic performance was consequently seen because approximately 86.7% of the DOX was released within 30 min in a manageable manner. The shift in the hydrophilic or hydrophobic character in reversibly photoresponding groups is typically significantly weaker than that for irreversible systems, which is one drawback of this approach. However, the photoactive moieties' utility in practical applications is severely constrained by the sluggish kinetics at which they react to light or return to their starting state in the absence of light [11].

10.2.2.2 Photoisomerization Based on Spiropyrans

Under UV light irradiation, colorless hydrophobic spiropyran derivatives can go through photoinduced ring-opening processes to create their isomeric hydrophilic merocyanine forms, which have strong visible absorption. Reversible photochemical conversion of the merocyanine isomer to the colorless spiropyrans is possible. Hence, adding spiropyran moieties to polymers can give them photoresponsive features that have a variety of uses, such as triggering the release of molecules from the spiropyran-containing polymers in response to light. The photoresponsive qualities of spiropyran-containing systems can, however, be reduced due to photoisomerization fatigue brought on by side reactions (the interaction of the excited triplet states of two spiropyran or with triplet oxygen). When the spiropyran moieties in the polymer nanoparticles are in other environments, like the surfaces of nanoparticles, they have significantly better reversibility [13, 14].

Wang et al. developed a nanogel of poly(acrylic acid-co-spiropyran methacrylate) cross-linked by N,Nbis(acryloyl)cystamine using the photoinduced hydrophobic-to-hydrophilic characteristic of SP to MC. This nanogel was transparent and pH and redox sensitive. The nanogels inflated when exposed to UV light (360 nm, 15 mW cm2) in an acidic environment because of the isomerization of hydrophobic SP to hydrophilic MC. When dithiothreitol, a reductive agent, was present, the disulfide cross-linkers broke down (DTT). Due to the disruption of the nanogels, 80% of the drug molecules (DOX was used as a model) were quickly released into the aqueous solution and cells. The unfilled nanogels had good biocompatibility before and after UV light irradiation (360 nm, 15 mW cm2, 1 min), even at higher concentrations (50 g mL1). This was demonstrated by in vitro cytotoxicity experiments. The cell viability of MCF-7 cancer cells for DOX-loaded nanogels (10 g mL1) reduced to 31% and 26%, respectively, before and after irradiation. The capacity of MC-based nanogels in imaging cancer cells (MCF-7 cells) was subsequently established, demonstrating the possibility of this material in applications of fluorescence imaging monitored medication delivery. This was made possible by the intense green emission of the MC in nanogels.

In a different instance, Li and colleagues proposed a new method for creating an amphiphilic dendrimer-star copolymer poly(caprolactone)-b-poly(methacrylic acid-co-spiropyran methacrylate) (DPCL-b-P(MAA-co-SPMA)) by combining ROP, ATRP, and postpolymerization. The SP isomerized to MC or merocyanine H+ (MCH) under UV irradiation or when the pH was decreased below 7.0, and DOX was released from the copolymer micelles as a result of altering the micelle morphologies [15, 16].

10.3 NEAR-INFRARED LIGHT-RESPONSIVE DRUG DELIVERY SYSTEMS

Due to deficiency of specificity, the conventional treatments of the cancer are not very effective. There are two types of conventional treatments: radiotherapy and chemotherapy. Novel drug delivery systems such as targeted DDSs are a recent topic for researchers. Novel DDSs also have some limitations, such as insufficient amount of drug release at the target site. So, it is essential to develop the drug delivery system such as stimuli drug delivery system which will overcome the limitations of the previously mentioned drug delivery systems. There are two types of stimuli: external and internal stimuli. Internal stimuli can be affected by individual variability, but external stimuli will overcome this drawback. One of external stimulus is light. Ultraviolet light has some limitations, such as poor penetration power and toxicity to cells and tissue. Near-infrared light with the wavelength of 650–900 nm has good penetration as well as not being toxic to cells and tissue [8].

10.3.1 Design Strategies and Mechanisms of Drug Release

There are three main mechanisms of drug release in NIRDDS:

1) Photothermal effect–based NIRDDS
2) Two-photon absorption-based NIRDDS
3) Up-converting nanoparticles

10.3.1.1 Photothermal Effect-Based NIRDDSs

The basic mechanism of drug release by photothermal effect is that upon exposure to near-infrared radiation, the carrier molecule (thermoresponsive material), which carries the drug moiety, is degraded or disrupted due to the increase in temperature. Various thermoresponsive materials are available, like carbon nanomaterials (graphene oxide and carbon nanotubes), gold nanomaterials (such as nanoparticles, nanorods, and nanocages), metallic oxides/sulfides, indocyanine green dye, melanin, and polyaniline [8].

10.3.1.1.1 Graphene-Based NIRDDSs

Graphene contains a sheet of two-dimensional (2D) single-layer sp2 hybridized carbon atoms in its aromatic structure. It contains strong carbon\carbon bonding, free π electrons, and reactive sites for surface reactions. This means graphene is a unique material which has mechanical, physicochemical, thermal, electronic, optical, and biomedical properties. Based on the number of layers in the structure, graphene-based material is divided into different categories: single-layer graphene, bilayer graphene, multilayer graphene, graphene oxide, and reduced graphene oxide [4, 17].

Graphene reacts with drugs by different mechanisms like adsorption or π-π hydrogen bonding by forming covalent bonds because it contains free hydroxyl and carboxylic groups. Graphene oxide also provides the functional groups to which DNA, protein, and RNA can be attached. Recently GO has been studied for various anticancer drug delivery methods, including chemotherapeutic agents, DNA, RNA, and genes. Graphene oxide absorbs NIR, produces heat, and causes hyperthermia [18].

Researcher has synthesized cisplatin multifunctional mesoporous silica nanoparticles. It contains fluorescent conjugates internally and polydopamine and graphene oxide externally. This structure enables sustained and controlled release of the cisplatin drug molecules. Along with this, the pH and NIR stimuli drug delivery system offers efficient cellular uptake and higher cytotoxicity to cancer cells [19].

There are also FA-GO (folic acid, graphene oxide) composite nanosheets, which work with manganese oxide nanoparticles. Manganese oxide nanoparticles decompose hydrogen peroxide into cancer cells, which leads to decreased hypoxic condition in cancer cells. Mno2 also acts as contrast for MRI imaging. A Mno2-FA-GO composite showed faster uptake in HeLa cells overexpressing FA-receptors. As the nanosheets have a large surface area, it showed fast degradation of hydrogen peroxide in cancer cells. Therefore, composite MnO2-FA-GO nanosheets could be a powerful carrier for cancer targeting and PTT applications [20].

10.3.1.1.2 Carbon Nanotube-Based NIRDDSs

Carbon nanotubes (CNT) are rolled graphite sheets in a cylindrical shape. The atoms are arranged in a hexagonal shape. CNTs have lengths in micrometers but diameters in nanometers. There are different types of carbon nanotubes available: single-walled, double-walled, and multi-walled CNTs (Figure 10.1) [21].

CNTs tend to form bundles in aqueous solution, so dispersibility into aqueous solution is the one of the major challenge in CNT, which can be overcome by surface functionalization. Doxorubicin was delivered through TAT-chitosan functionalized multi-walled CNTs. The system released doxorubicin for longer periods of time. MWCNT also retained optical properties for a high photothermal effect upon exposure to NIR radiation. So, through a combination of chemotherapy and photothermal therapy, the system exhibited superior anti-cancer efficacy [22].

Single-walled carbon nanotubes were used to deliver the anticancer drug doxorubicin. SWCNTs were incorporated into microcapsules with doxorubicin and evaluated for drug release patterns in the dark and under NIR irradiation. It was observed that there was no drug release in the dark, and in the presence of NIR, the specific release of doxorubicin was observed [23].

SWCNTs were coated with the polyethylene glycol (PEG) modified meso-porous silica (MS) and utilized as a multifunctional platform for imaging guided combination therapy of cancer. The

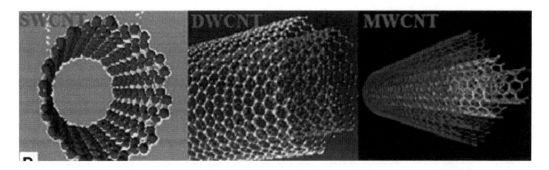

FIGURE 10.1 Structures of the three CNT types; SWCNT, DWCNT, and MWCNT, respectively. Adapted with permission from Aqel, A.; El-Nour, K. M. M. A.; Ammar, R. A. A.; Al-Warthan, A. Carbon Nanotubes, Science and Technology Part (I) Structure, Synthesis and Characterisation. Arab. J. Chem. 2012, 5 (1), 1–23. https://doi.org/10.1016/j.arabjc.2010.08.022.

anti-cancer drug doxorubicin was loaded into the structure. Irradiation of NIR led to release of the doxorubicin inside the cell, which has an anticancer effect [24].

10.3.1.1.3 Nanomaterial-Based NIRDDSs

Nowadays nano-scale drug delivery systems are gaining popularity for the treatment of cancer. Examples of nanomaterial are nanoparticles, nanocage, nanorods, and nanocubes. Gold nanoparticles have a better ability to convert radiation into heat than photothermal dyes. Gold nanoparticles have gained popularity in the treatment of cancer because they have a high optical absorption co-efficient, can be formulated on a large scale, and have a tunable absorption range [8].

Ultra-small gold nanoparticles were attached to the opening of meso-porous silica nanoparticles. This is an example of a combination of chemotherapy and photothermal therapy. Doxorubicin was loaded into MSN. Gold nanoparticles act as guards at the opening of MSN. Gold nanoparticles are irradiated with NIR and get heated. Studies showed that drug release occurred faster with NIR irradiation than without. So studies illustrate that MSN-Au nanoparticles can act as effective carriers for cancer therapy [25].

Resveratrol is loaded into chitosan-modified liposomes and surrounded by gold nanoshells. This structure is able to absorb NIR and convert it into heat. This will enable drug release. It also has high loading capacity, cellular uptake, and therapeutic effects of the drug [26].

10.3.1.1.4 Indocyanine Green Dye-Based NIRDDSs

Indocyanine green is a dye which absorbs near-infrared radiation and has the ability to convert light into heat. It absorption range is 740 to 800 nm. But it has some drawbacks, like poor solubility in water is poor, instability in aqueous medium, a short half-life (2–4 min), agglomeration in an aqueous medium, on-specific protein binding, and lack of targeting. Attempts have been made to reduce the aqueous instability as well to increase the half-life by formulating PLGA-coated indocyanine green dye nanoparticles [27].

The human epidermal growth factor receptor-2 (HER2) acts as a biomarker in ovarian cancer, so anti-HER2 antibody–loaded indocyanine green nanoparticles were formulated to target tumor cells in ovarian cancer. ICG nanoparticles will act as theranostic material to target the antibody to tumor cells [28].

Epirubicin is widely used as a chemotherapeutic agent in cancer treatment. Studies have been done on the formulation of ICG + epirubicin nanoparticles, which showed excellent stability, drug loading capacity, and pH/photo-responsive drug release and tumor targeting ability [29].

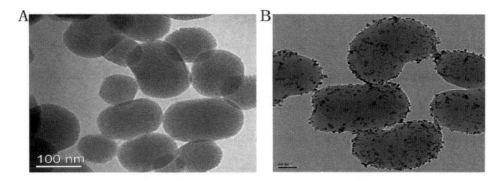

FIGURE 10.2 TEM images. (A) MSN (B) MSN–Au nanoparticles. Adapted with from permission Yang, Y.; Lin, Y.; Di, D.; Zhang, X.; Wang, D.; Zhao, Q.; Wang, S. Gold Nanoparticle-Gated Mesoporous Silica as Redox-Triggered Drug Delivery for Chemo-Photothermal Synergistic Therapy. J. Colloid Interface Sci. 2017, 508, 323–331. https://doi.org/10.1016/j.jcis.2017.08.050 [25].

10.3.1.1.5 Other Material-Based NIRDDSs

Some other materials also have the ability to convert NIR to heat. Copper sulfide, ferric oxide, polypyrrole, and black phosphorous are some examples. Copper sulfide (CuS) nanoparticles were used to deliver the anticancer drug doxorubicin (a CuS@mSiO$_2$-PEG core shell was prepared and was biocompatible and stable with high loading capacity). CuS will absorb NIR and convert it into heat, leading to release of the drug and giving a therapeutic effect.

Today ferric oxide (Fe$_3$O$_4$) nanoparticles are used for delivery of drugs because they have low toxicity, high drug loading capacity, and good magnetic behavior. Scientists have synthesized ferric oxide nanoparticles coated with Meso-2,3-dimercaptosuccinic acid (DMSA) for delivery of the anticancer drug doxorubicin for breast cancer. These synthesized nanoparticles have excellent pH- and NIR-induced drug release to target breast cancer cells (MDA-MB-231) [30].

Another study was done for the delivery of doxorubicin using polyacrylic acid-coated PPY/florescent mesoporous silica core-shell nanoparticles. These are also pH- and NIR-induced drug release systems [8].

10.3.2 Two-Photon Absorption-Based NIRDDSs

The need for UV or visible light and poor penetration power are limitations of light-responsive drug delivery systems. To overcome these drawbacks, two-photon excitation (TPE) in the NIR region is utilized for delivery of nanomedicines. They have good penetration power (up to 2 cm), lower scattering loss, and three-dimensional spatial resolution [31].

Croissant et al. developed mesoporous silica nanoparticles with fluorophore nanovalves. This system releases the drug due to two-photon excitation. First they developed a fluorophore, which was based on paracyclophane, which could absorb two photons in the NIR region and transfer energy through FRET to the azobenzene part of valves, which upon cis–trans isomerization opened valves to release the enclosed drug (camptothecin). Nanoparticles loaded with camtothecin showed 30% cell death upon irradiation with the 760-nm laser light, while no cell death was observed without irradiation [32].

10.3.3 Up-Converting Nanoparticles

Up-converting nanoparticles can transform NIR light into high-energy UV/visible light. This mechanism is used in various fields like drug delivery, photodynamic therapy, and biomedical imaging. For up-conversion of energy, lanthanide/rare earth metals are used. Liu et al. formulated NaYF4:

TmYb@NaYF4 core–shell nanoparticles coated with mesoporous silica. Then they installed an azo group that acts as a stirrer in the mesoporous silica. On nanoparticles, TAT peptide was attached to increase cellular uptake. Then the anticancer drug doxorubicin was loaded, which is called DoXUCNP@mSiO2-azo. When UV light irradiation occurs, the trans isomer of the azo compound is converted to a cis isomer, while on irradiation of visible light, the cis isomer is converted to a trans isomer. Nanoparticles absorb NIR and emit photons in the UV/visible region, which will be absorbed by the azo compounds in mesoporous silica. The reversible photoisomerization by simultaneous UV and visible light emitted by the UCNPs creates a continuous rotation–inversion movement. The back-and-forth wagging motion of the azo molecules will act as a molecular impeller that propels the release of doxorubicin.

10.4 APPLICATIONS OF PHOTOSENSITIVE NANOPARTICLES IN THE TREATMENT OF CANCER

Research has focused on developing novel therapeutic delivery systems with enhanced drug targeting to certain tissues in response to the growing demand for more efficient treatment techniques that are focused and safe. In particular, photo-triggered delivery systems provide a high degree of spatiotemporal control over where the medicine is given and the capacity to initiate the release of several drug payloads. All photoactivated medications used in clinical trials employ a photosensitizing agent for PDT, which is commonly used to treat cancer and has lately received more attention in research. These pharmaceuticals are supplied via conventional methods and are not inherently photosensitive [33–35].

10.4.1 PHOTOSENSITIVE NANOSYSTEMS FOR CANCER CHEMOTHERAPY

Clinical application of photosensitive delivery systems has previously been restricted by issues regarding the passive release of medications with exposure to sunlight after injection. However, recent studies have shown that these issues are immoderate. Targeted exposure of light results in optimal therapeutic release within a minute, whereas exposure to ambient sunlight results in very little drug leakage.

Second-degree selectivity is provided by the favorable deposition of nanoparticle drug delivery systems inside targeted tumor tissue, which results from changes in the leaky vasculature and accumulation of nanocarriers at the diseased site. Due to their adaptability, ease of fabrication, and enhanced time-dependent controlled release of drug, numerous efforts have since focused on developing nanoparticulate systems.

Various photosensitive nanoparticulate drug delivery systems have been developed using nano-sized structures like micelles, polymer-drug conjugates, dendrimers, and liposomes, including additional biorecognition components like antibodies[33].

10.4.2 CANCER THERAPIES BASED ON PHOTOTRIGGERED PHOTOSENSITIZER

PDT (photodynamic treatment) is a unique oncologic treatment that is tumor ablative and function sparing. It includes the delivery of a tumor-localizing photosensitizer (PS) and local activation of PS by illumination of tumors with specific-wavelength light [34]. Combining the PDT technique with the use of nanoparticles promoted theranostics, a phenomenon that was previously well known. This novel idea proposes a synergy between diagnosis and treatment, a method that has been demonstrated to be simple to implement in PDT employing nanoparticles as PSs photosensitizing vehicles.

An illustration of this would be the use of nanoparticles made with porphyrin (PPNs) and poly (vinyl alcohol). To be more precise, vehicles serving as PSs are capable of transporting antitumoral medications. If NIR light activates the PPNs, the antitumoral drugs would be released at the tumoral location. These innovative nanoplatforms not only released active substances at the targeted tumoral spot but also can modulate the biochemical reactions in the cancer cells leading to apoptosis[36]. Furthermore, gold nano-shells can be utilized in PDT as photosensitizers. They therefore are found to be able inhibit the growth of melanoma cells because of their capability to generate significant reactive oxygen species for the photodynamic method under NIR light [37]. In order to achieve a combined effect of chemo- and phototherapy in melanoma, Zhang et al. observed micelle production from block copolymer for the combined delivery of the anticancer drugs pheophorbide A and doxorubicin. The polymer chain included these chemicals, and the produced micelles were effectively absorbed into melanoma cells with light-induced ROS production seen both in vitro and in vivo. Micelles showed a significant increase in tumor growth suppression compared to treatments using just doxorubicin, nearly higher than that of micelles treated without irradiation [38].

Photo-based therapeutic methods like photodynamic and photothermal therapy have gained a lot of attention as cancer treatments due to their noninvasiveness, controllability, targeted therapy, high spatiotemporal accuracy, and low drug resistance. Continuously using tissue oxygen and the PDT-mediated consequences of shutting down vasculature would further worsen hypoxia and, in turn, reduce PDT efficiency since PDT solely relies on the availability of tissue oxygen to make O_2. PTT is an oxygen-free therapeutic approach that kills tumor cells by localized hyperthermia brought on by a photothermal chemical that is activated by near-infrared (NIR) light [39]. The key function of nano-agents in this process is to increase the selectivity of heat production at the lesional site [36]. PTT can increase intra-tumoral blood flow and dilate blood vessels to increase tumor reoxygenation in TME. External light sources (for instance, 600–800 nm light) that can more effectively penetrate the target tissue and can more thoroughly cover solid tumors may activate intracellular delivery as a result of photochemical internalization (PCI) caused by PDT, which can increase the permeability of the cell membrane and increase nanoparticle uptake by the cell. Paclitaxel (PTX) administration as a chemotherapeutic agent along with PTT for solid tumors, Sun et al. used Pluronic-b-poly(L-lysine) nanoparticles (Pluronic-PLL@AuNPs) coated with gold nanoparticles. After being exposed to 808-nm NIR laser irradiation, the nanoparticles demonstrated effective photothermal heating capabilities and a synergistic impact of chemo-photothermal therapy. The temperature of the Pluronic-PLL@Au NP tumors after PTX injection rose to 34 °C, which was sufficient to abolish tumors in vivo [40]. After 45 days of careful monitoring and therapy, it was ultimately possible to obtain a 100% survival rate in mice by combining PTT and PDT procedures. Just one out of all six mice had recurring malignancies as well. These nanoparticles can be utilized as imaging tools, which may be implicated in tumoral diagnoses and monitoring, and were shown to have a precision of about 95% [36].

10.4.3 Combined Application of Photodynamic or Photothermal Treatment with Chemotherapy

Chemotherapeutics can address the problem of light penetration in phototherapy and may also increase the sensitivity of cancer cells to hyperthermia or ROS, while broad-spectrum activity of photosensitive agents and lack of resistance to phototherapy provide activity against drug-resistant cancer cells. Combining photosensitive agents and chemotherapeutic drugs may result in synergistic therapeutic effects [41]. Chemotherapy treatment failure is often caused by MDR mechanisms involving cell membrane efflux pumps. By preventing or reducing drug efflux, many drug-delivery methods have been devised to enhance the accumulation of chemotherapeutic medicines in the cytoplasm or nucleus. Due to the suppression of drug-efflux P-glycoprotein pumps in MDR cells brought on by ROS production by the photosensitizer, chemotherapy and PDT may also improve the effectiveness of treatment. The observed effects on blood flow and oxygen saturation suggest that chemotherapy and PTT together may have synergistic efficacy against cancer hypoxia [42, 43].

TABLE 10.1
Examples of Recently Developed Photo-Triggered Drug Release Systems Loaded with Chemotherapeutic Agents, with and without Combination with PDT and PTT

Delivery System	Chemotherapeutic Agent	Therapy/Combination Therapy	Reference
Polymeric micelles	Doxorubicin	PDT	[44]
Hydrogel	Doxorubicin	PDT	[45]
Silica nanoparticles	Doxorubicin	PDT	[46]
Liposomes	Platinum (IV) chloride	PDT	[47]
Polymeric nanoparticles	DOX	PDT + chemotherapy	[48]
Micelles	Mitoxantrone (MX)	PDT + chemotherapy	[49]
Nanospheres	CP-TPP/Au/PEG	PDT + PTT	[50]
Nanocluster	Gold-hyaluronan	PTT + PDT	[51]
Nanorods	Gold	PTT+RT (radio therapy)	[52]

10.5 CONCLUDING REMARKS AND FUTURE PERSPECTIVES

Stimuli-responsive drug delivery systems achieve targeted delivery with controlled release of therapeutics for the treatment of chronic diseases like cancer. Light-responsive drug delivery systems are a futuristic type for the diagnosis and treatment of all types of cancers. The photochemical, photoisomerization, or photothermal mechanisms are common mechanisms of light-sensitive drug delivery. To date, a wide range of drug delivery systems has been developed with photo triggerable mechanisms, not limited to nanoparticles, liposomes, micelles, and hydrogels. These formulations that have been investigated are not limited to cancer therapy but are also used in the treatment of various dermatological conditions (usually UV-driven systems) and ocular diseases. In recent research, NIR light-responsive chemical moieties can be expanded to other areas such as immunotherapy and gene therapy for cancers and other diseases based on the results of this research. Combination treatments of chemotherapy and PDT and the potential for PTTs to combat multidrug resistance in cancer treatment is promising.

ACKNOWLEDGMENTS

The authors would like to thank Nirma University for providing financial assistance in the form of Nirma University fellowship-SRF to Atul Garkal (NU/Ph.D./IP/GAD/19–20/1496).

REFERENCES

[1] R. Jia, L. Teng, L. Gao, T. Su, L. Fu, Z. Qiu, Y. Bi, Advances in multiple stimuli-responsive drug-delivery systems for cancer therapy, Int. J. Nanomedicine. 16 2021) 1525–1551. https://doi.org/10.2147/IJN.S293427.

[2] N. Dhas, R. Kudarha, A. Pandey, A.N. Nikam, S. Sharma, A. Singh, A. Garkal, K. Hariharan, A. Singh, P. Bangar, D. Yadhav, D. Parikh, K. Sawant, S. Mutalik, N. Garg, T. Mehta, Stimuli responsive and receptor targeted iron oxide based nanoplatforms for multimodal therapy and imaging of cancer: Conjugation chemistry and alternative therapeutic strategies, J. Control. Release. 333 (2021) 188–245. https://doi.org/10.1016/j.jconrel.2021.03.021.

[3] F. Wang, Z. Yuan, P. McMullen, R. Li, J. Zheng, Y. Xu, M. Xu, Q. He, B. Li, H. Chen, near-infrared-light-responsive lipid nanoparticles as an intelligent drug release system for cancer therapy, Chem. Mater. 31 (2019) 3948–3956. https://doi.org/10.1021/acs.chemmater.9b00150.

[4] A. Garkal, D. Kulkarni, S. Musale, T. Mehta, P. Giram, Electrospinning nanofiber technology: A multifaceted paradigm in biomedical applications, New J. Chem. 45 (2021) 21508–21533. https://doi.org/10.1039/D1NJ04159B.

[5] T. Chanmee, P. Ontong, K. Konno, N. Itano, Tumor-associated macrophages as major players in the tumor microenvironment, Cancers (Basel). 6 (2014) 1670–1690. https://doi.org/10.3390/cancers6031670.

[6] H.S. El-Sawy, A.M. Al-Abd, T.A. Ahmed, K.M. El-Say, V.P. Torchilin, Stimuli-responsive nano-architecture drug-delivery systems to solid tumor micromilieu: Past, present, and future perspectives, ACS Nano. 12 (2018) 10636–10664. https://doi.org/10.1021/acsnano.8b06104.

[7] R.K. Jain, T. Stylianopoulos, Delivering nanomedicine to solid tumors, Nat. Rev. Clin. Oncol. 7 (2010) 653–664. https://doi.org/10.1038/nrclinonc.2010.139.

[8] A. Raza, U. Hayat, T. Rasheed, M. Bilal, H.M.N. Iqbal, "Smart" materials-based near-infrared light-responsive drug delivery systems for cancer treatment: A review, J. Mater. Res. Technol. 8 (2019) 1497–1509. https://doi.org/10.1016/j.jmrt.2018.03.007.

[9] S.A. Khayyat, L.S. Roselin, Recent progress in photochemical reaction on main components of some essential oils, J. Saudi Chem. Soc. 22 (2018) 855–875. https://doi.org/10.1016/j.jscs.2018.01.008.

[10] A. Chivate, A. Garkal, K. Hariharan, T. Mehta, Exploring novel carrier for improving bioavailability of itraconazole: Solid dispersion through hot-melt extrusion, J. Drug Deliv. Sci. Technol. 63 (2021) 102541. https://doi.org/10.1016/j.jddst.2021.102541.

[11] Y. Zhang, C. Xu, X. Yang, K. Pu, Photoactivatable protherapeutic nanomedicine for cancer, Adv. Mater. 32 (2020) 2002661. https://doi.org/10.1002/adma.202002661.

[12] P. Xiao, J. Zhang, J. Zhao, M.H. Stenzel, Light-induced release of molecules from polymers, Prog. Polym. Sci. 74 (2017) 1–33. https://doi.org/10.1016/j.progpolymsci.2017.06.002.

[13] M.-Q. Zhu, L. Zhu, J.J. Han, W. Wu, J.K. Hurst, A.D.Q. Li, Spiropyran-based photochromic polymer nanoparticles with optically switchable luminescence, J. Am. Chem. Soc. 128 (2006) 4303–4309. https://doi.org/10.1021/ja0567642.

[14] J.A. Delaire, K. Nakatani, Linear and nonlinear optical properties of photochromic molecules and materials, Chem. Rev. 100 (2000) 1817–1846. https://doi.org/10.1021/cr980078m.

[15] S. Chen, Q. Bian, P. Wang, X. Zheng, L. Lv, Z. Dang, G. Wang, Photo, pH and redox multi-responsive nanogels for drug delivery and fluorescence cell imaging, Polym. Chem. 8 (2017) 6150–6157. https://doi.org/10.1039/C7PY01424D.

[16] W. Yuan, X. Gao, E. Pei, Z. Li, Light- and pH-dually responsive dendrimer-star copolymer containing spiropyran groups: Synthesis, self-assembly and controlled drug release, Polym. Chem. 9 (2018) 3651–3661. https://doi.org/10.1039/C8PY00721G.

[17] S. Goenka, V. Sant, S. Sant, Graphene-based nanomaterials for drug delivery and tissue engineering, J. Control. Release. 173 (2014) 75–88. https://doi.org/10.1016/j.jconrel.2013.10.017.

[18] L. Liu, Q. Ma, J. Cao, Y. Gao, S. Han, Y. Liang, T. Zhang, Y. Song, Y. Sun, Recent progress of graphene oxide-based multifunctional nanomaterials for cancer treatment, Cancer Nanotechnol. 12 (2021) 18. https://doi.org/10.1186/s12645-021-00087-7.

[19] A.-V. Tran, K. Shim, T.-T. Vo Thi, J.-K. Kook, S.S.A. An, S.-W. Lee, Targeted and controlled drug delivery by multifunctional mesoporous silica nanoparticles with internal fluorescent conjugates and external polydopamine and graphene oxide layers, Acta Biomater. 74 (2018) 397–413. https://doi.org/10.1016/j.actbio.2018.05.022.

[20] J.H. Lim, D.E. Kim, E.-J. Kim, C.D. Ahrberg, B.G. Chung, Functional graphene oxide-based nanosheets for photothermal therapy, Macromol. Res. 26 (2018) 557–565. https://doi.org/10.1007/s13233-018-6067-3.

[21] A. Aqel, K.M.M.A. El-Nour, R.A.A. Ammar, A. Al-Warthan, Carbon nanotubes, science and technology part (I) structure, synthesis and characterisation, Arab. J. Chem. 5 (2012) 1–23. https://doi.org/10.1016/j.arabjc.2010.08.022.

[22] X. Dong, Z. Sun, X. Wang, X. Leng, An innovative MWCNTs/DOX/TC nanosystem for chemo-photothermal combination therapy of cancer, Nanomedicine Nanotechnology, Biol. Med. 13 (2017) 2271–2280. https://doi.org/10.1016/j.nano.2017.07.002.

[23] M.A. Correa-Duarte, A. Kosiorek, W. Kandulski, M. Giersig, L.M. Liz-Marzán, Layer-by-layer assembly of multiwall carbon nanotubes on spherical colloids, Chem. Mater. 17 (2005) 3268–3272. https://doi.org/10.1021/cm047710e.

[24] J. Liu, C. Wang, X. Wang, X. Wang, L. Cheng, Y. Li, Z. Liu, Mesoporous silica coated single-walled carbon nanotubes as a multifunctional light-responsive platform for cancer combination therapy, Adv. Funct. Mater. 25 (2015) 384–392. https://doi.org/10.1002/adfm.201403079.

[25] Y. Yang, Y. Lin, D. Di, X. Zhang, D. Wang, Q. Zhao, S. Wang, Gold nanoparticle-gated mesoporous silica as redox-triggered drug delivery for chemo-photothermal synergistic therapy, J. Colloid Interface Sci. 508 (2017) 323–331. https://doi.org/10.1016/j.jcis.2017.08.050.

[26] M. Wang, Y. Liu, X. Zhang, L. Luo, L. Li, S. Xing, Y. He, W. Cao, R. Zhu, D. Gao, Gold nanoshell coated thermo-pH dual responsive liposomes for resveratrol delivery and chemo-photothermal synergistic cancer therapy, J. Mater. Chem. B. 5 (2017) 2161–2171. https://doi.org/10.1039/C7TB00258K.

[27] V. Saxena, M. Sadoqi, J. Shao, Indocyanine green-loaded biodegradable nanoparticles: Preparation, physicochemical characterization and in vitro release, Int. J. Pharm. 278 (2004) 293–301. https://doi.org/10.1016/j.ijpharm.2004.03.032.

[28] B. Bahmani, Y. Guerrero, D. Bacon, V. Kundra, V.I. Vullev, B. Anvari, Functionalized polymeric nanoparticles loaded with indocyanine green as theranostic materials for targeted molecular near infrared fluorescence imaging and photothermal destruction of ovarian cancer cells, Lasers Surg. Med. 46 (2014) 582–592. https://doi.org/10.1002/lsm.22269.

[29] Y. Li, G. Liu, J. Ma, J. Lin, H. Lin, G. Su, D. Chen, S. Ye, X. Chen, X. Zhu, Z. Hou, Chemotherapeutic drug-photothermal agent co-self-assembling nanoparticles for near-infrared fluorescence and photoacoustic dual-modal imaging-guided chemo-photothermal synergistic therapy, J. Control. Release. 258 (2017) 95–107. https://doi.org/10.1016/j.jconrel.2017.05.011.

[30] Y. Oh, J.-Y. Je, M.S. Moorthy, H. Seo, W.H. Cho, pH and NIR-light-responsive magnetic iron oxide nanoparticles for mitochondria-mediated apoptotic cell death induced by chemo-photothermal therapy, Int. J. Pharm. 531 (2017) 1–13. https://doi.org/10.1016/j.ijpharm.2017.07.014.

[31] J. Croissant, M. Maynadier, A. Gallud, H. Peindy N'Dongo, J.L. Nyalosaso, G. Derrien, C. Charnay, J.-O. Durand, L. Raehm, F. Serein-Spirau, N. Cheminet, T. Jarrosson, O. Mongin, M. Blanchard-Desce, M. Gary-Bobo, M. Garcia, J. Lu, F. Tamanoi, D. Tarn, T.M. Guardado-Alvarez, J.I. Zink, Two-photon-triggered drug delivery in cancer cells using nanoimpellers, Angew. Chemie. 125 (2013) 14058–14062. https://doi.org/10.1002/ange.201308647.

[32] J. Croissant, A. Chaix, O. Mongin, M. Wang, S. Clément, L. Raehm, J.-O. Durand, V. Hugues, M. Blanchard-Desce, M. Maynadier, A. Gallud, M. Gary-Bobo, M. Garcia, J. Lu, F. Tamanoi, D.P. Ferris, D. Tarn, J.I. Zink, Two-photon-triggered drug delivery via fluorescent nanovalves, Small. 10 (2014) 1752–1755. https://doi.org/10.1002/smll.201400042.

[33] P. Pan, D. Svirskis, S.W.P. Rees, D. Barker, G.I.N. Waterhouse, Z. Wu, Photosensitive drug delivery systems for cancer therapy: Mechanisms and applications, J. Control. Release. 338 (2021) 446–461. https://doi.org/10.1016/j.jconrel.2021.08.053.

[34] S.S. Lucky, K.C. Soo, Y. Zhang, Nanoparticles in photodynamic therapy, Chem. Rev. 115 (2015) 1990–2042. https://doi.org/10.1021/cr5004198.

[35] C.S. Linsley, B.M. Wu, Recent advances in light-responsive on-demand drug-delivery systems, Ther. Deliv. 8 (2017) 89–107. https://doi.org/10.4155/tde-2016-0060.

[36] A. Crintea, A.G. Dutu, G. Samasca, I.A. Florian, I. Lupan, A.M. Craciun, The nanosystems involved in treating lung cancer, Life. 11 (2021) 1–21. https://doi.org/10.3390/life11070682.

[37] C. Beiu, C. Giurcaneanu, A.M. Grumezescu, A.M. Holban, L.G. Popa, M.M. Mihai, Nanosystems for improved targeted therapies in Melanoma, J. Clin. Med. 9 (2020). https://doi.org/10.3390/jcm9020318.

[38] C. Zhang, J. Zhang, Y. Qin, H. Song, P. Huang, W. Wang, C. Wang, C. Li, Y. Wang, D. Kong, Co-delivery of doxorubicin and pheophorbide A by pluronic F127 micelles for chemo-photodynamic combination therapy of melanoma, J. Mater. Chem. B. 6 (2018) 3305–3314. https://doi.org/10.1039/c7tb03179c.

[39] W. Zhang, C. Zhang, C. Yang, X. Wang, W. Liu, M. Yang, Y. Cao, H. Ran, Photochemically-driven highly efficient intracellular delivery and light/hypoxia programmable triggered cancer photo-chemotherapy, J. Nanobiotechnology. 21 (2023) 1–16. https://doi.org/10.1186/s12951-023-01774-w.

[40] Y. Sun, Q. Wang, J. Chen, L. Liu, L. Ding, M. Shen, J. Li, B. Han, Y. Duan, Temperature-sensitive gold nanoparticle-coated Pluronic-PLL nanoparticles for drug delivery and chemo-photothermal therapy, Theranostics. 7 (2017) 4424–4444. https://doi.org/10.7150/thno.18832.

[41] X. Huang, J. Wu, M. He, X. Hou, Y. Wang, X. Cai, H. Xin, F. Gao, Y. Chen, Combined cancer chemo-photodynamic and photothermal therapy based on ICG/PDA/TPZ-loaded nanoparticles (2019). https://doi.org/10.1021/acs.molpharmaceut.9b00119.

[42] X. Li, J.F. Lovell, J. Yoon, X. Chen, Clinical development and potential of photothermal and photodynamic therapies for cancer, Nat. Rev. Clin. Oncol. 17 (2020). https://doi.org/10.1038/s41571-020-0410-2.

[43] H.S. Han, K.Y. Choi, Advances in nanomaterial-mediated photothermal cancer therapies : Toward clinical applications Biomedicines (2021) 1–15.

[44] Q. Zheng, Y. He, Q. Tang, Y. Wang, N. Zhang, J. Liu, Q. Liu, S. Zhao, P. Hu, An NIR-guided aggregative and self-immolative nanosystem for efficient cancer targeting and combination anticancer therapy, Mol. Pharm. 15 (2018) 4985–4994. https://doi.org/10.1021/acs.molpharmaceut.8b00599.

[45] D.S.B. Anugrah, K. Ramesh, M. Kim, K. Hyun, K.T. Lim, Near-infrared light-responsive alginate hydrogels based on diselenide-containing cross-linkage for on demand degradation and drug release, Carbohydr. Polym. 223 (2019) 115070. https://doi.org/10.1016/j.carbpol.2019.115070.

[46] S. Li, F. Wang, Z. Yang, J. Xu, H. Liu, L. Zhang, W. Xu, Emulsifying performance of near-infrared light responsive polydopamine-based silica particles to control drug release, Powder Technol. 359 (2020) 17–26. https://doi.org/10.1016/j.powtec.2019.09.064.

[47] Y. Yang, X. Liu, W. Ma, Q. Xu, G. Chen, Y. Wang, H. Xiao, N. Li, X.J. Liang, M. Yu, Z. Yu, Light-activatable liposomes for repetitive on-demand drug release and immunopotentiation in hypoxic tumor therapy, Biomaterials. 265 (2021) 120456. https://doi.org/10.1016/j.biomaterials.2020.120456.

[48] Y. Zhang, F. Huang, C. Ren, L. Yang, J. Liu, Z. Cheng, L. Chu, J. Liu, Targeted chemo-photodynamic combination platform based on the DOX prodrug nanoparticles for enhanced cancer therapy, ACS Appl. Mater. Interfaces. 9 (2017) 13016–13028. https://doi.org/10.1021/acsami.7b00927.

[49] Y. Han, Y. An, G. Jia, X. Wang, C. He, Y. Ding, Q. Tang, Theranostic micelles based on upconversion nanoparticles for dual-modality imaging and photodynamic therapy in hepatocellular carcinoma, Nanoscale. 10 (2018) 6511–6523. https://doi.org/10.1039/c7nr09717d.

[50] X. Wei, H. Chen, H.P. Tham, N. Zhang, P. Xing, G. Zhang, Y. Zhao, Combined photodynamic and photothermal therapy using cross-linked polyphosphazene nanospheres decorated with gold nanoparticles, ACS Appl. Nano Mater. 1 (2018) 3663–3672. https://doi.org/10.1021/acsanm.8b00776.

[51] H.S. Han, K.Y. Choi, H. Lee, M. Lee, J.Y. An, S. Shin, S. Kwon, D.S. Lee, J.H. Park, Gold-nanoclustered hyaluronan nano-assemblies for photothermally maneuvered photodynamic tumor ablation, ACS Nano. 10 (2016) 10858–10868. https://doi.org/10.1021/acsnano.6b05113.

[52] Q. Sun, J. Wu, L. Jin, L. Hong, F. Wang, Z. Mao, M. Wu, Cancer cell membrane-coated gold nanorods for photothermal therapy and radiotherapy on oral squamous cancer, J. Mater. Chem. B. 8 (2020) 7253–7263. https://doi.org/10.1039/d0tb01063d.

11 Magnetic Field-Responsive Delivery Nanoplatforms in Cancer Theranostics

Manish P. Patel, Rutvi V. Patel, Mehul R. Chorawala, Avinash K. Khadela, Sandip P. Dholakia and Jayvadan K. Patel

CONTENTS

11.1	Epidemiology of Cancer	170
11.2	Introduction to Cancer	171
11.3	Treatment	172
11.4	Targeted Therapy	172
	11.4.1 Molecular Targeted Therapy	173
	11.4.2 Radiation Therapy	173
	11.4.3 Combination Therapy	174
	11.4.4 Targeted Therapy	174
11.5	Diagnosis	174
11.6	Cancer Theranostics	175
11.7	Nanoparticles in Cancer Theranostics	175
11.8	Magnetic Nanoparticles in Cancer Theranostics	176
11.9	Cancer Therapy with Magnetic Nanoparticle Drug Delivery	178
11.10	Magnetic Nanoparticles Used in Cancer Therapy and Diagnosis (Cancer Theranostics)	178
	11.10.1 Administration and Biodistribution of Magnetic Nanoparticles	178
11.11	Active Substances Attached to Magnetic Field Nanoparticles	179
	11.11.1 Chemotherapeutic Agents	179
	11.11.2 Antibodies	179
	11.11.3 Radiotherapy	180
	11.11.4 Gene Therapy	180
11.12	Synthesis of Superparamagnetic Iron Oxide Nanoparticles	180
11.13	SPION Functionalization	181
References		181

11.1 EPIDEMIOLOGY OF CANCER

The earliest known statement of cancer was written in an Egyptian papyrus around 1600 BC. It was thought to be untreatable until the late 1700s, when anaesthesia, better procedures, and pathological control improved the removal of tumours.[1] Cancer is a significant global health issue. A rise in cancer in the coming decades is foreseen by observing global demographic features; around 420 million new cancer cases will be encountered by 2025. Viewing GLOBACON data, 14.1 million new cancer patients were seen and 8.2 million patient deaths owing to cancer were estimated in 2021. According to a 2008 study by the International Agency for Research on Cancer, there could be 13–17 million disease-related deaths worldwide by 2030, vastly increasing since 1975.[2] The major contributor to cancer incidence and mortality globally is lung cancer. Treatments for cancer in the past 15 years have significantly changed owing to new developments in the fields of molecular and tumour biology. In the past, cancer was treated according to the organ it originates from or simplistic histomorphologic features [1]. Cancer also poses a critical public health issue in the united states. Statistically, one woman and two men out of three are likely to develop cancer throughout their lives. Due to the rise in cancer diagnoses, owing to the population ageing and growing and also rates of survival increasing, the number of cancer survivors has also grown. The phrase cancer survivor signifies an individual who has been diagnosed with cancer from the time of diagnosis through their life. Cancer survival includes three critical phases, the start of the diagnosis to the end of the first treatment, the extension of survival owing to the treatment, and lengthy survival.

Treatment focuses on lengthening the lives of cancer patients with advanced forms of sickness or curing cancer, all the while ensuring the highest possible quality of life. Patients free of cancer must handle abiding effects caused by the remedy and the psychological effects such as the fear of cancer's reappearance. These people who have survived the disease should also keep in mind other medical and social worries that may arise due to their age or other present diseases or socioeconomic status.[2] Blood cancer, cancers of the brain, and cancers of the lymph nodes, correspondingly, account for the highest percentage of malignancies in children.[1] In males, the highest occurrence of cancer is seen in the prostate, lung and bronchus, colon and rectum, and urinary bladder. For females, the frequent cancer types are breast, lung and bronchus, colon and rectum, uterine corpus, and thyroid. The data here signifies that prostate and breast cancer make up a high amount of cancer in females and males.[3]

11.2 INTRODUCTION TO CANCER

Cancer is a complex disorder that includes many tempo-spatial modifications in cell physiological behaviour that inevitably causes malignant tumours. The result of this disorder is excessive growth of tissue. Unconfined cell proliferation occurs during tumorigenesis and leads to the activation of oncogenes or the inactivation of tumour suppressor genes.[4] Neoplasm literally means "new growth" and can be defined as an abnormal mass of tissue that is uncoordinated with the normal tissues of the affected organ. Neoplasms are classified as either benign or malignant based on a number of important characteristics The most common forms of cancer occur in the epithelial tissues, which are separated from the underlying connective tissue, blood and lymphatic vessels, and nerves by a basement membrane. The cells and their nuclei show variation in size and shape, which is referred to as pleomorphism. (Philip c nasca) Cancerous cells occupy nearby tissues and organs, which is the main reason for death in cancer patients. The physiological process through which healthy cells convert into tumour cells has been the main focus of substantial studies in biomedical science for several decades.

Six necessary variations in cell biology can lead to cancer cell growth. All six modifications are used as characteristics of all types of cancer and are the following:

1) Self-reliance in signalling growth in tumour cells
2) Indifference to signals that discourage an increase in tumour cells
3) Escaping apoptosis
4) Immeasurable duplicate capability
5) Sustained vascularity
6) Permeation of tissues and spread of tumours

An unstable genome can cause a rise in variations, which was the authorised attribute to display six characteristics. But the metamorphosis speed of most genes is lower, making it unexpected to encounter multiple pathogenic mutations seen in cancer cells to exist irregularly.[5]

11.3 TREATMENT

The area of anticancer therapy has undergone substantial changes during the last 40 years. The time when radiotherapy and surgery were the sole options for halting the spread of cancer is long past.[5] The last century has shown that surgery, chemotherapy, and radiotherapy are effective cures for cancer. Usage of these either combined or individually can fruitfully combat the growth of the tumour and also eradicate it in some cases. For many cancers, such as colon cancer, early diagnosis and combined therapy have increased chances of survival. Curing cancer becomes complex after the tumour spreads to other parts but current treatments can be used to demote the tumour to resemble a chronic disease, although specific classes of tumours like glioblastoma still face considerable issues in prolonging life beneath 1 or 2 years.[6] Early diagnosis and treatment can effectively battle many cancers and reduce the chances of death so that less than half of the individuals with the disease will actually die. Treatments at this time include radiation therapy, surgery, and cytotoxic chemotherapy or hormonal therapy, based on the intensity and kind of cancer. As of now, few molecular targeting agents that have high chances of showing results have been green-lighted to be utilized in clinics.[7] Even though 90% of deaths from cancer are caused by metastasis, it is still the part of cancer pathogenesis about which we have the least amount of knowledge. Owing to the spread of the tumour by metastasis, a cancer cell undergoes the stages of occupying its neighbouring tissue, invading the microvasculature of the blood and lymph systems, transportation through the bloodstream to distant tissues, and survival by accommodating to the microenvironment of tissues far away from the initial source of the tumour. This accommodation in turn eases cell proliferation and colonization of the tumour.[8] Metastasis of cancer comprises epithelial-mesenchymal transition, a gathering of mutations taking place in stem cells.[9] The beginning stages of oncogenesis will be partially due to oncogenic drivers. ERBB2 amplification, EGFR mutations, and B-Raf mutations are examples of genetic changes that frequently produce oncogenic drivers. It is anticipated that tumour size will decrease as a result of "oncogene de-addiction", which addresses these carcinogenic promoters.[10] The majority of tumours will probably develop treatment resistance, maybe as a result of tumour cell heterogeneity and the choice of other molecular processes.[10] In the future, it will be beneficial to combine pharmaceuticals because administering one drug at a time will provide temporary effects. The development of individualized immunotherapeutic and/or anti-angiogenic drugs may alter the natural course of a number of malignancies.[10] Tumours demonstrate a variety of traits and capacities, including maintaining proliferation activity, avoiding growth inhibitors, avoiding apoptosis, enabling replicative immortality, and producing angiogenesis, along with stimulating invasion and spread. To get rid of certain traits and create therapeutic options, numerous investigations have been carried out. Operation; chemotherapy; radiation therapy; targeted therapy; hormone therapy; and other more recent therapeutics like immunotherapy, cell therapy, and palliative therapy are all regularly used in medical therapy. Surgery, the oldest and most fundamental treatment option, tries to remove the underlying abrasion. Even if a lymphadenectomy could enhance the impact of surgery, insufficient clearance remains the primary cause of cancer cell proliferation and metastasis. Particular medications and radiation are employed in radiotherapy and chemotherapy, correspondingly, to remove the abrasion. But in addition to the current issue of drug resistance with one or more medications, side effects and toxic effects are also significant.[11]

11.4 TARGETED THERAPY

Somewhere at the cellular and molecular level, personalized medicine can be precise, effective, and efficient. It lessens toxicity and harm to healthy cells. Additionally, it is also progressively gaining popularity as a treatment, but its price and range of applications restrain its growth.[11]

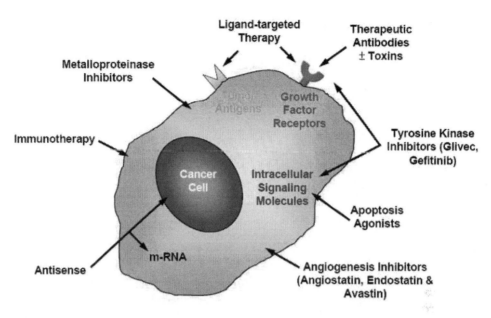

FIGURE 11.1 Types of targeted therapy in cancer therapy [12].

11.4.1 MOLECULAR TARGETED THERAPY

In order to stop the proliferation and dissemination of cancer cells, a treatment method known as "molecular targeted therapy" uses medications or other chemicals that specifically target certain molecules. For the effective development of molecularly targeted medicines for tumours, obvious target selection is crucial. Treatment of cancer that uses molecularly targeted therapeutics may display a variety of features and functionalities. Depending on the targets, they may affect receptors, growth factors, signalling pathways, or cell surface antigens that control the formation of new blood vessels, the spread of cancer cells, and cell division cycle development and induce apoptosis. Small molecules, monoclonal antibodies, immunotherapeutic cancer vaccines, and gene therapy are the different categories of substances employed in molecular personalized medicine.[13]

11.4.2 RADIATION THERAPY

Radiotherapy, surgeries, chemotherapy, immunotherapeutic, and hormonal therapies are all examples of treatment techniques. About 50% of cancer patients undergo radiotherapy at some point during their illness, and it accounts for 40% of cancer cures. Radiotherapy is still a crucial part of cancer management. Radiation therapy's primary objective is to rob cancer cells of their capacity for cell replication and proliferation.[8] To kill tumour cells, radiation is utilized as a destroying agent. Because it generates ionized particles and stores energy with in cellular region the radiation employed in it is known as ionizing radiation. This stored energy has the ability to harm tumour cells or alter their genetic makeup so they die.[14] The DNA or RNA of cells is damaged by radiating high energy, which prevents future cell division and proliferation.[15] While both cancerous and normal cells are destroyed by radiation, the aim of radiation treatment is to increase the dose of radiation to aberrant tumour cells while decreasing exposure to healthy cells that are nearby or in the radiation's route.[16]

In contrast to tumour cells, healthy cells can typically heal themselves more quickly and maintain their regular activity. Tumour cells are typically less effective at mending radiation-induced destruction than healthy cells, which leads to uneven cancer cell deaths.[17] In order to treat a range of disease conditions except for operations, vascular radiology, also known as IR, employs X-rays and other computer tomography to "see" within the body while directing catheters and other extremely small tools through the body to the area of a problem. Interventional radiology has grown at an extraordinary rate over the past 20 years.[18]

11.4.3 Combination Therapy

A key component for the treatment of cancer is combination therapy, a mode of care that includes multiple chemotherapeutic drugs. Even though conventional therapy is currently the most effective therapy for various types of cancer, combination therapy is usually thought to be more effective than monotherapy. Radiotherapy could be harmful to the cancer patient and come with a number of hazards and adverse effects. It can also significantly lower the patient's immune system by weakening progenitor cells and making them more vulnerable to host illnesses.[19] The idea for combination treatment was first conceived in 1965 by Emil Frei; James F. Freireich suggested that the very first combination therapy for acute leukaemia might exist. Methotrexate, 6-mercaptopurine, vincristine, and prednisone, commonly referred to as the POMP regimen, were employed to treat children and adolescents with acute lymphocytic leukaemia. This therapy was found to be effective in lowering disease severity and extending recovery.[19] Combined drug therapy with phytochemicals is found to be more beneficial and helpful than treatment with one drug. The reactions that can occur when utilizing combined drug therapy might be helpful (additive or synergistic), neutral, or negative (by raising harm and/or reducing efficiency).[20]

11.4.4 Targeted Therapy

The idea of personalized medication delivery appeals to people because it replicates some of the benefits of topical drug treatment, such as elevated concentration and less systemic exposure.[21] Medicines encapsulated in nanocarriers (NCs) are normally administered systemically, or therapeutics are locally delivered to the sick tissue in approaches for targeted therapeutic delivery in cancer.

Small molecule inhibitors, chemotherapy, RNAi, and other medicinal compounds can all be enclosed in nanocarriers to increase their solubility and bioavailability, change their bio-distribution, and make it easier to penetrate into the target cell.[21]

11.5 DIAGNOSIS

In epidemiological and intervention research, it has been demonstrated that timely diagnosis and effective cancer therapy reduce mortality and pain.

Obtaining tissue or a cancer sample via surgical intervention or percutaneous aspiration, or collecting a cancer sample via surgical intervention, percutaneous aspiration, or the collecting of highly porous cells from bodily fluids, among other methods, is typically necessary to confirm a cancer diagnosis. A pathologist next examines the sample and assesses whether the tissues are cancerous, notes the activity (benign or malignant), and records the types of cells. The doctor will also assess any lymph nodes provided by the doctor for signs of malignancy and assess the cancer stage (as a measure of the extent of heterogeneity).[22]

A tool known as a microtome is used to slice tiny slices of cancer from a paraffin block if the specimen will be seen under microscopic examination.[23] Traditional chemotherapy approaches to cancer treatment are constrained by toxicities and limited selectivity, and existing medical testing must be enhanced to enable early detection capabilities.[24]

11.6 CANCER THERANOSTICS

Cancer continues to be one of the most pressing health concerns, including over 10 million cases annually, and is a challenging condition to treat. Effective diagnosis and treatment methods with few side effects are needed for effective treatment.[25] The existing methods for diagnosing and treating cancer have limitations, which may be part of the reason cancer mortality is rising.[26] When compared to chemotherapeutic drugs, tumour cell biomarkers, and cell imaging, early-stage therapeutic and diagnostic agents, including quantum dots, radioactive elements, nanostructured lipid carriers, and plasmonic nanobubbles, have the ability to accelerate the identification, management, and treatment of cancer patients.[10] Theranostics are growing in popularity since they are tailored treatments and can be utilized for diagnostic imaging with few or minor modifications to support personalized medicine.[27] Magnetic resonance imaging, optical imaging, and photothermal imaging adjuvants can all be made of nanomaterials. By extending circulation durations, shielding encapsulated pharmaceuticals from deterioration, and also by improving tumour absorption through drug diffusion and sustained release in addition to receptor-mediated endocytosis, nanoformulations can improve treatment outcomes when utilized as drug carriers. Additionally, multiple therapeutic modalities such as chemotherapy and heat can be co-administered to get the benefits of collective effects. Likewise, scanning and therapeutic agents can be co-delivered to offer the interconnection of diagnostics, treatment, and follow-up.[24] A number of platforms can merge scanning techniques and treatments to increase effectiveness and reduce the risk of drug therapies.[16] The term "theranostics" combines the words "therapeutics" and "diagnostics". It is also possible to observe initial treatment response and forecast treatment effectiveness by using treatment followed by diagnostics. Co-development of diagnostics and treatment is another alternative.[16] Cancer theranostics, itself suggests it is a combined therapeutic and diagnostic approach to the disorder with the goal of facilitating patient care and lowering treatment delays. It appears to be crucial for individualized treatment for cancer. The benefits of highly precise time- and space-based control of compound release have drawn further attention to the utilization of lasers as a remote-activation method for medication delivery. A developing externally activated potential treatment for several disorders is photochemotherapy, also known as photodynamic therapy (PDT).[28] PDT is the process of administering a non-toxic medication or dyes designated as a photocatalyst (PS) systemically, regionally, or externally to an individual who has a lesion—often but not always a cancerous growth. PDT is a photochemistry-based method that generates reactive molecular species in response to light activation of a chemical compound known as a photocatalyst, also known as photosensitizer (PS), in order to convey toxicity.[28] The photosensitizer is often injected into the bloodstream or used externally in therapeutic settings, after by luminance with a laser delivery device appropriate for the topographical region which is getting cured.[29] PDT, also known as photothermal treatment (PTT), for malignancy includes heating malignant tissue by exposing it to infrared energy (VIS-NIR light). In contrast to PDT, where the activation of a photosensitizer produces reactive oxygen species (ROS), in photothermal therapy (PTT), the photon's infrared energy is taken up by the photon and transformed into heat. PTT can result in various biological modifications, including tissue carbonization and alterations to the structural proteins. Similar to PDT, PTT offers limited intrusiveness and regional selectivity, enabling just the diseased tissue to be exposed to radiation, whereas the adjacent benign tissue sustains only minor harm.[16]

11.7 NANOPARTICLES IN CANCER THERANOSTICS

The three main pillars of anticancer therapy are operations, radiation, and chemo (chemotherapy).

To kill cancerous growth in anatomical structures, chemo utilizes lethal chemicals. Furthermore, chemo is costly and has severe adverse effects because it damages healthy cells in contrast to malignant cells. A cancerous cell can sometimes be precisely treated by utilizing nanoimaging and nano-drug delivery methods, eliminating undesirable systemic toxicity. Pre-clinical studies are still being conducted to create nanomaterial-based medicine delivery systems, particularly liposomes, which have already received USFDA approval to cure some tumours. Maghemite–ferric oxide and magnetite ferrous oxide are the two magnetic nanoparticles that are most frequently employed. MNPs may also be made using precious metals such as iron, nickel, and cobalt, as well as ferrites of the formula MeO@Fe2O3 (Me = Mg, Zn, Mn, Ni, Co, etc.). An option for anti-tumour medication therapy for malignancy is gene therapy. This treatment could target specific genes and control the abnormal morphology of genes implicated in the development of cancer. The administration of regenerative medicine is highly difficult due to the extremely short half-lives of deoxyribonucleic acid (DNA) and or ribonucleic acid(RNA) in animals.[30] Therefore, when employed for drug administration, nanomaterials must do two tasks: first, they must shield a medication from deterioration, and, second, they must carry it to tumour cells. One other benefit of magnetic nanoparticle (MNP) technology is that it draws NPs to the cancer sites and speeds up the transport of the genome into the deep tissues[20] Gene transfer is enhanced by magnetofection (MF) as DNA-loaded particles are guided and maintained in close contact with the target cells. The process, known as magnetofection (MF), was originally developed primarily to enhance genome transport into deep tissues, a situation that is more straightforward and simple to manage than animal models.[31] Therapeutic agents, protein and peptide delivery, nanoparticle tailoring, neuroelectronic interfaces, operations, and imaging are some of the medical applications of nanotechnology.

Nanobiotechnology seems to be a new field of study that combines nanoscale technology and molecular biology, while the phrase "medical and biological (biomedical) nanotechnology" is used to describe the application of nanotechnology in the healthcare profession. Investigation in biomedical nanotechnology has concentrated on identifying units joined to incurable diseases, such as malignancy, type of diabetes, and neurodegenerative illnesses, as well as the identification of microbes and viruses linked to illness (e.g. pathogenic bacteria, fungi, and HIV).

Cancer was one of the five preliminary disorders that contributed to thousands of fatalities in the early twentieth century, and it continues to be challenging to cure and a key cause of suffering and/or death, as there are over 10 million new cases yearly.[26] Nanoparticles have the ability to play an important role in osseous tissue engineering therapeutically.[32] To increase the effectiveness of chemo and lessen its adverse effects, the use of nanostructured materials for the administration of anti-tumour drugs has been actively researched.[7] Nanoscience has a lot of potential for use in medicine delivery. The medicinal agent may be carried by a bifunctional nanostructure tailored to particular sick tissues or areas in the body. The domain of treating malignancy is where effective drug tailoring has mostly been investigated. Chemotherapy's effectiveness is limited by the substantial symptoms and the propensity for tumour cells to exhibit multi-drug resistance. One of the fastest-growing prominent domains of cancer investigation right now is the hunt for therapeutic approaches.[33]. In numerous papers, a wide range of various tailoring tactics for tumour therapy have already been studied in depth. A logical strategy is to couple tumour chemotherapy drugs with agonists that are particular towards the area of the cancerous cells in order to facilitate their localization in tumours.[33] In order to limit the accessibility of normal tissue to medications or genes, nanomaterials might also be created to transport pharmaceuticals and genetic mutations selectively to specific anatomical structures. Additionally, thermoelectric ablation employs certain nanoparticles. In fact, there are ongoing human studies for tumour thermoelectric treatment using a variety of nanomaterial categories, namely metallic, magnetic, and liposomal nanomaterials. By focusing on tumour stem cells, investigation in the struggle against tumours advances still further (CSCs). In fact, CSCs have not only the potential to perform a significant role in the development, progression, and treatment response of cancer, but anti-cancer agents may also enhance the proportion of CSCs in the disease, enabling those cells to persist and escape to other locations.[2]

11.8 MAGNETIC NANOPARTICLES IN CANCER THERANOSTICS

Nanotechnology, which holds a lot of promise to help with the treatment and identification of cancers, has recently undergone extensive study and is progressively emerging as a new form of tumour treatment. The nanoparticle-based sector is now expanding quickly.

Magnetic Field-Responsive Delivery Nanoplatforms

Numerous nanoparticle-based products are already in various preclinical studies and clinical research stages, and an increasing number of medicinal products are available in the market. In this chapter, we condense the most current advancements in MNPs for therapeutic uses and talk about the possibilities of magnetic nanoparticles in cancer diagnostics and therapeutics and upcoming research directions [11]. Another strategy is magnetic drug targeting (MDT), which makes magnetically lipid nanoparticles or magnetic polymeric aggregates holding drug ingredients an appealing technique since it enables the localized use of an externally applied magnetic field to localize the drug at a specified target region.

Because of the distinct magnetic properties and low toxic effects, iron oxide nanomaterials, particularly magnetite and maghemite, have been proven useful.[33] Magnetic nanoparticles are thought to be an ideal delivery vehicle of drugs. Numerous sizes of magnetic particle can be manufactured, and they can also be synthesized and characterized to transport different molecules. Several issues need to be taken into account when using magnetic nanoparticles as a transporter, along with physical stability and biocompatibility.[20] The concept behind targeting drugs through magnetic nanoparticles is straightforward: following the administration of drugs through the bloodstream, MNPs also can be given through the intravascular route and targeted directly at the cancer site with the help of a magnetic field targeting the area being treated. Properties like microscopic particle size, high surface area, magnetostrictive response, and superparamagnetism make magnetic nanoparticle nanomagnetic material.[34] MNPs can be prepared and placed in a magnetic flux that is always present, and the electromagnetic radiation in an external magnetic field absorbs the heat generated [35]. Magnetic Nano Particles are often present in a superparamagnetic state, when use in Biomedical research [34]. Fe3o4 (Magnetite) and γ-Fe_2O_3 (Maghemite) are the examples of Iron Oxide Nanoparticles which are most often used Nano Materials [36]. It is widely acknowledged that MNPs play a significant part in cancer detection, administration, and management [37] .Tumour imaging technologies make it possible to diagnose cancer before symptoms appear. Popular imaging techniques such as magnetic resonance imaging (MRI), magneto-acoustic tomography (MAT), computed tomography, and near-infrared imaging are used to diagnose late and early-stage cancer.[21] MRI strongly influences the prior diagnosis of tumours and SPIONs, commonly known as superparamagnetic iron oxide nanoparticles mostly represented as a contrast agent for MRI.[4] High

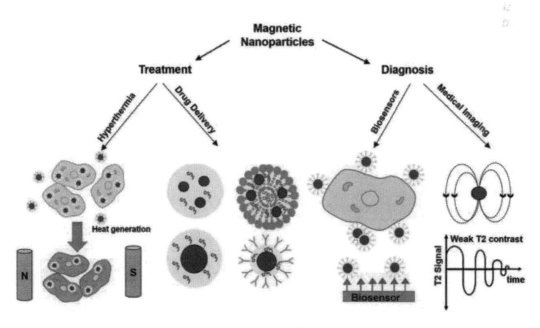

FIGURE 11.2 Different forms of magnetic nanoparticles [39].

saturation magnetism, superior biocompatibility, and high stability in acidic or basic media are all features of carbon-protected magnetic nanoparticles.[38]

11.9 CANCER THERAPY WITH MAGNETIC NANOPARTICLE DRUG DELIVERY

Serious side effects such as nausea, hair loss, bone marrow suppression, liver toxicity, and kidney toxicity have been observed because the conventional chemotherapy route of administration is systemic. These factors set the chemotherapeutic agent dose and restrict the tumour-affecting actions of the drugs.[36] The prospect of using local enhancement techniques with magnetic nanoparticle drug delivery allows the drug to aggregate and act in a predetermined location. This approach was initially introduced in 1978.[40] Three main elements are used in magnetic nanoparticles: iron (Fe), cobalt (Co), and nickel (Ni). Under physiologic settings, all elements are ferromagnetic, even though their magnetization patterns differ. Co 160, Ni 55, and Fe 218 emu/g are the values. Most often, they are combined with O_2 (oxygen), CO_2 (carbon dioxide), or other metal ions to form hybrids.[41] The biomedical applications of iron compounds are a major factor in their use. They exhibit the least cytotoxicity and are even applied medically to replace iron.[42] The coating of the nanomaterials helps to promote stability, reduce agglomeration, and improve bioavailability of drugs. A wide variety of materials, including fatty acids, polyethylene glycol (PEG), dextran, and chitosan may be used.[43] Saturated fats, (polyethylene glycol) PEG, dextran, and chitosan are just a few of the many components that can be used.[43] This method can, in theory, be applied to any tumour, regardless of its size, degree of differentiation, or location.[44]

11.10 MAGNETIC NANOPARTICLES USED IN CANCER THERAPY AND DIAGNOSIS (CANCER THERANOSTICS)

Nanoparticle Composition	Size (nm)	Loaded Drug	Application	References
Composite polymeric nanoparticle composed of $MnFe_2O_4$ and poly-N isopropylacrylamide-co-poly glutamic acid	17 ± 2 nm	Curcumin	pH-sensitive and thermally responsive for hyperthermia and targeted drug release	[45]
Core: iron oxide nanoparticle Shell: polyethyleneglycol and luteinizing hormone-releasing hormone (LHRH) peptide	36.5 nm	Doxorubicin	Chemotherapeutic drug and mild hyperthermia	[46]
Core: iron oxide Shell: chitosan	50 nm	Temozolomide	Delivery of drug to the cells of brain	[38]
Poly (N-isopropylacrylamide-co-acrylamide)-block-poly (ε-caprolactone) copolymer micelles with loaded iron oxide	70 nm	Doxorubicin	High efficiency of direct energy heating (hyperthermia) and temporal and spatial drug release	[45]
Core: iron oxide Shell: polyvinyl alcohol (PVA)	75 nm	Doxorubicin paclitaxel	Drug releasing vehicle under an external magnetic field for treating breast and cervical cancer models	[39]

11.10.1 ADMINISTRATION AND BIODISTRIBUTION OF MAGNETIC NANOPARTICLES

When employing magnetic nanocomposites in treatment of cancer, the method of administration is crucial. Nanoparticles can be administered directly into the blood (intravenously or intra-arterially), via the oral route (aerosol), or directly into the tissue (interstitially) depending on the intended use and target structure. However, the parenteral route is typically recommended. However, the majority of these particles are held in the liver and spleen and eliminated by the kidneys.[47] Therefore, circulatory clearance by excretion and metabolic processes may be problematic. In mice with subcutaneously injected hepatocellular cancers, nanoparticles managed to considerably slow down the tumour growth. To accomplish localized amplification, researchers used magnetism and intravenously injected magnetic nanocomposites connected to doxorubicin.[48] Lung cancer may be treated by the atomizer administration of nanocomposites made by magnetism. In both cell culture and animal investigations, researchers created a product called iron oxide (Fe_3O_4) nanoparticles encapsulated with a monomer of polylactic-co-glycolic acid (PGLA) that was successfully sustained (mice).[37] The study examined if there was any buildup in the axillary lymphatic nodes of mice by injecting different magnetic nanostructured materials formulations by transdermal route into their chest walls.[49] Remsen et al., in comparison, failed to detect any appreciable differences in the development of human pulmonary tumour tissues following iv(intravenous), intra-arterial, or intra-tumoral treatment in a rat model. Particularly in the event of overheating (hyperthermia), immediate insertion into the tumour is employed.[50] Magnetic nanocomposites can always be distributed directly through active or passive processes. Following intravenous administration, inactive dispersion mostly happens by diffusion. Although the increased blood artery porosity of tumours is advantageous, it is frequently offset by the high pressure found in tumour interstitial tissue. By joining the nanomaterials with the right tumour-specific ligands, nanomaterials are also able to be effectively delivered to the tumour. (e.g. antibodies). Magnetization can be used to direct materials, especially magnetized nanomaterials. Its absorption, allocation, and durability must be enhanced by combining active chemicals using nanomaterials or embedding them in nanostructures. This ought to make it easier for substances to get through cellular membranes, easily penetrate, and raise people's endurance overall.[22]

11.11 ACTIVE SUBSTANCES ATTACHED TO MAGNETIC FIELD NANOPARTICLES

Loaded nanoparticle technologies in combination with pharmaceutical delivery methods have drawn ever more attention. The ability to combine qualities boosts the effectiveness of medicine distribution in addition to immune response and reticuloendothelial system (RES) avoidance.[51]

11.11.1 Chemotherapeutic Agents

Anti-cancer drugs bound to magnetised nanomaterials have been the subject of numerous studies. Silicate nanomaterials with a structure made of y (yttrium), V (vanadium), or Eu (europium) and the chemotherapeutic medication doxorubicin were used by Shanta Singh et al. to cure carcinoma cell lines. An extra-lethal impact might be produced by simply introducing a rotating magnetic field following introducing of iron oxide particles.[52] Yallapu et al. improved the clinical effectiveness towards malignancy by altering the distribution of both the enclosed medicine by covering magnetite nanoparticles using -cyclodextrin and F127 polymer.[53] Guo et al. employed Ni (nickel)-based nanocomposites to increase the toxicity of daunorubicin in vitro towards cell lines.[54] At our facility, MDT, a unique method for delivery of drugs that makes use of targeted therapies closely linked to superparamagnetic iron oxide nanoparticles, is being created to treat malignancy. By injecting the resultant suspension—ferrofluid—directly into the artery, the tumour is given immediate nutrition. An electromagnetic force is used to focus the nanoparticles as well as the medicine connected to them in the tumour.[55]

11.11.2 Antibodies

Immunotherapy occasionally has significant drawbacks; hence there have been efforts to enhance this form of therapy with the aid of medication delivery. Wang et al. identified tumour cells that circulated within the blood of patients suffering from non-small-cell lung cancers by combining immunoglobulin with magnetic iron oxide nanoparticles.[48] Nano-immunoliposomes are a different approach. Nanosized ionic liposomes containing exterior components of immunoglobulin to the transferrin receptor, which exhibits enhanced expression in tumour cells, are used to enclose superparamagnetic iron oxide nanoparticles.[56] Although not completely medicinal, these techniques may help in early detection and repetition control.[57]

11.11.3 Radiotherapy

Iron oxide nanoparticles were employed by Klein et al. to in vitro sensitize cancer cells to radiation. ROS were identified in increased concentrations in the cells following the integration of the nanomaterials, which improved the cells' responsiveness to radiation.[58]

11.11.4 Gene Therapy

Medical care using recombinant DNA has a lot of promise. Poorly functioning genes are ultimately responsible for a cell's malignancy. Gene therapy can intervene at the site where a tumour first develops.[59] The biggest challenge is delivering genetic medicine, or genetic information, to the altered cells. The use of nanomaterials can make up for the lack of adequate and effective vectors in this situation. The process is frequently called "magnetofection", which refers to the controlled translocation of cells using nanocomposites with nucleotide bases attached.[30] Fe_3O_4 magnetic nanomaterials were used by Qi et al. to deliver siRNA to tumour cells effectively. They achieved a vastly greater efficiency of gene transfer with the method and a magnetic nanomaterials than they had with previous transfection agents.[60]

11.12 SYNTHESIS OF SUPERPARAMAGNETIC IRON OXIDE NANOPARTICLES

The coprecipitation of magnetite represents one of the most widely used chemical processes for producing superparamagnetic iron oxide nanoparticles.[61] Two methods could be employed; the first uses oxidizing substances like nitrates to partially oxidize ferrous hydroxide, while the latter uses an O_2-free atmosphere and involves adding an oxidizing sedimentation to an aqueous solution of ferrous (Fe^{2+}) and ferric (Fe^{3+}) ions.[62] Massart produced magnetized nanoparticles of about 10 nm by utilizing the second method. However, Sugimoto et al. were able to create bigger magnetic nanoparticles with diameters ranging from 30 to 200 nm utilizing the first method. Amongst variables influencing these techniques, pH and zeta potential have the greatest effects on the effectiveness and calibre of the chemical processes.[63] These factors have an impact on nanomaterial size, which is a measure of the durability of their dispersal. The microemulsion technique produces better homogenous and finer nanomaterials because the features of aquatic microparticles enable the containment of development and coagulation processes.[64] As Fe2O3 precipitates inside the constricted space, surfactant-stabilized watery molecules assembled into an oil colloidal suspension metal origins provide regulated scale superparamagnetic iron oxide nanoparticle synthesis.[65] Depending on hydroxyl, there is a second effective way of creating homogenous nanomaterials. Ionic compounds precipitate as nanocrystals with a size of less than 100 nm when they are decreased in a hydroxyl solution.[66] Sun et al. observed comparable outcomes after creating 3- to 20-nm-diameter narrow-size distribution nanomaterials from iron (III) acetylacetonate in phenyl ether in the presence of alcohol, oleic acid, and oleylamine at 265°C.[67]

Additional techniques involve the utilization of polymer backbone platforms where SPION can evenly deposit, as well as elevated ultrasonic waves that produce hot open spaces that then arbitrarily generate energetic waves that finally result in a decrease in nanomaterial size.[68] Parallel to hydroxyl-based synthesis, magnetized nanoparticles were also created electrochemically by reducing metal ions that had already been absorbed from the electrode and depositing them on an electrode.[51]

11.13 SPION FUNCTIONALIZATION

To provide colloid suspend ability, SPIONs are covered with polymeric materials. Water-soluble polymeric materials, for example, operate as stabilizers or give nanomaterials bioavailability. The observable traits of the polymer matrices and the superparamagnetic core may also be coupled. This connection is reliant on an outside magnetism, the power of which is passed to the local matrix and causes structural changes. Thus, it is interesting to combine magnetized nanoparticles with polymeric materials that are susceptible to thermal effects brought on by an external magnetic field. Based on the coating's hydrophobicity, magnetized nanoparticles may be incorporated into the lipid bilayer or the liposome.[68–73] To prevent additional instability of the magneto-liposome, nanomaterial size must be regulated, and a suitable covering is essential.[33]

REFERENCES

1. Papac RJ, "Origins of cancer therapy." *Yale J. Biol. Med.* **2001**,74(6),391.
2. Gobbo OL, Sjaastad K, Radomski MW, Volkov Y, Prina-Mello A, "Magnetic nanoparticles in cancer theranostics." *Theranostics* **2015**,5(11),1249.
3. Hassanpour SH, Dehghani M, "Review of cancer from perspective of molecular." *J. Cancer Res. Pract.* **2017**,4(4),127–129.
4. Sarkar S, Horn G, Moulton K, Oza A, Byler S, Kokolus S, et al., "Cancer development, progression, and therapy: An epigenetic overview." *Int. J. Mol. Sci.* **2013**,14(10),21087.
5. Seyfried TN, Shelton LM, "Cancer as a metabolic disease." *Nutr. Metab. 2010 71* **2010**,7(1),1–22.
6. Shewach DS, Kuchta RD, "Introduction to cancer chemotherapeutics." *Chem. Rev.* **2009**,109(7),2859–2861.
7. Croker AK, Allan AL, "Cancer stem cells: Implications for the progression and treatment of metastatic disease." *J. Cell. Mol. Med.* **2008**,12(2),374–390.
8. Chaffer CL, Weinberg RA, "A perspective on cancer cell metastasis." *Science (80).* **2011**,331(6024), 1559–1564.
9. Seyfried TN, Huysentruyt LC, "On the origin of cancer metastasis." *Crit. Rev. Oncog.* **2013**,18(1–2),43–73.
10. Arnedos M, Soria JC, Andre F, Tursz T, "Personalized treatments of cancer patients: A reality in daily practice, a costly dream or a shared vision of the future from the oncology community?" *Cancer Treat. Rev.* **2014**,40(10),1192–1198.
11. Li X, Li W, Wang M, Liao Z, "Magnetic nanoparticles for cancer theranostics: Advances and prospects." *J. Control. Release* **2021**,335,437–448.
12. Tsimberidou AM, "Targeted therapy in cancer." *Cancer Chemother. Pharmacol.* **2015**,76(6),1113–1132.
13. Lee YT, Tan YJ, Oon CE, "Molecular targeted therapy: Treating cancer with specificity." *Eur. J. Pharmacol.* **2018**,834,188–196.
14. Baskar R, Lee KA, Yeo R, Yeoh KW, "Cancer and radiation therapy: Current advances and future directions." *Int. J. Med. Sci.* **2012**,9(3),193.
15. Jackson SP, Bartek J, "The DNA-damage response in human biology and disease." *Nat. 2009 4617267.* **2009**,461(7267),1071–1078.
16. Chen X, Wong STC, "Cancer theranostics: An introduction." *Cancer Theranostics* **2014**,3–8.
17. Waks AG, Winer EP, "Breast cancer treatment: A review." *JAMA* **2019**,321(3),288–300.
18. Goswami T, Jasti BR, Li X, "Sublingual drug delivery." *Crit. Rev. Ther. Drug Carr. Syst.* **2008**,25(5),449–484.
19. Mokhtari RB, Homayouni TS, Baluch N, Morgatskaya E, Kumar S, Das B, et al., "Combination therapy in combating cancer." *Oncotarget* **2017**,8(23),38022.
20. Sauter ER, "Cancer prevention and treatment using combination therapy with natural compounds." **2020**,13(3),265–285. https://doi.org/10.1080/17512433.2020.1738218.
21. Rosenblum D, Joshi N, Tao W, Karp JM, Peer D, "Progress and challenges towards targeted delivery of cancer therapeutics." *Nat. Commun. 2018 91* **2018**,9(1),1–12.
22. Dürr S, Janko C, Lyer S, Tripal P, Schwarz M, Zaloga J, et al., "Magnetic nanoparticles for cancer therapy." *Nanotechnol. Rev.* **2013**,2(4),395–409.
23. Yap TA, Omlin A, De Bono JS, "Development of therapeutic combinations targeting major cancer signaling pathways." *J. Clin. Oncol.* **2013**,31(12),1592–1605.
24. Siegel R, Desantis C, Virgo K, Stein K, Mariotto A, Smith T, et al., "Cancer treatment and survivorship statistics, 2012." *CA. Cancer J. Clin.* **2012**,62(4),220–241.

25. Shanbhag PP-, Jog S V., Chogale MM, Gaikwad SS, "Theranostics for cancer therapy." *Curr. Drug Deliv.* **2013**,10(3),357–362.
26. Jeyamogan S, Khan NA, Siddiqui R, "Application and importance of theranostics in the diagnosis and treatment of cancer." *Arch. Med. Res.* **2021**,52(2),131–142.
27. Hapuarachchige S, Artemov D, "Theranostic pretargeting drug delivery and imaging platforms in cancer precision medicine." *Front. Oncol.* **2020**,10,1131.
28. Rai P, Mallidi S, Zheng X, Rahmanzadeh R, Mir Y, Elrington S, et al., "Development and applications of photo-triggered theranostic agents." *Adv. Drug Deliv. Rev.* **2010**,62(11),1094–1124.
29. Celli JP, Spring BQ, Rizvi I, Evans CL, Samkoe KS, Verma S, et al., "Imaging and photodynamic therapy: Mechanisms, monitoring, and optimization." *Chem. Rev.* **2010**,110(5),2795–2838.
30. Plank C, Zelphati O, Mykhaylyk O, "Magnetically enhanced nucleic acid delivery. Ten years of magnetofection—Progress and prospects." *Adv. Drug Deliv. Rev.* **2011**,63(14–15),1300–1331.
31. I. Schwerdt J, F. Goya G, Pilar Calatayud M, B. Herenu C, C. Reggiani P, G. Goya R, "Magnetic field-assisted gene delivery: Achievements and therapeutic potential." *Curr. Gene Ther.* **2012**,12(2),116–126.
32. Ahmed N, Fessi H, Elaissari A, "Theranostic applications of nanoparticles in cancer." *Drug Discov. Today* **2012**,17(17–18),928–934.
33. Douziech-Eyrolles L, Marchais H, Hervé K, Munnier E, Soucé M, Linassier C, et al., "Nanovectors for anticancer agents based on superparamagnetic iron oxide nanoparticles." *Int. J. Nanomedicine* **2007**,2(4),541.
34. Wu M, Huang S, "Magnetic nanoparticles in cancer diagnosis, drug delivery and treatment." *Mol. Clin. Oncol.* **2017**,7(5),738.
35. Mariappan L, Shao Q, Jiang C, Yu K, Ashkenazi S, Bischof JC, et al., "Magneto acoustic tomography with short pulsed magnetic field for in-vivo imaging of magnetic iron oxide nanoparticles." *Nanomedicine Nanotechnology, Biol. Med.* **2016**,12(3),689–699.
36. Rosenberger I, Strauss A, Dobiasch S, Weis C, Szanyi S, Gil-Iceta L, et al., "Targeted diagnostic magnetic nanoparticles for medical imaging of pancreatic cancer." *J. Control. Release* **2015**,214,76–84.
37. Verma NK, Crosbie-Staunton K, Satti A, Gallagher S, Ryan KB, Doody T, et al., "Magnetic core-shell nanoparticles for drug delivery by nebulization." *J. Nanobiotechnology* **2013**,11(1),1–12.
38. Lu AH, Zhang XQ, Sun Q, Zhang Y, Song Q, Schüth F, et al., "Precise synthesis of discrete and dispersible carbon-protected magnetic nanoparticles for efficient magnetic resonance imaging and photothermal therapy." *Nano Res. 2016 95* **2016**,9(5),1460–1469.
39. Farzin A, Etesami SA, Quint J, Memic A, Tamayol A, "Magnetic nanoparticles in cancer therapy and diagnosis." *Adv. Healthc. Mater.* **2020**,9(9),e1901058.
40. Klostergaard J, Seeney CE, "Magnetic nanovectors for drug delivery." *Nanomedicine Nanotechnology, Biol. Med.* **2012**,8(SUPPL. 1),S37–S50.
41. Widder KJ, Senyei AE, Scarpelli DG, "Magnetic microspheres: A model system for site specific drug delivery in vivo1." **2016**,158(2),141–146. https://doi.org/10.3181/00379727-158-40158.
42. Spinowitz BS, Kausz AT, Baptista J, Noble SD, Sothinathan R, Bernardo M V., et al., "Ferumoxytol for treating iron deficiency anemia in CKD." *J. Am. Soc. Nephrol.* **2008**,19(8),1599–1605.
43. Veiseh O, Gunn JW, Zhang M, "Design and fabrication of magnetic nanoparticles for targeted drug delivery and imaging." *Adv. Drug Deliv. Rev.* **2010**,62(3),284–304.
44. Mukherjee S, Liang L, Veiseh O, "Recent advancements of magnetic nanomaterials in cancer therapy." *Pharmaceutics* **2020**,12(2).
45. Kim DH, Vitol EA, Liu J, Balasubramanian S, Gosztola DJ, Cohen EE, et al., "Stimuli-responsive magnetic nanomicelles as multifunctional heat and cargo delivery vehicles." *Langmuir* **2013**,29(24),7425–7432.
46. Taratula O, Dani R, Schumann C, . . . HX-I journal of, 2013 undefined, "Multifunctional nanomedicine platform for concurrent delivery of chemotherapeutic drugs and mild hyperthermia to ovarian cancer cells." *Int. J. Pharm.* **2013**,458(1),169–180.
47. Jain TK, Reddy MK, Morales MA, Leslie-Pelecky DL, Labhasetwar V, "Biodistribution, clearance, and biocompatibility of iron oxide magnetic nanoparticles in rats." *Mol. Pharm.* **2008**,5(2),316–327.
48. Chao X, Zhang Z, Guo L, Zhu J, Peng M, Vermorken AJM, et al., "A novel magnetic nanoparticle drug carrier for enhanced cancer chemotherapy." *PLoS One* **2012**,7(10),e40388.
49. Hiraiwa K, Ueda M, Takeuchi H, Oyama T, Irino T, Yoshikawa T, et al., "Sentinel node mapping with thermoresponsive magnetic nanoparticles in rats." *J. Surg. Res.* **2012**,174(1),48–55.
50. Wilson PWF, Polonsky TS, Miedema MD, Khera A, Kosinski AS, Kuvin JT, "Systematic review for the 2018 AHA/ACC/AACVPR/AAPA/ABC/ACPM/ADA/AGS/APhA/ASPC/NLA/PCNA guideline on the management of blood cholesterol: A report of the american college of cardiology/american heart association task force on clinical practice guidelines." *J. Am. Coll. Cardiol.* **2019**,73(24),3210–3227.

51. Mulens V, Morales M del P, Barber DF, "Development of magnetic nanoparticles for cancer gene therapy: A comprehensive review." *ISRN Nanomater.* **2013**,2013,1–14.
52. Shanta Singh N, Kulkarni H, Pradhan L, Bahadur D, "A multifunctional biphasic suspension of mesoporous silica encapsulated with YVO4:Eu3+ and Fe3O4 nanoparticles: Synergistic effect towards cancer therapy and imaging." *Nanotechnology* **2013**,24(6),065101.
53. Yallapu MM, Othman SF, Curtis ET, Gupta BK, Jaggi M, Chauhan SC, "Multi-functional magnetic nanoparticles for magnetic resonance imaging and cancer therapy." *Biomaterials* **2011**,32(7),1890–1905.
54. Guo D, Wu C, Hu H, Wang X, Li X, Chen B, "Study on the enhanced cellular uptake effect of daunorubicin on leukemia cells mediated via functionalized nickel nanoparticles." *Biomed. Mater. (Bristol. Online)* **2009**,4(2).
55. Alexiou C, Tietze R, Schreiber E, Lyer S, "[Nanomedicine: Magnetic nanoparticles for drug delivery and hyperthermia—new chances for cancer therapy]." *Bundesgesundheitsblatt. Gesundheitsforschung. Gesundheitsschutz* **2010**,53(8),839–845.
56. Wang Y, Zhang Y, Du Z, Wu M, Zhang G, "Detection of micrometastases in lung cancer with magnetic nanoparticles and quantum dots." *Int. J. Nanomedicine* **2012**,7,2315.
57. Yang C, Rait A, Pirollo KF, Dagata JA, Farkas N, Chang EH, "Nanoimmunoliposome delivery of superparamagnetic iron oxide markedly enhances targeting and uptake in human cancer cells in vitro and in vivo." *Nanomedicine Nanotechnology, Biol. Med.* **2008**,4(4),318–329.
58. Klein S, Sommer A, Distel LVR, Neuhuber W, Kryschi C, "Superparamagnetic iron oxide nanoparticles as radiosensitizer via enhanced reactive oxygen species formation." *Biochem. Biophys. Res. Commun.* **2012**,425(2),393–397.
59. Li C, Li L, Keates AC, "Targeting cancer gene therapy with magnetic nanoparticles." *Oncotarget* **2012**,3(4),365.
60. Qi L, Wu L, Zheng S, Wang Y, Fu H, Cui D, "Cell-penetrating magnetic nanoparticles for highly efficient delivery and intracellular imaging of siRNA." *Biomacromolecules* **2012**,13(9),2723–2730.
61. Sugimoto T, Matijević E, "Formation of uniform spherical magnetite particles by crystallization from ferrous hydroxide gels." *J. Colloid Interface Sci.* **1980**,74(1),227–243.
62. Massart R, "Preparation of aqueous magnetic liquids in alkaline and acidic media." *IEEE Trans. Magn.* **1981**,17(2),1247–1248.
63. Jun YW, Choi JS, Cheon J, "Heterostructured magnetic nanoparticles: Their versatility and high performance capabilities." *Chem. Commun.* **2007**,(12),1203–1214.
64. Fitzgerald PF, Butts MD, Roberts JC, Colborn RE, Torres AS, Lee BD, et al., "A proposed CT contrast agent using carboxybetaine zwitterionictantalum oxide nanoparticles: Imaging, biological, and physicochemicalperformance." *Invest. Radiol.* **2016**,51(12),786.
65. Okoli C, Sanchez-Dominguez M, Boutonnet M, Järås S, Civera C, Solans C, et al., "Comparison and functionalization study of microemulsion-prepared magnetic iron oxide nanoparticles." *Langmuir* **2012**,28(22),8479–8485.
66. Shen YF, Tang J, Nie ZH, Wang YD, Ren Y, Zuo L, "Tailoring size and structural distortion of Fe3O4 nanoparticles for the purification of contaminated water." *Bioresour. Technol.* **2009**,100(18),4139–4146.
67. Sun S, Zeng H, Robinson DB, Raoux S, Rice PM, Wang SX, et al., "Monodisperse MFe2O4 (M = Fe, Co, Mn) nanoparticles." *J. Am. Chem. Soc.* **2004**,126(1),273–279.
68. Choi SK, Myc A, Silpe JE, Sumit M, Wong PT, McCarthy K, et al., "Dendrimer-based multivalent vancomycin nanoplatform for targeting the drug-resistant bacterial surface." *ACS Nano* **2013**,7(1),214–228.

12 Ultrasound-Responsive Delivery Nanoplatforms in Cancer Theranostics

Ruchi Tiwari, Vaseem Ahmad Ansari, Juhi Mishra and Namdev L. Dhas

CONTENTS

12.1 Introduction 184
12.2 Barriers in Targeting Tumours with Nanocarriers 186
12.3 Physics of Ultrasound Delivery 187
    12.3.1 Sonoporation 188
    12.3.2 Cavitation 188
    12.3.3 Hyperthermia 189
12.4 Theranostic Nanocarriers 190
    12.4.1 Mesoporous Silica Nanoparticles 190
    12.4.2 Microbubbles 194
    12.4.3 Micelles 195
    12.4.4 Gold Nanoparticles 196
    12.4.5 Hydrogels 197
    12.4.6 Liposomes 198
    12.4.7 Niosomes 200
    12.4.8 Polymer Nanoparticles 201
12.5 Conclusion 202
12.6 Abbreviations 203
References 204

12.1 INTRODUCTION

With 17.6 million new cases and 9.6 million deaths recorded from cancer in 2018, it is the second-biggest cause of death worldwide. Despite ongoing advancements in contemporary medicine, effective tumour identification and therapy remain difficult issues. Chemotherapy is a crucial therapeutic approach that is frequently used to treat cancer (1). Due to numerous barriers (anatomical, such as membranes; physiological, such as kidney filtration; and pathophysiological, such as tumour heterogeneity) and drug characteristic restrictions (such as solubility, stability), the delivery of chemotherapeutic agents is frequently ineffective, which results in unsatisfactory anticancer efficacy with serious toxic side effects. It is difficult to provide a correct diagnosis and, consequently, effective treatment since malignant tumours are varied in nature. The enormous diversity of cancer subtypes; distinctive genetics and epigenetics; and dynamic factors including age, environment, lifestyle, and medical history all contribute to the interpatient tumour heterogeneity (2). Therefore, patient classification must be at its best for the development and confirmation of successful medicines. Conventional cancer therapeutic techniques, such chemotherapy and radiotherapy, do not, however, have the patient-specific, personalized platforms needed to effectively treat cancer. The distribution of chemotherapeutic medications in patient-specific clinical settings is one example of such personalized platforms (3).

Anticancer drug delivery to solid tumours has always been extremely difficult. The lack of tumour selectivity in anticancer medicines makes them potentially harmful to both cancer and healthy cells (4). Additionally, aberrant vasculature and a disturbed extracellular matrix in tumour tissue may serve as pharmacological barriers and reduce the absorption of medications (5). The introduction of nanotechnology has demonstrated promising outcomes in overcoming the erratic distribution and pharmacokinetics of anticancer medicines. Through the so-called increased permeability and retention effect, the use of nanoscale formulations can aid to boost medication stability, limit degradation, and increase in vivo pharmacokinetics (6). Utilizing nanosystems that respond to stimuli is another method of attacking cancer cells. This "smart" technology responds to both internal and external stimuli that take advantage of the properties of the tumour microenvironment, such as a lower pH value and greater temperature.

It is common knowledge that administering many anticancer medications can result in significant systemic toxicity, which in some circumstances may be dose limiting. Whether intravenous injection or oral administration is used, the medication frequently builds up in healthy, normal tissues and destroys them (7). To reduce their systemic adverse effects and boost their therapeutic effectiveness, it is therefore vital to target and release these medications at the desired areas in a regulated manner. Drug delivery systems (DDSs) like liposomes, polymeric nanoparticles, or nanoemulsions (NEs) have been intensively researched to address the shortcomings and downsides of conventional medications, such as uncontrolled release and nonspecific biodistribution. However, even standard DDSs frequently struggle to release the cargo in a well-controlled way at the specified location. Smart DDSs have therefore been created to give drug release at the target site in a spatially and temporally regulated manner, preserve the drug/agent in the target site for a longer time, boost therapeutic efficacy, and reduce unwanted systemic side effects (8, 9).

Another important category used for cancer diagnosis is ultrasound (US) nanotheranostics. Most ultrasound contrast materials are gas-filled microbubbles (MBs) with a high degree of echogenicity. As a real-time, portable, nonionizing, and deep tissue-penetrating technique for tumour diagnosis, ultrasound offers several benefits (10). Additionally, it can be employed in sonodynamic therapy and high intensity focused ultrasound for cancer treatment. Contrast-enhanced ultrasonography may boost ultrasound backscatter, or the reflection of the ultrasound waves, which might result in a distinctive sonogram with improved contrast because of the wide echogenicity range. More precisely, ultrasonography may be utilized to image tumour receptor density, blood flow, and perfusion. Through the thermal and mechanical effects that ultrasound produces, medications are released from the nanosystems (11). Thermo-sensitive liposomes or polymer micelles were developed to release their drug payload upon exposure to mild heat (usually between 40 and 42°C) on the tumour, allowing for faster drug release and deeper tumour penetration. For thermal drug administration, 1,2-dipalmitoylsn-glycero-3-phosphocholine (DPPC) is the most often utilized phospholipid. Applying mild heat to the formulation will result in a more flexible and permeable liposome that can release the medication since its melting point is just above body temperature (41.5°C) (12, 13). A study was conducted on the first thermosensitive liposomes based on DPPC. Additionally, the release rate of traditional thermosensitive liposomes can be altered by adding lysolipids such 1-myristoyl-2-palmitoyl-sn-glycero-3-phosphocholine (MPPC) or 1-myristoyl-2-stearoyl-sn-glycero-3-phosphocholin (MSPC) in varying ratios. Lysolipid thermosensitive liposome technology is used in ThermoDox, which is mentioned in numerous clinical trials, to encapsulate doxorubicin (14).

Due to the physical properties of US and the efficient, safe, and non-invasive way in which it is used, US is frequently used for diagnostic imaging, therapeutic reasons (such as kidney stone comminution, physiotherapy, or tumour tissue ablation), or both ("theranostics"). Acoustic waves, a type of pressure wave, with a frequency above the range of human hearing (more than 20 kHz) make up the US and require a medium to pass through (unlike light or electromagnetic waves). In gases or liquids, acoustic waves primarily propagate longitudinally (15). Transversal waves resulting from shear stress have also been discovered to arise in solids. US waves often have inherent physical characteristics of all waves, including attenuation, reflection, refraction, amplification, absorption, and scattering. Most frequently, a transducer with a piezoelectric crystal, which can translate an electrical signal into mechanical pressure waves, serves as the source of ultrasound (16). Ultrasound-responsive particles either expand and contract in response to low-pressure ultrasound pulses that massage the vascular wall or grow and collapse in response to high-pressure ultrasound that releases drug payload through microjet. This improves the extravasation of co-administered medicines and blood vascular permeability. The commercially available phospholipid micro bubble was used in phase 1 clinical research to increase the therapeutic effectiveness of the chemotherapy drug gemcitabine. The direct interaction between endothelial cells and the micro bubbles made possible by ultrasonic application increased the cancer patient's response to gemcitabine by enhancing intracellular stress signalling and medication sensitivity. This chapter provides an overview of the various hybrid nanoplatforms now in use and the recent developments that have led to their usage in cancer theranostics (17, 18).

12.2 BARRIERS IN TARGETING TUMOURS WITH NANOCARRIERS

Recent developments in the synthesis and characterization of modified nanoparticles for cancer imaging and treatment have produced several interesting candidates for testing in human clinical trials. Despite these developments, biological, immunological, and translational obstacles continue to restrict the clinical applications of nanoparticle-based medicinal and imaging agents. The high likelihood of being cleansed by blood circulation is the first challenge for nanoparticles right after intravenous delivery, well before they reach the tumour microenvironment. It is possible because blood proteins may opsonize nanoparticles, which are then recognized by cells of the mononuclear phagocyte system (MPS) and ultimately removed from circulation (19). The populations of nanoparticles that the MPS cannot remove must disperse out of circulation. Nanoparticles must efficiently interact with their environment when they are in motion. The second hurdle for nanoparticles is their efficient extravasation through the tumour microenvironment. Compared to normal tissues, tumour tissue has a unique structural makeup. Typically, aberrant vasculature, overexpression, and a high level of extracellular matrix (ECM) are present in the tumour structure. The aberrant characteristics of tumours are the main causes of ineffective nanoparticle delivery to tumours. The crosslinked network of collagen and elastin fibres, proteoglycans, and hyaluronic acid that makes up the ECM of tumours creates a structure that resembles a cross-linked gel. Therapeutic nanoparticles that are diffused across the interstitium encounter severe resistance due to the tumour's highly developed and overexpressed ECM (Figure 12.1). In addition to this, high interstitial fluid pressure (IFP), which is caused by fast cell growth in a small space; high tumour vascular permeability; and the lack of a lymphatic drainage system all reduce the force needed for nanoparticles to enter tumours (20–23).

The vasculature of some organs, like the brain and retina, is far more impervious, acting as a barrier to practically all therapeutic access. The BBB, which is made up of tight connections between endothelial cells in the brain, is the primary illustration. Large and hydrophilic molecules find it challenging to pass over this barrier since it is a complicated biological system made up of numerous unique proteins and receptors. Since 2001, numerous investigations have demonstrated that the BBB can be safely and reversibly disrupted by ultrasound in conjunction with the injection of MBs. Several chemotherapeutic drugs, including herceptin, liposomal DOX, methotrexate, cytarabine, and DOX, have been demonstrated to be more effectively delivered over the BBB in healthy brains when using ultrasonography in cancer treatment. Additionally, it was shown how to transfer and transfect genes in healthy brain tissue, and gene therapy for cancer treatment was successful with the help of ultrasonography (24, 25).

Drug and nanoparticle distribution is significantly hampered by mucus. Delivery of nanoparticles to the diseased area through a dense mucosal structure is made more difficult by the presence

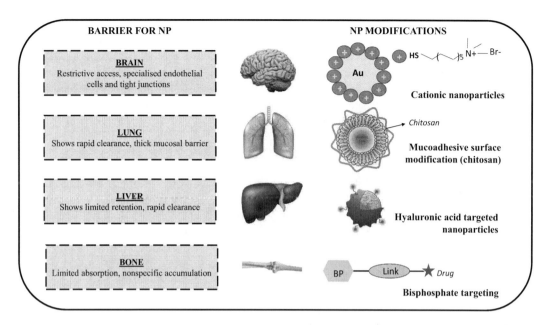

FIGURE 12.1 Image showing variety of tumour targeting obstacles and their adjustments.

of mucus in the bladder, lungs, and vagina. For instance, it is difficult to transfer drugs containing nanoparticles to the lung epithelium because of a variety of extracellular barriers, including mucuciliary clearance and the existence of thick mucosal coatings around the tissues (26). Low-frequency ultrasound was said to be capable of quickly delivering a therapeutic agent to the intestinal mucosa. In Franz diffusion cells, it was initially investigated whether low-frequency ultrasound could enhance the delivery of the medications to the colon tissues. According to the results of the ex vivo experiment, low-frequency ultrasound (20 kHz) therapy was able to boost the delivery of dextran (3 kDa) labelled with Texas red up to seven times more than control. The success of conventional drug delivery methods in treating several bladder disorders, such as interstitial cystitis, hyperactive bladder, bladder cancer, and urinary tract infections, is similarly constrained by mucous barriers. When mucous creates a physical or chemical barrier that prevents a medicine or drug carrier from reaching the target, ultrasonic radiation may be used to enhance transmucosal drug delivery (9, 26, 27).

12.3 PHYSICS OF ULTRASOUND DELIVERY

Light has been frequently used in nanomedicine to both initiate medication release and probe the sick tissue to gather clinical data. This stimulus has benefits, including great biocompatibility and ease of focusing its impact in certain zones. Due to the physical properties of US and the efficient, safe, and non-invasive way in which it is used, US is frequently used for diagnostic imaging, therapeutic reasons (such as kidney stone comminution, physiotherapy, or tumour tissue ablation), or both ("theranostics"). Acoustic waves, a type of pressure wave, with a frequency above the range of human hearing (more than 20 kHz) make up US and require a medium to pass through (unlike light or electromagnetic waves). In gases or liquids, acoustic waves primarily propagate longitudinally (28). Transversal waves resulting from shear stress have also been discovered to arise in solids. US waves often have inherent physical characteristics of all waves, including attenuation, reflection, refraction, amplification, absorption, and scattering. Most frequently, a transducer with a piezoelectric crystal, which can translate an electrical signal into mechanical pressure waves, serves as the source of ultrasound. When these pressure waves pass through the transmitting medium, they create local oscillatory motion of particles, which causes a change in local density in the medium (succession of compression and decompression events) (11, 29). The quantity of energy received by the targeted tissue is directly correlated with the applied acoustic pressure (measured in Pa). By adjusting various factors like frequency, intensity, or exposure period, one might affect the biological consequences caused by US application. Additionally, the use of US might be either continuous or discontinuous (pulsed mode). Tissue overheating and injury result from prolonged exposure to US, which is applied continuously. For the clinical ablation of tumour tissue, this effect is used.

12.3.1 Sonoporation

The process of sonoporation is the development of pores in a cell membrane as a result of ultrasonic exposure. It is a biological process caused by cavitation that is frequently used for drug administration. From a therapeutic perspective, sonoporation creates flaws in the cell membrane that allow chemicals that are not permeable to membranes to enter the cells. The cell membrane can be pierced by high-speed microjets created by asymmetric inertial cavitation (Figure 12.2). Cells can undergo sonoporation by means of the cavitation process. It was shown that ultrasound can cause sonoporation, improving DNA transport to mammalian cells. Live slime mould amoeba suspensions were treated with fluorescein-labelled dextran, which is generally impermeable to cells of their size, and exposed to ultrasound in this investigation. The ultrasound-treated cells had a 40% increase in fluorophore absorption, and the idea was later used for the delivery of DNA in mammalian cells (12, 30).

12.3.2 Cavitation

Cavities or bubbles either already exist in the liquid or are produced as a result of the pressure dropping to a level below the liquid's vapour pressure. Acoustic cavitation refers to the development, oscillation, expansion, and collapse of these cavities in an ultrasonic field. Cavitation can be used to increase the effect of sonoporation, and for this reason, exogenous MBs can be injected into the cellular microenvironment. In order to improve the contrast of ultrasound images, MBs are frequently utilized in diagnostic imaging. On the other hand, when imaging, ultrasonic frequency and intensity are controlled so as not to disturb MBs, which otherwise may result in undesirable side effects. The ability of ultrasound to disrupt MBs in specific regions allows for the on-demand

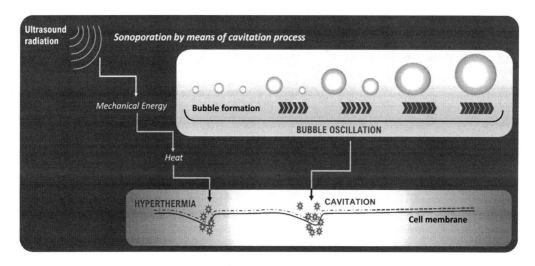

FIGURE 12.2 Physics of ultrasound-assisted theranostic approach.

release of drugs. MBs have also been used in drug delivery. According to how MBs collapse in reaction to ultrasonic energy (Figure 12.2), cavitation can be divided into two categories: stable and inertial cavitation (31).

A non-linear, sustained, and periodic expansion and contraction of a gas bubble is referred to as stable cavitation. Stable cavitation occurs when the acoustic pressure is at a level for which bubble response is generally linear. In each acoustic cycle, the bubbles oscillate around a value for the equilibrium radius. Microstreaming in the surrounding fluid, fluid shear stresses, and localized heat creation are all effects of this oscillation. While the converse occurs, albeit to a smaller amount, during the positive pressure phase of the cycle, dissolved gas diffuses into the bubbles during the negative pressure phase and continues to build inside these bubbles, leading to an increase in size. As a result of the net movement of gas into these bubbles, the size of the bubbles increases with repeated exposure. Stable cavitation is primarily used to promote the extravasation (penetration) of nanoparticles through vascular permeability, which enhances medication, gene, and nanoparticle delivery and deposition to whole tissues. Additionally, it stimulates ion channels and receptors, which concurrently alters cell permeability and action potential and makes it easier to deliver drugs to cells (32–35).

When ultrasound is applied to MBs, it causes them to collapse violently, a process known as inertial cavitation. When the bubble reaches its natural resonance size or at higher peak negative pressures, inertial or collapse cavitation takes place, producing oscillations with a considerable amplitude (14, 15). The bubbles fluctuate nonlinearly in these circumstances, resulting in subharmonic and higher harmonic oscillations. In the positive pressure phase, the surrounding liquid's compression squeezes the gas bubble until it becomes a supercritical fluid, producing pressures greater than 100 atm. High-pressure shock waves and high local temperatures are produced by this collapse. High-speed liquid microjets are created at the tiny level if the collapse is not symmetrical. Cavitation can affect nearby tissues mechanically, thermally, and chemically. Inertial cavitation's primary use is to alter cellular permeability to improve the transport of therapeutic agents (genes or drugs) to certain cellular platforms. Compared to steady cavitation, inertial cavitation causes membrane holes of bigger diameters. Pore diameters can range from a few micrometres to hundreds of nanometres (36).

12.3.3 Hyperthermia

In comparison to other external sources of energy, ultrasound is a suitable source of high energy for clinical therapy because of its good ability to penetrate deeply within the human body and its focus. It has been demonstrated that both thermal and non-thermal effects follow the entry of ultrasonic waves into the body. The increase in temperature (Figure 12.2) caused by the tissue's absorption of ultrasonic energy as a result of mechanical compression and decompression is known as ultrasound-induced hyperthermia. Friction effects use a percentage of the mechanical energy, which is ultimately converted into heat. Targeted tissues experience localized hyperthermia and frictional heat as a result of the rotation or vibration that ultrasound waves generate in the tissue's molecules. Drugs are delivered to the tumour site using local hyperthermia-induced drug delivery, which is produced by the thermal effects of US that are spatially constrained. This technique aims to increase the therapeutic impact of chemotherapeutic medications in order to prevent adverse effects brought on by their unintended diffusion into nearby healthy tissues. With satisfactory acceptable safety and few side effects, this method has been adopted into clinical practice as an adjuvant therapy for the treatment of several human malignancies (37). The phospholipid bilayer of the cell membrane becomes more fluid during hyperthermia, altering and frequently increasing the permeability of the cell membrane for medicines or nanoparticles. The structure of nanoparticles can be deformed or completely disrupted by swelling as a result of a slight temperature increase due to the absorption of ultrasonic energy. The eventual burst of nanoparticles aids in improving therapeutic medication distribution at the desired site of action. Another thermal effect of ultrasound is localized thermal ablation, which is produced with the use of high intensity focused ultrasound (HIFU). Higher intensities of ultrasound (>5 W/cm^2) are employed for local heating of focused tissues for accomplishing ablation. In order to provide localized hyperthermia within the tumour while protecting healthy tissue from the unfavourable effects of heating, tumour tissue might be loaded with NPs and then subjected to US waves. Heat generated locally can release medications, amplify their cytotoxic effects, and ultimately kill tumour cells. Therefore, using a hyperthermia method based on nanotechnology can improve the overall purpose of cancer therapy. Thermosensitive nanoparticles are nanoparticles used for this purpose and are made of components (usually polymers or lipids) that react to heat. In order to improve the targeted drug release, thermosensitive nano/micro particles can be activated using ultrasound (19, 38).

12.4 THERANOSTIC NANOCARRIERS

In the world of contemporary medicine, nanocarriers have found a place for themselves. Nanocarriers are suitable for various delivery tasks due to their unique physical and chemical properties. Liposomes, micelles, biodegradable polymer nanoparticles, dendrimers, carbon nanotubes, metal nanoparticles, and others are examples of nanocarriers (Table 12.1). These nanocarriers are perfect for this purpose because they may be made to deliver therapeutic and diagnostic chemicals to the desired region without causing unwanted side effects. Based on individual variability, these theranostic nanocarrier systems can be customized for personalized medicine (Figure 12.3). Therefore, these theranostic nanocarriers allow for the modification of dose regimen and real-time monitoring of a patient's disease status and treatment (39, 40).

12.4.1 Mesoporous Silica Nanoparticles

Mesoporous silica nanomaterials (MSNs) are classified as a subclass of nanomaterials with pore sizes ranging from 2 to 50 nm, which fall between microporous and macroporous materials. Due to their advantageous features, MSN nanostructures have received the greatest attention in studies focusing on silica (SiO_2) nanostructures, which have been widely employed for US-responsive drug delivery and US contrast enhancement. MSNs have exceptional optical qualities, low density, excellent adsorption capacity for several cargo molecules, great thermal and chemical stability, and resilience to corrosion in challenging environments. A wide variety of morphologies and textures may be synthesized on a large scale, and the mesopores' large surface area and volume enable them to have a high drug loading capacity (21, 55). Their physicochemical characteristics can be tailored to offer sensitivity to different stimuli, targeting capability, biocompatibility, biodegradability, and controlled release of encapsulated cargos. Drug administration, diagnostic imaging, biocatalysis, biosensors, enzyme supports, protein adsorption and separation, and nucleic acid detection and purification are just a few of the uses for MSNs. MSNs have a large surface area and clearly defined mesopores, making them attractive platforms for multi-component drug delivery (56).

Metalloproteinase-sensitive fluorophores that are exclusively active in tumoral tissue have been added by Zhang et al. to the exterior of MSN. In order to dissolve the extracellular matrix and facilitate the colonization of nearby tissues by cancerous cells, some proteolytic enzymes, such as metalloproteinases MMP-2, are overproduced by tumoral tissues. A particular peptide called Pro-Leu-Gly-Val-Arg that is cleavable by MMP-2 was coupled to the TAMRA fluorophore. Dabcyl was also placed on the opposite end of this peptide to inhibit the TAMRA fluorescence. These nanocarriers contained camptothecin, an effective cytotoxic agent, which was kept inside the pore network by the presence of these moieties on pore outputs that serve as pore blockers. When the nanoparticle enters the tumoral tissue, MMP-2 cleaves this peptide, causing the medication to start to exit the body and activating the TAMRA fluorescence (Figure 12.4), which makes the nanoparticle visible through fluorescence imaging (22).

TABLE 12.1
Ultrasound-Assisted Theranostic Approaches to Cancer Treatment

Drug	Delivery System	Hz	Effects	Reference
Curcumin	Nanoparticles	5–12 MHz	FA-FCP combined with the external LIFU and the endogenic acidic environment can have powerful theranostic functions and provide a novel type of non-invasive and integrated tumour theranostic option.	(41)
Indocyanine	Microbubble	12 MHz	The versatility of this drug delivery system with dual-imaging and ultrasound-triggered drug release characteristics for potential future applications in cancer theranostics.	(42)
Microneedle based	Microbubble	20 KHz	This method would be a promising complementary approach to reduce the side effects and increase efficacy of chemotherapy, especially in patients with superficial local solid tumours with high proliferation and low migration rates.	(43)
Gold	Nanoparticle	13.56 MHz	The AuNP capabilities on producing photoacoustic (PA) signals and photothermal effects have been used to image and treat tumour progression, respectively.	(44)
7-ethyl-10-hydroxycamptothecin (SN-38)	Nanocrystal	40 kHz	The SN-38 nanocrystals exhibited a considerably higher cytostatic activity against cancer cells than that of irinotecan hydrochloride.	(45)
Molybdenum disulphide (MoS$_2$)	Nanosheets	–	MoS$_2$ could facilitate tracking biodistribution of MoS$_2$ at the tumour site and showed enhanced biocompatibility.	(46)
Polydopamine (PDA)	Nanoparticle	12 MHz	Used in radiotherapy, photothermal therapy combined with chemotherapy, photothermal therapy combined with immunotherapy, photothermal therapy combined with photodynamic/chemodynamic therapy, and cancer theranostics.	(47)
Porphyrin	Microbubble	5–14 MHz	The increased preferential accumulation and penetration of PTX-HSA-NPs suppressed tumour growth 10-fold more than without exposure to ultrasound. In conclusion, the developed porphyrin-MB-NPs establish a new paradigm in simultaneous bi-functional ultrasound/photoacoustic imaging diagnosis and locally triggered release of nanomedicine and enhanced chemotherapy efficiency.	(48)
Titanium dioxide (TiO$_2$)	Nanoparticle	–	TiO$_2$ significantly reduced the side effects of DOX, augmented the levels of ROS, and achieved effective and safe therapy, indicating its potential for the multi-mechanism therapy of prostate cancer.	(49)
Vanadium-doped titanium dioxide, V-TiO$_2$	Nanospindles	–	It tends to quickly recombine with the nanoparticles, thus suppressing the production of reactive oxygen species that could attack the tumour cells.	(50)
Graphene	Nanoparticle	–	It enhances therapeutic effects, reducing the toxicity of chemotherapeutic drugs, overcoming multidrug resistance (MDR), and inhibiting tumour metastasis.	(51)
Gold (Au)	Nanoparticles	1 MHz	A higher effect was observed at the highest intensities, and increased anti-cancerous activity was observed.	(52)
Graphene	Nanosheet	–	Graphene has a good biocompatibility (if functionalized), biodegradability and multi-functionalities, which makes them suitable candidates for cancer nanotheranostics.	(53)
CaCO3	Nanoparticle	–	It enables efficient amplification of tumour oxidative stress for effective cancer treatment.	(54)

MESOPOROUS SILICA NPs
Active drug delivery systems for cancer treatments and allows their doping with diverse materials (such as fluorophores, nanoparticles etc.) for various applications in cancer therapy.

MICROBUBBLES
Using MBs to deliver genes or immune-stimulatory substances to boost cell-based immunotherapy could be a potential way to promote cancer immunotherapy.

MICELLES
Micelles have been widely employed as drug delivery vehicles for cancer therapy due to their good physical, chemical, and biological characteristics (nontoxicity, lack of protein adsorption, and diminished reticuloendothelial system (RES) uptake) following intravenous injection.

GOLD NANOPARTICLES
Gold nanoparticles are used in photothermal therapy (PTT), which uses electromagnetic radiation to generate heat for the thermal death of cancer cells.

HYDROGELS
Hydrogels can be effectively used to deliver radioisotope and chemotherapeutic agents for cancer treatment.

LIPOSOMES
Liposomes overcome the limitations of chemotherapy by improving the bioavailability and stability of the drug molecules and minimizing side effects by site-specific targeted delivery of the drugs.

NIOSOMES
Niosomes help in targeting the drug to the cancer cells, increasing the treatment duration by reducing the severe side toxic effects and improving the drug stability.

POLYMERIC NANOPARTICLES
Polymeric nanoparticles show controlled mode of action, different administration methods to the tumor site, both organic and inorganic drug delivery.

FIGURE 12.3 Various nanocarriers and their function in the treatment of cancer.

The pores are loaded with the intended medicine, and to prevent the pores from opening too early and releasing unwanted cargo, cap should be grafted onto the pores. These caps may respond to stimuli in order to achieve scheduled or controlled cargo discharge. For MSN preparation, there are four general approaches: chemical etching methods, template-directed procedures, sol-gel approaches, and microwave-assisted approaches. For theranostic applications with multifunctional features, silica NPs can be created using a core-shell formulation and by surface functionalization with suitable polymeric or lipidic material. Opsonization is one of the most significant issues MSNs face during in vivo delivery, and it would be avoided by PEGylation, zwitterions, or lipid coatings (57). On the other hand, electrostatic charge modification or active targeting may be employed to improve cargo uptake and distribution within cells. Protein corona formation can be avoided in

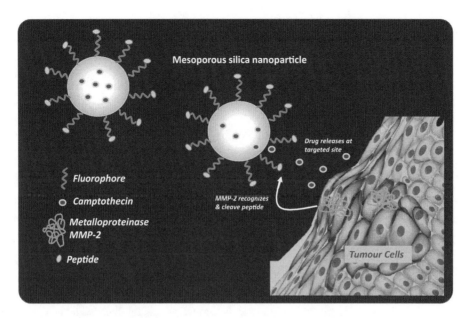

FIGURE 12.4 Ultrasound-assisted microporous silica nanoparticles towards a theranostic approach.

physiological conditions by increasing the hydrodynamic diameter and altering the surface electrical charge. This phenomenon may have a positive or negative impact on how helpful MSN is for biological applications. There is no widely accepted agreement on the cytotoxicity of MSNs and possible carcinogenesis (58).

The preparation technique, particle size, particle shape, surface chemistry, and mode of administration all affect how MSNs are distributed in the body. Furthermore, the dosage, cell type, and incubation period all affect cellular absorption. The effectiveness of MSN clearance is influenced by particle size, surface functionalization, electrostatic charge, and shape. In comparison to spherical NPs, non-spherical structures such as short and long rod-shaped MSNs exhibit varying degrees of biocompatibility, bio-distribution, and clearance. Rod-shaped MSNs in particular exhibit greater cell internalization. MSNs are inorganic substances that do not react to changes in temperature. As a result, temperature-sensitive polymers can be employed to give MSN-polymer hybrids temperature-responsive drug release capabilities (59, 60).

Targeting the glypican 3 protein was reported as a biocompatible technique for in vitro molecular imaging of hepatocellular cancer cells (GPC3). To improve ultrasound contrast, a brand-new GPC3-targeting peptide was coupled to fluorescent silica nanoparticles. To improve mesenchymal stem cell survival and permit simultaneous US imaging, a cell-penetrating peptide to MSNs that were loaded with the Wnt3a protein was attached. It was possible to create a dual-functional MSN by emulsion-templating from the silica precursors bis(triethoxysilyl)ethane (BTSE) or bis(3-trimethoxysilylpropyl) amine (TSPA) in order to produce exosome-like silica NPs and increase cell survival by releasing insulin-like growth factors continuously from mesenchymal stem cells. Theranostic compounds made of silica can be used well for simultaneous medication delivery and ultrasound imaging. Numerous novel biological applications, such as US imaging, sonodynamic treatment, HIFU tumour ablation, and US-triggered drug administration, have been made possible by the synergistic combination of silica NPs and US irradiation. Targeted US imaging can use ligand-conjugated MSNs as contrast agents. Additionally, silica nanostructures can be employed in imaging, cargo delivery, and US-based treatments (61–64). The TSPA structures were enabled for US contrast enhancement and had 40% exosome-like shape. Additionally, these elements generated a positively charged structure (zeta potential) that was used to identify negatively charged cells and enhance cell absorption. Due to its unique benefits including customizable and uniform pore size, large surface area and inner pore volume, nontoxicity, and biocompatibility, silica-based nanoparticles play a crucial role in cancer therapy (65). As active drug delivery systems for cancer therapies, silica nanoparticles' mesoporous structure also enables them to be doped with a variety of materials (such as fluorophores, nanoparticles, etc.) for a variety of cancer therapy applications. Although gold nanoparticles have made good progress in the treatment of cancer, optical imaging using gold nanoparticles has a limited practical future because gold nanoparticles have a lower optical signal than some fluorescent dyes or quantum dots (QDs). More research must be done on the toxicity consequences before QDs are widely used in clinical diagnosis and therapy, even if they may be thought of as multifunctional nanoparticles in biomedical applications including in situ optical imaging and drug delivery. Contrarily, silica nanoparticles have large loading capacities for a variety of substances and are biocompatible, non-toxic, and biodegradable, making them potentially safer and more effective for diagnostic techniques (66, 67).

12.4.2 Microbubbles

Typically, "MBs" refers to a hollow particle filled with a particular gas and encased in a particular layer that acts as a shell. The finding of a connection between gas bubbles in the bloodstream and the potent US echo observed after a US irradiation marks the beginning of MB development. In the US, MBs are frequently used as contrast agents in medical imaging. Due to their compressible gas-filled core, they efficiently respond to US pressure waves and disperse the incident US energy. As a result, they can generate successive waves, amplify US signals, and eventually raise the contrast of the image. MBs have been utilized as carriers in US-based drug and gene delivery systems. They can be loaded with a therapeutic material, followed or traced to the target region using low-intensity US imaging, and then destroyed with a high-intensity burst of US (68). As a result, they can use microstreaming and ARF to locally release the loaded medications and increase the therapeutic agents' penetration depth into the targeted tissue. Gene transfection, medicinal substances, and anticancer medications can all benefit from the use of MBs. By changing their surface with various ligands, they can also be directed to tissues. Azmin et al. provided a theoretical framework for the creation of MB-based theranostic systems by reviewing MB dynamics and the underlying physical concepts of MBs (51). MBs can be altered to enhance their performance, effectiveness, and characteristics. For cargo transportation, many areas of the MB structure can be used. For cargo loading, it is feasible to establish an oil layer within the MB. Both cargo and particular ligands can be added to the surface of MBs to make them more targetable. There may be more opportunities if MBs are used in conjunction with other kinds of NPs. These hybrids might improve how quickly, deeply, and effectively nanomedicines are absorbed. These mixtures have been utilized in theranostics for multimodal imaging or for simultaneous imaging and medication delivery (69). MBs can influence cell signalling pathways to improve the endocytosis process. According to one study, MBs and US irradiation together had a synergistic effect in initiating exocytosis, which resulted in the release of extracellular vesicles. MBs were initially created as contrast agents before being applied to the transportation of payload. They now function as theranostic agents. Numerous studies have demonstrated that using MBs and US radiation simultaneously improves the efficacy of drug delivery or imaging contrast in conditions such as cancer, infectious diseases, brain disorders, cardiovascular diseases, and vaccines. Dong and colleagues created plasma MBs by emulsifying plasma gas with surfactant to create the plasma. In response to US irradiation, these MBs released the loaded medication and produced active free radicals (such as nitric oxide and hydrogen peroxide). Additionally, MBs have been used to deliver oxygen to the tumour microenvironment. The pO_2 of the tumour tissue increases nearly six-fold following exposure to US in an oxygen-loaded lipid-coated preparation of MBs with mixed gas (O_2/C3F8 5:1 v/v). The use of MBs in cell or gene therapy has been covered in numerous articles (28, 70).

A viable strategy for enhancing cancer immunotherapy might involve the delivery of genes or immune-stimulatory substances via MBs. Due to internal rapid gas diffusion and the instability of conventional lipid shells, MBs have poor in vivo stability and a relatively short circulation half-life of 5–20 min. Additionally, the ability of MBs to load drugs is constrained, and it is difficult to add functional molecules to their surface to facilitate targeted drug delivery. The MBs may also permanently harm normal tissues that are off target. Therefore, tumour endothelial and cardiovascular targets may be the only ones to which MBs and US-triggered medication delivery is directed. Fortunately, modifying the stiffness of the MB shell could help to overcome these restrictions (fabricated from synthetic polymers, phospholipids, or albumin). PVA has been utilized as a shell to enhance the physicochemical characteristics of MBs. PVA-based MBs exhibit remarkably long-term stability. PVA-based MBs are susceptible to being absorbed by the reticuloendothelial system due to their biocompatibility, yet they remain in the blood circulation long enough to function as an effective contrast agent (71). Liquid perfluoro hexane (PFH), phosphatidylethanolamine, and halocarbons are some innovative nano systems that have been created as probes for US molecular imaging applications and as carriers for drug/gene delivery, ultimately leading to highly effective tumour growth prevention. A new method for delivering drugs to the brain, called FUSIN (focused ultrasound combined with MB-mediated intranasal administration), was developed by Dezhuang et al (30). For dual-imaging guided tumour targeted therapy, Yan et al. employed indocyanine green–conjugated lipid MBs as an ultrasound-responsive drug delivery method (72). To further these efforts, several related theranostic strategies are currently being implemented in our lab and other labs. For instance, chemotherapy-loaded MBs that are targeted to tumour blood vessels can be destroyed locally to cause drug release, drug delivery across biological barriers, and drug efficacy only at the target site. Theranostic MB formulations would also enable simultaneous feedback on the therapeutic efficacy of the prior administration via both functional perfusion monitoring and molecular ultrasound imaging when such agents are administered more than once, as will likely be the case clinically (73).

12.4.3 Micelles

Micelles are self-assembling colloidal structures in aqueous solutions consisting of amphiphilic molecules. They have a hydrophilic exterior coating that helps them escape being ingested by the immune system, which makes them a very valuable vehicle for targeted drug administration, particularly in the treatment of cancer. Due to their propensity to enter tumour vasculature and their ability to effectively treat patients with breast and lung cancer, polymeric micelles have attracted a lot of attention. Micelles have a huge prospect for medication administration as controlled systems because they are sensitive to biological stimuli like ultrasound. In order to cure canine cancer, Horise et al. used high-intensity focused ultrasound and sonodynamic therapy with anticancer micelles (34). They were successful in demonstrating the method's anticancer efficacy and demonstrating its potential to become a standard treatment for human cancer. This work and others suggest that ultrasound waves may be able to dislodge micelle nanocarriers, allowing medicines to diffuse into cancer sites. One of the most beneficial types of nanocarriers for effective medication delivery is micelles. The observation that some surfactants generated particles with a size range of 10–200 nm when the concentration was raised led to the discovery of micelles. Micelles are a type of colloidal dispersion made up of amphiphilic molecules with hydrophilic tails pointing in the direction of the surrounding water, forming a shell, and hydrophobic heads pointing in the direction of the structure's centre (typically made up of hydrocarbon chains). Van der Waals bonds are used to generate the micelle core. While hydrophilic molecules can bind to the micelle's surface, hydrophobic cargos can be found inside the micelle's core (74). The hydrophilic component also contributes significantly to the integrity of the structure and safeguards the micelles against external deterioration or elimination. Micelles can be created using synthetic polymer components or naturally occurring surfactants. The surfactant molecules or the polymer building blocks begin to form the micelle structure when their concentration rises above a particular threshold. The "critical micellar concentration" (CMC) is what is meant by this. The micelles will remain dissolved in the medium if the CMC value is below it. Polymeric micelles often exhibit greater thermodynamic stability than surfactant micelles because they have lower CMCs. Micellar stability is generally increased by using polymers with more hydrophobic blocks, shorter hydrophilic blocks, and longer hydrophobic chains. The CMC value for polymeric micelles is typically 106 to 107 M, whereas it ranges from 103 to 104 M for surfactant micelles. Diblock copolymers, triblock copolymers, and graft polymers are parts of polymeric micelles. The most popular materials are diblock or triblock copolymers that are amphiphilic. Biocompatibility, biodegradability, solubility, release rate, and hydrophobic core composition are taken into consideration while choosing polymers (75). The core of the micelles serves as a drug reservoir. One of the most often used forms of micelles in various investigations is a pluronic copolymer micelle. Hydrophobic poly (ethylene oxide) and hydrophobic poly (propylene oxide) triblock copolymers make up pluronic micelles. Micelles made of pluronic P105 have been used more frequently than other types of pluronic polymers as US-triggered drug delivery agents. Making the micelles responsive to various stimuli (US, heat, light, or a reduced pH in a tumour setting) might potentially be a goal of modifications in order to stimulate medication release. The mechanical effects of US heating can be used for micelle-based drug delivery in addition to thermo-responsive micelles. Stabilized doxorubicin-loaded pluronic P-105 micelles (Plurogel) were used in conjunction with low-intensity focused ultrasound by Nelson et al. to treat colon cancer in rats. When compared to non-insonated controls, a substantial decrease in tumour size was found (76). Three synergistic US effects are believed to have taken place in this instance: an increase in micelle extravasation, the release of DOX, and an increase in intracellular drug absorption. Polyesters and poly (amino acids) are two of the most extensively studied types of polymers among the polymeric micelle-based DDS. The delivery of anticancer medications and contrast agents for tumour therapy and diagnosis has led to the development of numerous types of micelles formulations based on PEG-PLA, PEG-PLGA, and PEG-PCL. The drug-encapsulated polyester micelles, however, frequently struggle with size management and burst cargo release, which can result in unfavourable side effects and worse therapeutic performance (77, 78). Theranostic polyester micelles could be used to support the imaging modalities, such as optical imaging, magnetic resonance imaging (MRI), ultrasound, photoacoustic, positron emission tomography (PET), single-photon emission computed tomography (SPECT), and X-ray computed tomography (CT), for monitoring the distribution and accumulation of nanocarriers in the body and providing visible and noninvasive feedback on the therapeutic response. Due to their superior physicochemical and biological characteristics, such as nontoxicity, lack of protein adsorption, and decreased absorption by the reticulo-endothelial system (RES) after intravenous injection, PEG-PLA micelles have been widely employed as drug delivery vehicles for cancer therapy (79).

12.4.4 Gold Nanoparticles

Larger gold nanoparticles (AuNPs) (10 or 16 nm) can only pass through the cell membrane and are only located in the cytoplasm. In contrast, smaller gold NPs (less than 6 nm) can efficiently reach the cell nucleus. This finding pertains to the possibility of increased toxicity for tiny NPs that penetrate the nucleus. Due to their fascinating properties, including their tiny size, high reactive surface area, targetability to cells, and capability of functionalization, metal NPs can be used as both diagnostic and medication delivery instruments. Owing to their unique chemical and physical characteristics, AuNPs have generated a great deal of interest. Due to their large internal pore volume, tunable size, ease of synthesis and surface modification, good thermal stability, and advantageous biocompatibility, prussian blue nanoparticles (PBNPs) have attracted growing research interest in the fields of immunosensors, bioimaging, drug delivery, and application as therapeutic agents. Due to their outstanding biocompatibility and versatile physicochemical features, AuNPs are gaining popularity. Since Faraday's 1857 report of the first scientific paper linking the red colour of AuNPs to their colloidal nature, AuNPs have been developed as stable nano-cargoes, metal catalysts, and organic photosensitizers (PSs) or to directly generate ROS and heat under near-infrared light irradiation, promising biologically friendly nanoparticles for medical treatment and diagnostics (80–83).

There are five different ways that the therapeutic and diagnostic uses of AuNPs in conjunction with US irradiation may operate. AuNPs have the potential to cause hyperthermia, raise the attenuation coefficient of US waves in the medium, act as a nucleation point for cavitation bubble formation, lower the cavitation threshold, and produce free radicals and ROS. It was determined that the mechanical and thermal impacts of ultrasonication were responsible for the sono-sensitizing features of AuNPs. Moradi et al. investigated hyperthermia using AuNPs while exposed to US radiation (84). When exposed to laser light, AuNPs catalysed the photolysis of the azide groups on their surface, resulting in the formation of N_2 MBs. As a result of backscattered US signals, these MBs may improve the contrast of US imaging. These nanostructures demonstrated improved penetration into the target tissue since they were smaller than MBs. Additionally, their tiny size and negative post-gas production zeta potential resulted in a longer blood residence time and better clearance. This technology could be controlled, was responsive across a wide spectrum, and could be applied to various imaging modalities. AuNPs are of great interest because of their distinctive optical properties. The important optical property of AuNPs—localized surface plasmon resonance—is caused by the external specific wavelength laser excitation of a collective and coherent oscillation of the conduction electrons close to the gold nanoparticle surface (83). This oscillation is in resonance with the incident light frequency and results in scattering peaks, spectral absorption, and local field enhancements (LSPR) (55, 80). The LSPR of AuNPs is confirmed to be highly sensitive to the size, material shape, dimension, and dielectric characteristics of the surrounding media in numerous investigations. Additionally, LSPR is used in a variety of imaging techniques, including dark-field microscopy (DFM), two-photon luminescence (TPL), and photoacoustic imaging, which frequently uses the optical absorption of AuNPs. The distributions of localized surface plasmons in nanoparticles within the NIR range have been visualized using two-photon luminescence imaging, which involves scanning a focused laser beam across the target tissues. This TPL is incredibly sensitive to the plasmonic excitations in gold nanostructures that result in greatly increased electro-magnetic fields. Strong TPL signal enhancements are thus produced by the electrons' collective oscillations inside the gold nanoparticles. Additionally, gold nanorods can strongly scatter to detect cancer cells under stimulation in the near-infrared region where biological tissues exhibit weak extinction coefficients, making them desirable TPL imaging substrates (57, 70–73).

12.4.5 Hydrogels

The hydrogel skeleton, the therapeutic substance, and the imaging moiety make up the three fundamental parts of a theranostic hydrogel. Additionally, during manufacturing, cross-linking and gelation techniques allow for the control and modification of gel formation, drug release, and degradation in vivo. Monomers, oligomers, premade polymers, tiny molecules, and even combinations of these can be used to build the skeleton of polymer hydrogels. Polymer-based hydrogels, such as those based on synthetic polymers like poly (lactic-co-glycolic acid) (PLGA) and polyethylene glycol (PEG), as well as natural polymers like chitosan, gelatin, and polypeptides, are biocompatible, have sturdy structures and a variety of physiochemical and mechanical properties that can be tailored to meet various needs in vivo. In addition to organic molecules, hydrogel (58) networks can also contain inorganic components such metallic nanoparticles, silica, and carbon materials. By doing so, the platform's mechanical properties will be improved and it will become more versatile. Hydrogels, a new class of nanomaterials with three-dimensional networks capable of absorbing substantial amounts of water, are crucial for medical rehabilitation. A high degree of biocompatibility is required for the use of hydrogels as DDS in the human body, and this biocompatibility can be attained by employing biomass polymers like cellulose, chitin, and chitosan as well as other polysaccharides that retain substantial amounts of water. Chemical cross-linking can be used to create hydrogels from biomass polymers for tissue engineering, and freezing-thawing and phase-inversion processing can be used to produce physical effects. A hydrogel typically releases the medicine as it swells, but when it contracts, the release rate drops or even stops (43, 59). However, occasionally the reverse behaviour is seen because potent drug-hydrogel interactions might restrict release. In this instance, as the hydrogel contracts, the medication is released along with the water. An unusual natural remedy that has been utilized to heal skin wounds is mimosa pudica root extract. This formulation was placed onto cellulose hydrogel films and activated by US exposure, which broke the hydrogen bonds and released the medication. mPEG-PLGA-BOX (BOX = 2,2'-bis(2-oxazoline)) block copolymer, an injectable, biocompatible, and thermosensitive hydrogel system for US-triggered drug release, has been disclosed. A small molecule medication, DOX, and a big molecule, FITC-dextran, had their in vitro release profiles at rest and after being activated by the US, respectively, evaluated. Over the course of seven days, an extended baseline release rate was gauged in vitro. The release rate substantially tripled when US (1 MHz, CW, 0.4 W/cm^2) triggered the DDS. The release rate restored to baseline when the US was turned off. After giving mice and rats a subcutaneous injection into their backs, the in vivo release profile of DOX was evaluated (44, 45, 60). The outcomes demonstrated that the hydrogels maintained their position and offered a consistent release for at least seven days. The in vivo release from the hydrogel was boosted by around 10-fold after US application. The temperature was increased to 40°C in vivo after exposure to US (0.4 W/cm^2), which led to the suggestion that thermal effects were the hypothesized mechanism. After receiving US therapy, the blood's DOX levels were assessed. With and without US irradiation, the blood DOX levels did not differ in a statistically significant way. As a result, local release to the nearby muscle was shown, confirming localized US-responsive drug release. The fact that there was an increase in DOX concentration in muscle but not in blood was explained by the possibility that the increase in blood DOX concentration was too modest in comparison to the baseline to be detected, making systemic toxicity improbable (61).

In theranostics, where injectable hydrogels provide non-invasive treatment and monitoring with a single injection, providing higher patient comfort and effective therapy, polymer hydrogels have been used to give well-controlled drug release and targeted therapy. Hydrogels can also be utilized to deliver chemotherapeutic drugs and radioisotopes for the treatment of cancer (46, 62). For instance, one team of researchers was successful in using a single vehicle to deliver a radioisotope along with chemotherapy drugs to cancer tissues. They created a macroscale thermosensitive injectable micellar hydrogel that could work as a drug reservoir for radioactive and chemotherapeutic agents. Rapid intracellular DOX release and radioisotope immobilization at the internal irradiation hot centre allowed the tumour cells to quickly ingest this hydrogel. With few side effects, a peritumoral injection of this system greatly reduced tumour development (47, 63).

12.4.6 Liposomes

The vesicle equivalent of cellular membranes, liposomes are made up of two encased layers of phospholipids. Phospholipids are amphiphilic molecules with a hydrophilic head and one or more lengthy hydrophobic hydrocarbon chains. Liposomes mostly consist of phospholipids, although they can also contain cholesterol and other polymeric building components. Phospholipids self-assemble and form a variety of forms in the presence of water because of their nature. The bending of the lipid bilayer and vesicle formation lower the edge interaction energy that results from the partial exposure of nonpolar hydrocarbon chains to the aqueous phase, making vesicles the most stable structures. The physical characteristics of liposomes are influenced by the techniques of manufacture, lipid type and charge, lipid composition, surfactant, organic solvent, and ionic strength of the suspension media (48, 64). Depending on the preparation technique, liposomes can be created in a multilayer configuration. The most common techniques for creating vesicles include reverse phase evaporation, detergent depletion, lipid hydration, freeze-thawing, and alcohol injection. A lipid bilayer shell surrounds the internal aqueous compartments of liposomes, which are tiny carriers. Hydrophobic medications can be effectively entrapped within the lipid bilayer of liposomes, whereas hydrophilic pharmaceuticals can be contained within their cores. It is possible to create monodisperse liposomes with an optimum diameter (less than 200 nm), allowing them to extravasate through solid tumours' leaky blood vessels (65). Liposomes that have had their surfaces modified with polymers such as polyethylene glycol, also known as stealth liposomes, exhibit less absorption into RES cells and longer circulation durations than liposomes without these modifications. As opposed to free drugs, drug encapsulation inside liposomes can greatly boost the accumulation of the loaded medications at the locations of solid tumours. However, as the medications must first be released from the carriers before they can affect the surrounding cells, this does not necessarily represent an equally great increase in therapeutic effects. Several research have combined liposomes and ultrasound. Low-frequency ultrasound (LFUS) increases cell membrane permeability, which improves medication and gene uptake. The administration of LFUS also improves the permeability of liposomes, speeding up the pace at which content is released since the phospholipid bilayer structure surrounding the liposomes is comparable to that of cells. LFUS (20 kHz) was found to improve drug release from liposomes in several in vitro tests (49, 66). Combining US with liposomal cytostatic medications has been shown to have increased therapeutic benefits when used to treat tumours in in vivo trials. The lipid content of the liposomes influences how sensitive they are to acoustic waves. For instance, HSPC-based (solid) liposomes have lower son sensitivity than DOPE-based (fluid) liposomes, which are highly son sensitive and effectively release their payload as a result of irreversible membrane breakage during sonication. Additionally, after six minutes of insonation, DSPE-based liposomes emitted more DOX than DSPC-based liposomes (69% vs. 9%, respectively). Following six minutes of sonication, the contents of liposomes made from DOPE/DSPC/DSPE-PEG2000/cholesterol (25:27:8:40 mol%) were released in 70% of the total amount. For targeting purposes, ligands and antibodies can be added to liposomes to functionalize them. The tactic minimizes potential systemic negative effects while improving therapeutic or diagnostic efficacy (57, 67). In a study, human epidermal growth factor receptor 2 (HER2)–positive breast cancer cells were the target of calcein and doxorubicin- and trastuzumab-functionalized liposomes. The likelihood of a relationship between trastuzumab-functionalized liposome and these cancer cells would be higher than that of an interaction with other healthy cells since these cancer cells overexpress the HER2 receptor on their surface (68). The release of medicines from liposomes was triggered using low-intensity focused ultrasound (LIFU). When LIFU was added, the therapeutic results were significantly enhanced. Trastuzumab-functionalized liposomes demonstrated increased cellular toxicity and higher drug uptake by the HER2-positive cell line. Similar research used the use of transferrin-functionalized calcein-loaded liposomes to target HeLa cells. In this way, a synergistic impact between the targeting qualities and the administration of LIFU boosted the therapeutic efficacy of the procedure. Similar outcomes were seen in a study that combined the use of LIFU with calcein-loaded liposomes functionalized with human serum albumin for site-specific breast cancer therapy (69). These studies demonstrate the beneficial synergy between targeted LIFU delivery and treatment outcomes. US-responsive liposomes could be created by conjugating liposomes with MB. Recently, liposome-MB conjugates were presented as platforms for cancer therapy that are US-responsive. They have only been studied as passive carriers, since they lack good tumour penetration due to their size. In 2018, Banerjee et al. discovered that paclitaxel-liposome-nanobubble conjugates, which are submicron sized (756 ± 180.0 nm), have a pro-apoptotic anticancer impact in addition to enabling image guidance (70). US-induced cavitation produced medicine release in the paclitaxel-liposome-nanobubble conjugates. A 10-fold increase in cellular internalization was seen in in vitro studies when compared to a control sample. Moreover, the great anticancer activity both in vitro and in vivo (98.3 ± 0.8% tumour growth suppression) was explained by the substantial synergism between polystyrene (PS) and paclitaxel (combination index, CI < 0.1). When exposed to mild hyperthermia, thermo-responsive liposomes, which are sensitive to temperature increases, can be induced to release their load. Phase III clinical studies for Thermodox, temperature-sensitive liposomes (TSLs) encapsulating DOX, were completed; however, the main endpoint of a 33% increase in progression-free survival was not met. The TSLs release their payload when the local temperature reaches their membrane's solid to fluid transition temperature (Tm), which is the point at which their fluidity transforms from having a solid-like phase/structure to having a liquid-crystalline phase/structure. Since tumours usually have a higher body temperature than healthy tissue, this is an example of triggered release (51, 71).

Liposomes can deliver drugs because of their many benefits. They play a part in improving drug solubility, acting as a sustained release system, delivering targeted drug delivery, lowering drug toxicity, offering protection against drug degradation, extending the half-life of APIs in circulation, effectively combating multidrug resistance, raising the therapeutic index of the entrapped drug, and shielding APIs from their environment. Targeted drug delivery and controlled drug release have both been made possible by advancements in liposomal vesicle formation (disease-specific localization). Since the first-line treatments for cancer are surgery, radiation therapy, and chemotherapy, this feature is generally beneficial for cancer treatment. Chemotherapy must be given systemically in some malignant conditions (72). Many APIs used in chemotherapy to date have been extremely cytotoxic to both cancerous and healthy cells. As a result of the free medication being administered directly into the bloodstream that circulates the body, people experience a wide range of adverse effects and restrictions. The chemotherapeutic agent may subsequently be absorbed by both cancerous and healthy tissues, resulting in severe organ damage in the heart, kidneys, liver, and other bodily organs. In order to maximize the amount of chemotherapy taken up by the cancer cells, patients are occasionally given the greatest dose of the drug feasible. The capacity of a cancer treatment to shrink tumours and eliminate them without harming healthy tissues is fundamental to its success in extending patients' survival times and improving their quality of life (52, 73).

12.4.7 NIOSOMES

Niosomes are non-ionic bilayered surfactant vesicles like liposomes. These multilayered, unilayered, or bilayered vesicles are all possible. Concentric vesicles that are nested inside of one another make up multilayered vesicles. Amphiphilic compounds such alkyl ethers, alkyl glyceryl ethers, terpenoids, polysorbates, and polyoxyethylene ethers are employed as non-ionic surfactants in niosomes. These stop the aggregation of vesicles and the change from the gel to the liquid phase, which lessens niosome leakage. Three criteria—intended function, preparation method, and vesicle size—are used to categorize niosomes. Multilamellar vesicles (MLVs), large unilamellar vesicles (LUVs), and small unilamellar vesicles (SUVs) are the three primary forms of niosomes based on the size and number of layers. Niosomes are typically submicron sized. SUV particles range in size from 10 to 100 nm, LUV particles range from 100 to 3000 nm, and MLV particles range in size from 5 micrometres to more. The reports also include some enormous vesicles. Niosome formation is fundamentally analogous to liposome formation. The vesicle structure is created by the self-assembly of amphiphilic molecules (53, 74). However, controlling outside energy would make this procedure easier. For thermodynamically stable niosomes, the right surfactant and charge-inducing chemical compositions are necessary. The most crucial elements that must be considered for the preparation of niosomes with desirable properties include monomer concentration, hydration temperature, time of hydration, pH of the hydration medium, cosurfactant, cholesterol, aqueous interlayer, lipid chain length, chain-packing, membrane asymmetry, and the type of drug. The preparation technique must be taken into consideration because it affects the vesicles' physicochemical characteristics and pharmacokinetics (75). Thin-film hydration, ether injection, sonication, reverse phase evaporation, freeze and thaw, heating, and dehydration/rehydration are a few techniques used to create niosomes. The creation of niosomes using microfluidics would result in more uniform niosomes with predetermined sizes as opposed to the bulk approaches. The desired entrapment efficacy, size, preferred materials, drug-loading procedures, homogeneity, and number of layers can all be taken into consideration when choosing the preparation method.

The drawbacks of liposomes led to interest in niosomes. Niosomes provide several benefits over traditional liposomes, including greater chemical stability; increased osmotic activity; a longer shelf life; easier surface modification; less toxicity; and more compatibility, biodegradability, and reduced immunogenicity. Other benefits of niosomes include osmotic action, a lengthy storage time, controlled properties, and a reasonably straightforward production procedure (39, 76). However, niosomes also experience stability and cargo leakage, just as liposomes do. Drugs that are hydrophilic, lipophilic, or both can be loaded into niosomes simultaneously. Some factors to consider in drug delivery applications include the shape, vesicle size, vesicle charge, encapsulation efficacy, stability, permeability, and release profile. Niosomes made of surfactants are nonimmunogenic, biocompatible, and degradable. Through their closed bilayer structure, they operate as a drug depot in the body, releasing medications under strict control to the target site over an extended period. Reduced clearance and targeted delivery of medicines inside niosomes enhance their therapeutic benefits. Niosomes may tolerate a large variety of pharmaceuticals with a wide range of solubility because of their hydrophilic, amphiphilic, and lipophilic characteristics. Niosomes may increase the bioavailability of normally poorly soluble medications and boost the effectiveness of topical treatments. Niosomes also shield the encapsulated active medicinal substances from harmful conditions both inside and outside the body, making it easier to administer labile and sensitive medications. The type of surfactant, characteristics of the medication enclosed, temperature of hydration, detergent, membrane-spanning lipids, polymerization of surfactant monomers, and charged molecules are the key factors affecting niosome stability. A hydrophilic head and a hydrophobic tail must be present in the surfactant used to make niosomes (55, 77).

Niosome-based drug delivery systems via transdermal, parenteral, and ocular routes have received extensive research. The slow penetration rate of traditional transdermal techniques can be overcome via niosomal administration via transdermal channels. Drugs like diclofenac, flurbiprofen, and nimesulide are made more bioavailable and therapeutically effective by being included in niosomal formulations. When compared to commercially available formulations, the chitosan-coated niosomal version of timolol maleate has a better impact on lowering intraocular pressure while having fewer adverse cardiovascular effects. Niosomal formulations have been used in numerous additional therapeutic applications because of their numerous favourable qualities, as detailed in the following sections. Cancer chemotherapy frequently has negative side effects and poor therapeutic efficiency. The anticancer drug doxorubicin, a broad-spectrum anthracycline, has a dose-dependent, permanent cardiotoxic impact. However, mice with the S-180 tumour who received this medication by niosomal administration showed longer survival times and reduced sarcoma growth (78). Because niosomes are so effective at encasing drugs, their circulation is extended and their metabolism is altered. Daunorubicin hydrochloride, a different well-known anticancer medicine, had enhanced antitumor activity as compared to the free drug. In comparison to unencapsulated cisplatin, niosomal cisplatin demonstrated a 1.5-fold increase in cytotoxic impact against BT-20 breast cancer cells. The ascitic lymphoma cells of Dalton were rapidly and destroyed by the niosomal formulation. When bleomycin, a potent anticancer agent, was encapsulated in niosomes containing 47.5% cholesterol as opposed to its free drug form, it was discovered that larger levels of the drug were deposited at the tumour site. Cancer chemotherapy, HIV/AIDS treatment, vaccine and antigen administration, lung delivery, transdermal distribution, and the transport of proteins and peptides have all benefited from the usage of niosomes. Despite their resemblance to liposomes, niosomes have unique properties that should be considered in the development of more affordable and effective medicinal formulations. It is crucial to effectively examine the true potential of these vesicular systems, and it will soon be necessary to invest more money in their research (58, 69).

12.4.8 Polymer Nanoparticles

Nanospheres, nanocapsules, and polymersomes are examples of polymeric NPs. The most often used polymers are PS, PLGA, PLA, and polycaprolactone (PCL). Polymeric nanoparticles can be created via polymerization of monomers or by employing premade polymers (59, 78). Synthesis methods can be divided into one-step and two-step procedures. Emulsification is not necessary to produce nanoparticles in one-step operations; however, in two-step procedures, an emulsification system must first be prepared before nanoparticle formation may take place. Polymeric nanoparticles are thought to be biocompatible, biodegradable, and non-toxic since they are created from natural polymers like chitosan, dextran, heparin, and hyaluronan or biodegradable synthetic polymers like PLA, poly (glycolic acid), and PLGA. To increase the activity of polymers, several chemicals can be added. For example, copolymerized polymeric NPs containing PEG can evade detection by mononuclear phagocytic cells (79). The polymeric shell may also increase the stability of polymeric NPs, allowing them to withstand pressure in US pressure fields. However, a variety of circumstances, including an increasing size, may affect the characteristics of the nanocapsules. Another typical class of polymeric nanocarriers that have been investigated for medication and gene delivery are polymersomes (80). These polymers have a synthetic vesicle membrane made of amphiphilic block copolymers, which resembles the lipid bilayers in a cell membrane in terms of structure. Polymersomes are known for their ability to self-assemble in an aqueous solution, and this trait has allowed them to be widely used as DDSs. These synthetic vesicles are stable, have a changeable membrane, and can enclose both kinds of substances. They also have a huge internal compartment (i.e., hydrophilic, and lipophilic molecules) (81). As NPs offer several benefits over traditional therapeutic techniques, their potential as putative DDSs is becoming more and more apparent. In the creation of NPs, biocompatible and biodegradable polymers are employed. Following the release of the medication from its encapsulation, the polymer matrix is broken down into innocuous molecules like water, nitrogen, and hydrogen before being eliminated from the body. Different formulations of nanoparticles need to be used to address the issue of regulated, sustained release of anticancer medications (63–65). To attain the optimum therapeutic level in target tissues for the necessary periods for best therapeutic efficacy and release of a constant amount of medication per unit time, the polymer properties can be modulated. The blood-brain barrier makes it challenging for many medications to reach the treatment site in the case of central nervous system malignancies (65, 82). This barrier can be overcome by drug-loaded NPs, which have also been demonstrated to significantly raise the therapeutic concentrations of anti-cancer medicines in brain tumours. Chemotherapy causes adverse effects on healthy tissues because pharmacologically active cancer medicines have limited tumour tissue selectivity and dose-limiting toxicity. An intriguing strategy that could aid in overcoming these difficulties involves employing NPs laden with anticancer medicines to target cancer cells (66, 84).

12.5 CONCLUSION

A variety of techniques have been used to generate ultrasound-responsive hydrogels for delivery applications, including cancer treatments and bone regeneration. As an external trigger, ultrasound has a lot to offer. Nanostructures and their extraordinary features are used in smart medication delivery systems, a novel strategy, to increase control over the distribution process (68). One of the biggest obstacles in cancer diagnosis is the precise drug administration to a particular tumour spot. Despite the huge potential of the existing theragnostic nanomaterials, it is necessary to develop next-generation design concepts and their efficient implementation methods. Numerous different nano-platforms have been tested for medication delivery to tumours, and several them have been successful in obtaining FDA approval and are now frequently used in clinical settings (69). However, the bulk of them have fallen short in either early or late-stage clinical studies. Making more nanomaterials will be necessary in the future, but it will also be crucial to understand how cancer cells interact with nanomaterials at the subcellular level. US-responsive materials have included MBs, nanoemulsions, polymeric structures, lipid vesicles, surfactant-based micelles, and inorganic nanoparticles such gold nanoparticles, titania nanostructures, carbon nanostructures, and silica nanostructures (83). Future ultrasound-responsive drug delivery systems have significant safety issues, particularly about high-intensity focused ultrasound. It is always advantageous to operate a system with the least amount of energy necessary in order to minimize any potential harm. Future ultrasound-responsive hydrogels will probably be logically constructed to reduce the energy needed to cause the release and lessen the possibility of causing harm to the tissues in the vicinity. Overall, ultrasonography has a great deal of promise to support the development of on-demand drug delivery platforms and to enhance the clinical effectiveness of numerous high-tech drug delivery applications (84).

12.6 ABBREVIATIONS

AuNPs	gold nanoparticles
BBB	blood brain barrier
BTSE	bis(triethoxysilyl)ethane
TSPA	3-trimethoxysilylpropyl)amine
CMC	critical micellar concentration
CT	X-ray computed tomography
DDS	drug delivery system
DFM	dark-field microscopy
DNA	deoxyribonucleic acid.
DPPC	1,2-dipalmitoylsn-glycero-3-phosphocholine
DSPE	distearoylphosphatidylethanola
ECM	extracellular matrix
FITC	fluorescein isothiocyanate
GNPs	gold nanoparticles
GPC3	Glypican 3
HER2	human epidermal growth factor receptor 2
HIFU	high-intensity focused ultrasound
IFP	interstitial fluid pressure
LFUS	low-frequency ultrasound
LSPR	local field enhancements
LUV	large unilamellar vesicles
MBs	microbubbles
MLV	multilamellar vesicles
MPPC	1-myristoyl-2-palmitoyl-sn-glycero-3-phosphocholine
MPS	mononuclear phagocyte system
MRI	magnetic resonance imaging
MSNs	mesoporous silica nanomaterials
MSPC	1-myristoyl-2-stearoyl-sn-glycero-3-phosphocholin
NEs	nanoemulsions
PCL	polycaprolactone
PEG	polyethylene glycol
PET	positron emission tomography
PFH	perfluoro hexane
PLGA	poly lactic-co-glycolic acid
PS	photosensitizers
PSPLBC	paclitaxel-liposome-nanobubble conjugates
PVA	polyvinyl alcohol
QDs	quantum dots
RES	reticulo-endothelial system
SPECT	single-photon emission computed tomography
SUV	small unilamellar vesicles
TPL	two-photon luminescence
TSLs	Thermodox, temperature-sensitive liposomes
US	ultrasound

REFERENCES

1. Alves CG, Lima-Sousa R, de Melo-Diogo D, et al. IR780 based nanomaterials for cancer imaging and photothermal, photodynamic and combinatorial therapies. Int. J. Pharm. 2018; 542: 164–175.
2. Boman NL, Bally MB, Cullis PR, et al. Encapsulation of vincristine in liposomes reduces its toxicity and improves its anti-tumor efficacy. J. Liposome Res. 1995; 5: 523–541.
3. Canavese G, Ancona A, Racca L, et al. Ultrasound using nanoparticles: Special attention to sonodynamic therapy against cancer. Chem. Eng. J. 2018; 340: 155–172.
4. Cao ZQ, Wang GJ. Multi-stimuli-responsive polymer materials: Particles, films, and bulk gels. Chem. Rec. 2016; 16: 1398–1435.
5. Dancy JG, Wadajkar AS, Connolly NP, et al. Decreased nonspecific adhesivity, receptor-targeted therapeutic nanoparticles for primary and metastatic breast cancer. Sci. Adv. 2020; 6: 3931.
6. Evjen TJ, Hagtvet E, Nilssen EA, et al. Sonosensitive dioleoyl phosphatidyl ethanolamine-containing liposomes with prolonged blood circulation time of doxorubicin. Eur. J. Pharm. Sci. 2011; 43: 318–324.
7. Guo Q, Zhang L, He M, et al. Doxorubicin-loaded natural daptomycin micelles with enhanced targeting and anti-tumor effect in vivo. Eur J Med Chem. 2021; 222: 113582.
8. Hervouet E, Simonnet H, Godinot C. Mitochondria and reactive oxygen species in renal cancer. Biochimie. 2007; 89: 1080–1088.
9. Hirabayashi F, Iwanaga K, Okinaga T, et al. Epidermal growth factor receptor-targeted sonoporation with MBs enhances therapeutic efficacy in a squamous cell carcinoma model. PLoS One. 2017; 12: 0185293.
10. Horise Y, Maeda M, Konishi Y, et al. Sonodynamic therapy with anticancer micelles and high-intensity focused ultrasound in treatment of canine cancer. Front Pharmacol. 2019; 10: 545.
11. Horise Y, Maeda M, Konishi Y, et al. Sonodynamic therapy with anticancer micelles and high-intensity focused ultrasound in treatment of canine cancer. Front Pharmacol. 2019; 545.
12. Hossen S, Hossain MK, Basher MK, et al. Smart nanocarrier-based drug delivery systems for cancer therapy and toxicity studies: A review. J. Adv. Res. 2019; 15: 1–8.
13. Hu Z, Yang XY, Liu Y, et al. Investigation of HIFU-induced anti-tumor immunity in a murine tumor model. J. Transl. Med. 2007; 5: 1–1.
14. Hua S, De Matos MB, Metselaar JM, et al. Current trends and challenges in the clinical translation of nanoparticulate nanomedicines: Pathways for translational development and commercialization. Front Pharmacol. 2018; 9: 790.
15. Huang C, Ding S, Jiang W, et al. Glutathione-depleting nanoplatelets for enhanced sonodynamic cancer therapy. Nanoscale. 2021; 13: 4512–4518.
16. Hurwitz M, Stauffer P. Hyperthermia, radiation and chemotherapy: The role of heat in multidisciplinary cancer care. Semin. Oncol. 2014; 41: 714–729.
17. Husseini GA, El-Fayoumi RI, O'Neill KL, et al. DNA damage induced by micellar-delivered doxorubicin and ultrasound: Comet assay study. Cancer Lett. 2000; 154: 211–216.
18. Husseini GA, Pitt WG, Martins AM. Ultrasonically triggered drug delivery: Breaking the barrier. Colloids Surfaces B. 2014; 123: 364–386.
19. Husseini GA, Pitt WG, Williams JB, et al. Investigating the release mechanism of calcein from eliposomes at higher temperatures. J. Colloid Sci Biotechnol. 2014; 3: 239–244.
20. Husseini GA, Rapoport NY, Christensen DA, et al. Kinetics of ultrasonic release of doxorubicin from pluronic P105 micelles. Colloids Surfaces B. Biointerfaces. 2002; 24: 253–264.
21. Javadi M, Pitt WG, Tracy CM, et al. Ultrasonic gene and drug delivery using eLiposomes. J. Control Release. 2013; 167: 92–100.
22. Hu JJ, Liu LH, Li ZY, et al. MMP-responsive theranostic nanoplatform based on mesoporous silica nanoparticles for tumor imaging and targeted drug delivery. J. Mater Chem B. 2016 Mar 21; 4(11): 1932–1940.
23. Kang J, Wu X, Wang Z, et al. Antitumor effect of docetaxel-loaded lipid MBs combined with ultrasound-targeted MB activation on VX2 rabbit liver tumors. J. Ultrasound Med. 2010; 29: 61–70.
24. Kaplun AP, Bezrukov DA, Shvets VI. Rational design of nano- and micro-size medicinal forms of biologically active substances. Appl. Biochem Microbiol. 2011; 47: 711–717.
25. Kim D, Han J, Park SY, et al. Antitumor efficacy of focused ultrasound-MFL nanoparticles combination therapy in mouse breast cancer xenografts. Materials. 2020; 13: 1099.
26. Li D, He S, Wu Y, et al. Excretable lanthanide nanoparticle for biomedical imaging and surgical navigation in the second near-infrared window. Adv. Sci. 2019; 6: 1902042.
27. Li P, Zheng Y, Ran H, et al. Ultrasound triggered drug release from 10-hydroxycamptothecin-loaded phospholipid MBs for targeted tumor therapy in mice. J. Control. Release. 2012; 162: 349–354.

28. Yang C, Xiao H, Sun Y, et al. Lipid MBs as ultrasound-stimulated oxygen carriers for controllable oxygen release for tumor reoxygenation. Ultrasound Med. Biol. 2018; 44(2): 416–425.
29. Li Y, Huang W, Li C, et al. Indocyanine green conjugated lipid MBs as an ultrasound-responsive drug delivery system for dual-imaging guided tumor-targeted therapy. RSC Adv. 2018; 8: 33198–33207.
30. Liu D, Yang F, Xiong F, et al. The smart drug delivery system and its clinical potential. Theranostics. 2016; 6: 1306–1323.
31. Ye D, Zhang X, Yue Y, et al. Focused ultrasound combined with MB-mediated intranasal delivery of gold nanoclusters to the brain. J. Control Release. 2018; 286: 145–153.
32. Li Y, Huang W, Li C, et al. Indocyanine green conjugated lipid MBs as an ultrasound-responsive drug delivery system for dual-imaging guided tumor-targeted therapy. RSC Adv. 2018; 8(58): 33198–33207.
33. Liu L, Chang S, Sun J, et al. Ultrasound-mediated destruction of paclitaxel and oxygen loaded lipid MBs for combination therapy in ovarian cancer xenografts. Cancer Lett. 2015; 361: 147–154.
34. Horise Y, Maeda M, Konishi Y, et al. Sonodynamic therapy with anticancer micelles and high intensity focused ultrasound in treatment of canine cancer. Front Pharmacol. 2019; 10: 545.
35. Wu P, Jia Y, Qu F, et al. Ultrasound-responsive polymeric micelles for sonoporation-assisted site-specific therapeutic action. ACS Appl. Mater. Interfaces. 2017; 9(31): 25706–25716.
36. Liu M, Zhang P, Deng L, et al. IR780-based light-responsive nanocomplexes combining phase transition for enhancing multimodal imaging-guided photothermal therapy. Biomater Sci. 2019; 7: 1132–1146.
37. Liu Y, Liu J, Chen D, et al. Quinoxaline-based semiconducting polymer dots for in vivo NIR-II fluorescence imaging. Macromolecules. 2019; 52: 5735–5740.
38. Lyon PC, Gray MD, Mannaris C, et al. Safety and feasibility of ultrasound-triggered targeted drug delivery of doxorubicin from thermosensitive liposomes in liver tumours (TARDOX): A single-centre, open-label, phase 1 trial. The Lancet Oncol. 2018; 19: 1027–1039.
39. Maloney E, Hwang JH. Emerging HIFU applications in cancer therapy. Int. J. Hyperther. 2015; 31: 302–309.
40. Moradi S, Mokhtari-Dizaji M, Ghassemi F, et al. Increasing the efficiency of the retinoblastoma brachytherapy protocol with ultrasonic hyperthermia and gold nanoparticles: A rabbit model. Int. J. Radiat. Biol. 2020; 96(12): 1614–1627.
41. Guo X, Mei J, Jing Y, et al. Curcumin-loaded nanoparticles with low-intensity focused ultrasound-induced phase transformation as tumor-targeted and pH-sensitive theranostic nanoplatform of ovarian cancer. Nanoscale Res. Lett. 2020; 15: 1–1.
42. Li Y, Huang W, Li C, et al. Indocyanine green conjugated lipid microbubbles as an ultrasound-responsive drug delivery system for dual-imaging guided tumor-targeted therapy. RSC Adv. 2018; 8: 33198–33207.
43. Zandi A, Khayamian MA, Saghafi M, et al. Microneedle-based generation of microbubbles in cancer tumors to improve ultrasound-assisted drug delivery. Adv. Healthc. Mater. 2019; 8: 1900613.
44. Zhao N, Pan Y, Cheng Z, et al. Gold nanoparticles for cancer theranostics—A brief update. J. Innov. Opt. Health Sci. 2016; 9: 1630004.
45. Koseki Y, Ikuta Y, Taemaitree F, et al. Fabrication of size-controlled SN-38 pure drug nanocrystals through an ultrasound-assisted reprecipitation method toward efficient drug delivery for cancer treatment. J. Cryst. Growth. 2021; 572: 126265.
46. Song C, Yang C, Wang F, et al. MoS 2-Based multipurpose theranostic nanoplatform: Realizing dual-imaging-guided combination phototherapy to eliminate solid tumor via a liquefaction necrosis process. J. Mater. Chem. B. 2017; 5: 9015–9024.
47. Zhu M, Shi Y, Shan Y, et al. Recent developments in mesoporous polydopamine-derived nanoplatforms for cancer theranostics. J. Nanobiotechnol. 2021; 19: 1–22.
48. Moon H, Kang J, Sim C, et al. Multifunctional theranostic contrast agent for photoacoustics-and ultrasound-based tumor diagnosis and ultrasound-stimulated local tumor therapy. J. Control Rel. 2015; 218: 63–71.
49. Yuan P, Song D. MRI tracing non-invasive TiO2-based nanoparticles activated by ultrasound for multi-mechanism therapy of prostatic cancer. Nanotechnol. 2018; 29: 125101.
50. Wang X, Wang X, Zhong X, et al. V-TiO2 nanospindles with regulating tumor microenvironment performance for enhanced sonodynamic cancer therapy. Appl. Phys. Rev. 2020; 7: 041411.
51. Cui G, Wu J, Lin J, et al. Graphene-based nanomaterials for breast cancer treatment: Promising therapeutic strategies. J. Nanobiotechnology. 2021; 19: 1–30.
52. Shanei A, Akbari-Zadeh H, Fakhimikabir H, et al. The role of gold nanoparticles in sonosensitization of human cervical carcinoma cell line under ultrasound irradiation: An in vitro study. J. Nano Res. 2019; 59: 1–14.
53. Tabish TA, Zhang S, Winyard PG. Developing the next generation of graphene-based platforms for cancer therapeutics: The potential role of reactive oxygen species. Redox. Biol. 2018; 15: 34–40.

54. Dong Z, Feng L, Hao Y, et al. Synthesis of CaCO$_3$-based nanomedicine for enhanced sonodynamic therapy via amplification of tumor oxidative stress. Chem. 2020; 6: 1391–1407.
55. Mangraviti A, Gullotti D, Tyler B, et al. Nanobiotechnology-based delivery strategies: New frontiers in brain tumor targeted therapies. J. Control. Release. 2016; 240: 443–453.
56. Mishra B, Patel BB, Tiwari S. Colloidal nanocarriers: A review on formulation technology, types and applications toward targeted drug delivery. Nanomedicine Nanotechnology. Biol Med. 2010; 6: 9–24.
57. Mitragotri S. Healing sound: The use of ultrasound in drug delivery and other therapeutic applications. Nat. Rev. Drug Discovery. 2005; 4: 255–260.
58. Morey M, Pandit A. Responsive triggering systems for delivery in chronic wound healing. Adv. Drug Delivery Rev. 2018; 129: 169–193.
59. Paris JL, Cabañas MV, Manzano M, et al. Polymer-grafted mesoporous silica nanoparticles as ultrasound-responsive drug carriers. ACS Nano. 2015; 9: 11023–11033.
60. Paunovska K, Gil CJ, Lokugamage MP, et al. Analyzing 2000 in vivo drug delivery data points reveals cholesterol structure impacts nanoparticle delivery. ACS Nano. 2018; 12: 8341–8349.
61. Paunovska K, Sago CD, Monaco CM, et al. A direct comparison of in vitro and in vivo nucleic acid delivery mediated by hundreds of nanoparticles reveals a weak correlation. Nano Lett. 2018; 18: 2148–2157.
62. Silva R, Ferreira H, Little C, et al. Effect of ultrasound parameters for unilamellar liposome preparation. Ultrason. Sonochem. 2010; 17: 628–632.
63. Singh AP, Biswas A, Shukla A, et al. Targeted therapy in chronic diseases using nanomaterial-based drug delivery vehicles. Signal Transduction Targeted Ther. 2019; 4: 1–21.
64. Tan T, Hu H, Wang H, et al. Bioinspired lipoproteins-mediated photothermia remodels tumor stroma to improve cancer cell accessibility of second nanoparticles. Nat. Commun. 2019; 10: 1–7.
65. Thorat ND, Bauer J, Tofail SA, et al. Silica nano supra-assembly for the targeted delivery of therapeutic cargo to overcome chemoresistance in cancer. Colloids Surfaces B Biointerfaces. 2020; 185: 110571.
66. Tian Y, Liu Z, Tan H, et al. New aspects of ultrasound-mediated targeted delivery and therapy for cancer. World Int. J. Nanomed. 2020; 15: 401–418.
67. Toraya-Brown S, Fiering S. Local tumour hyperthermia as immunotherapy for metastatic cancer. Int. J. Hyperther. 2014; 30: 531–539.
68. Turk MJ, Reddy JA, Chmielewski JA, et al. Characterization of a novel pH-sensitive peptide that enhances drug release from folate-targeted liposomes at endosomal pHs. Biochim Biophys Acta—Biomembr. 2002; 1559: 56–68.
69. ud Din F, Aman W, Ullah I, et al. Effective use of nanocarriers as drug delivery systems for the treatment of selected tumors. Int. J. Nanomedicine. 2017; 12: 7291–7291.
70. van der Meel R, Sulheim E, Shi Y, et al. Smart cancer nanomedicine. Nat. Nanotechnology. 2019; 14: 1007–1017.
71. Wang L, Wu B, Chen J, et al. Monolayer hexagonal boron nitride films with large domain size and clean interface for enhancing the mobility of graphene-based field-effect transistors. Adv. Mater. 2014; 26: 1559–1564.
72. Wang N, Liu C, Yao W, et al. Endogenous reactive oxygen species burst induced and spatiotemporally controlled multiple drug release by traceable nanoparticles for enhancing antitumor efficacy. Biomater Sci. 2021; 9: 4968–4983.
73. Wang TY, E Wilson K, Machtaler S, et al. Ultrasound and MB guided drug delivery: Mechanistic understanding and clinical implications. Curr. Pharm. Biotechnol. 2013; 14: 743–752.
74. Wang Y, Wang B, Zhang L, et al. Mitochondria-targeted nanospheres with deep tumor penetration for photo/starvation therapy. J. Mater Chem. B. 2020; 8: 7740–7754.
75. Wu P, Dong W, Guo X, et al. ROS-responsive blended nanoparticles: Cascade-amplifying synergistic effects of sonochemotherapy with on-demand boosted drug release during SDT process. Adv. Healthcare Mater. 2019; 8: 1900720.
76. Xu H, Zhang X, Han R, et al. Nanoparticles in sonodynamic therapy: State of the art review. RSC Adv. 2016; 6(56): 50697–50705.
77. Yuan Y, Zhang J, Qi X, et al. Furin-mediated intracellular self-assembly of olsalazine nanoparticles for enhanced magnetic resonance imaging and tumour therapy. Nat. Mater. 2019; 18: 1376–1383.
78. Zhang H, Cui W, Qu X, et al. Photothermal-responsive nanosized hybrid polymersome as versatile therapeutics codelivery nanovehicle for effective tumor suppression. Proc. Natl. Acad. Sci. USA. 2019; 116: 7744–7749.
79. Zhang H, Xia H, Wang J, et al. High intensity focused ultrasound-responsive release behavior of PLA-b-PEG copolymer micelles. J. Control Release. 2009; 139: 31–39.

80. Zhang L, Gu FX, Chan JM, et al. Nanoparticles in medicine: Therapeutic applications and developments. Clin. Pharmacol. Ther. 2007; 83: 761–769.
81. Zhang L, Yi H, Song J, et al. Mitochondria-targeted and ultrasound-activated nanodroplets for enhanced deep-penetration sonodynamic cancer therapy. ACS Appl. Mater. Interfaces. 2019; 11: 9355–9366.
82. Zhang Q, Zhou H, Chen H, et al. Hierarchically nanostructured hybrid platform for tumor delineation and image-guided surgery via NIR-II fluorescence and PET bimodal imaging. Small. 2019; 15: 1903382.
83. Zhang X, He S, Ding B, et al. Cancer cell membrane-coated rare earth doped nanoparticles for tumor surgery navigation in NIR-II imaging window. Chem Eng J. 2020; 385: 123959.
84. Zhao C, Tong Y, Li X, et al. Photosensitive nanoparticles combining vascular-independent intratumor distribution and on-demand oxygen-depot delivery for enhanced cancer photodynamic therapy. Small. 2018; 14: 1703045.

13 Exogenous Dual/Multi-Responsive Delivery Nanoplatforms in Cancer Theranostics

Mershen Govender and Yahya E. Choonara

CONTENTS

13.1 Introduction 208
13.2 Dual-Responsive Therapy 210
    13.2.1 Combination Photothermal Therapies 210
    13.2.2 Combination Radiotherapy 212
    13.2.3 Magnetic Resonance-NIR-II Platforms 214
    13.2.4 Ferrous-Based Nanoplatforms 214
    13.2.5 Magnetic Resonance-Ultrasound Therapy 217
13.3 Multi-Responsive Theranostic Platforms 218
    13.3.1 MRI-NIRFI/PET-Dual Modal PET-MRI/PTT-PDT Nanoplatforms ... 218
    13.3.2 Chemotherapy-Radiotherapy-Phototherapy 218
    13.3.3 Chemotherapy-Photodynamic-Photothermal Therapy 219
    13.3.4 Metal-Based Targeted Antimicrobial Nanoplatforms 220
13.4 Conclusion 222
References 223

13.1 INTRODUCTION

Cancer is associated with the abnormal, uncontrolled growth of cells, with patient mortality a significant concern in the field of medicine. The diagnosis and treatment of the various forms of cancer have consequently been subjects of intense focus, with many innovative and novel treatment strategies being implemented over the last few decades. Despite the significant advancement in the field, diagnostic limitations and often limited treatments still exist, warranting continuous research to ensure an increased disease prognosis and patient survival (1). Conventional cancer treatments such as chemotherapy, surgery, and radiotherapy, while having varying degrees of efficacy, are often limited due to their systemic toxicities, indiscriminate mechanisms of action, and variations between patients (2–5). Additionally, chemotherapeutic treatment methods often results in substandard delivery to the affected tissues or cells (3). As a result of these shortcomings, nanomedical approaches to cancer treatment have since been developed to include early cancer detection, targeted drug delivery, and personalized treatment approaches. These approaches have been proposed as major developments in the field of cancer theranostics.

Theranostics, through the combining of specific targeted therapies for diagnostic tests (3, 6), has opened avenues for the simultaneous observation and monitoring of posttreatment chemotherapeutic drug efficacy, personalized and targeted treatments, and precise diagnosis and treatment for cancer patients (1, 6). Using nanomedicine platforms, which include lipid-based systems, polymeric particles, inorganic materials, and dendrimers (7) that integrate the disciplines of engineering, physics, biology, and chemistry, the potential for effective theranostic treatment has been highlighted. This is due to their stellar biocompatibility, biodegradability, relative functionalization ease, small size (1–100 nm), increased drug loading capabilities, and enhanced permeation and retention (EPR) effects. Nanoparticles also offer excellent multifunctional applications such as bio-sensing, bio-imaging, therapeutics, and diagnostics (7, 8). Furthermore, drugs, imaging agents, and nucleic acids can be incorporated into a single, integrated, functional platform (7). Table 13.1 provides a summary of the nanotheranostic systems that are currently undergoing clinical trials.

Theranostic nanomedicines may be prepared through various techniques, including (i) the loading of the imaging agent into nanoparticles or conjugation of therapeutic agents and (ii) tagging of radioisotopes, contrast imaging agent, and fluorescent dyes with therapeutic nanoparticles. Examples of effective theranostic platforms include biocompatible polymeric nanoparticles, lipid nanoparticles (nanostructured lipid carriers, liposomes, solid lipid nanoparticles), and metal-based nanosystems (gold, ferrite, and silver nanoparticles), which have been used to encapsulate the processing and visualization agents and provide its theranostic effect (3, 10).

The delivery of nanosystems with no site-specific functionality may, however, exert the same adverse effects commonly associated with conventional systemically administered chemotherapy. To overcome this potential therapeutic concern, site-specific stimuli-responsive systems have been

TABLE 13.1
Nanotheranostic Systems Currently under Clinical Trials, Adapted with Permission from Tan et al. (9) © 2020 Elsevier B.V.

Clinical Trial ID	Nano-Dosage Form	Cancer Type	Description
NCT02106598	Silica nanoparticles	Head and neck melanoma, breast cancer, colorectal cancer	Real-time image-guided intraoperative mapping of nodal metastases
NCT03409198	Liposomes	Breast cancer	Evaluating immunogenic chemotherapy combined with ipilimumab and nivolumab in breast cancer (ICON)
NCT03350945	Carbon nanoparticles	Colorectal cancer	Evaluating the effectiveness of application of carbon nanoparticles for tumor localization and lymph node mapping in laparoscopic colorectal surgery
NCT02495896	Recombinant EphB4-HSA fusion protein	Solid tumors	Study of the side effects and best dose of recombinant EphB4-HSA fusion protein when given together with standard chemotherapy regimens in treating patients with solid tumors
NCT02181075	Lipid-based nanoparticles	Liver tumors	A targeted delivery to achieve enhanced intra-tumor doxorubicin concentrations via a specially formulated lipid-based nanoparticles which is activated by mild hyperthermia using focused ultrasound
NCT01895829	Ferumoxytol–iron oxide nanoparticles	Head and neck cancer	Feasibility study of iron oxide nanoparticle magnetic resonance dynamic contrast enhanced MRI for primary and nodal tumor imaging in locally advanced head and neck squamous cell carcinomas
NCT03246659	CCK-2/gastrin receptor-localizing radiolabelled peptide probe	Medullary thyroid carcinoma	A novel CCK-2/gastrin receptor-localizing radiolabelled peptide probe for personalized diagnosis and therapy of patients with progressive or metastatic medullary thyroid carcinoma

developed. These stimuli-responsive systems are functional only in the presence of a required stimuli and therefore prevent unwanted drug release in areas of the body where the cancer is not present. Using such systems, however, can be inconsistent due to changes in the microenvironment due to the presence of tumors. External or exogenous stimuli may therefore be more suited to drug release from nanocarriers in cases of improperly functioning biological stimuli.

The application of ultrasound, magnetic field, visible light, γ-ray irradiation, UV light, and near-infrared (NIR) light are examples of constitute exogenous stimuli that are effective in providing an exogenous theranostic effect (11). These regulated exogenous stimuli allow for pulsatile drug release (12) as well as spatial and temporal control, therefore minimizing toxicity in a particular organ (13, 14), overcoming inter-patient variability (15), and increasing the half-life of drugs at the target site (16). These properties ultimately allow for greater treatment control, decreased dosing, and a lower side effect profile. With the potential and success of single stimuli-responsive platforms noted to provide enhanced patient treatment, dual- and multi-responsive systems have been developed to provide greater control of treatment. This chapter therefore provides for the numerous types of dual/multifunctional nano-based exogenous theranostic agents for cancer treatment recently reported. Additionally, herein an overview of the nanoparticle system has been presented with a focus on their effectiveness in cancer treatment and potential to decrease side effects and increase patient quality of life.

13.2 DUAL-RESPONSIVE THERAPY

13.2.1 Combination Photothermal Therapies

Numerous responsive systems have been noted to provide improved cancer treatment using a single stimulus. Laser-triggered photothermal therapy (PTT) treatment is an example of such a platform that has been noted to significantly improve tumor-specific treatment with fewer side effects (17). PTT, however, also has the potential to be used in combination with theranostic agents for increased anti-cancer efficacy. The near-infrared (NIR) dye IR780 iodide is such an agent which has been shown to non-invasively kill cancer cells with a high efficiency following laser irradiation (18). This has been achieved in the NIR imaging of breast tumors (19), as well as in cancer imaging and photodynamic therapy (PDT) (19). A disadvantage of using IR780 iodide, however, is its low lipophilicity and high toxicity. To overcome these challenges, Jiang et al. (19) loaded IR780 into human serum albumin nanoparticles (HSA-IR780 NPs) via protein self-assembly for the PDT and PTT of adenocarcinoma. The HSA-IR780 NP platform resulted in a significantly increased (1000-fold) solubility with a decreased toxicity (from 2.5 to 25 mg/kg). The photothermal effects were additionally evaluated in male tumor-bearing BALB/c mice and showed substantial tumor inhibition after injection the of HSA-IR780 NPs (minimal tumor size after 14 days vs. a placebo group displaying a 12-fold increase in tumor volume).

In another study aiming to enhance the delivery of IR780, Li et al. (17) used CXCR4-targeted nanostructured lipid carriers (NLCs) containing a CXCR4 antagonist (AMD3100). The dual-function system for the PTT of breast cancer exhibited reduced cancer cell invasiveness, as well as simultaneous tumor-targeting (Figure 13.1). Cumulative in vitro IR780 release from the nanosystem (particle size of ± 135 nm) was additionally noted to be low (<9%) after 24 hours. Concentration-dependent laser-induced temperature increases were also observed for the dual-function system when compared with free IR780 during the irradiation cycles (>60ºC vs. <40ºC). Furthermore, the free IR780 was rapidly degraded as evident by its low concentration (evaluated by HPLC), while after three cycles of irradiation, the IR780-loaded NLCs were >50% and did not reach zero until cycle 8, indicating the protective properties of the IR780-AMD-NLCs. Cell viability studies in the 4T1-Luc cell line noted a slight toxicity of both blank AMD-NLCs and the IR780-loaded NLCs without laser irradiation, with the cells exposed to the IR78-loaded NLCs and free IR780 after laser irradiation determined to be 20% and 92.2%, respectively. Furthermore, in vivo, the tumor

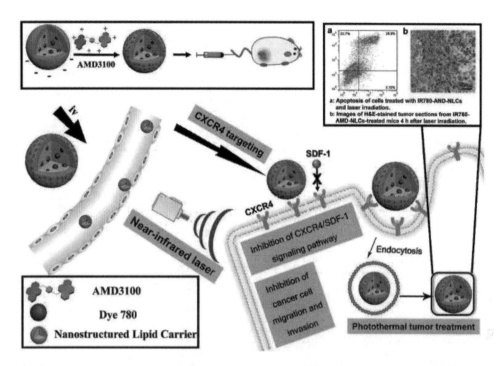

FIGURE 13.1 Images of (i) the schematic design of IR780-AMD-NLCs. AMD3100 can be used as an antagonist of CXCR4 to inhibit breast tumor lung metastasis. AMD3100 also increases the ability of NLCs to target 4T1 cells overexpressing CXCR4 and facilitates laser-induced photothermal antitumor treatment after the NLCs have accumulated at the tumor site. Scale bar: 100 μm. (ii) (A) the tumor temperature after exposure to laser irradiation in mice treated with saline, IR780 and IR780-AMD-NLCs ($n = 3$); (B) images of H&E-stained sections of tumors collected from mice treated with saline, IR780 solution, IR780-NLCs, or IR780-AMD-NLCs at 4 h after laser irradiation. Scale bar: 100 μm; (C) the growth of 4T1-Luc tumors (*$P < 0.05$ vs. free IR780/laser group, **$P < 0.01$ vs. saline group); (D) tumor weight (*$P < 0.05$ vs. free IR780/laser group, **$P < 0.01$ vs. free IR780/laser group); (E) photographs of tumors; and (F) body weight in the different groups of mice after various treatments. Reproduced with permission from Li et al. (17) © 2017 Acta Materialia Inc.

temperature of the group administered with the dual-function system increased from 33.5°C to 49.3°C compared with the control group (IR780 solution only), where the temperature was determined to 43.2°C. This increase in temperature was noted to promote tumor death (17).

Ricciardi et al. (20) further developed a core–shell gold–silica nanoplatform coupled with a photosensitizer iridium(III) complex for multimodal cellular imaging, photodynamic, and photothermal therapies. Ir_1-embedded silica nanoparticles (Ir_1-SiO_2) and Ir_1-embedded silica nanoparticles with a gold core (Ir_1-$AuSiO_2$) were compared to feature the photodynamic activity of Ir_1-SiO_2 and the combined PDT–PTT Ir_1-$AuSiO_2$ effect. Temperature variations for Ir_1-SiO_2 were observed after excitation with blue laser ($\lambda = 405$ nm) but not green ($\lambda = 532$ nm), with a 13 ± 1°C change seen with both $AuSiO_2$ and Ir_1-$AuSiO_2$ upon excitation with the green laser. Additionally, exposure to the blue laser resulted in an approximate 6 ± 1°C temperature variation after 10 min for $AuSiO_2$, while Ir_1-$AuSiO_2$ had a temperature variation of 9 ± 1°C. This allowed for the transition metal complexes' luminescence properties during cellular uptake and intracellular distribution to be determined. Phosphorescent emission was also exhibited with in vitro studies in human glioblastoma cells (U87MG) displaying remarkable phototoxicity at photosensitizer concentrations <0.5 μM, due to the synergistic PDT and PTT effects. Ir_1-$AuSiO_2$ furthermore showed a cell viability of 10% and had the most efficient phototherapy activity when the photothermal and photodynamic effects were combined.

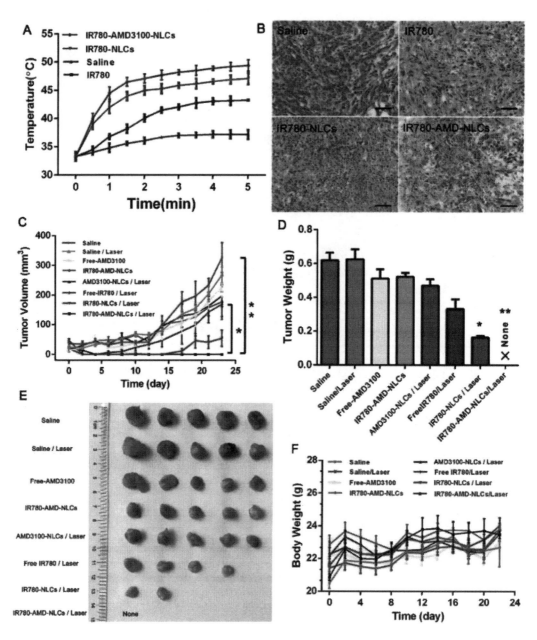

FIGURE 13.1 (Continued)

13.2.2 Combination Radiotherapy

The use of radiotherapy in combination with theranostic agents has been investigated for anticancer applications. In a study undertaken by Luna-Gutiérrez and co-workers evaluated $^{177}Lu_2O_3$-iPSMA and $^{177}Lu_2O_3$-iFAP nanoparticles (36 ± 7 nm), activated by neutron irradiation, for both radiotherapy and SPECT imaging in colorectal cancer metastasis (21). The biokinetic evaluation of both nanosystems in nude HCT116 tumor-bearing mice displayed statistically significant tumor retention ($p <$ 0.05). Tumor size progression additionally correlated with the delivered doses: $^{177}Lu_2O_3$-iFAP (105 ± 14 Gy), $^{177}Lu_2O_3$-iPSMA (99 ± 12 Gy), and $^{177}Lu_2O_3$ (58 ± 7 Gy) nanoparticles.

Results of this study showed that tumor radiation absorbed doses and tumor metabolic activity had no significant difference ($p < 0.05$) between the $^{177}Lu_2O_3$-iPSMA and $^{177}Lu_2O_3$-iFAP nanoparticles. Standard uptake values of 0.421 ± 0.092, 0.375 ± 0.104 and 1.821 ± 0.891 were also observed in $^{177}Lu_2O_3$-iPSMA, $^{177}Lu_2O_3$-iFAP, and $^{177}Lu_2O_3$, respectively. Tumor progression was significantly reduced (7.5 times) by the $^{177}Lu_2O_3$-iPSMA and $^{177}Lu_2O_3$-iFAP groups compared to the controls, with tumors 3-fold smaller than in the $^{177}Lu_2O_3$ group. This was attributed to the prolonged retention times, as well as the iPSMA or iFAP molecular recognition/radiotherapy combination. It was also noted that there was no evidence of renal and liver toxicity and uptake in non-target tissues was negligible. Additionally, a "compassionate use" protocol was implemented for one patient with multiple colorectal liver metastases (prostate-specific membrane antigen-positive), as described in Figure 13.2. In the study, doses of 42–210 Gy successfully demonstrated the potential use of $^{177}Lu_2O_3$-iPSMA for colorectal liver metastases.

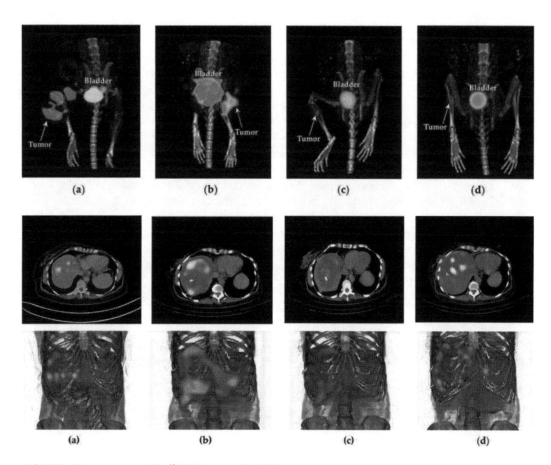

FIGURE 13.2 Images of (i) [^{18}F]FDG-microPET/CT imaging of (a) the control (untreated mouse), (b) the $^{177}Lu_2O_3$, (c) the $^{177}Lu_2O_3$-iPSMA, and (d) the $^{177}Lu_2O_3$-iFAP groups at 21 days of treatment of mice bearing HCT116 tumors, and (ii) the different molecular images (top: axial sections; bottom: PET/CT or SPECT/CT images) of a patient with multiple colorectal liver metastases (PSMA-positive): (a) [18F]FDG PET/CT [Day 0]; [^{99m}Tc]Tc-iPSMA SPECT/CT images at (b) 3 h and (c) 24 h post-injection [administration on Day 13]; and (d) $^{177}Lu_2O_3$-iPSMA SPECT/CT images acquired on Day 22. It can also be observed that over almost three months, the metastases moved due to the appearance of bilomas and the presence of necrotic tumors. The $^{177}Lu_2O_3$-iPSMA SPECT/CT image was contrast-enhanced to highlight the uptake of the radio-nanoparticles in tumor lesions as proof of concept. Reproduced with permission from Luna-Gutiérrez et al. (21) (distributed under a Creative Commons Attribution [CC BY 4.0] license).

13.2.3 MAGNETIC RESONANCE-NIR-II PLATFORMS

The use of near-infrared II (NIR-II) in combination with other exogenous stimulus has also been researched and includes the use of magnetic resonance (MR) imaging with PTT. In a study by Li et al (22), a phototheranostic urokinase plasminogen activator receptor (uPAR)-targeted nanoprobe (CH4T@MOF-PEGAE) was developed by applying NIR-II/MR dual modal image-guided PTT for glioblastoma (Figure 13.3). The nanoprobe was designed with the aim of combining the high detection sensitivity of the MR imaging with the spatiotemporal resolution of NIR-II fluorescence imaging. The intravenously delivered NIR-II fluorophore (CH4T)-loaded Fe-based metal-organic framework (MOF) nanoprobe (CH4T@MOF-PEGAE) was additionally modified for tumor-targeting using the AE105 peptide.

The octahedral CH4T@MOF-PEGAE (~60 nm) exhibited temperature increases to 53.8°C with 808-nm laser irradiation, with a concentration and laser-power–dependent photothermal effect. The study noted that no significant hemolysis was exhibited from the in vivo hemolysis assay, while cell viability (>90%) was observed in L929 normal cells and U87MG glioblastoma cancer cells after a 24-hour incubation. Furthermore, the nanoprobe exhibited no inflammation, apparent organ damage, or toxic side effects on the liver and kidney, while serum biochemical index results were similar to the control. NIR-II fluorescence intensity was significantly stronger (2.4-fold) as compared with the CH4T@MOF-PEG-SCM (nanoparticles without AE105 modification) after six hours in the uPAR-expressing human glioblastoma U87MG cell line. Tail vein injection in glioblastoma (GBM) xenograft mice model was used to assess the dual-mode imaging properties, with the NIR-II fluorescence signal detected at the tumor site 3 hours after injection, with the highest tumor-to-background ratio (TBR) value (4.34 ± 0.16) being reached 12 hours post injection. The TBR and fluorescence signals from the developed platform were also noted to be consistently higher than the control group. Passive targeting was almost three times less than that of the active targeting via the AE105 modification. In vivo, excellent tumor accumulation ability and primary liver clearance was exhibited by the CH4T@MOFPEG-AE group. A tumor temperature of 50°C was reached after laser irradiation at 808 nm for 5 min, eventually reaching 56°C, demonstrating a significant PTT effect. The results of this study noted that the nanoprobes demonstrated both the successful reduction of the U87MG tumors along with the effective guidance of the glioblastoma surgical resection via real-time intraoperative NIR-II imaging.

13.2.4 FERROUS-BASED NANOPLATFORMS

A ferrous-supply-regeneration core-corona nanosystem comprising a sorafenib (SRF, ferroptosis inducer) nanocrystal, Fe^{3+}, and tannic acid used in combination with PDT was developed by Liu et al. (23) for the treatment of breast cancer (Figure 13.4). In the developed system, methylene blue (MB) was loaded onto a core–corona SRF@Fe^{III}TA (SFT), whereby upon passive tumor accumulation and subsequent corona erosion, fluorescence recovery would be induced by the concomitant MB release. This allowed for complete elimination of the tumor through the multimodal effect of ferroptosis therapy and PDT. SFT and SFT-MB sizes were 207 and 220 nm, respectively, with a drug-loading of ~50%. Both apoptosis and ferroptosis were suggested to be possibly involved in the cellular death pathway determined using 4T1 murine mammary cancer cells, pre-treated with ferrostatin-1 and Ac-DEVD-CHO, before the cytotoxicity assay. At pH 4.0, after 72 hours of incubation, SRF release of 68% was achieved compared with <20% at pH 7.4. In 4T1 tumor-bearing female BALB/c mice, red fluorescent signals were seen to rapidly accumulate at the tumor sites just at one hour post-injection with no tumor-targeting effect seen when free MB was administered. Furthermore, tumor proliferation was almost completely inhibited with low levels of acute systematic toxicity after SFT-MB treatment.

Another study undertaken by Sang et al. (24) also focused on ferroptosis induction using co-loaded R780-Hex, magnetic iron oxide nanoparticles (MIONPs) and sorafenib (Sor) into a black

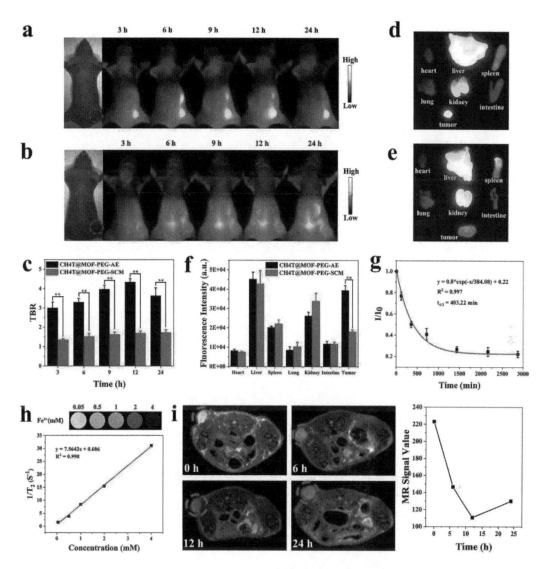

FIGURE 13.3 NIR-II fluorescence images of the xenograft GBM model at 3, 6, 9, 12, and 24 h after tail vein injection of the (a) CH4T@MOF-PEG-AE or (b) CH4T@MOF-PEG-SCM, and (c) TBR of the xenograft GBM model overtime after tail vein injection of the CH4T@MOF-PEG-AE or CH4T@MOF-PEG-SCM with the irradiation of 808 nm laser. *Ex vivo* NIR-II fluorescence images of major organs and tumors after treated with (d) CH4T@MOF-PEG-AE or (e) CH4T@MOF-PEG-SCM, (f) quantification analysis of the fluorescence signals of major organs and tumors at 24 h post-injection, (g) blood circulation half-life curve of CH4T@MOF-PEG-AE in mice fitted with a first-order exponential decay ($n = 3$) and, (h) T2-weighted MR images and T2 relaxation curves of CH4T@MOF-PEG-AE versus Fe concentration (mM) and (i) T2-weighted MR images and the corresponding MR signal values of the xenograft GBM model before and after CH4T@MOF-PEG-AE treatment (tumors are highlighted by red circles). **$p < 0.01$. Reproduced with permission from Li et al. (22) © 2020 The Authors.

hole quencher (BHQ)-based fluorescence "off–on" nano-photosensitizer complex assembly (CSO-BHQ-IR780-Hex/MIONPs/Sor). The novel multi-stimuli responsive nanosystem released lipid peroxidation producing payloads when disassembled by GSH attack and triggered by NIR irradiation. The MIONPs exhibited particle sizes of 50–100 nm with a maximum absorption at 780 and 825 nm, respectively. Good magnetic targeting ability was additionally shown in 4T1 cells. After

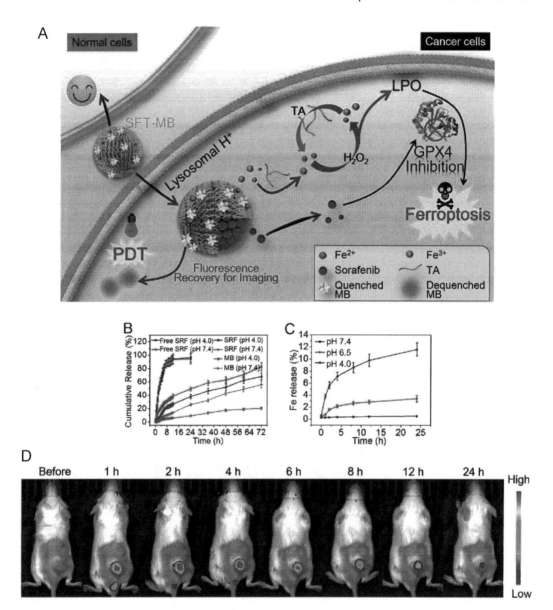

FIGURE 13.4 (A) Schematic illustration of SFT-mediated combination of ferroptosis and image-guided PDT, (B) cumulative release of SRF from free SRF or SFT-MB and cumulative release of MB from SFT-MB under different pH values, (C) cumulative release of Fe under different pH, and (D) in vivo fluorescence images of 4T1 tumor-bearing mice at different time points after intravenous injection of SFT-MB. Adapted with permission from Liu et al. (23). Copyright (2018) American Chemical Society.

administration of the developed nanosystem into the caudal vein of female Wistar breast tumor rat model, the sorafenib plasma $t_{1/2}$ was 31.25 ± 0.12 h, 26-fold higher than that of the free drug. Fluorescence signal when exposed to a magnetic field was also stronger than when not exposed, demonstrating the self-assembly's magnetic nature to enhance the nanoparticle/tumor-targeting EPR effect by accurately accumulating in the tumors, as monitored by the IR780-Hex photosensitizer (Figure 13.5).

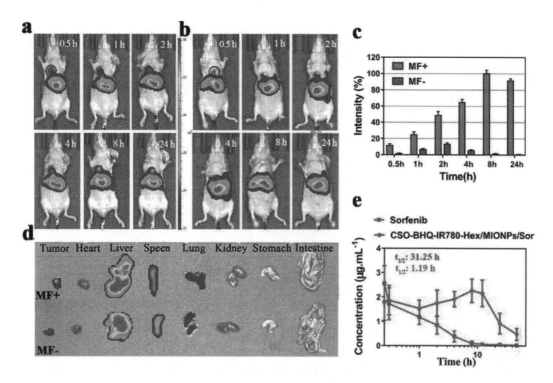

FIGURE 13.5 In vivo imaging of NIR nanophotosensitizer and pharmacokinetic study. The tumor of mice exposed to a magnetic field (a), and not exposed to a magnetic field (b) after treating with CSO-BHQ-IR780-Hex/MIONPs/Sor from 0 to 24 h; the in vivo images were obtained; (c) fluorescence intensity of MF+ and MF− groups; (d) mice injected with the complex self-assembly for 24 h; then, the organs (heart, liver, lung, kidney, spleen, lung, kidney, stomach, and intestine) and tumor tissue were harvested; and (e) Mean blood concentration–time curve of rats after injecting with CSO-BHQ-IR780-Hex/MIONPs/Sor self-assembly or sorafenib ($n = 6$) for 72 h (24). Reproduced with permission from Sang et al. (24). Copyright (2019) American Chemical Society.

13.2.5 Magnetic Resonance-Ultrasound Therapy

The use of magnetic resonance in combination with ultrasound therapy has also been investigated for its anticancer potential. In a study undertaken by Wang and co-workers (25) a nanoscale superparamagnetic iron oxide (SPIO) platform was shown to be able to induce the sensitivity of imaging and efficiency of energy deposition by a clinical MR-guided focused ultrasound surgery (MRgFUS) system. The SPIO surfaces were functionalized with anti-epidermal growth factor receptor (EGFR) monoclonal antibodies. Results from this study showed that the PEGylated SPIO and EGFR-targeted (anti-EGFR-PEG-SPIO) nanoparticles had diameters of 38.5 and 45.7 nm, respectively, while the theranostic agents for MRI and MRgFUS ablation were evaluated in a H460 cell-xenografted rat model and exhibited enhanced targeting and an improved contrast imaging at the tumor site. H460 cell viability was also observed to be above 96% at 5 mg/mL, decreasing to 85% at 80 mg/mL, for the anti-EGFR-PEG-SPIO and PEGylated SPIO, respectively, demonstrating no obvious cytotoxicity for both.

Results from the in vitro MRgFUS lung tumor ablation showed significantly lower sonication energy levels were used in the targeted and non-targeted groups (32 and 54 W, respectively) compared to the positive control group (76 W), although the peak temperatures for both nanoparticles were similar to the positive control group. The negative control group (54 W) also displayed a significantly lower peak temperature (low power, 54 W). Significant decreases in tumor size were additionally observed in the positive control, non-targeted, and targeted group, with increases observed in the negative control group.

13.3 MULTI-RESPONSIVE THERANOSTIC PLATFORMS

13.3.1 MRI-NIRFI/PET-Dual Modal PET-MRI/PTT-PDT Nanoplatforms

With the development and success of dual-responsive theranostic platforms, multi-responsive platforms with three or more mechanisms of actions of have been developed for enhanced anticancer treatment. In a study by Li and co-workers (26) a smart and highly versatile "all-in-one" nanoporphyrin nanoplatform capable of integrating near-infrared fluorescent imaging (NIRFI), positron emission tomography (PET), MRI, dual modal PET-MRI, PTT, and PDT was developed. This agent also integrated ^{64}Cu for PET, Gd (III) for MRI, and chemotherapy drugs in a single system.

Disulfide-crosslinked nanoparticles (CNPs) displayed sizes of 32 ± 8 nm, and cell uptake was evaluated in the SKOV3 ovarian cancer, the PC3m prostate cancer and MB49 mouse bladder cancer cell lines, with PDT efficacy noted to be light dose and nanoparticle dose dependent. Tissue distribution was determined using intravenously administered Gd(III)-chelated CNPs and Gd(III)-chelated non-crosslinked nanoparticles into SKOV3-xenografted nude. The particles dissociated at the tumor site, ultimately resulting in significant MRI contrast, beginning four hours post injection (Figure 13.6A).

Lukianova-Hleb et al. (27) additionally developed a four-component nanosystem, comprising colloidal gold, encapsulated drugs, low-energy short laser pulses, and X-rays. A receptor-mediated endocytosis process resulted in self-assembly of the antibody-functionalized gold nanoparticles and drug nanocarriers, which remained stable until laser pulse and X-ray stimulation. Upon activation, a plasmonic nanobubble was generated resulting in the mechanical destruction of the cancer cell or drug ejection into the cytoplasm. The ejection of the drug functioned by disrupting the cancer cell's liposome and endosome. The results of the study noted a 100-fold improvement in anticancer efficacy in resistant and aggressive head and neck cancer. Additionally, the entry doses of the respective drugs and X-rays were reduced to 2–6% of their clinical doses, thereby sparing healthy cells (27) (Figure 13.6B).

13.3.2 Chemotherapy-Radiotherapy-Phototherapy

The use of chemotherapy in combination with other synergistic therapies has been explored in an effort to enhance anticancer treatments while decreasing systemic side effects and increasing patient outcomes. In a study undertaken by Mirrahimi et al. (28) a synergistic multimodal nanocomplex (ACA) consisting of an alginate hydrogel co-loaded with cisplatin and gold nanoparticles was developed. The ACA was designed to exert a unique combination of chemotherapy, radiotherapy, and PTT and was evaluated in CT26 colon adenocarcinoma tumor-bearing mice. The trimodal thermo-chemo-radio therapy utilized in this study involved the targeted delivery of cisplatin in conjunction with laser irradiation (532 nm) and X-ray (6 MV).

The results of this study displayed distinct antitumor effectiveness, with complete tumor regression and no evidence of relapse during the 90-day follow up period, when compared to its dual therapy (thermo-chemotherapy and chemoradiation) counterparts. This study also investigated the effect of heat generation on drug release and determined that in response to temperature increases from the thermo- and radiotherapy, the amount of cisplatin released after two hours varied drastically between 25°C (23.4%) and 56°C (94.7%). This result highlighted the importance of considering the effect of additional stimuli on drug release to prevent unwanted adverse drug effects.

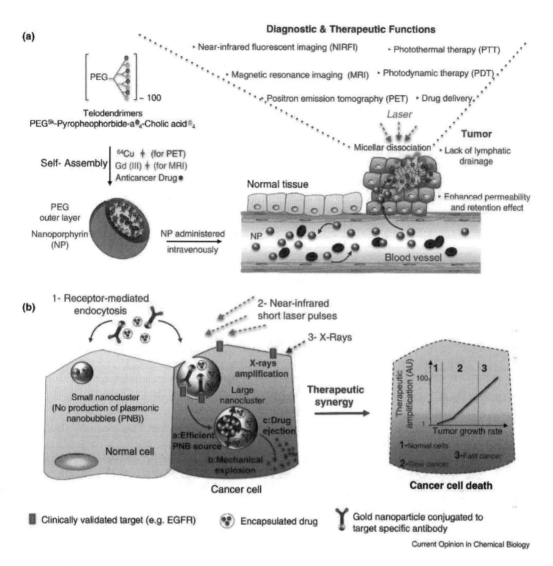

FIGURE 13.6 Images of the (a) multifunctional theranostic nanomedicine (26), and (b) a schematic illustration of quadrapeutics (27). Reproduced with permission from Kojima et al. (7) © 2015 Elsevier Ltd.

13.3.3 CHEMOTHERAPY-PHOTODYNAMIC-PHOTOTHERMAL THERAPY

The use of chemotherapy in conjunction with PDT and PTT has been investigated to enhance the anticancer activity of the respective individual therapies. In a study undertaken by Luo and coworkers (29), a tri-synergistic tumor therapy of PDT, PTT, and chemotherapy was developed using mesoporous silica-coated gold nanorods (MMSGNRs) to target to hepatoma cells through the lactobionic acid targeting moiety and was loaded with the photosensitizer Al(III) phthalocyanine chloride tetrasulfonic acid, $AlPcS_4$. The developed platform was additionally loaded with a redox-responsive Pt(IV) complex that converted to the cytotoxic Pt(II) drug within the within the tumor environment, inducing cell death. The synergy behind using PDT and PTT in the MMSGNR system was due to thermal effects of PTT (>41°C) increasing blood flow, which allows for greater oxygen supply, promoting the effects of PDT. Additionally, the thermal effects of PTT promote blood permeability, which has an increased effect on cytotoxic drug permeation, which in the case of the MMSGNR-AlPcS$_4$ system was the Pt(IV) complex.

The results of the study displayed an appreciable tumor cell selectivity and responsiveness by the MMSGNR-AlPcS$_4$ nanocomposite with the triggered release of AlPcS$_4$ and cisplatin. Interestingly, no tumor growth was seen after 12 days in the in vivo group administered with the nanocomposite platform and exposed to PDT and PTT treatment, while the control group of PBS administration were sacrificed after their tumors grew 18.4-fold. Additionally, the group administered with the nanocomposite but no exposed to PDT and PTT did not show any obvious tumor suppression. The results of this study therefore displayed the increased potential of using multi-responsive platforms combining synergistic therapies for increased anti-cancer treatment.

13.3.4 Metal-Based Targeted Antimicrobial Nanoplatforms

The treatment of cancer through thermal therapy often results in the induction of inflammation, which can lead to further metastasis and infection. The use of transitional platforms that are delivered prior to induction into an active dual anticancer-antimicrobial form using an exogenous stimulus has been researched to overcome this therapeutic concern. These platforms are usually composed of metal compounds that exert an antimicrobial effect in addition to having anti-cancer properties. The platforms can additionally be CT guided to ensure that platforms exert their actions at the required area to ensure maximum therapeutic benefit with minimal side effects. Silver (Ag) nanoparticles are an example of metal-derived nanosystems and have a variety of applications, including (i) drug delivery, (ii) therapeutics, and (iii) diagnostics/imaging (30).

In a study by Wu et al. (31), trifunctional PEG-IL/ZrO$_2$-Ag@SiO$_2$ nanorattles were developed for CT imaging-guided, simultaneous tumor microwave thermal therapy. The nanorattles additionally had an antibacterial effect at the site of action. The results of this study noted that the prepared nanorattles (145 nm with a hydrodynamic size of 181 nm) resulted in excellent bacterial resistance and microwave thermal properties after intratumor and intravenous injections (50 mg/kg), exhibiting an effective reduction in inflammatory responses during the microwave thermal therapy (1.8 W, 450 MHz). The anti-tumor effects were determined to be 96.4%, with an overall survival rate of 80%, compared to a placebo group (40% survival), in H22-tumorbearing mice (Figure 13.7). Tissue analysis after administration also detailed the biocompatibility of the prepared nanorattles. Further MTT assays on HepG2 cells detailed the biocompatibility of the prepared nanorattles with an 84% cell survival at a concentration of 12.5 mg/mL. Antimicrobial efficacy tests were also undertaken by the researchers and displayed activity against resistant *E. coli* and methicillin-resistant *S. aureus* at all test concentrations. In each of these studies, ZrO$_2$-Ag@SiO$_2$ prior to IL loading and modification with PEG was also evaluated and displayed a higher zone of inhibition during antimicrobial studies and a lower cellular activity (70%) during MTT analysis.

In another study by Zhu et al. (32), the researchers demonstrated the anticancer and antibacterial effects of *in situ* NIR-light-assisted reduced antimicrobial peptide (AMP)-protected gold nanoclusters on HT-29 human colon cancer cells and *E. coli* respectively. The AMP used in this study was tachyplesin-I. The prepared system (CNP@AMPAuNCs), which was designed to possess a synergistic combination of antibacterial, fluorescent, and PTT properties and was noted to have increased anticancer activity after irradiation with 100% killing of the cancer cells achieved at 0.3 mg/mL. Cytotoxicity analysis of the developed nanosystem against HT-29 cells however detailed that >89% were active after exposure to 0 to 0.5 mg/mL of CNP@AMPAuNCs for 24 h without irradiation, confirming that irradiation of the system was required for its anticancer activity. The CNP@AMPAuNCs were additionally highly active against *E. coli*, noting its antibacterial properties.

Other metal-based theranostic nanoplatforms have also been developed for cancer treatment and include IL@ZrO$_2$ nanostructures for the image-guided treatment of tumors (33), tantalum sulfide nanosheets for computed tomographic imaging-guided chemo-photothermal therapy (34), and

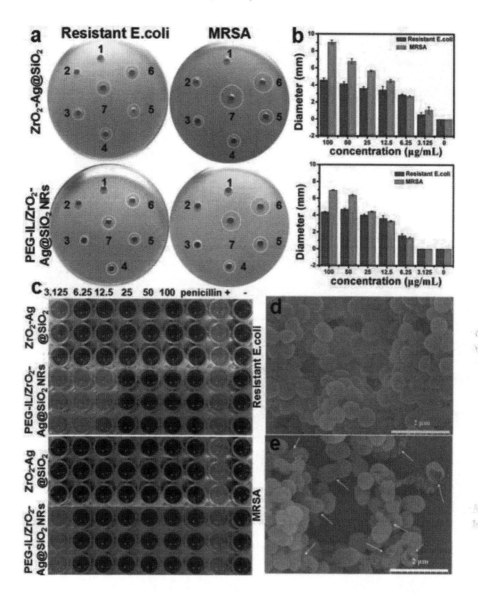

FIGURE 13.7 (i) (a) The anti-bacterium-circles of agarose diffusion assays on MRSA and resistant *E. coli* using ZrO_2-Ag@SiO_2 and PEG-IL/ZrO_2-Ag@SiO_2 NRs; (b) the diameter of the anti-bacterium-circle based on (a); (c) the minimal inhibition concentrations of ZrO_2-Ag@SiO_2 and PEG-IL/ZrO_2-Ag@SiO_2 NRs on MRSA and resistant *E. coli*; (d) SEM image of normal MRSA; (e) SEM image of MRSA which was treated with PEG-IL/ZrO_2-Ag@SiO_2 NRs, the arrow indicated that membrane perforation, cell membrane rupture and content exudate after 6 h of incubation induced by the PEG-IL/ZrO_2-Ag@SiO_2 NRs. (ii) (a) Different concentrations of ZrO_2@SiO_2 and ZrO_2-Ag@SiO_2 NRs CT imaging in vitro (1, 2, 3, 4, 5, 6, 8, 10, and 20 mg mL^{-1}); (b) CT imaging in vivo prior to and post subcutaneous injection of PEG-IL/ZrO_2-Ag@SiO_2 NRs (left) and the volume rea construction image (right); (c) CT imaging in vivo at different time after intravenous injection of PEG-IL/ZrO_2-Ag@SiO_2 NRs. From left to right: 0 h, 2 h, 4 h, 6 h, 12 h (the dose was 50 mg kg^{-1}); (d) FLIR images of the mice in microwave and PEG-IL/ZrO_2-Ag@SiO_2 NRs + microwave groups at 1 min intervals; (e) temperature change values in microwave and PEG-IL/ZrO_2-Ag@SiO_2 NRs + microwave groups. Reproduced with permission from Wu et al. (31) © 2018 Elsevier Ltd.

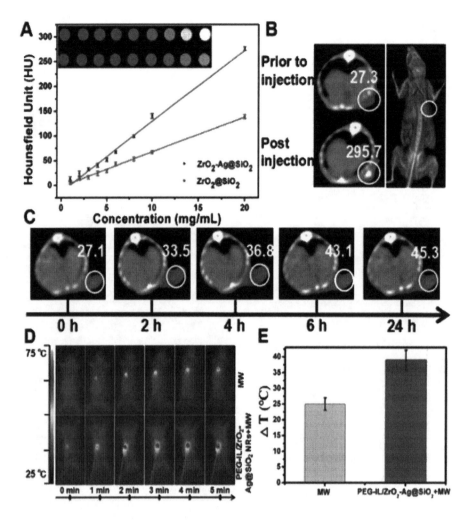

FIGURE 13.7 (Continued)

magnetically targeted iron (III) oxide–gold core–shell nanoparticles (35). In each of these platforms, the in vivo anticancer activity of the respective platforms was evaluated and noted to be highly effective. The metals which were utilized have been determined previously to have potential antibacterial properties; however, antimicrobial efficacy studies were not undertaken in these studies.

13.4 CONCLUSION

The use of synergistic dual- and multi-responsive exogenous nanoplatforms, receptive to different stimuli, has resulted in a significant improvement in the treatment of various cancers when compared to the individual therapies. These stimuli include magnetic fields, ultrasound, light (NIR, visible, and UV) and radiation and have been used in varying combinations with the aim of improving treatment efficacy while increasing biocompatibility and safety. Additionally, the platforms can be designed to provide further health benefits such as anti-inflammatory and antimicrobial effects to assist in healing of the treated tissue. While research into dual- and multi-responsive nanoplatforms is still limited, the studies that have been performed have displayed their significant potential in advancing anticancer treatment. This will ultimately address the need for effective treatment platforms that will increase disease prognosis and increase patient quality of life.

REFERENCES

1. Jeyamogan S, Khan NA, Siddiqui R. Application and importance of theranostics in the diagnosis and treatment of cancer. Arch Med Res. 2021;52(2):131–142.
2. Moradi Kashkooli F, Soltani M, Souri M. Controlled anti-cancer drug release through advanced nano-drug delivery systems: Static and dynamic targeting strategies. J Control Release Off J Control Release Soc. 2020;327:316–349.
3. Rizwanullah M, Ahmad MZ, Garg A, Ahmad J. Advancement in design of nanostructured lipid carriers for cancer targeting and theranostic application. Biochim Biophys Acta—Gen Subj. 2021;1865(9):129936.
4. Ko C-N, Li G, Leung C-H, Ma D-L. Dual function luminescent transition metal complexes for cancer theranostics: The combination of diagnosis and therapy. Coord Chem Rev. 2019;381:79–103.
5. Zugazagoitia J, Guedes C, Ponce S, Ferrer I, Molina-Pinelo S, Paz-Ares L. Current challenges in cancer treatment. Clin Ther. 2016;38(7):1551–1566.
6. Kang SJ, Jeong HY, Kim MW, Jeong IH, Choi MJ, You YM, et al. Anti-EGFR lipid micellar nanoparticles co-encapsulating quantum dots and paclitaxel for tumor-targeted theranosis. Nanoscale. 2018;10(41):19338–19350.
7. Kojima R, Aubel D, Fussenegger M. Novel theranostic agents for next-generation personalized medicine: Small molecules, nanoparticles, and engineered mammalian cells. Curr Opin Chem Biol. 2015;28:29–38.
8. Madamsetty VS, Mukherjee A, Mukherjee S. Recent trends of the bio-inspired nanoparticles in cancer theranostics. Front Pharmacol. 2019;10:1264.
9. Tan YY, Yap PK, Xin Lim GL, Mehta M, Chan Y, Ng SW, et al. Perspectives and advancements in the design of nanomaterials for targeted cancer theranostics. Chem Biol Interact. 2020;329:109221.
10. Ahmad J, Rizwanullah M, Amin S, Warsi HM, Ahmad ZM, Barkat AM. Nanostructured lipid carriers (NLCs): Nose-to-brain delivery and theranostic application. Curr Drug Metab. 2020;21:1136–1143.
11. Chen H, Zhao Y. Applications of light-responsive systems for cancer theranostics. ACS Appl Mater Interfaces. 2018;10(25):21021–21034.
12. Satarkar NS, Hilt JZ. Magnetic hydrogel nanocomposites for remote controlled pulsatile drug release. J Control Release. 2008;130(3):246–251.
13. Du L, Jin Y, Zhou W, Zhao J. Ultrasound-triggered drug release and enhanced anticancer effect of doxorubicin-loaded poly(D,L-lactide-co-glycolide)-methoxy-poly(ethylene glycol) nanodroplets. Ultrasound Med Biol. 2011;37(8):1252–1258.
14. Hoare T, Santamaria J, Goya GF, Irusta S, Lin D, Lau S, et al. A magnetically triggered composite membrane for on-demand drug delivery. Nano Lett. 2009;9(10):3651–3657.
15. Raza A, Hayat U, Rasheed T, Bilal M, Iqbal HMN. "Smart" materials-based near-infrared light-responsive drug delivery systems for cancer treatment: A review. J Mater Res Technol. 2019;8(1):1497–1509.
16. Han L, Tang C, Yin C. Dual-targeting and pH/redox-responsive multi-layered nanocomplexes for smart co-delivery of doxorubicin and siRNA. Biomaterials. 2015;60:42–52.
17. Li H, Wang K, Yang X, Zhou Y, Ping Q, Oupicky D, et al. Dual-function nanostructured lipid carriers to deliver IR780 for breast cancer treatment: Anti-metastatic and photothermal anti-tumor therapy. Acta Biomater. 2017;53:399–413.
18. Yue C, Liu P, Zheng M, Zhao P, Wang Y, Ma Y, et al. IR-780 dye loaded tumor targeting theranostic nanoparticles for NIR imaging and photothermal therapy. Biomaterials. 2013;34(28):6853–6861.
19. Jiang C, Cheng H, Yuan A, Tang X, Wu J, Hu Y. Hydrophobic IR780 encapsulated in biodegradable human serum albumin nanoparticles for photothermal and photodynamic therapy. Acta Biomater. 2015;14:61–69.
20. Ricciardi L, Sancey L, Palermo G, Termine R, De Luca A, Szerb EI, et al. Plasmon-mediated cancer phototherapy: The combined effect of thermal and photodynamic processes. Nanoscale. 2017;9(48):19279–19289.
21. Luna-Gutiérrez M, Ocampo-García B, Jiménez-Mancilla N, Ancira-Cortez A, Trujillo-Benítez D, Hernández-Jiménez T, et al. Targeted Endoradiotherapy with Lu2O3-iPSMA/-iFAP nanoparticles activated by neutron irradiation: Preclinical evaluation and first patient image. Pharmaceutics. 2022;14:720.
22. Li Z, Wang C, Chen J, Lian X, Xiong C, Tian R, et al. uPAR targeted phototheranostic metal-organic framework nanoprobes for MR/NIR-II imaging-guided therapy and surgical resection of glioblastoma. Mater Des. 2021;198:109386.
23. Liu T, Liu W, Zhang M, Yu W, Gao F, Li C, et al. Ferrous-supply-regeneration nanoengineering for cancer-cell-specific ferroptosis in combination with imaging-guided photodynamic therapy. ACS Nano. 2018;12(12):12181–12192.

24. Sang M, Luo R, Bai Y, Dou J, Zhang Z, Liu F, et al. BHQ-cyanine-based "off—on" long-circulating assembly as a ferroptosis amplifier for cancer treatment: A lipid-peroxidation burst device. ACS Appl Mater Interfaces. 2019;11(46):42873–42884.
25. Wang Z, Qiao R, Tang N, Lu Z, Wang H, Zhang Z, et al. Active targeting theranostic iron oxide nanoparticles for MRI and magnetic resonance-guided focused ultrasound ablation of lung cancer. Biomaterials. 2017;127:25–35.
26. Li Y, Lin T, Luo Y, Liu Q, Xiao W, Guo W, et al. A smart and versatile theranostic nanomedicine platform based on nanoporphyrin. Nat Commun. 2014;5(1):4712.
27. Lukianova-Hleb EY, Ren X, Sawant RR, Wu X, Torchilin VP, Lapotko DO. On-demand intracellular amplification of chemoradiation with cancer-specific plasmonic nanobubbles. Nat Med. 2014;20(7):778–784.
28. Mirrahimi M, Beik J, Mirrahimi M, Alamzadeh Z, Teymouri S, Mahabadi VP, et al. Triple combination of heat, drug and radiation using alginate hydrogel co-loaded with gold nanoparticles and cisplatin for locally synergistic cancer therapy. Int J Biol Macromol. 2020;158:617–626.
29. Lùo G-F, Chen W-H, Lei Q, Qiu W-X, Liu Y-X, Cheng Y-J, et al. A triple-collaborative strategy for high-performance tumor therapy by multifunctional mesoporous silica-coated gold nanorods. Adv Funct Mater. 2016;26(24):4339–4350.
30. Chugh H, Sood D, Chandra I, Tomar V, Dhawan G, Chandra R. Role of gold and silver nanoparticles in cancer nano-medicine. Artif Cells, Nanomedicine, Biotechnol. 2018;46(Sup1):1210–1220.
31. Wu Q, Yu J, Li M, Tan L, Ren X, Fu C, et al. Nanoengineering of nanorattles for tumor treatment by CT imaging-guided simultaneous enhanced microwave thermal therapy and managing inflammation. Biomaterials. 2018;179:122–133.
32. Zhu S, Wang X, Li S, Liu L, Li L. Near-infrared-light-assisted in situ reduction of antimicrobial peptide-protected gold nanoclusters for stepwise killing of bacteria and cancer cells. ACS Appl Mater Interfaces. 2020;12(9):11063–11071.
33. Shi H, Niu M, Tan L, Liu T, Shao H, Fu C, et al. A smart all-in-one theranostic platform for CT imaging guided tumor microwave thermotherapy based on IL@ZrO(2) nanoparticles. Chem Sci. 2015;6(8):5016–5026.
34. Liu Y, Ji X, Liu J, Tong WWL, Askhatova D, Shi J. Tantalum sulfide nanosheets as a theranostic nanoplatform for computed tomography imaging-guided combinatorial chemo-photothermal therapy. Adv Funct Mater. 2017;27(39):1703261.
35. Abed Z, Beik J, Laurent S, Eslahi N, Khani T, Davani ES, et al. Iron oxide-gold core-shell nanotheranostic for magnetically targeted photothermal therapy under magnetic resonance imaging guidance. J Cancer Res Clin Oncol. 2019;145(5):1213–1219.

14 Aptamer-Based Targeting of Nanoplatforms for Cancer Theranostics

*Gaurav Tiwari, Ancha Kishore Babu,
Charul Khatri and Namdev L. Dhas*

CONTENTS

14.1 Introduction 225
14.2 Structure of Aptamers 227
14.3 Screening of Aptamers 228
14.4 Aptamer-Mediated Drug Delivery for Targeted Cancer Therapy 228
    14.4.1 Nucleic Acid Aptamer-Based Dual Responsive Nanoplatforms 228
    14.4.2 Aptamers for Targeted Drug Delivery 230
    14.4.3 Targeted Chemotherapy 230
    14.4.4 Aptamers Modified with Drugs for Therapy 232
    14.4.5 Targeted Biotherapy 232
14.5 Aptamer-Conjugated Nanomaterials 232
    14.5.1 Aptamer-Conjugated Organic Nanomaterials 234
    14.5.2 Aptamer-Conjugated Inorganic Nanomaterials 235
    14.5.3 Aptamer-Conjugated Nanomaterials for PDT 236
    14.5.4 Aptamer-Conjugated Nanomaterials for PTT 237
14.6 Aptamer-Mediated Antitumour Effects 239
14.7 Conclusion 240
14.8 Abbreviations 240
References 242

14.1 INTRODUCTION

Due to rapid development in several areas, significant advancements in the treatment of diseases, including cancer, have been made recently. Due to difficulties in the clinical management and diagnosis of cancer, the high mortality rate of malignant tumours continues to seriously harm people's lives and physical well-being on a global scale (1). Chemotherapy, radiation, and surgery are common forms of conventional tumour treatment, but each of these modalities has limits in real-world clinical settings. Unfortunately, surgical treatment is not as effective as non-solid tumours like leukaemia and is less effective for advanced solid tumours than it is for early-stage solid tumours (2). Currently, chemotherapy is a crucial component of the cancer treatment process. After intravenous injection, it is typically disseminated throughout the body to produce therapeutic effects. However, this kind of treatment lacks selectivity and will kill tumour tissues together with normal tissues, resulting in non-specific harm. Additionally, prolonged usage of chemotherapy-related medications might result in patients developing drug resistance and lessen the efficacy of treatment (3). Ionizing radiation is utilized in radiotherapy to treat malignant tumours. Serious side effects from local and systemic radiotherapy are frequently seen, including systemic responses, radiation pneumonia, and radiation osteonecrosis. Therefore, it is critical to investigate efficient techniques for cancer diagnosis, early detection, and treatment approaches in order to drastically lower cancer mortality. Prognosis would be improved by a therapeutic approach that combined early clinical diagnosis with ideal therapy regimens (4). A technology that delivers medications to cancer cells only could do this. In the expanding field of biotechnology, the special qualities of aptamers that are compatible with many platform designs have aided in the quick analysis, detection, and treatment of cancer. The systematic evolution of ligand by exponential enrichment (SELEX) method produces aptamers, which are single-stranded DNA, RNA, or altered nucleic acid sequences with a high affinity for specific target binding (5). Aptamers can bind to the target with great specificity and affinity due to their distinctive three-dimensional spatial structure. Some aptamers can influence the function of target proteins by interacting with them because they serve as regulatory proteins (4, 6). Due to their automated synthesis, long-term stability as dry powder or in solution, capacity to sustain reversible denaturation, well-established selection process, easy and controllable modification to fulfil various diagnostic and therapeutic purposes, slow degradation kinetics, non-toxicity and lack of immunogenicity, and quick tissue penetration, aptamers have emerged as one of the most promising tools for adding target specificity to nanomaterials for intracellular applications. Aptamers have many different targets, such as proteins, ions, and cells. Aptamers are superior to antibodies because they have greater target affinity, greater specificity, ease of synthesis and modification, good stability, and ease of storage (Table 14.1). A wide range of aptamers aimed at various targets have so far been examined. Aptamers are employed frequently in tumour detection, diagnostics, imaging analysis, and drug targeted therapy because of their special properties (6–10).

Aptamers undergo routine alterations with a specified goal as synthetic compounds. Aptamers have been changed in a variety of ways to increase their adaptability up to this point, such as adding biotin, fluorescent dyes, or radionuclides. The biodistribution or pharmacokinetics of an aptamer is often influenced by its physical and biological features (9). The biocompatibility, stability, and bioavailability of aptamers must therefore be improved via grafting modification in order to fulfil therapeutic needs and improve aptamer features. Additionally, aptamers can successfully transfer proteins, drugs, or nucleic acids into cell structures by conjugating to small interfering RNAs (siRNAs), drug molecules, or nanoparticles, reducing toxicity and negative consequences. Biocompatible and biodegradable materials are used to create nanoscale particles and materials (10). It can effectively transfer nucleic acids, proteins, or peptides to cells. The improved permeability and retention properties of solid tumours allow for the non-specific accumulation of the nanomaterial as a medication carrier in the cancer tissue location. The enhanced permeability and retention (EPR) effect's passive targeting effectiveness, however, is modest. Nanomaterials can be actively targeted for aggregation in tumour tissue while reducing damage to healthy tissues by combining

TABLE 14.1
Analysing the Differences between Aptamers and Antibodies

Characteristics	Aptamers	Antibodies
Targets	Widely	Mostly immunogenic macromolecular targets
Size	15–30 kD	50–100 kD
Affinity	High	High
Immunogenicity	Low	High
Tissue penetration	Fast	Slow
Cost	Low	Expensive
Toxicity	Low	High
Stability	High	Low
Production process	Chemical	In vivo production

them with aptamers that can selectively recognize tumour cell antigens or biomarkers. A new avenue for tailored cancer treatment is made possible by the identification of circulating tumour cells (CTCs), which allows for early diagnosis and targeted therapy. A compound of aptamers-based nanomaterials that has a high affinity for tumour biomarkers (11, 12).

Aptamer-conjugated nanomaterials may offer a more effective and risk-free solution to fulfil the growing demand for novel approaches in the battle against cancer by fusing the intrinsic properties of nanomaterials with the specialized recognition capacity of aptamers. Here, we concentrate on the development of innovative aptamer nanomaterial-based techniques for targeted cancer therapy as well as aptamer-coupled nanomaterials for specific cancer cell detection. This chapter first looks at recent developments in the utilization of aptamer-conjugated nanoparticles and aptamer-tethered DNA/lipid nanostructured materials for selective cancer cell identification. The use of aptamer-conjugated nanomaterials in novel methods including photodynamic therapy (PDT) and photothermal therapy (PTT) is also reviewed. Aptamer-conjugated nanomaterials are intriguing prospects for use in future cancer treatments since this aptamer-targeted approach has excellent efficacy and few side effects (12–15).

14.2 STRUCTURE OF APTAMERS

The word "aptamer", which comes from the Latin "aptus" and means "fit", was first used in a publication in 1990. Because they may reassemble to bind related targets with a high degree of specificity and affinity, aptamers are regarded as chemicals or synthetic "antibodies". They are single-stranded DNA or RNA sequences known as short-stranded oligonucleotides, which show a great deal of potential for targeted tumour diagnostics and treatment. They often fold into a distinctive three-dimensional shape and can detect targets including cells, proteins, and tiny chemical compounds (14). To determine the three-dimensional structure of an aptamer, look for a stem, an inner ring, a bulge, a hairpin, a pseudo-knot, a triple structure, or a G-quadruplex structure. The contact between the aptamer and its target is facilitated by the complementary geometry, stacking interaction of the aromatic ring with the nucleobase of the aptamer, electrostatic link between the charged groups, van der Waals interaction, and hydrogen bond (15). Since their discovery, they have drawn more attention and are now employed as therapeutic and diagnostic targeted ligands. They differ from antibodies in that they have a low molecular weight, a stable structure, chemical group flexibility, quick blood clearance, and non-immunogenicity. Due to their propensity to reassembly bind related targets with high affinity and specificity, aptamers are regarded as chemicals or synthetic "antibodies". Aptamers, however, promise to have remarkable benefits over antibodies due to several distinctive properties (1, 4). First, the limits imposed by the requirement for cell lines or animals can theoretically be overcome by selecting aptamers in vitro for any given target. As a result, since the technique does not need the introduction of an animal immune system, which is required for the formation of antibodies, aptamers may also be chosen against hazardous or non-immunogenic targets. Second, once chosen, aptamers may be produced in huge quantities with good repeatability and purity. Third, it is considerably simpler to further change aptamers with functional groups while maintaining the aptamer's affinity once the key sequence for target binding of an aptamer is understood. Examples of such functional groups include fluorophores, nanoparticles, or enzymes. Fourth, unlike antibodies, aptamers are very stable and may regain their active conformation following heat denaturation. Dissociation constants for some RNA aptamer target complexes can approach picomolar values, suggesting that aptamers have a greater affinity for binding targets than antibodies. They are very selective and can tell a target molecule from non-target molecules or even a group from other amino acids. They serve as both medicinal drugs and particular ligands for drug delivery. By binding to relevant protein targets and cancer cells, they can increase the effectiveness of tumour treatment while minimizing harmful and side effects. The development of biosensors that use aptamers as recognition components has promise for the early detection of cancer in clinical settings. In the meantime, aptamers, which can lessen side effects of most chemotherapeutic drugs, have a bright future in targeted therapeutic drug delivery to cancer cells and tissues (16, 17).

14.3 SCREENING OF APTAMERS

The SELEX method is used to choose aptamers, and it starts with an oligonucleotide library made of 1013–1016 distinct sequences that were chemically generated (18). The SELEX technology involves three steps: To find the binding sequence, first build a random screening library of DNA or RNA using molecular biology technologies. Then, incubate the random library with the target. The PCR amplification product can be improved and used for the subsequent round of screening after several rounds of screening have been completed. A nucleic acid library with high affinity to the target can be cloned and sequenced to produce a specific aptamer sequence. Aptamers can be successfully screened in vitro using the usual SELEX, although the process can take weeks or even months to complete. Automation technology, capillary electrophoresis technology, microfluidic chip technology, nanotechnology, high throughput sequencing technology, and so on are but a few instances of the novel, successful, and high-throughput screening techniques that have evolved in recent years. These techniques have considerably increased the effectiveness of screening and the number of objects that can be screened (19). As an illustration, consider the SELEX capillary electrophoresis technology, which has the advantages of low sample infusion, high separation efficiency, economy, and high automation. For material separation and analysis, it is frequently employed. The first-time capillary electrophoresis technology was used to screen aptamers was in 2004 by a research group. With just a few rounds of screening, capillary electrophoresis technology effectively separates the bound and unbound aptamer molecules, which has considerably increased screening efficiency and drastically reduced the screening cycle (20). By creating a non-equilibrium capillary electrophoresis of equilibrium mixtures (NECEEM) for aptamer screening, researchers advanced capillary electrophoresis SELEX even further. The low background of this technology, which is 100–1000 times lower than that of the standard method, is its main advantage. Toggling SELEX, expression cassette SELEX, photo SELEX, and automated SELEX, as well as customized or primer-free capillary electrophoresis SELEX, have all considerably enhanced aptamer selection and screening (21, 22). As cancer cells may be identified by cell surface molecules that are overexpressed or altered as a result of multiple oncogenic mutations, scientists have also developed cell-SELEX aptamer screening technology. The cell-SELEX screening procedure's incubation, isolation, and amplification phases are comparable to those of traditional SELEX. The goal of cell-SELEX is complete living cells, which keep the regular conformation of cell surface proteins, as opposed to ordinary SELEX (Figure 14.1), which targets target molecules, and the screened aptamers are more suited for biological applications. Due to the lack of a necessity to initially identify the molecular markers on the cell surface, the aptamer selection process is incredibly straightforward (16–22).

14.4 APTAMER-MEDIATED DRUG DELIVERY FOR TARGETED CANCER THERAPY

14.4.1 Nucleic Acid Aptamer-Based Dual Responsive Nanoplatforms

One strategy is to use aptamers, which may be constructed from peptides or nucleic acids. Peptide aptamers are exceedingly simple combinatorial protein structures where a variable peptide sequence that has affinity for a particular target protein is coated with an inert, constant scaffold protein. A few artificial combinatorial proteins are now undergoing clinical trials, while most are being examined in preclinical research. Most peptide aptamers are used in in vivo diagnostics and medical treatment (23, 24).

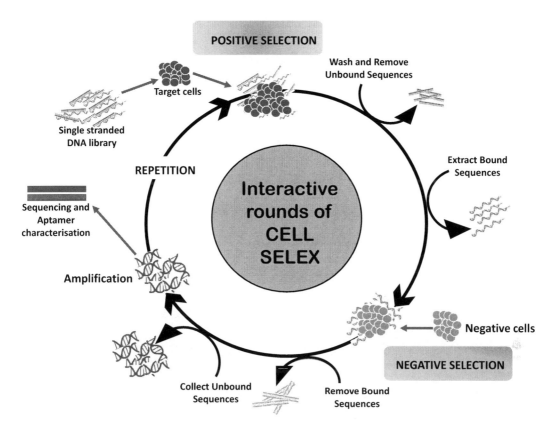

FIGURE 14.1 Production of aptamers through systematic evolution of ligand by exponential enrichment (SELEX) method.

Single-stranded oligonucleotide (DNA/RNA) molecules called nucleic acid aptamers fold into distinctive secondary or tertiary structures that enable them to bind to their respective targets with high affinity and precision. With several advantages over antibodies, aptamers are emerging as promising therapeutic agents (Table 14.2). These benefits include more efficient and affordable chemical synthesis, straightforward and controllable modification with functional moieties to meet various clinical needs, large-scale commercial production, excellent stability, non-toxicity, the availability of antidotes for agonistic aptamers, quick tissue penetration, and minimal immunogenicity (19–22). Since their initial publication in the 1990s, a wide range of aptamers have been created, each with a unique target that can range from tiny molecules and metal ions to proteins. As the target for in vitro selection, the Tan group published cell-SELEX in 2006 utilizing a cultivated precursor T cell acute lymphoblastic leukaemia (ALL) cell line. Cell-SELEX DNA aptamers identify a specific target molecule's relatively natural form on the cell membrane, which is often a cell surface transmembrane protein, and target complete living cells (23, 25).

Aptamers produced by cell-SELEX can be used in applications for cancer detection and treatment since they can be obtained without prior knowledge of the molecular fingerprints of proteins located on the cell surface. The three main phases of cell-SELEX are partitioning, amplification, and incubation (26). DNA aptamers have been created using this technique to target a variety of cancerous cells, including those that cause lung, liver, ovarian, colorectal, B-cell lymphoma, breast, and acute myeloid leukaemia (AML) cancers. Aptamers have demonstrated excellent potential in the discovery of biomarkers, cell sorting, detection, imaging, and clinical therapy in addition to their exceptional binding profile (27).

TABLE 14.2
Examples of Nucleic Acid Aptamers That Bind to Molecular Targets of Therapeutic Interest

Aptamer	Molecular Target	Associated Disease
Macugen	Vascular endothelial growth factor (VEGF)	Age-related macular degeneration
AS1411	Nucleolin	Cancer
Sgc8	Protein tyrosine kinase 7 (PTK-7)	Cancer
TD05	Immunoglobulin μ heavy chains (IGHM)	Lymphoma
ARC1779	A1 domain of von Willebrand factor (vWF)	Thrombotic microangiopathies and carotid artery disease
TBA	α- thrombin	Thrombosis

14.4.2 Aptamers for Targeted Drug Delivery

Aptamers can be used to create platforms for targeted medication administration because of their high binding affinity and specificity for cancer cells and related proteins. Through greater efficacy, decreased toxicity, and decreased side effects, targeted therapy can enhance the treatment of cancer. Researchers have made tremendous progress in creating forms of targeted therapy, including as chemotherapy and phototherapy, thanks to the use of cell-SELEX to produce several cancer cell–specific aptamers (27, 28).

Aside from acting as medicines in and of themselves, aptamers have also been thoroughly investigated as targeting ligands for aptamer drug conjugates-based drug delivery. Aptamer drug conjugates have several potential advantages over antibody-drug conjugates (ADCs), some of which have previously received United States Food and Drug Administration (USFDA) approval for the treatment of cancer (29). Aptamers, for instance, can be chemically or enzymatically altered with a variety of functional groups to increase biostability or bioconjugation with medicines. Additionally, aptamers can be retrieved from extremely hot or chemically corrosive environments for possible effective bioconjugation (30). Finally, aptamers and aptamer drug conjugates have relative molecular weights that make them more likely than aptamer drug conjugates to permeate bodily tissues swiftly and profoundly. Chemotherapy, immunotherapy, radiation, and phototherapy are just a few of the therapeutic modalities for which aptamer drug conjugates have been investigated. Aptamer drug conjugates can be used in a variety of disorders, including cancer and acquired immune deficiency syndrome (AIDS), depending on disease-related biomarkers. Additionally, by enhancing the drug loading capacity and passive drug delivery effectiveness of aptamer drug conjugates, nanotechnology can help them evolve even further (31).

14.4.3 Targeted Chemotherapy

One of the most significant therapeutic techniques is chemotherapy. Unfortunately, chemotherapeutic medicines show decreased efficacy and increasing adverse effects due to their lack of specificity and propensity to produce cytotoxicity in both malignant and healthy cells. Aptamers made from cell-SELEX, however, exhibit great specificity and affinity for their targets. As a result, two drug delivery platforms, aptamer nanomaterial conjugates and aptamers physically intercalated with medication, address the drawbacks of systemic chemotherapy (32).

Aptamer drug conjugates can be designed in a variety of ways thanks to the programmability, simplicity of synthesis, and modification of aptamers. For instance, site-specific chemical conjugation or physically complexing medications with aptamers can be used to create aptamer drug conjugates with a 1:1 ratio of aptamer:drug since a well-defined aptamer/drug ratio is necessary to consistently achieve high efficiency of drug delivery. For instance, by studying the distinct aptamer conformation, the chemotherapy drug doxorubicin (Dox) has been physically complexed with a prostate specific membrane antigen (PSMA) targeting aptamer at a 1:1 ratio. Dox was intercalated into an inherent double-stranded 5'-(GC)-3' sites of this aptamer (33). A second method for creating aptamer-drug conjugates is covalent conjugation. In one instance of Dox, an ApDC was created by coupling Dox with the DNA aptamer sgc8, which has a strong affinity for the biomarker tyrosine-protein kinase-like 7 (PTK7). To conditionally release the drug at tissues or subcellular organelles, linkers used to connect aptamers and medicines can be created. The sgc8-Dox aptamer drug conjugates, for instance, related to an acid-labile hydrazone linker, allowing Dox to be conditionally released in an acidic tumour environment or in an acidic endosome or lysosome (28, 34). It is also significant to note that the aptamer sgc8's capacity to internalize into cancer cells allowed the matching aptamer drug conjugates to do the same, enabling efficient intracellular drug delivery. In order to increase the effectiveness of drug delivery, aptamer drug conjugates can also be created with an aptamer-to-drug ratio of 1:n, allowing one aptamer to carry numerous copies of the drug. Highly toxic chemicals are probably rejected during traditional chemotherapeutic drug screening because of their extreme toxicity to healthy tissues. The therapeutic potency may be modest as a result; thus, a relatively high dose of these medications is needed for the best therapeutic efficacy (35). This restriction can be partially bypassed and therapeutic efficacy increased by using an aptamer drug conjugate design that can disperse several doses of medication. For instance, numerous aptamer drug conjugate derivatives have been developed by either utilizing the numerous intrinsic drug intercalating sites in aptamers or by purposefully engineering a predetermined number of such drug intercalating sites on aptamers to benefit from physical drug complexation with the tandem 5'-(GC)-3' or 3'-(GC)-5' sites. In one instance, a lengthy double-stranded DNA (dsDNA) that contains about 100% of drug-intercalating sites was attached to one aptamer to create an aptamer drug conjugate, also known as a nano train (36). This aptamer drug conjugate was created by a hybridization chain reaction (HCR). Such a design increased the drug delivery effectiveness by maximizing the drug loading capacity. In a mouse xenograft tumour model, these nano trains were effective and precise in delivering Dox to the target cancer cells. Additionally, numerous drug copies can be chemically coupled with a single aptamer molecule to create aptamer drug conjugates.

Using programmed automated nucleic acid synthesis and a phosphoramidite from a prodrug, multiple drug copies were site-specifically conjugated onto one aptamer molecule in an example of an aptamer drug conjugate (37). To build an aptamer drug conjugate and deliver 5-fluorouracil (5-FU) for colon cancer treatment, a phosphoramidite prodrug for the drug was developed. At any preset locations in aptamer drug conjugates, the resulting phosphoramidite prodrug may then be created on an automated DNA synthesizer utilizing traditional solid-phase DNA synthesis chemistry. As a result, it is possible to create aptamer drug conjugates by packing several drug molecules onto a single aptamer drug conjugate molecule. Notably, a photocleavable linker connecting the 5-FU moiety to the phosphoramidite prodrug's backbone was also developed, allowing for regulated drug release. For delivering the 5-FU prodrug to the intended HCT116 colon cancer cells, the generated aptamer drug conjugate has demonstrated its efficacy and selectivity. Using natural aptamers and no other chemical alterations, aptamer-drug adducts were created, yet another illustration of how to create aptamer drug conjugates with multiple drug copies on each aptamer (38). Aptamer drug conjugate adducts can be created utilizing aptamers, anthracycline medicines (like Dox), and cisplatin in a bio-friendly reaction environment and in the presence of formaldehyde, much like natural drug-genome adduct formation during drug-induced apoptosis. A conditionally cleavable crosslinker between the Dox and the nucleotides on aptamers was given by the formaldehyde. Additionally, it was shown that most of these anthracyclines medications were conjugated on guanosine, indicating a strategy for the programmable and predictable synthesis of aptamer drug conjugates. Another differentiating characteristic that makes chemically specified aptamer drug conjugates programmable and synthesizable makes aptamer drug conjugates attractive for industrial manufacturing as well as clinical translation. Remember that the drug conjugation sites on aptamer structures are crucial for ensuring that aptamer drug conjugates operate properly and that drugs are delivered specifically; this is a crucial factor to consider (39–41).

14.4.4 Aptamers Modified with Drugs for Therapy

Aside from aptamer nanomaterial conjugates, poisons like gelonin or cytotoxic medications like Dox, docetaxel, daunorubicin, and cisplatin have also been added to aptamers to make them more effective as pharmaceuticals. These aptamer drug conjugates are expected to have fewer side effects while also improving therapeutic efficacy due to the aptamers precise targeting. The Tan group, for instance, has investigated the covalent bonding of Dox to the DNA aptamer sgc8c (29, 33, 42).

With little harm to healthy cells, this aptamer drug conjugate was able to deliver the drug precisely. Drug aptamer cross-linking turned out to be laborious, even though the technique might successfully reduce the cytotoxicity of chemotherapy and boost its therapeutic efficacy. Researchers created a universal phosphoramidite that has an anticancer drug moiety and a photocleavable linker to solve this challenge. They also developed an automated modular synthesis of aptamer drug conjugates for targeted drug delivery (43). This module effectively integrated various medicines into aptamer drug conjugates at predetermined locations. The aptamer drug conjugates were able to specifically recognize the target cancer cells and release the medications in a photo controllable manner as a result. The outcomes showed that automated and modular aptamer drug conjugate technology has potential for use in targeted cancer therapy (44).

14.4.5 Targeted Biotherapy

Scientists have created new cancer treatments in addition to conventional cancer therapies like chemotherapy and phototherapy. These include immune cells that have been engineered with an aptamer so they can be directed to destroy cancer cells in vivo while sparing healthy ones. For instance, the direct therapy of cancer with natural killer lymphocytes (NK cells) is now being researched. An artificial diacyllipid tail with two stearic acids acts as the membrane anchor and connects the aptamers to it through a polyethylene glycol (PEG) segment. It has been shown that immune effector cells modified with aptamers may identify leukaemia cells via MHC nonrestricted structures, leading in enhanced cancer cell targeting and killing. This was the first account of the use of aptamer-controlled T-cell redirection to specifically attack cancer cells (Table 14.3) (45–47).

14.5 APTAMER-CONJUGATED NANOMATERIALS

Nanomaterials shine as medication carriers and signal reporters in the treatment of cancer because of their special qualities. It is true that aptamers conjugated with nanomaterials can increase the effectiveness of cancer treatment. A variety of nanomaterials, including graphene or fullerenes as well as silica and gold nanoparticles (AuNPs), have been employed to create aptamer nanomaterial conjugates for cancer treatment (48). For instance, gold nanorods (AuNRs) conjugated to the aptamer Chlorine 6 (Ce6) could precisely bind to and destroy certain cells via PDT and PTT. In this way, a smart multifunctional nanostructure (SMN) made of porous hollow magnetite nanoparticles that can incorporate Dox, polyethylene glycol (PEG) ligands, and aptamers could deliver SMN-Dox complexes into target cells and reduce the proliferation of those cells (49). High-energy radiation beams are focused on the tumour during radiation therapy (RT), a localized form of cancer treatment. More than half of cancer patients benefit from it, making it one of the most prevalent and important treatments for cancer treatment. However, while fighting the cancer cells, RT causes harm to the healthy tissues. These unintended side effects continue to exist and impair the overall effectiveness of this main cancer therapy strategy (50). Radiosensitizer usage is one method for improving the safety and efficacy of RT. High atomic number materials, such as gold nanoparticles (GNPs), are some of the most well-liked radiosensitizers that have recently garnered a lot of interest.

TABLE 14.3
Theranostic Approach of Aptamer Conjugates towards Cancer

Conjugate Name	Target Cell	Cancer Type
DNA aptamer conjugates		
Wy-5A	Prostate cancer cells (PC-3)	Prostate cancer
XL-33	Metastatic colon cancer cells (SW620)	Colon cancer
XQ-2d	Pancreatic ductal adenocarcinomas (PDAC) cells (PL45)	Pancreatic cancer
C-2	Liver cancer cells (HepG2)	Liver cancer
JHIT2	Liver cancer cells (HepG2)	Liver cancer
AGC03	Gastric cancer cells (HGC-27)	Gastric cancer
Cy-apt20	Gastric carcinoma cells (AGS)	Gastric cancer
LXL-1	Metastatic breast cancer cells (MDA-MB-231)	Breast cancer
GMT3	Glioblastoma multiforme cells (A172)	Brain tumour
Myo040–7–27	Myoglobin	Breast cancer
RNA aptamer conjugates		
A30	Human epidermal growth factor receptor 3 (HER 3)	Lung cancer, breast cancer, gastric cancer, pancreatic cancer
A9-g	Prostate specific membrane antigen (PSMA)	Prostate cancer
AFP-apt	Alpha fetoprotein (AFP)	Liver cancer
A-P50	Nuclear factor kappa light chain enhancer of activated B cells (NF-KB)	Lung caner
AS1411	Nucleolin	Lung cancer, renal cancer, breast cancer, pancreatic cancer
E0727/CL428/ KD1130/ TuTu2231	Epidermal growth factor receptor (EGFR)	Squamous cell carcinoma, breast cancer, glioblastoma multiform, lung cancer
NOX-A12	C-X-C motif chemokine ligand 12 (CXCL12)	Multiple myeloma, chronic lymphocytic leukaemia, etc.
OPN-R3	Osteopontin (OPN)	Breast cancer
YJ-1	Carcinoembryonic antigen (CEA)	Colorectal cancer
Drug aptamer conjugates		
CD4	RAR-related orphan receptor gamma (RORγt)	Gene therapy
sgc8	Tyrosine protein kinase like 7 (PTK7)	Gene therapy
SL$_2$-B	Human liver cancer cell line (Hep G2)	Liver cancer
RNV-L7	Low-density lipoprotein receptor (LDL-R)	Liver cancer
CD28Apt7	B16	Skin cancer
P19	Pancreatic cancer cell line (PANC-1)	Pancreatic cancer
AP-1	CD133	Anaplastic thyroid carcinoma
AptABCG2	ATP-binding cassette G2 (ABCG2)	Mammary cancer
MUC-1	Michigan Cancer Foundation7 (MCF-7)	Mammary cancer
S30	CD33	Blood cancer
AS1411	Nucleolin	Mammary cancer Skin cancer photothermal therapy
MP7	Programmed cell death protein 1 (PD-1)	Immunotherapy
RCA II	Cluster of differentiation 40 (CD40)	Immunotherapy

Following the collision of radiation beams at the tumour location, radiosensitizers enhance radiation effects. These nanoparticles are the best candidates for reticuloendothelial system escape and adequate accumulation in tumour cells because of their tiny size (51). Quantum dots (QDs), a new class of nanomaterial, have drawn a lot of attention as prospective medication delivery systems. For the chemotherapy of ovarian cancer, researchers described the creation and delivery of a tumour-targeted, pH responsive quantum dot-mucin1 aptamer-doxorubicin (QD-MUC1-DOX) combination. This approach indicated the proposed conjugates significant potential for treating multi-drug resistant ovarian cancer (49, 51).

The use of nanomaterials in bioanalysis and biomedicine is essential. Nanomaterials have overcome many of the limits of traditional therapeutic and diagnostic approaches thanks to their distinctive physicochemical features, including an ultra-small size, a huge surface area, and the capacity to load. The improvement of the specific recognition capability for sick tissues is essential for the development of nanomedicine. Aptamers and nanomaterials working together is a potential development for targeted medicine delivery. The effects of photodynamic and photothermal therapy, as well as aptamer-based inorganic and organic nanomaterials, on the treatment of cancer, will be covered in this part (52–54).

14.5.1 Aptamer-Conjugated Organic Nanomaterials

As the first drug delivery technology to be studied, liposomes have several attractive qualities, including excellent drug loading efficiency, good biocompatibility, low toxicity, and low immunogenicity. The first liposomal medication for the treatment of solid tumours to receive food and drug administration (FDA) approval is doxil, which is PEGylated liposomal Doxorubicin. With the quick advancement of biotechnology, it has been possible to successfully create liposomal systems with precise targeting capabilities by adding various molecular recognition components, such as folate, peptides, antibodies, and aptamers (55). Aptamer-based liposomes have gotten a lot of attention among them. Researchers first described therapeutic liposome nanoparticles modified with therapeutic aptamers for targeted drug delivery. A sgc8 aptamer that demonstrated strong binding and internalization capacity for the targeted cells was conjugated with this approach (56). Researchers developed an aptamer-based lipopolymer for tumour-specific delivery of clustered regularly interspaced short palindromic repeats (CRISPR) to control vascular endothelial growth factor A (VEGFA) in osteosarcoma, for example, showing that sgc8-modified liposomes could deliver loaded drug to targeted cancer cells with high specificity and excellent efficiency. In this scenario, the LC09 aptamer might make it easier to distribute CRISPR plasmids in a targeted manner to reduce VEGFA expression, hence preventing lung metastasis and orthotopic osteosarcoma malignancy. The micelle structure is a different form of promising aptamer-based organic nanomaterial. The multivalent effect of this drug delivery method results in good aptamer binding to target (57–60). It can therefore be developed for a variety of bio applications. A cross-linked aptamer lipid micelle system was created in 2018 for outstanding stability and selectivity in target cell recognition. In this simple method, a methacrylamide branch was connected to aptamer and lipid segments by means of a successful photo-induced polymerization procedure. This described solution offered greater bio stability in a cellular context than conventional aptamer lipid micelles, thus enhancing the targeting capabilities for imaging applications (61). In another study, hydrophobic prodrug bases were combined with bio-orthogonal chemistry for hydrogen peroxide and pH-independent cancer chemodynamic therapy to create a new aptamer prodrug conjugate micelle. In-depth mechanistic studies show that intracellular Fe^{2+} might activate this mechanism, which would then cascade bio-orthogonal processes to produce hazardous C-centred free radicals (62).

Target-responsive DNA hydrogels were among the aptamer-based organic nanomaterial systems that were extensively used in biomedical and pharmaceutical applications because of their exceptional mechanical properties and programmable characteristics. The first adenosine sensitive hydrogel for potential medication release was created in 2008. The DNA nanohydrogels in this study were created using cross-linking oligonucleotides that were rationally designed and two oligonucleotides that included polyacrylamide. The aptamer sequence for adenosine was present in the DNA linker (63). The aptamer will bind to target molecules when they already have adenosine molecules, which will cause the cross-links to break down and the hydrogel to dissolve. This technology may therefore be investigated for target-responsive medication release. Another notable example was the discovery of a physically cross-linked DNA network to isolate bone marrow mesenchymal stem cells (BMSCs) from a variety of non-target cells (64). The Aptamer19S sequence-containing nanomaterial produced a 3D microenvironment that allowed the imprisoned stem cells to continue to perform at a high level. According to the complementary base pairing approach, aptamers may also be simply coupled to produce various DNA nanostructures for targeted cancer cell detection and subsequent uses. As a prominent DNA-based nanostructure, DNA origami has been changed by numerous small molecule drugs, functional NA sequences, and nanomaterials. Li's team revealed a smart DNA nanorobot for intelligent medicine delivery in cancer therapy in 2018. (65). By functionalizing on the outside with aptamer AS1411, this DNA nanorobot may specifically distribute thrombin to tumour-associated blood vessels to restrict the development of the tumour. This led to the development of the Apt-Dox-origami-ASO aptamer-functionalized DNA origami, which concurrently delivers Dox and antisense oligonucleotides (ASOs) to cancer cells. Utilizing molecular engineering, many aptamer-based DNA nanostructures have been researched as cargo carriers for targeted drug delivery in cancer treatment (59–65).]

14.5.2 Aptamer-Conjugated Inorganic Nanomaterials

AuNPs, a significant inorganic nanomaterial, have drawn a lot of interest in biomedicine due to their high surface-to-volume ratio, low toxicity, exceptional stability, and biological compatibility. First, variations in surface plasmon resonance (SPR) in AuNPs show that they have substantial distance-dependent optical features, which are often shown by a colour shift in AuNPs brought on by an uneven interval distance. Second, AuNPs have a large capacity for DNA loading, excellent intracellular stability, and simple surface modification. Third, the robust interaction between thiol and gold allows for the effective and economical strategy for AuNP alteration. Furthermore, rapid development of size- and shape-controlled metallic NPs has led to their application in the targeting and therapy of cancer cells (66). The detection and treatment of cancer regularly make use of aptamer-conjugated gold nanomaterials, which combine the advantageous properties of aptamers with gold nanoparticles. A polymeric network of nanoparticles was constructed using target DNA molecules in a conventional investigation to allow for the exact detection of polynucleotides. A few of the numerous corresponding studies that are currently being conducted include aptamer functionalized AuNPs-Fe_3O_4-GS capture probe for monitoring circulating tumour cell in whole blood, apt-AuNPs with enzyme responsiveness for mucin 1 protein (MUC1) detection, aptamer-conjugated gold nanomaterials combined with graphene oxide for photothermal therapy of breast cancer, and apt-AuNPs with enzyme responsiveness for MUC1 detection (67).

One of the inorganic nanomedicines that has been shown to be beneficial as a drug delivery system carrier is silica nanoparticles. These particles successfully altered pH and temperature, as well as some photochemical and redox processes, to regulate drug release in vivo and in vitro. When combined with certain elements like aptamers, they can enhance cancer therapy outcomes with a lower amount of drug (68). A new redox-sensitive nanocontainer for triplex cancer-targeted treatment based on mesoporous silica nanoparticles was reported by Cai's team as an early example (MSN). The cytochrome C (CytC) sealed MSNs used in their investigation were altered using the AS1411 aptamer. Additionally, this system can trigger a specific release of Dox into tumour cells by disrupting the S-S connections (5). Researchers developed two photon MSNs based on fluorescence resonance energy transfer (FRET) in 2021 for multiplexed intracellular imaging and customized drug delivery. The MSNs may display varied two-photon multicolor fluorescence by varying the doping ratio of the three dyes. The nanosystem can also successfully integrate into cancer cells since it contains the anticancer drug Dox and an aptamer cap on top. Additionally, aptamer-targeted MSNs are commonly used for gene-targeted delivery, guarding against nuclease degradation of gene therapy molecules (56). For usage in the biomedical industry, traditional carbon nanomaterials including fullerene, graphene, carbon dots/nanobots/nanotubes, and hybrids offer unique advantages. Aptamer-functionalized carbon nanomaterials create ideal nanoplatforms for the treatment and detection of cancer. Researchers developed a multifunctional substance that has both sustained release and heat-stimulating properties. With the inclusion of MUC1 aptamers, this nanoparticle can detect targeted Michigan Cancer Foundation-7 (MCF-7) breast cells with outstanding identification capacity. Aptamer-based graphene nanostructures have also enabled some significant developments in cancer gene therapy (10, 57). Aptamer-based graphene quantum dots loaded with porphyrin derivatives photosensitizer were first disclosed in 2017 for fluorescence-guided photothermal/photodynamic synergetic treatment. These multifunctional theranostic nanomaterials have shown high practicality for detecting intracellular cancer-related miRNA, and their intrinsic fluorescence may be used to distinguish cancer cells from somatic cells (43, 58).

14.5.3 Aptamer-Conjugated Nanomaterials for PDT

Since PDT is noninvasive, regionally selective, and has low toxicity, it has gained popularity as a therapeutic strategy for the treatment of tumours in recent years. Following the buildup of photosensitizers at the tumour site, reactive oxygen species (ROS) are produced by NIR laser irradiation. Toxic ROS injure tumour cells by the oxidation of protein, DNA, or RNA. To combat the hypoxic environment and growing problems of the tumour, many photosensitizers have been created. An activatable ROS-producing system, for instance, was developed by researchers by manipulating the metabolic interaction between linoleic acid hydroperoxide and catalytic iron (II) ions. Through the creation of ROS, the designed nanoparticles were able to cause cancer cell death through apoptosis both in vitro and in vivo. For synergistic photothermal and increased photodynamic therapy, researchers created a Pt nano enzyme with functionalized nanoplatform black phosphorus/Pt-Ce6/PEG nanosheets. The Pt nano enzyme would break down H_2O_2 into oxygen to increase the photodynamic impact (68).

Due to its strong loading capabilities and practical surface fictionalization, manganese dioxide (MnO_2) is being used more frequently in biological fields. MnO_2 has been employed as a photosensitizer carrier to create unique activatable PDT systems because of its quenching capability. MnO_2 is a special kind of nanomaterial that responds to the tumour microenvironment and can interact with glucocorticoid suppressible hyperaldosteronism (GSH) to solve the issues with PDT therapy. A novel method was put out by researchers that made use of the photosensitizer hematoporphyrin monomethyl ether (HMME), with mesoporous MnO_2 ($mMnO_2$) serving as the HMME carrier (69). When loaded onto the surface of $mMnO_2$ nanoparticles and sealed by aptamers on the particle surface, the photosensitizers were in the quenching condition. The photosensitizers were released as a result of the aptamers' ability to preferentially target the membrane protein MUC1 on the tumour cell. When it interacted with healthy cells deficient in MUC1, neither the HMME nor the PDT was released. However, the aptamer was liberated from the MUC1 protein and HMME when MUC1 overexpression was present in breast cancer cells. After that, cancer cells were destroyed by ROS that were produced as a result of laser radiation. The researchers used Michigan Cancer Foundation-7 (MCF-7) to investigate the tumour-targeting release of HMME (55–60). Confocal laser scanning microscopy (CLSM) findings revealed that the fluorescence of HMME in MCF-7 cells was incredibly brilliant, whereas it was incredibly low in Hs578bst cells. Additionally, MCF-7 (56.4%) showed a higher degree of ROS after laser exposure than Hs578Bst (1.24%), which confirmed the results of the HMME imaging (68). In PDT therapy, up-conversion nanoparticles (UCNPs) are often utilized as energy donors and delivery vectors for photosensitizers. This created system offers a simple but effective strategy for the selective death of tumour cells, with low harm to normal cells, and creates a novel path for using PDT in targeted cancer treatment, as opposed to the usual PDT method.

Additionally, to improve the efficacy of PDT, UCNPs with recognition moieties functionalized can be given the capacity to precisely administer a photosensitizer to target cells (45–54).

Using rolling circle amplification (RCA), a long single-stranded DNA (ssDNA) with an AS1411 aptamer and a DNA enzyme was created. UCNPs served as the vehicle for loading the ssDNA. The multivalence of the ssDNA contributed to the high recognition and drug-loading capabilities of the up conversion nanoplatform, while DNA enzymes employed gene-silencing strategies to prevent the expression of survivin. A protein called nucleolin is involved in several cellular activities, including the metabolism of DNA and the regulation of RNA processes including transcription, translation, and ribosome assembly (14). Along with being present in the cytoplasm and cell membrane, it is mostly found in the nucleolus. Numerous malignancies are characterized by the dysregulation of nucleolin expression, which results in the accumulation of nucleolin mRNAs and proteins as well as its cell surface overexpression as compared to normal cells. A nucleolin-targeting aptamer called AS1411 is regarded as the most thoroughly studied aptamer in the field of cancer research. This finding emphasizes its excellent potential and safe usage in therapeutic settings (34). The AS1411 aptamer served as the targeting agent in this nanosystem to identify the overexpressed nucleolin on breast cancer cells. Near-infrared radiation (NIR) irradiation activated PDT, which produced ROS that killed the cancer cells. To evaluate the targeting of this nanosystem, flow cytometry and CLSM were used to measure the uptake of UCNP-ApDz-TMPyP4 in MCF-7 cells, the target cancer cell, and BRL 3A cells, and the control cell. The CLSM results confirmed the flow cytometry investigations in that the MCF-7 cells had much greater fluorescence intensities than the BRL 3A cells. The cytotoxicity of MCF-7 cells was assessed using the (3-[4,5-dimethylthiazol-2-yl]-2,5 diphenyl tetrazolium bromide) MTT test and cytotoxicity assay. The results showed that UCNP was not cytotoxic, although cell survival rates in the UCNP-ApDz-TMPyP4 groups were 36.3%. According to recent data, the emergence of cancer cells that are resistant to PDT always has a negative impact on its effectiveness (68). However, when treated to tumour tissues, this multifunctional up conversion nanoplatform exhibits an impressive anticancer response both in vivo and in vitro, cooperating with PDT and DNA enzyme-based gene therapy, and may be a praiseworthy alternative treatment for cancer (68).

14.5.4 Aptamer-Conjugated Nanomaterials for PTT

Since Goldman removed tumours with laser irradiation in 1966, PTT has attracted a lot of interest in the world of cancer therapy. Due to its low toxicity, noninvasive nature, ease of use, and low recurrence, PTT demonstrated significant promise as a cancer therapy. A NIR laser is utilized when photosensitizing chemicals have accumulated in the tumour location. The target tissues are partially or completely burned away as a result of the high photothermal conversion efficiency of photosensitizing chemicals, which transforms the optical energy that is absorbed into thermal energy (65). PTT frequently needs the tumour centre to reach high temperatures (50°C) for the tumours to be effectively abated. NIR light may penetrate deep into tissues, leading to the development of numerous organic and inorganic phototherapeutic agents. The creation of sensitive and precise targeted imaging and diagnostic modalities for cancer is the consequence of the surface modification of QDs with aptamers that might bind to antigens present on the target cells (Figure 14.2). B quantum dots (BQDs), for instance, were developed by scientists and are biocompatible. When exposed to NIR light, BQDs can successfully produce photothermal effects. BQDs-PEG has been shown through experimental research done in vivo and in vitro to dramatically kill cancer cells and stop tumour development by photothermal effects. Another researcher made dumbbell-shaped gold nanorods and core–shell Cu7S4 heterostructures. The photothermal stability and conversion efficiency of both gold nanorods-Cu7S4 heterostructures (= 56% and 62%, respectively) were very good. The in vitro photothermal ablation of cancer cells was carried out using low cytotoxicity and effective photothermal treatments. Researchers developed a unique dual-targeted small-molecule organic photothermal agent that can target both biotin and mitochondria for enhanced photothermal treatment.

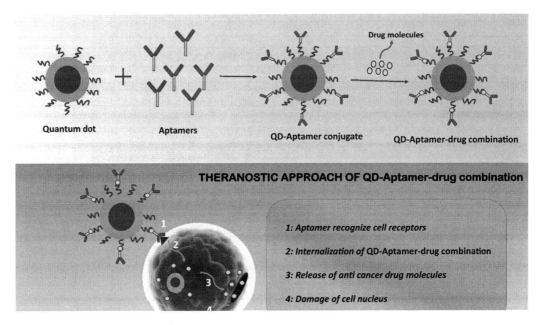

FIGURE 14.2 Image showing theranostic mechanism of QD-aptamer-drug conjugates for cancer.

According to in vivo photothermal treatment experiments, the dual-targeted photothermal agent performed much better in tumour inhibition (66).

Due to its straightforward synthesis, strong stability, huge surface area, great water dispersibility, and biocompatibility, molybdenum disulphide (MoS_2) is a typical transition metal disulphide, which is a class of two-dimensional nanomaterials. Especially high NIR absorption is found in MoS_2. It has a high photothermal conversion efficiency as a result. Numerous studies have used it in photothermal treatment as a result of this. However, it must be enhanced in terms of identifying certain tumour cells. For instance, researchers created a novel technique that employed MoS_2 as the photothermal agent. First, scientists created MoS_2-BSA nanosheets by combining MoS_2 and bovine serum albumin (BSA). The free carboxyl group on the MoS_2-BSA surface was then activated using ethyl dimethyl aminopropyl carbodiimide (EDC) and (n-hydroxy succinimide) NHS. Aptamers were then modified to create composite MoS_2-BSA-Apt nanosheets, which were stable and biocompatible (58–60).

Due to the strong affinity between aptamers and receptors on the cell surface, the composition would distinguish MCF-7 human breast cancer cells from other cells and then enter the cells by endocytosis. When cancer cells are subjected to an 808-nm laser, which is based on MoS_2 nanosheets, they are destroyed by the heat that is created. The results of the fluorescence inverted microscope revealed that after MoS_2-BSA-Apt was cultured with MCF-7 and MCF-10A human breast cancer cells, MCF-7 human breast cancer cells had uniformly distributed green fluorescence signals, while MCF-10A human breast cancer cells had almost no fluorescence signal. The results suggested that MoS_2-BSA-Apt might target MCF-7 human breast cancer cells since it had a greater impact on killing cells than MoS_2-BSA under the same laser irradiation period. MoS_2 nanosheets also show magnetic, fluorescence, and other properties. These traits enable the development of a platform for accurate diagnosis and therapy that integrates photothermal therapy with imaging diagnostics (60, 61).

Graphene oxides (GOs) and gold nanoparticles (AuNPs) have been utilized extensively in cancer therapy because of their excellent photothermal conversion efficiencies and biocompatibilities. GOs are desirable nanomaterials for drug delivery because they have a large loading capacity. GOs and

AuNPs combined to create a nanomatrix may thereby enhance the photothermal effects on malignancies. Researchers used GOs' high loading capabilities to attach AuNPs on them in order to boost the photothermal impact (10). Strong Au-S bonds were used to immobilize thiolated MUC1 aptamers, which can selectively detect breast cancer cells, on the surface of AuNPs in order to increase the targeting efficiency of this nano matrix. The GOs subsequently absorbed the aptamer gold nanoparticle conjugates. The cancer cells are subsequently heated by the laser's radiation, which kills them. To evaluate the targeting of aptamer gold nanoparticles graphene oxide conjugates, researchers employed MCF-7 cells (MUC1-positive cell lines) and EA.hy926 cells (MUC1-negative cell lines). The results demonstrated that even though MCF-7 cells exhibited substantial red fluorescence, this technology may be developed to produce Apt-AuNPs-GOs loaded with a heat shock protein inhibitor to improve the efficacy of photothermal treatment (62, 63).

14.6 APTAMER-MEDIATED ANTITUMOUR EFFECTS

In comparison to other malignancies of the female reproductive system, ovarian cancer frequently resists chemotherapy, leading to greater mortality rates in ovarian cancer patients. In order to enhance the prognosis of ovarian cancer patients, it is crucial to administer precise drug-targeted therapy. Treatments for epithelial ovarian cancer depend increasingly on biomarkers, and numerous laboratories have lately started to produce these tools (64).

Most studies suggests that CD44 and epithelial cell adhesion molecule (EpCAM) both contribute to chemoresistance and play significant roles in the development and incidence of ovarian cancer. Due to the co-expression of CD44 and EpCAM in ovarian cancer cells (OCs), it is possible to overcome chemoresistance by using bispecific medicines that target both molecules. A bispecific CD44-EpCAM aptamer, for instance, was developed by researchers to concurrently block CD44 and EpCAM. The tumour-inhibiting effects of this bispecific CD44-EpCAM aptamer were superior to those obtained with single aptamers. The CD44-EpCAM aptamer may therefore be a useful medication for treating advanced ovarian cancer (64).

Major cell cycle regulators are inhibited by the DNA binding protein inhibitors Id1–4, and overexpression of the Id protein encourages cancer cell proliferation and prevents apoptosis. Additionally, Id1 expression is linked to higher malignancy and more aggressive clinical signs. Using the yeast mammalian two-hybrid technique, a peptide aptamer called Id1 was screened from a random combination expression pool in order to block Id1 and Id3. Id1 was produced and purified after being coupled with a cell penetrating protein transduction domain. According to experimental findings, Id1 biological effects are caused by the functional blockage of Id1 and Id3. These results indicate that additional research on Id1/3-PA7 is necessary to learn more about its stability, functional specificity, and toxicity in vivo. As they play crucial roles in regulating tumour spread, anxelekto receptor tyrosine kinases (AXL-RTK) are important targets for the treatment of ovarian cancer. It has been demonstrated that AXL inhibition increases chemotherapy sensitivity, making it a viable option for ovarian cancer patients (60–65). The anticancer effectiveness of a nuclease AXL aptamer that targets AXL-RTK was tested in several intraperitoneal injection animal models. Combining paclitaxel with AXL aptamer therapy might significantly enhance the anticancer effects of paclitaxel in mice. These findings suggest that AXL aptamer has the potential to be an effective treatment for ovarian cancer by inhibiting AXL activity and tumour growth in vivo. A drawback of aptamer-based therapy is that the therapeutic substance may only be delivered intracellularly if aptamers are not endocytosed into the target cell and stick to the cell surface (66). Researchers found an aptamer, R13, that can attach to and internalize OCs through caveolae- and clathrin-mediated endocytosis to address this problem. The membrane protein that R13 targets is one that is expressed on the surface of tumour cells and may serve as a biomarker to distinguish between cancer and healthy cells. R13 appears to be a promising new tool for detection and drug administration in ovarian cancer, according to imaging studies conducted on a mouse tumour model to ascertain its targeting capacity (67).

Aside from having the ability to inhibit biomarkers, aptamers (such as aptamer siRNA chimaeras) can also be employed on their own to treat other diseases, including prostate cancer (68). A unique aptamer siRNA chimeric delivery method, mediated by cationic Au-Fe$_3$O$_4$ nanoparticles, was described in a study on ovarian cancer. Small interference RNA (siRNA), microRNA (miRNA), and short hairpin RNA (shRNA) are the three types of RNA-based aptamers used in practice. They are thought to be cutting-edge treatments for malignant tumours because of their precise cytotoxic effects on cancer cell lines in both in vitro and in vivo studies. Single-stranded RNA molecules with a high level of specificity and selectivity are known as RNA aptamers. Due to malfunctions, they are trained to adversely regulate unchecked cell division and necrosis. Through selection, separation, and amplification, the practical approach of systematic evolution of ligands by exponential enrichment (SELEX) is used to produce RNA aptamers from various RNA library sequences. The VEGF-RNA aptamer and Notch3 siRNA were combined to create the aptamer siRNA chimaera, which binds Au-Fe$_3$O$_4$ nanoparticles through electrostatic contact. The produced chimeric complexes enhance the anti-tumour action and have a better effectiveness in silencing the Notch3 gene. The effective transport of Au-Fe$_3$O$_4$ nanoparticles may also overcome the multidrug resistance of cisplatin, indicating that Au-Fe$_3$O$_4$ nanoparticles may be employed in cancer targeted treatment to overcome multidrug resistance (68). Researchers created an EpCAM-siPKC aptamer (EpCAM-siRNA chimaera) in a different investigation. They found that protein kinase c iota (PRKCI), a member of the protein kinase (PKc) family associated with increased PKC expression in ovarian cancer, can be significantly amplified. Additionally, PKC silencing triggers ovarian cancer cell death that is PRKCI amplified. These findings suggest that PKC is necessary for PRKCI-amplified cells to be carcinogenic. The EpCAM-siPKC aptamer was able to significantly reduce the formation of intraperitoneal tumours in xenotransplantation mice and successfully cause apoptosis in PRKCI-amplified OCs. This study indicated that PKC siRNA supplied by EpCAM aptamer can be utilized to inhibit such tumours and showed a specific medical strategy aimed at a subpopulation of OCs possessing PRKCI amplification (69–71).

14.7 CONCLUSION

Aptamers have a wide range of applications in the study of cancer diagnosis and treatment, as demonstrated by the advancements made in recent decades. When aptamers were being developed as nucleic acid functionalized substitutes for folic acid, peptides, and antibodies for targeted drug administration, several of their distinctive features received a lot of attention. This succinct review highlights some recent developments in aptamer-based cancer therapy systems. Aptamer and aptamer-drug conjugate research is still in its infancy. The clinical application obstacles should be addressed with serious effort. The delivery of effective therapeutic drugs to cancer cells in their target tissues was another aspect of aptamers' most notable triumphs. This was made possible by their fusion with nanomaterials, which increased the specificity of the diagnostic signal. In conclusion, the previous explanation highlighted the therapeutic relevance and adaptability of aptamers as well as recent developments in this exciting area of aptamers.

14.8 ABBREVIATIONS

5-FU	5-fluorouracil
ADCs	Antibody-drug conjugates
AIDS	Acquired immune deficiency syndrome
ALL	Acute lymphoblastic leukaemia
AML	Acute myeloid leukaemia
ASOs	Antisense oligonucleotides
AuNPs	Gold nanoparticles (AuNPs)
AuNRs	Gold nanorods
AXL-RTK	Anxelekto receptor tyrosine kinases

BMSCs	Bone marrow mesenchymal stem cells
BQDs	Quantum dots
BSA	Bovine serum albumin
Ce6	Chlorine 6
CLSM	Confocal laser scanning microscopy
CRISPR	Clustered regularly interspaced short palindromic repeats
CTCs	Circulating tumour cells
Cyt C	Cytochrome C
Dox	Doxorubicin
DsDNA	Double-stranded DNA
EDC	Ethyl dimethyl aminopropyl carbodiimide
EpCAM	Epithelial cell adhesion molecule
EPR	Enhanced permeability and retention
FDA	Food and drug administration
FRET	Fluorescence resonance energy transfer
GOs	Graphene oxides (GOs)
GSH	Glucocorticoid suppressible hyperaldosteronism
HCR	Hybridization chain reaction
HMME	Hematoporphyrin monomethyl ether
Id1	Inhibition of differentiation
MCF	Michigan Cancer Foundation
MCF-7	Michigan Cancer Foundation-7
MHC	Major histocompatibility complex
$mMnO_2$	Mesoporous MnO2
$MnO2$	Manganese dioxide
MoS2	Molybdenum disulfide
MSN	Mesoporous silica nanoparticles
MTT	(3-[4,5-dimethylthiazol-2-yl]-2,5 diphenyl tetrazolium bromide)
MUC1	Mucin 1 protein
NECEEM	Non-equilibrium capillary electrophoresis of equilibrium mixtures
NHS	N-hydroxysuccinimide
NIR	Near-infrared radiation
NK cells	Natural killer lymphocytes
PDT	Photodynamic therapy
PEG	Polyethylene glycol
PEG	Polyethylene glycol
PKc	Protein kinase
PSMA	Prostate specific membrane antigen
PTK7	Tyrosine-protein kinase-like 7
PTT	Photothermal therapy
QD-MUC1-DOX	Quantum dot-mucin1 aptamer-doxorubicin
RCA	Rolling circle amplification
ROS	Reactive oxygen species
SELEX	Systematic evolution of ligand by exponential enrichment
SiRNAs	Small interfering RNAs
SMN	Smart multifunctional nanostructure
SS	Single stranded
UCNPs	Upconversion nanoparticles
USFDA	United States Food and Drug Administration
VEGF	Vascular endothelial growth factor
VEGFA	Vascular endothelial growth factor A

REFERENCES

1. Vermes A, Guchelaar HJ, Dankert J. Flucytosine: A review of its pharmacology, clinical indications, pharmacokinetics, toxicity and drug interactions. J. Antimicrob. Chemother. 2000; 46(17): 1–9.
2. Reuther T, Schuster T, Mende U, et al. Osteoradionecrosis of the jaws as a side effect of radiotherapy of head and neck tumour patients-a report of a thirty-year retrospective review. J. Oral Maxillofac. Surg. 2003; 32: 289–295.
3. Bates PJ, Kahlon JB, Thomas SD, et al. Antiproliferative activity of G-rich oligonucleotides correlates with protein binding. J. Biol. Chem. 1999; 274: 26369–26377.
4. Bates PJ, Laber DA, Miller DM, et al. Discovery and development of the G-rich oligonucleotide AS1411 as a novel treatment for cancer. Exp. Mol. Pathol. 2009; 86: 151–164.
5. Cai B, Yang X, Sun L, et al. Stability and bio activity of thrombin binding aptamers modified with D-/L-isothymidine in the loop regions. Org. Biomol. Chem. 2014; 12: 8866–8876.
6. Chen K, Liu B, Yu B, et al. Advances in the development of aptamer drug conjugates for targeted drug delivery. Wiley Interdiscip. Rev. Nanomed. Nanobiotechnol. 2017; 9: e1438.
7. Chi-hong BC, George AC, Van QH, et al. Inhibition of heregulin signaling by an aptamer that preferentially binds to the oligomeric form of human epidermal growth factor receptor-3. Proc. Natl. Acad. Sci. USA. 2003; 100: 9226–9231.
8. Dong Y, Yao C, Zhu Y, et al. DNA functional materials assembled from branched DNA: Design, synthesis, and applications. Chem. Rev. 2020; 120: 9420–9481.
9. Wu X, Chen J, Wu M, et al. Aptamers: Active targeting ligands for cancer diagnosis and therapy. Theranostics. 2015; 5(4): 322–344.
10. Dou B, Xu L, Jiang B, Yuan R, et al. Aptamer-functionalized and gold nanoparticle array-decorated magnetic graphene nanosheets enable multiplexed and sensitive electrochemical detection of rare circulating tumor cells in whole blood. Anal. Chem. 2019; 91: 10792–10799.
11. Kong R, Chen Z, Ye M, et al. Cell-SELEX-based aptamer-conjugated nanomaterials for enhanced targeting of cancer cells. Sci. China Chem. 2011; 54: 1218.
12. Elghanian R, Storhoff JJ, Mucic RC, et al. Selective colorimetric detection of polynucleotides based on the distance-dependent optical properties of gold nanoparticles. Sci. 1997; 277: 1078–1081.
13. Esposito CL, Passaro D, Longobardo I, et al. A neutralizing RNA aptamer against EGFR causes selective apoptotic cell death. PLoS One. 2011; 6: e24071.
14. Safarzadeh Kozani P, Safarzadeh Kozani P, Rahbarizadeh F. Flexible aptamer-based Nucleolin-targeting cancer treatment modalities: A focus on immunotherapy, radiotherapy, and phototherapy. Trends in Med. Sci. 2021; 1(3): e113991.
15. Liu Q, Jin C, Wang Y, Fang X, Zhang X, Chen Z, Tan W. Aptamer-conjugated nanomaterials for specific cancer cell recognition and targeted cancer therapy. NPG Asia Mater. 2014; 6:e95.
16. Fan X, Sun L, Wu Y, et al. Bioactivity of 2′-deoxyinosineincorporated aptamer AS1411. Sci. Rep. 2016; 6: 25799.
17. Ferreira CS, Matthews CS, Missailidi S. DNA aptamers that bind to MUC1 tumour marker: Design and characterization of MUC1-binding singlestranded DNA aptamers. Tumor Biol. 2009; 27: 289–301.
18. Razlansari M, Jafarinejad S, Rahdar A, et al. Development and classification of RNA aptamers for therapeutic purposes: An updated review with emphasis on cancer. Mol. Cell Biochem. 2022; 478(7): 1573–1598.
19. Fu Z, Xian GJ. Aptamer-functionalized nanoparticles in targeted delivery and cancer therapy. Int. J. Mol. Sci. 2020; 21: 9123.
20. Gefen T, Castro I, Muharemagic D, et al. A TIM-3 oligonucleotide aptamer enhances T cell functions and potentiates tumor immunity in mice. Mol. Ther. 2017; 25: 2280–2288.
21. Gong Y, Tian S, Xuan Y, et al. Lipid and polymer mediated CRISPR/Cas9 gene editing. J. Mat. Chem. B. 2020; 8: 4369–4386.
22. Gragoudas ES, Adamis AP, Cunningham ET, et al. Pegaptanib for neovascular age-related macular degeneration. N. Engl. J. Med. 2004; 351: 2805–2816.
23. He S, Gao F, Ma J, et al. Aptamer PROTAC conjugates (APCs) for tumor-specific targeting in breast cancer. Angew. Chem. Int. Ed. 2021; 60: 23299–23305.
24. Herrmann A, Priceman SJ, Swiderski P, et al. CTLA4 aptamer delivers STAT3 siRNA to tumor-associated and malignant T cells. J. Clin. Invest. 2014; 124: 2977–2987.
25. Hu Q, Wang S, Wang L, et al. DNA nanostructure-based systems for intelligent delivery of therapeutic oligonucleotides. Adv. Healthc. Mat. 2018; 7: 1701153.

26. Hu R, Wen W, Wang Q, et al. Novel electrochemical aptamer biosensor based on an enzyme-gold nanoparticle dual label for the ultrasensitive detection of epithelial tumour marker MUC1. Biosens. Bioelectron. X. 2014; 53: 384–389.
27. Huang BT, Lai WY, Chang YC, et al. A CTLA-4 antagonizing DNA aptamer with antitumor effect. Mol. Ther.—Nucleic Acids. 2014; 8: 520–528.
28. Ireson CR, Kelland LR. Discovery and development of anticancer aptamers. Mol. Cancer Ther. 2006; 5: 2957–2962.
29. Ishizaki J, Nevins JR, Sullenger BA. Inhibition of cell proliferation by an RNA ligand that selectively blocks E2F function. Nat. Med. 1996; 2: 1386–1389.
30. Jeong H, Lee SH, Hwang Y, et al. Multivalent aptamer—RNA conjugates for simple and efficient delivery of doxorubicin/siRNA into multi drug-resistant cells. Macromol. Biosci. 2017; 17: 1600343.
31. Kang H, O'Donoghue MB, Liu H, et al. A liposome-based nanostructure for aptamer directed delivery. Chem. Commun. 2010; 46: 249–251.
32. Lai WY, Huang BT, Wang JW, et al. A novel PD-L1-targeting antagonistic DNA aptamer with antitumor effects. Mol. Ther.–Nucleic Acids. 2016; 5: e397.
33. Lebruska LL, Maher LJ. Selection and characterization of an RNA decoy for transcription factor NF-κB. Biochemistry. 2016; 38: 3168–3174.
34. Li F, Lu J, Liu J, et al. A water-soluble nucleolin aptamer-paclitaxel conjugate for tumor specific targeting in ovarian cancer. Nat. Commun. 2017; 8: 1390.
35. Li J, Mo L, Lu CH, et al. Functional nucleic acid-based hydrogels for bioanalytical and biomedical applications. Chem. Soc. Rev. 2016; 45: 1410–1431.
36. Li N, Nguyen HH, Byrom M, et al. Inhibition of cell proliferation by an anti-EGFR aptamer. PLoS One 2011; 6: e20299.
37. Li S, Jiang Q, Liu, S, et al. A DNA nanorobot functions as a cancer therapeutic in response to a molecular trigger in vivo. Nat. Biotechnol. 2018; 36: 258–264.
38. Li SL, Jiang P, Jiang FL, et al. Recent advances in nanomaterial-based nanoplatforms for chemodynamic cancer therapy. Adv. Funct. Mat. 2021; 31: 2100243.
39. Li X, Figg CA, Wang R, et al. Cross-linked aptamer—lipid micelles for excellent stability and specificity in target-cell recognition. Angew. Chem. Int. Ed. 2018; 130: 11589–11593.
40. Li X, Zhao Q, Qiu L. Smart ligand: Aptamer-mediated targeted delivery of chemotherapeutic drugs and siRNA for cancer therapy. J. Control. Release. 2013; 171: 152–162.
41. Liang C, Li F, Wang L, et al. Tumor cell-targeted delivery of CRISPR/Cas9 by aptamer-functionalized lipopolymer for therapeutic genome editing of VEGFA in osteosarcoma. Biomaterials. 2017; 147: 68–85.
42. Liu H, Zhang, Q, Wang S, et al. Bacterial extracellular vesicles as bioactive nanocarriers for drug delivery: Advances and perspectives. Bioact. Mat. 2021; 14: 169–181.
43. Liu YJ, Dou XQ, Wang F, et al. IL- 4Rα aptamer-liposome-CpG oligodeoxynucleotides suppress tumour growth by targeting the tumour microenvironment. J. Drug Target. 2017; 25: 275–283.
44. Lozano T, Soldevilla MM, Casares N, et al. Targeting inhibition of foxp3 by a CD28 2'-flfluro oligonucleotide aptamer conjugated to P60-peptide enhances active cancer immuno therapy. Biomaterials. 2016; 91: 73–80.
45. Lupold, SE, Hicke BJ, Lin Y, et al. Identification and characterization of nuclease stabilized RNA molecules that bind human prostate cancer cells via the prostate-specific membrane antigen. Cancer Res. 2002; 62: 4029–4033.
46. Mahmoudpoura M, Dingd S, Lyud Z, et al. Aptamer functionalized nanomaterials for biomedical applications: Recent advances and new horizons. Nano Today. 2016; 39: 101177.
47. Mallikaratchy P, Tang Z, Kwame S, et al. Aptamer directly evolved from live cells recognizes membrane bound immunoglobin heavy mu chain in Burkitt's lymphoma cells. Mol. Cell. Proteomics 2007; 6: 2230–2238.
48. Martell RE, Nevins JR, Sullenger BA. Optimizing aptamer activity for gene therapy applications using expression cassette SELEX. Mol. Ther. 2002; 6: 30–34.
49. Martin DF, Maguire MG, Ying GS, et al. Ranibizumab and bevacizumab for neovascular age-related macular degeneration. N. Engl. J. Med. 2011; 364: 1897–1908.
50. McNamar JO, Andrechek ER, Wang Y, et al. Cell type–specific delivery of siRNAs with aptamer-siRNA chimeras. Nat. Biotechnol. 2006; 24: 1005–1015.
51. McNamara JO, Kolonias D, Pastor F, et al. Multivalent 4–1BB binding aptamers costimulate CD8+ T cells and inhibit tumor growth in mice. J. Clin. Invest. 2008; 118: 376–386.
52. Mi J, Zhang X, Giangrande PH, et al. Targeted inhibition of αvβ3 integrin with an RNA aptamer impairs endothelial cell growth and survival. Biochem. Biophys. Res. Commun. 2005; 338: 956–963.

53. Miao Y, Gao Q, Mao M, et al. Bispecific aptamer chimeras enable targeted protein degradation on cell membranes. Angew. Chem. Int. Ed. 2021; 60: 11267–11271.
54. Moosaviana SA, Sahebkar A. Aptamer-functionalized liposomes for targeted cancer therapy. Cancer Lett. 2019; 448: 144–154.
55. Ng EWM, Shima DT, Calias P, et al. Pegaptanib, a targeted anti-VEGF aptamer for ocular vascular disease. Nat. Rev. Drug Discov. 2006; 5: 123–132.
56. Ni S, Zhuo Z, Pan Y, et al. Recent progress in aptamer discoveries and modifications for therapeutic applications. ACS Appl. Mat. Interfaces. 2021; 13: 9500–9519.
57. Ni X, Castanares M, Mukherjee A, et al. Nucleic acid aptamers: Clinical applications and promising new horizons. Curr. Med. Chem. 2011; 18: 4206–4214.
58. Nimjee S M, White RR, Becker RC, Sullenger BA. Aptamers as therapeutics. Annu. Rev. Pharmacol. Toxicol. 2017; 57: 61–79.
59. Pan Q, Nie C, Hu Y, et al. Aptamer functionalized DNA origami for targeted codelivery of antisense oligonucleotides and doxorubicin to enhance therapy in drug-resistant cancer cells. ACS Appl. Mat. Interfaces. 2020; 12: 400–409.
60. Pratico ED, Sullenger BA, Nair SK. Identification and characterization of an agonistic aptamer against the T cell costimulatory receptor, OX40. Nucleic Acid. Ther. 2013; 23: 35–43.
61. Prodeus A, Abdul-Wahid A, Fischer NW, et al. Targeting the PD-1/PD-L1 immune evasion Axis with DNA aptamers as a novel therapeutic strategy for the treatment of disseminated cancers. Mol. Ther.– Nucleic Acids. 2015; 4: e237.
62. Seeman N C, Sleiman, HF. DNA nanotechnology. Nat. Rev. Mat. 2018; 3: 17068.
63. Shangguan D, Cao Z, Meng L, et al. Cell-specific aptamer probes for membrane protein elucidation in cancer cells. J. Proteome Res. 2008; 7: 2133–2139.
64. Soldevilla MM, Hervas S, Villanueva H, et al. Identification of LAG3 high affinity aptamers by HT-SELEX and conserved motif accumulation (CMA). PLoS One. 2017; 12: e0185169.
65. Soldevilla MM, Villanueva H, Bendandi M, et al. 2-fluoro-RNA oligonucleotide CD40 targeted aptamers for the control of B lymphoma and bone-marrow aplasia. Biomaterials. 2015; 67: 274–285.
66. Leng C, Zhang X, Xu F, Yuan Y, Pei H, Sun Z, Li L, Bao Z. Engineering gold nanorod-copper sulfide heterostructures with enhanced photothermal conversion efficiency and photostability. Small. 2018; 14(12): e1703077.
67. Song XR, Liu C, Wang N, et al. Delivery of CRISPR/cas systems for cancer gene therapy and immunotherapy. Adv. Drug Deliv. Rev. 2021; 168: 158–180.
68. Song Y, Zhu Z, An Y, et al. Selection of DNA aptamers against epithelial cell adhesion molecule for cancer cell imaging and circulating tumor cell capture. Anal. Chem. 2013; 85: 4141–4149.
69. Soundararajan S, Chen W, Spicer EK, Courtenay-Luck N, Fernandes DJ. The nucleolin targeting aptamer AS1411 destabilizes bcl-2 messenger RNA in human breast cancer cells. Cancer Res. 2008; 68: 2358–2365.
70. Sun S, Liu H, Hu Y, et al. Selection and identification of a novel ssDNA aptamer targeting human skeletal muscle. Bioact. Mat. 2022; 20: 166–178.
71. Tan W, Donovan MJ, Jiang J. Aptamers from cell-based selection for bioanalytical applications. Chem. Rev. 2013; 113: 2842–2862.

15 Peptide-Based Targeting of Nanoplatforms for Cancer Theranostics

*Samson A. Adeyemi, Philemon N. Ubanako,
Pavan Walvekar and Yahya E. Choonara*

CONTENTS

15.1 Introduction 245
15.2 Peptide-Based Targeting for Cancer Theranostics 246
15.3 Albumin-Based Nanoformulations 247
      15.3.1 General Characteristics of Albumin 247
15.4 Application of HSA for Cancer Theranostics 248
15.5 Application of BSA for Cancer Theranostics 250
15.6 Ferritin-Based Nanoformulations 252
      15.6.1 Characteristics of Ferritin 252
15.7 Application of Ferritin for Cancer Theranostics 252
15.8 Gelatin and Its Applications for Cancer Theranostics 253
      15.8.1 Characteristics of Gelatin 253
15.9 Theranostics Applications of Gelatin in Cancer 253
15.10 Transferrin for Cancer Therapy 254
      15.10.1 Characteristics of Transferrin 254
15.11 Transferrin and Its Applications for Cancer Theranostics 255
15.12 Therapeutic Peptides as Theranostic Agents in Cancer Therapy 256
15.13 Tumour-Specific Peptides for Cancer Theranostics 257
      15.13.1 Tumour Cell Surface Receptor Targeting 258
      15.13.2 Tumour Microenvironment–Targeting Peptides 261
      15.13.3 Subcellular Organelle–Targeting Peptides 261
15.14 Nanoparticle-Based Drug Delivery Systems for Cancer Theranostics 262
      15.14.1 Nanoparticulate Theranostic Systems for Tumour Targeting 262
      15.14.2 Exosomes as Nanoplatforms for Cancer Theranostics 262
15.15 Theranostics: The Paradigm Shift in Cancer Therapy 264
15.16 Conclusion and Future Directions 265
References 266

15.1 INTRODUCTION

Cancer remains a deleterious disease and global burden, taking many lives. Based on the available data, it accounted for nearly 10 million deaths in 2020, and the global cancer burden has been projected to increase for the next two decades.[1] Within the last decade (starting in approximately 2015), the US government came up with an ambitious initiative on precision medicine with the main objective to explore and make tailor-made and precise patient care a clinical reality, where treatment is individualized based on each person's genetic information, lifestyle and environment.[2] To achieve this promising objective, cutting-edge nanotechnological devices and therapeutic nanoplatforms with potential for one or more clinical applications are to be produced.[3]

Peptides are made up of amino acids coupled together by amide linkage usually within 50 amino acids sequence and they are mainly present in bioorganisms.[4] As protein fragments, their intrinsic properties confer on them high biocompatibility and biodegradability. More importantly, their homogeneous nature enhances their bioactivity, including targeted binding, response to stimulus and therapeutic efficacy. These viable features pointed to peptides as perfect fundamental structures for biomedical nanoplatforms with potential clinical applications.[5] Over the past decades, advances in synthetic techniques and peptide screening libraries provided for the enormous utilization of peptides as targeting ligands, bioresponsive moieties and therapeutic agents in biomedical systems, including theranostic nanosytems.[6]

The three therapeutic approaches of radiation, chemotherapy and biological therapies still face numerous challenges, including drug resistance and negative side effects and low or no selectivity, that limit their efficacies. With the advent of molecular biology and immunity, the use of peptide drugs has gained tremendous momentum and applications due to their unique potential such as reduced toxicity, enhanced selectivity, low drug–drug interaction and effective biocompatibility.[7] More importantly, promising antitumour peptide-based drug candidates cause apoptosis and prevent the metastasis of tumour cells through the enhancement of immune function. Nonetheless, clinical application of peptide drugs is still faced with enormous limitations such as short half-life and enzymatic degradation as well as their inability to cross the cell membrane.[8]

The use of inorganic nanoparticulate drug delivery strategies has received much attention and progress due to their ease of synthesis, regulated shape and size and ease of surface functionalization.[9] In this way, it is possible to design and fabricate nanoparticles with multifunctional properties by the incorporation and conjugation of diverse components into and onto their surfaces such as therapeutic drugs, targeted moieties and imaging molecules, which confer on them multifunctional properties for bioimaging and cancer therapy[10, 11]

Biocompatible endogenous peptide-based nanoformulations are presently fabricated to simultaneously deliver anticancer drugs and diagnostic agents (Figure 15.1). This phenomenon, named theranostics, has gained tremendous and widespread applications, as peptide and protein nanoformulations are usually encapsulated with dyes, inorganic nanoclusters, contrasting agents and drug payloads for combinatorial imaging-guided cancer therapeutics. Using this design, theranostics employs the natural properties of peptides/proteins to evade clearance by reticuloendothelial cells and display prolonged circulation in the blood.[12] Similarly, the nanoscale size of the nanoplatform enables them to advance deeply into tumour tissues for enhanced imaging and deliver their therapeutic payloads at the active sites for optimal efficacy.[13] Interestingly, theranostics has been proposed as a new and revolutionary therapeutic approach in cancer therapy, which foster simultaneous diagnosis and treatment response monitoring using personalized medicine with improved accuracy and specificity.[14] Also, it is possible to encapsulate a number of various anticancer drugs into a single theranostics nanoplatform through intelligent design to achieve synergistic treatment of cancer.[15, 16] In this chapter, we elucidate the need to develop an efficient delivery system for peptide drugs to fully harness their effectiveness as cancer therapeutics. We outline their incorporation into nanoparticulate platforms for enhanced efficacy and discuss their engineering for targeted and tailor-made theranostics applications for cancer therapy. In a nutshell, peptide-based targeting of nanoplatforms for cancer theranostics represents the next paradigm for research orientation that will stimulate promising therapeutic interventions in the field of cancer treatment.

15.2 PEPTIDE-BASED TARGETING FOR CANCER THERANOSTICS

Quite often, natural proteins are employed in most experiments; meanwhile, there are a number of limiting factors that endanger these protein-based bio-platforms. Ease of degradation, poor

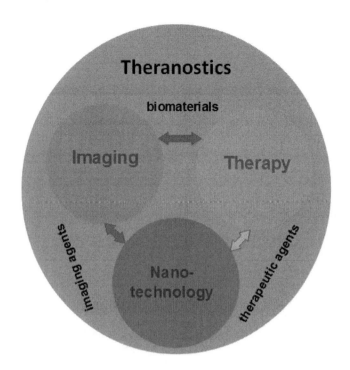

FIGURE 15.1 Schematic representation of theranostics.

permeability and potential immune responses are among the disadvantages of protein-based bio-platforms.[17] More importantly, it is possible to alter the protein function after their conjugation to or functionalization with other molecules. As such, in situ polymerization techniques for protein encapsulation have been developed in recent years as a viable alternative methods for the modification of proteins.[18] In this way, through the incorporation of cross-linkers to the surface of the protein and subsequent polymerization, protein nanogels are obtained. Notably, it is possible for such nanogels to keep the protein properties and functions within a complex biological or chemical environment and thereby escape the immunogenicity of proteins. Based on this modification, the outer layer polymers are cleaved and degraded, while the inner proteins remain intact and can be released by both internal and external stimulus. Using this strategy, a large number of therapeutic proteins have been fabricated for use in cancer therapy.[19, 20] In this section, some of the widely use proteins/peptides and their theranostics applications in cancer therapy will be discussed. Albumin-based, ferritin-based, gelatin-based and transferrin-based nanoformulations are briefly discussed.

15.3 ALBUMIN-BASED NANOFORMULATIONS

15.3.1 General Characteristics of Albumin

On a commercial platform, albumins are produced in large quantities from a number of sources, including human serum albumin (HAS), bovine serum albumin (BSA), rat serum (RSA) and egg white (ovalbumin).[21] For the sake of this chapter, BSA and HAS will be reviewed. BSA and HSA share similar homology sequence of about 76%.[22] BSA has a net negative potential charge of 18 mV with an isoelectric point (pl) of 4.7. Structurally, BSA is a heart-shaped molecule made up of three repeating clusters which are further sub-divided into two domains each. Its unique characteristics, including its abundance, cost effectiveness, drug delivery capability and ease of purification, enhance its extensive application in cancer theranostics.[23] On the other hand, HSA, as the most ubiquitous plasma protein, 35–50 mg/mL human serum, is produced in the liver. It is a single-chain polypeptide without glycosidic bond having a molecular weight of 66,500 Da. It is heart shaped and consists of 585 amino acid residues.[24] Structurally, using X-ray crystallography, the main difference between BSA and HAS is the presence of two tryptophan amino acid residues in BSA (Trp-135 and Trp-212), whereas only one of this amino acid is found in HAS (Trp-214). A unique property of HAS is its high solubility and robust thermal durability (available at 60°C for 10 h), with a stable pH range between 4–9. Additionally, its preferential absorption into tumour interstitium through the SPARC glycoprotein and gp60 glycoprotein transcytosis pathways, non-toxic nature, biodegradability and immunogenicity made it a viable candidate with increased applications for cancer theranostics.[17]

15.4 APPLICATION OF HSA FOR CANCER THERANOSTICS

A number of organic and inorganic molecules have been isolated and delivered using HAS as a natural delivery carrier. Superparamagnetic iron oxide, chlorin e6 (Ce6), IR780 and IR825, among other theranostic agents, have been synthesized using HAS. As a natural peptide with multiple binding domains, HAS is able to bind with several organic dyes without a covalent bond to produce high fluorescence HAS-dye complexes with improved quantum yield.[17] Near-infrared (NIR) dyes such as IR780, IR825 and indocyanine (ICG) have been extensively applied for cancer theranostics investigations in recent times due to their potential to penetrate deeply into living systems with less interference. For instance, it is possible to bind IR825 to the hydrophobic pocket of HAS through a hydrophobic bonding in a molar ratio of 1:1, to produce a HAS-IR825 complex.[25] This HAS-IR825 complex displayed a high fluorescence quantum yield at an excitation wavelength of 600 nm with high absorbance. Meanwhile, the same complex with a low fluorescence quantum yield under 808 nm excitation showed improved performance in NIR imaging and photothermal therapy (PTT) at diverse wavelengths. Subsequently, Chen and his colleagues (2014) synthesized a gadolinium-loaded HAS-IR825 complex used as a dual-modal imaging-guided PTT of tumour.[26] A Gd(III) compound of diethylenetriamine pentaacetic acid was conjugated with HAS to produce a HAS-IR825 conjugate that was further bound to IR825 dye in a HAS-Gd-IR825 complex. Interestingly, this HAS-Gd-IR825 complex showed high fluorescence and NIR absorbance with improved T1 relaxivity of 4.82 mM^{-1} s^{-1}. In another recent study, Han and co-workers (2017) fabricated a gemcitabine functionalized HSA-IR780 agent employed for chemotherapy and image-guided PTT of tumour.[27] Through hydrophobic interactions and the use of cathepsin B cleavage peptide GFLG, HAS was primarily linked with gemcitabine and subsequently mixed with IR780 at a molar ratio of 1:1.

In order to circumvent the challenge of dye leakage into in vivo circulation, observed with the HAS-NIR dyes complexes discussed previously, possibly due to non-covalent interactions, Rong et al. (2015) developed a covalently linked HAS-NIR cyanine dye complex using heptamethine CySCOOH dye conjugated to the lysine residues of HSA through an advanced EDC/NHS chemistry for enhanced NIR fluorescence, photoacoustic (PA), thermal multimodality imaging and PTT.[28] Compared to the native dye without CySCOOH, the fabricated heptamethine CySCOOH-HAS complex showed increased PTT potential, prolonged in vivo circulation and enhanced tumour accumulation under similar conditions. More importantly, it is possible to preferentially bind the Cys34 residue of HSA via a Michael addition reaction to the maleimide group of heptamethine CySCOOH dye. Lisitskiy and co-workers (2017) employed the Michael addition reaction technique to conjugate fluorescent dye Cy7 to Cys34 residue of HAS.[29] Meanwhile, the lysine residues of HSA were linked to pTFT (5-trifluoromethyl-2'-deoxyuridine 5'-monophosphate) as a chemotherapeutic drug through a pH and redox dual-sensitive linker. In this way, the fabricated drug-loaded Cy7-HAS complex could serve dual purposes as optical and fluorescence 19F MR imaging as well as delivery cargo for chemotherapeutics.

It has been well established that nanocarriers within the size range of 100–200 nm accumulate efficiently in tumour tissues through the enhanced permeability and retention (EPR) effect.[30] Also, in order to expand the application of theranostic agents, more functional ingredients are incorporated into their design and fabrication. With HSA having an effective diameter of 7.2 nm, more effort has been devoted to designing and fabricating HAS-based complexes with desirable sizes and enhanced properties for efficient multifunctional HAS-based theranostics. Based on these modalities, paclitaxel- and melanin-loaded HAS nanoparticles (HMP-NPs) using a desolvation-crosslinking technique with an average size of 192 nm were synthesised.[31, 32] In vivo experimental results showed that the HMP-NPs exhibited an efficient photoacoustic imaging signal intensity in the tumour site and enhanced chemotherapeutic efficacy to the tumour with prolonged circulation time. Additionally, this technique was employed by Li and co-workers (2015) to developed platinum-decorated HAS-based nanoparticles (Pt(IV)-probe@HSA) for theranostic usage.[33] The researchers surface-functionalized HSA nanoparticles with NIR fluorophore Cy5, Pt(IV) as an antitumour prodrug and quencher Qsy21. This synthesized Pt(IV)-probe@HSA was efficient in the selective activation and release of Pt(IV) prodrug locally and allowed for enhanced real-time imaging of tumour cells with high resolution.

For most of the HSA nanoparticulate systems discussed, a major disadvantage in their formation is the possible side effects caused by aldehyde residue in in vivo application.[34] Meanwhile, the development of some easy techniques employing exogenous cross-linkers or non-toxic chemicals has been widely accepted for their synthesis as alternatives to circumvent this problem.[35, 36] For instance, intermolecular disulphide HAS internal cross-linking was employed to synthesis HAS-ICG nanoparticles with an average hydrodynamic size of about 75 nm for imaging-guided photodynamic (PD) and photothermal (PT) cancer therapies.[36] Using this approach, the first step was to break down the intramolecular disulphide bonds within HSA using glutathione as an endogenous reducing agent and thereafter develop HSA nanoparticles using the desolvation technique. In another development, the same method was employed to synthesize HAS nanoassemblies (HSA-Ce6 NAs) using chlorin e6 for multi-modal imaging-guided PD therapy.[35] Interestingly, these nanoassemblies, with an average size of 100 nm, exhibited intelligent preference for tumour selectivity and efficient PD therapeutic potential with promising triple-modal (fluorescence, PA and MR) imaging.

Without making use of exogenous cross-linkers, Chen and his colleagues developed multifunctional HAS-based theranostic nanoparticles using a drug-induced protein assembly technique.[37] In this experiment, three clinically approved agents, ICG, PTX and HSA, were incorporated to formulate a multifunctional theranostics agent similar to Abraxane. In this formulation network, chemotherapy, NIR fluorescence imaging, PTT and thermal imaging were succinctly and efficiently put together. In turn, this formulation exhibited a synergistic therapeutic efficacy in treating both subcutaneous and metastatic breast tumours. This technique was modified to design and fabricate tumour-specific theranostics agents for multimodal imaging-guided therapy of tumours.[38] The strategy employed the pre-modification of HSA with either tumour-specific tri-amino acyclic Arg-Gly-Asp (RGD) peptide to form (HAS-RGD) conjugate or the conjugation of HSA with photosensitizing agent Ce6 to produce (HAS-Ce6). Thereafter, PTX as an anticancer drug was linked to the network to facilitate the self-assembly of HAS-RGD and HAS-Ce6 to yield two separate sets of nanoparticles. Subsequently, the incorporation of manganese (II) will confer on both nanoparticles MR and fluorescence imaging properties that could be tracked and applied for bimodal cancer PDT and chemotherapy. In a similar study, iron (II) phthalocyanine (FePc), as a photosensitizer agent, was used to synthesize multifunctional HAS-FePc nanoparticles for PA imaging-guided PTT.[39] These fabricated HAS-FePc nanoparticles showed improved PA imaging potential, enhanced stability, effective PTT quality and reduced prolonged toxicity in vivo.

Furthermore, Moon and colleagues (2015) developed PTX-loaded HSA nanoparticles which were surface-functionalized with porphyrin microbubbles for application as cancer theranostics.[40] In order to produce a system to simultaneously intensify ultrasound and PA signal, porphyrin microbubbles were produced by conjugating porphyrin to a phospholipid. This fabricated nanoparticles with multifaceted theranostic potentials, exhibited improved sensitivity in PA and ultrasound imaging and enhanced delivery of PTX as an anticancer drug to tumour sites. Similarly, superparamagnetic iron oxide–loaded HSA-decorated nanoparticles were fabricated enhanced absorption of pulsed microwave energy and transform efficiency into shockwave with the thermoelastic potential in addition to its application as MR contrast agents.[41]

15.5 APPLICATION OF BSA FOR CANCER THERANOSTICS

With several binding domains that are hydrophobic, bovine serum albumin (BSA) is a natural polypeptide that can serve as a delivery vehicle for several small NIR dye agents. Through hydrogen bonding and hydrophobic interactions, the hydrophobic binding site of BSA was employed to selectively bind squaraine (SQ) with improved fluorescence intensity up to 80% efficiency.[42] Subsequently, a dual-purpose polymeric complex capable of serving as a PTT agent and effective bioimaging probe was fabricated as a supermolecular adducts of SQ and BSA (SA@BSA). Similarly, folic acid was conjugated to the surface of the supermolecular adducts (SQ@BSA-FA) with enhanced characteristics to monitor the biodistribution of the adducts in a time-dependent manner with tumour-specific targeting. In another experiment, Chen and Liu (2016) synthesized a BSA-NIR complex for NIR fluorescence imaging, PTT and PDT with a single excitation wavelength using ICG, an FDA-approved NIR.[43] The development of BSA-based systems for theranostics applications have been reported. These systems are covalently modified with NIR dyes. Lee and co-workers (2016) fabricated a pH-responsive theranostics agents consisting of zinc as the coordinating agent. Using both BHQ-3 quencher (NIR dark quencher) or Cy5.5 dye (donor NIR dye), modified polyethylene glycol-BSA-imidazole covalently linked complexes were produced.[44] Interestingly, the synthesized BSA-based complexes rapidly disassembled and emitted strong NIR fluorescence at endo/lysosome pH of 5 within cancer cells.

Aside from the conjugation of BSA to NIR dyes, it is possible to engineer BSA with other functional agents through complexation, including manganese,[45] gadolinium[46] and graphene derivatives.[47] For instance, BSA complexes with manganese as the contrasting agent have been produced for cancer theranostics.[48] By mimicking the drug–substrate interaction technique, a BSA-MnO_2 theranostics agent with multifunctional properties was synthesized in which BSA was applied a reducing agent and carrier.[45] It should be noted that both PTX and ICG can be incorporated into BSA-MnO_2 using complimentary easy and energy-saving methods. The PTX/ICG-loaded BSA-MnO_2 theranostics agent can be applied for chemotherapy and MR imaging-guided PTT in both in vitro and in vivo experiments.

Using gadolinium, a size-tunable Gd_2O_3@BSA complex was produced by employing hollow BSA conjugated with Ce6 as a theranostics agent for application for MR imaging-guided PDT and PTT.[46] Interestingly, the BSA nanoreactor was able to readily conjugate photosensitizers and monitor the longitudinal relaxivity of Gd_2O_3 effectively. Similarly, Yang et al. (2016) fabricated Gd:CuS@BSA complexes using a simple method in which BSA was employed as a biotemplate at physiological temperature and incorporated with Gd and CuS for multifunctional theranostics application.[49] This synthesized Gd:CuS@BSA theranostics agent exhibited an ultrasmall size (about 9 nm), enhanced longitudinal relaxivity, acceptable temperature increase and improved PA signals under NIR irradiation.

In a recent experiment, the amino groups of BSA-Gd conjugate and the carboxyl groups of MoS_2 nanoflakes were employed through an amine reaction to fabricate a multifunctional MoS_2-Gd-BSA theranostics agent. The excellent photothermal effect of MoS_2 nanoflakes with the improved longitudinal proton relaxivity of BSA-Gd complex were employed in this multifunctional theranostics agent.[50] This novel MoS_2-Gd-BSA theranostics agent showed a potent NIR absorbance and increased relaxivity, which are useful for PA and T1-weighted MR dual-modal imaging-guided PTT of cancers. In addition, carmustine as a chemotherapeutic drug was incorporated into FITC-BSA nanoparticles using the desolvation/denaturation technique. This was subsequently conjugated with Gd (III) salt as a MR agent to produce a potent nanomedicine with dual imaging modalities.[51]

Sheng and co-workers (2013) developed a BSA-assisted synthesis technique to fabricate a graphene oxide (nano-rGO) theranostic agent.[47] The surfaced functionalized BSA nano-rGO complex exhibited an increased stability with reduced cytotoxicity, with enhanced PA imaging and PTT application. Based on this novelty, a pH-BSA-pheophorbide-a (PheoA) photosensitizer, loaded with GO and functionalized for active targeting using folate receptor (PheoA + GO:FA-BSA-c-PheoA), was fabricated as a nanocarrier for theranostic applications.[52] Interestingly, this theranostic agent displayed viable pH-responsive photosensitive potential and exhibited synergistic dual PTT and PDT imaging-guided effects by NIR irradiation against tumours.

In addition to its multifunctional role as a delivery vector to incorporate different types of functional molecules, including dyes and chemotherapeutic drugs, BSA can also serve as a coating agent for other nanovectors to improve blood circulation and retention time, enhance solubility in water and increase physiological stability for theranostics applications. Based on this unique property of BSA, NaGdY4-based upconversion nanoparticles (UCNPs) were coated using BSA to produce UCNP@BSA nanoparticles with advanced stability and enhanced solubility in water within physiological environments.[53] Subsequently, two photosensitizer dyes, IR825 and rose bengal (RB), can be efficiently incorporated into the BSA outer shell of the UCNP@BSA system. Worthy of note is the compliance of the characteristic absorbance peak of RB with the green emission peak of UCNPs (980 nm excitation), which displayed a deleterious effect on cancer cells by PDT. As such, this theranostic agent loaded with two dyes can be employed for upconversion optical imaging, MR diagnostics and bimodal PDT and PTT treatments in vitro and in vivo. Wang and colleagues (2015) fabricated a multifunctional theranostics agent with a magnetite Fe3O4 core-shell, coated with BSA and functionalized with Anti-EGFR mab C225 as an active targeting agent and loaded with gemcitabine as an anticancer drug.[54] The multifunctional structure displayed dual-targeted thermochemotherapy against pancreatic cancer, specifically targeting different pancreatic tumour cell EGFR-expressions and regulating diverse cellular targeting by MR imaging. Using this modality, doxorubicin (DOX), as an antitumoral drug, was encapsulated with BSA and loaded with magnetite Fe_5C_2 for theranostics applications.[55] Interestingly, under an acidic environment, the DOX is released and irradiated by NIR and serves as an intelligent nanoplatform for effective chemotherapy, PTT and MR imaging. Furthermore, superparamagnetic magnetite Fe_3O_4-coated BSA–poly(ethylene glycol) (PEG) and DOX as a pH-responsive protein-polymer bioconjugate were synthesized for combined chemotherapy and MRI diagnostics.[56] Using the same technique, poly(ε-caprolactone) was coated with BSA to produce a BSA–poly(ε-caprolactone) bioconjugate produced as an upconversion theranostics agent for simultaneous application in chemotherapy and cancer cell imaging using PDT.[57] Through the use of a biocompatible and straightforward technique, Prussian blue (PB) nanoparticles coated with BSA and non-covalently surface-functionalized with ICG molecules were produced.[58] Within this network, PB was incorporated as the MR contrast enhancer while BSA conferred high stability. Notably, the multifunctional theranostics nanoplatform has the potential for bimodal MR signalling and NIR fluorescence imaging and also provides for combined treatment with PDT and PTT. Last, a multi-purpose theranostics nanoplatform was recently fabricated with a gold (Au) inner core, which was encapsulated with an outer BSA shell conjugated with DOX and folate. This theranostics agent was fabricated and applied for computed tomography (CT) imaging and targeted cancer therapy.[59]

In summary, it is possible to produce albumin-based nanoparticles by engaging albumin as a template, scaffold or stabilizer. Albumin can be conjugated to drugs, polymers and contrast molecules. Strategic techniques such including covalent and non-covalent conjugation or assembly methods can be used. While a large number of scientists prefer to use the non-covalent self-assembly technique for the synthesis of albumin-based nanoparticles following the breakthrough achieved with Abraxane, nanoparticles produced using the non-covalent method will be different from batch to batch. This anomaly has to be handled and overcome in future research.[60]

15.6 FERRITIN-BASED NANOFORMULATIONS

15.6.1 CHARACTERISTICS OF FERRITIN

As an abundant protein found in the intracellular and extracellular matrixes in circulation, ferritin has a molecular weight of 450 kDa, with a hollow nanostructure with internal and external dimensional measurements of 8 and 12 nm, respectively.[61] As it name connotes, ferritin possess an approximation of about 4500 iron atoms which are bioavailable and non-toxic.[62] In advanced eukaryotes like mammals, a single ferritin protein is made up of 24 subdomains perfectly assembled to form a spherical symmetrical protein shell.[62] Principally in eukaryotes, the ferritin proteins are formed by the self-assembly of two subdomain types, the L and H ferritin chains, with molecular weights of 19 and 21 kDa, respectively. The L-chain is without ferroxidase and activates iron through the activity of nucleases, while the H-chain contains ferroxidase, which oxidizes Fe(II) to Fe(III).[63] As one of the multifunctional proteins for the storage and metabolism of iron, ferritin plays crucial roles in angiogenesis, proliferation and immunosuppression.[62]

With ferritin as a robust and abundant protein, it is disassembled at equilibrium under extreme pH conditions from acidic (pH 2–3) or basic (pH 10–12) and can withstand high temperatures of up to 80°C without being denatured.[64] Notably, these characteristics, in addition to its biodegradability and biocompatibility, made ferritin an idea candidate for cancer theranostics.[65]

15.7 APPLICATION OF FERRITIN FOR CANCER THERANOSTICS

Ferritin, as a macromolecule, serves as a dual-purpose protein both as a reactor for the production of several non-native metallic nanoparticles and nanovector for diverse applications.[65] A typical example is the encapsulation of Zn hexadecafluorophthalocyanine (ZnF16Pc), a potent hydrophobic photosensitizer, into the matrix of the Cys-Asp-Cys-Arg-Gly-Asp-Cys-Phe-Cys (RGD4C) of modified ferritins (P-RFRTs). The loading efficiency was above 60 wt% for effective PDT treatment.[66] Subsequently, the P-RFRTs were surface-functionalized with ZW800, which is a NIR dye molecule, for improved tracking of the P-RFRT particles. Using the same method, RGD4C-modified ferritin was also loaded with DOX by Zhen and co-workers.[67] In order to produce a multimodal imaging-guided fluorescence/PA/PTT theranostic agent, a step-wise change in pH self-assembly technique was adopted to synthesize ferritin nanoparticles loaded with IR820 as a NIR dye.[68] The fabricated nanocages have the capacity to diagnose and treat cancer effectively using two varied excitation wavelengths (808 nm for PA and PTT treatments and 550 nm for high quantum-yield fluorescence imaging). Also, it is possible to produce biomimetic ferritin nanocages loaded with CuS nanoparticles using a direct synthesis technique.[68] The produced CuS-ferritin theranostics agent possessed vibrant NIR absorbance, increased photothermal conversion efficiency, improved biocompatibility and excellent PA contrast. Of note is the ability of the ^{64}CuS-ferritin theranostic system to serve as a distinct PET imaging agent when incorporated with the irradiated ^{64}Cu.

In some malignant tissues such as colon cancer, pancreatic cancer, breast cancer and testicular seminoma, an increased level of L-ferritin in tumour tissue compared to normal tissue is quite obvious.[62] Quite often, an increased L-ferritin level is associated with increased expression of its receptor, which facilitates its cellular internalization through endocytosis. As observed by Geninatti Crich and colleagues (2015), high expression of L-ferritin receptors corroborates increased uptake of L-ferritin in breast cancer MCF-7 cells.[69] Based on this modality, a curcumin-loaded ferritin-based theranostic nanoplatform was developed to deliver curcumin as a natural anticancer drug and GdHPDO3A, a MR contrast agent.[69] Notably, the theranostic nanoplatform selectively delivered curcumin and GdHPDO3A to breast cancer cells. Additionally, ferritin was loaded with curcumin and GdHPDO3A for MR contrast imaging in breast cancer stem cells.[70] Turino and co-workers (2017) coated poly (lactic-co-glycolic acid) (PLGA) nanoparticles with L-ferritin to produce a multifunctional theranostic nanoplatform for enhanced targeting to breast cancer MCF-7 cells.[71] In this work, amphiphilic Gd and PTX were added to the functionalized PLGA nanoparticles for simultaneous MR contrast analysis and MRI-guided chemotherapy, respectively. Interestingly, the outer ferritin shell provided more stability to the PLGA nanoparticles, thereby enhancing their ability to avoid the non-specific and fast release of PTX and Gd before reaching the action sites.

In concise terms, regardless of the unique hollow structure of ferritin that makes it a suitable vector, slight conformational changes occur upon its loading with therapeutic agents, which may alter its activity. Also, the toxicity of ferritin-based nanoparticles should be thoroughly investigated. Last, it is possible for surface-decorated ferritins to be identified as foreign agents. Based on this disadvantage, the grafting density of ferritin should be considered when surface targeting ligands are needed.

15.8 GELATIN AND ITS APPLICATIONS FOR CANCER THERANOSTICS

15.8.1 Characteristics of Gelatin

As a polyampholytic protein, gelatin is a biomacromolecule possessing both anions and cations in addition to having hydrophobic groups.[72] Gelatin is made of the triple amino acids proline, alanine and glycine woven together in a repeating unit.[73] Gelatins are produced from the hydrolysis of collagen either though acidic, basic or enzymatic degradation. Depending on the chemical hydrolysis involved, their isoelectric points vary from 4.5–6.0 for alkaline treatment and 7–9 for acid hydrolysis.[74] Additionally, it is possible to optimize and control the release of anticancer agents from gelatin by changing the source of gelatin, its molecular weight and the degree of its crosslinking.[75] Other characteristic properties of gelatin include its natural composition, biodegradability, biocompatibility, water solubility and non-toxicity.[76] These multifaceted properties confer on gelatin its application as delivery vehicle for several therapeutic/diagnostic agents and its usage in the design and development of viable theranostics.

15.9 THERANOSTICS APPLICATIONS OF GELATIN IN CANCER

A unique property of gelatin that made it a viable application in coating various metal nanoparticles is its native hydrophobic and hydrophilic segments inherent in each single polypeptide chain.[77] For instance, self-assembled gelatin was used to coat iron oxide nanoparticles to produce a theranostic agent abbreviated as AGIO.[78] Meanwhile, using the electrolytic co-deposition method, it is possible to encapsulate both anticancer DOX and calcium phosphate (CaP) in gelatin. The subsequent theranostics nanoplatform produced from their complexation (AGIO@CaP-DOX) displayed excellent MR contrast, enhanced cytocompatibility and effective cellular internalization toward HeLa cells. In a work by Cheng and co-workers (2014), a gelatin-based multifunctional nanoplatform was fabricated.

In this nanosystem, oleylamine-coated Fe_3O_4 nanoparticles formed the inner core upon their encapsulation with amphiphilic gelatin as the outer shell. FITC molecules were incorporated as the fluorescence labelling agent, and platinum(IV) prodrug was employed as the anticancer drug. The fabricated theranostics nanoplatform was adopted for efficient fluorescence and MR imaging-guided chemotherapy.[79] In another study, silica-coated iron oxide magnetic nanoparticles were modified by encapsulating them using both gelatin and oleic acid. In this way, the solubility of iron oxide nanoparticles is enhanced with increased biocompatibility to facilitate effective treatment-response monitoring of tumours.[77] Specifically, hydrophobic PTX as an anticancer drug was successfully entrapped into the lipophilic oleic acid-gelatin shell. Interestingly, the fabricated theranostic nanoplatform exhibited high R-squared value, reduced cellular toxicity, increased PTX delivery and potent anticancer efficacy in vitro. In a subsequent experiment, the biodistribution, pharmacokinetics and antitumoral activity as well as the tumour diagnostic efficiency of the fabricated theranostic nanoplatform were evaluated in an in vivo study to ascertain its potential clinical applications.[80]

Quite often, aside from iron oxide nanoparticles, gelatin is employed to coat gold nanoparticles to optimize their biocompatibility and improve their stability. For instance, a gelatinized gold-based nanostructure was formed for application in fluorescence imaging-guided chemotherapy.[81] Within this nanoplatform, DOX was first covalently linked with gelatin and thereby employed to encapsulate preformed epigallocatechin gallate (EGCG)-functionalized Au nanoparticles. In this network, DOX serves dual purposes as an anticancer agent as well as an indicator for fluorescence. The fabricated theranostics nanoplatform (DOX-gelatin/EGCG Au) showed an enhanced anti-proliferation efficacy against human prostate cancer cells (PC-3) and can monitor the intracellular enzyme-induced release of DOX through the measurement of the recovery fluorescence signal of DOX. Studies have shown that the covalent conjugation of DOX to nanovectors may be problematic due to low in vivo release and reduced drug activity.[82] To circumvent this anomaly, a new dual-responsive pH and temperature theranostic system was established to fabricate non-covalently bound DOX to biosynthesized gelatin-coated gold nanoparticles (DOX-AuNPs@gelatin).[82] This synthesized multifunctional nanoplatform serves as a suitable agent for cancer theranostics because of its excellent biocompatibility, enhanced DOX loading capacity due to the non-covalent complexation and efficient release of DOX within the nanosystem, especially in the tumour microenvironment.

Not too long ago, Hu and colleagues (2015) designed and fabricated angiopep-2–functionalized gelatin-based nanoparticles (Angio-DOX-DGL-Gel-NP).[83] They were designed to improve the efficiency of tumour targeting, increase drug retention at tumour sites and enhance tumour penetration. Within this theranostic network, the outer shell was made up of dendrigraft poly-lysine conjugated with angiopep-2 and DOX, while the inner core consisted of gelatin nanoparticles degraded by matrix metalloproteinase-2. When evaluated for fluorescence imaging-guided chemotherapy, the results were excellent for antitumour therapy both in vitro and in vivo.

Summarily, the use of gelatins as nanoplatforms in drug delivery networks and their ability to produce nano- and micro-particles are well established.[75] It is possible to encapsulate small drug molecules or large bioactives into gelatins and have them released in a regulated manner. Despite this promising potential, gelatin remains underutilized in cancer theranostics platforms. Based on available data, we propose the use of gelatin-based nanoparticles will gain more momentum and applications in cancer theranostics, especially in brain delivery research and oral administration.

15.10 TRANSFERRIN FOR CANCER THERAPY

15.10.1 CHARACTERISTICS OF TRANSFERRIN

With a molecular weight of about 79 kDa, transferrin, as a monomeric glycoprotein, has 679 amino acids.[84] There are three carbohydrate side chains that surround the transferrin molecule, two N-linked (Asn-413 and Asn-611) and one O-linked (Ser-32)[85] Structurally, the polypeptide chain of transferrin is divided into two similar units, the C-domain with 343 amino acids and the N-terminal with 336 amino acids. Connecting the two structurally similar subunits of the polypeptide chain is a short linear spacer sequence.[86] Within each domain is a reversible binding site specific for binding ferric iron with an affinity of 10^{22} M^{-1} at physiological pH (7.4).[87] One critical biological role of transferrin is the biodistribution and control of iron (Fe) circulation. Interestingly, a number of biological processes depend on Fe for their functioning, including DNA synthesis, cellular metabolism and proliferation, oxygen transport and electron transfer, among others.[88] Specifically, transferrin targets transferrin receptors on the cell surface to produce a transferrin-transferrin receptor complex which in turn is internalized through receptor-mediated endocytosis.[9,89] Importantly, transferring receptors have been shown to overexpress on the surfaces of many cancer types, such as colon cancer, melanoma, breast carcinoma, ovarian carcinoma and glioblastoma.[90]

15.11 TRANSFERRIN AND ITS APPLICATIONS FOR CANCER THERANOSTICS

Transferrin has been employed by several investigations for active delivery of theranostic agents due to its targeting ability as a promising ligand. For instance, in order to track and target tumour cells that express transferrin receptors on their surfaces, transferrin was functionalized with graphene quantum dots for tumour imaging.[91] Within this system, DOX, as an anticancer drug, was employed to encapsulate graphene via π–π stacking and hydrophobic interactions, with a potential application for fluorescence imaging-guided chemotherapy. In another development, PEGylated transferrin loaded with DOX was used to encapsulate diamond nanoparticles. The fabricated multifaceted theranostics nanoplatform was investigated for active targeting and chemotherapeutic efficacy on human hepatoma (HepG2) cell lines, with an overexpression of transferrin receptors against normal cell lines (L-02) with minimal expression of transferrin receptors.[92] The dual-purpose multifunctional theranostic micelles of D-alpha-tocopheryl PEG 1000 succinate (TPGS) conjugated to transferrin was produced by Muthu and co-workers (2015).[93] Within this system, ultra-bright gold clusters were incorporated for imaging purposes, while DOX was employed as an ideal anticancer agent. The fabricated theranostic system was adopted for simultaneous imaging and chemotherapy. Interestingly, the transferrin-functionalized micelle outperformed the native theranostic micelle-based nanoformulation in terms of higher cellular uptake, with increased cytotoxicity in a breast cancer cell line (MDA-MB-231-luc). Following this investigation, a liposome-based nanosystem was fabricated in which transferrin was employed to surface-modified TPGS for theranostic application. Both quantum dots and DOX were incorporated into the system for imaging-guided chemotherapy of brain cancer.[94] This results of the theranostic liposomes decorated with transferrin showed enhanced and prolonged targeting of quantum dots and DOX within the brain compared to the native liposomal formulation without transferrin. Also, NIR dye (Cy7) and hydrophobic anticancer drug PTX were incorporated into transferrin-decorated mesoporous silica-coated Fe_3O_4 nanoparticles for NIR/MR bimodal imaging-guided chemotherapy.[95]

In a recent study, nuclear-targeted TAT peptide (YGRKKRRQRRR) was conjugated to transferrin and employed to encapsulate magnetic nanoparticles proposed to be applied for PTT treatment.[96] First, pre-conjugated TAT peptide and transferrin were linked to the magnetic nanoparticles and subsequently bound to NIR dye Cy7. The fabricated theranostic system has the potential to specifically target the cancer cell nucleus and instigate bimodal guided-imaging of NIR and MR for PTT treatment. Wang and co-workers (2017) developed a theranostic system with a core shell of UCNP ($NaYF_4:Gd^{3+}, Yb^{3+}, Er^{3+}$). The fabricated system has the ability to upconvert NIR to visible light and displayed increased active tumour targeting with enhanced biocompatibility.[97] Within this network, protoporphyrin IX, as a clinically approved PTT agent, was carefully incorporated into the transferrin shell, with the potential of being recognized by cancer cells for effective PDT with NIR irradiation and luminescence bio-imaging. In another investigation, Hou and colleagues (2017) explored a new method called the diffusion molecular retention tumour targeting effect to synthesize a theranostic active targeting system for PA imaging-guided synergistic application for photothermic and chemotherapeutic interventions in tumours.[98] In the meantime, iron-dependent artesunate (AS), as an effective anticancer drug, and transferrin were linked in hollow mesoporous CuS nanoparticles (HMCuS NPs) to produce AS/transferrin-HMCuS NPs. Notably, the fabricated AS/transferrin-HMCuS NPs instigated the accumulation and retention of AS, facilitated it specifically targeting breast cancer MCF-7 cells through transferrin receptor-mediated endocytosis, improved the synergistic efficiency of chemotherapy-phototherapy and enhanced anticancer effects overall.

The application of transferrin is not limited to its usage for ligand targeting. It is also versatile for the encapsulation of various imaging and therapeutic agents. For instance, with its self-assembling ability, transferrin nanoparticles (transferrin-IR780 NPs) loaded with NIR bye IR780 were developed for application in cancer targeted imaging and photothermy.[99] Interestingly, this fabricated system showed viable photostability, acceptable and narrow size distribution, efficient photothermal conversion and promising targeting and theranostics capability. Similarly, cytotoxic chelating agent (NNE3TA:2,2'-(7-(2-((carboxymethyl)(4-nitrobenzyl)amino)ethyl)-1,4,7-triazonane-1,4-diyl)diacetic acid) and NIR dye Cy5.5 were coated with transferrin to produce (NNE3TA-transferrin-Cy5.5) theranostics agent.[100] Notably, the NNE3TA-transferrin-Cy5.5 theranostics nanoplatform showed improved active iron chelation for cancer therapy, and excellent NIR imaging was exhibited in vitro. Zhu and co-workers (2017) developed a simple technique in which a drug-induced transferrin self-assembly occurred to produce tumour-targeted nanoparticles for fluorescence and bi-modal imaging-guided PTT of glioma.[101] Within this network, it is possible for transferrin to simultaneously conjugate and encapsulate ICG using both hydrophobic interaction and hydrogen bonding. Of particular interest is the environmentally friendly nature of the technique, which is safe, straightforward, simple and mild without the application of any toxic reagents. Notably, the synthesized transferrin-ICG nanoparticles exhibited efficient direct tumour-targeting, improved biocompatibility, pronounced bimodal imaging-guided efficiency and effective PTT treatment which could be employed for enhanced theranostics application for both orthotopic and subcutaneous brain tumour treatment.

In simple terms, transferrin has been shown in this work to exhibit dual potential applications as a delivery vector as well as a targeting moiety. Nonetheless, reports have shown that the targeting potential of transferrin-modified silica nanoparticles was completely lost in serum-rich media as a result of the shielding effects of the surfaced modified transferrin proteins.[102] As such, the targeting functionality of transferrin within advanced biological media depends largely on the bioavailability of large transferrin molecules. Therefore, keeping the targeting efficacy of transferrin-based nanoparticles remains a hurdle to overcome in in vivo experiments.

15.12 THERAPEUTIC PEPTIDES AS THERANOSTIC AGENTS IN CANCER THERAPY

A number of characteristics position therapeutic peptides as natural biopolymers for tumour therapy against other bioplatforms. Among such advantages are their favourable pharmacokinetic profiles, enhanced solubility, excellent tissue penetration, easy modification, high specificity and viable cost effectiveness.[103] Usually, therapeutic peptides have their source in the natural world, including animal, plants and microbial pools.[104] Nonetheless, there are other therapeutic peptides that are obtained though the screening of peptide libraries.[105] In general terms, the use of naturally occurring peptides as therapeutic agents is limited due to their limited in vivo circulation, reduced membrane permeability and low physical and chemical stability as well as their rapid clearance by the immune system. Thus, several attempts have been made to develop multifunctional approaches for peptide-based nanoformulations for cancer therapy.[6]

Abbreviated as (KLAKLAK)$_2$, the 14–amino acid helical-shaped peptide KLAKLAKKLAKLAK is a proapoptosis peptide with viable antimicrobial efficacy and reduce cytotoxicity in mammals (193). Its antimicrobial property was investigated against both Gram-negative and Gram-positive bacteria such as *Escherichia coli*, *Pseudomonas aeruginosa* and *Staphylococcus aureus*.[106] By employing chiral-independent mechanisms, (KLAKLAK)$_2$ demonstrates targeted disruption of the mitochondrial membranes in eukaryotes against plasma membranes.[107] Unfortunately, its application for biomedicines is restricted due to its poor ability to penetrate the cellular membrane. Interestingly, researchers have fabricated multifunctional nanoplatforms for the delivery of (KLAKLAK)$_2$ across the cell membrane for improved and efficacious tumour therapy.[108] For example, self-delivery system PpIX-PEG-(KLAKLAK)$_2$ was developed by Han and co-workers (2015) to achieve mitochondria-targeted photodynamic tumour therapy.[109] Within this system, by using a short PEG linker, Protoporphyrin IX (PpIX) was conjugated to (KLAKLAK)$_2$ for efficient cellular uptake. Upon its cellular internalization, the (KLAKLAK)$_2$ proapoptotic peptide targets and disrupts the mitochondrial membrane to induce cell death by apoptosis. Similarly, PpIX as a photosensitizer produces reactive oxygen species (ROS) to destroy mitochondria for optimum antitumoral effects. Using this strategic mitochondria-targeted self-delivery system with the proapoptotic peptide (KLAKLAK)$_2$, a new horizon has opened in the fabrication of therapeutic peptides for the eradication of tumours.

Another peptide with therapeutic ability that has gained wide application is an amphipathic melittin peptide made up of 26 amino acids. This cationic peptide (GIGAVLKVLTTGLPALISWIKRKRQQ-NH2) has its source from bee venom.[110] By attaching itself to the phospholipid bilayer on the membrane, the melittin peptide is able to disrupt the cell membrane and produce transmembrane pores to enhance the permeability of the membrane and thereby facilitate cell lyses.[111] While melittin exhibits a viable and effective anticancer property, its in vivo applications are often restricted by the haemolysis effect.[112] A strategic way to circumvent melittin's haemolytic effect prior to its arrival to the tumour site is to deeply hide the cationic amino acid of melittin by conjugating it with an amphipathic α-helical peptide.[113] The use melittin as an anticancer agent in an in vivo experiment showed its therapeutic efficacy but with some side effects. Notably, upon its encapsulation into diverse nanoplatforms, its therapeutic efficiency improved for treating tumour growth.[114]

Another potent therapeutic peptide is the autophagy-inducing peptide beclin-1. This peptide has its source from the coiled-coil beclin-1 protein with a 60 kDa. Interestingly, its precursor, coiled-coil beclin-1 protein, has the potential to interact with the apoptosis inhibitor Bcl-2, which plays a critical role in autophagy regulation.[115] Previous reports have shown that beclin-1 peptide, when bound to P13Ks, can effectively stimulate autophagy. Interestingly, P12Ks are key principal players as initiators in autophagosome formation.[116] Thus, the autophagy-inducing peptide beclin-1 can be employed for the induction of tumour cells autophagy thereby facilitating autophagic cell death. Based on this characteristic, it was possible to conjugate beclin-1 peptide to several nanoparticulate systems to induce autophagic tumour cell death.[117]

Quite often, most therapeutic peptides exert their therapeutic potentials by perturbing tumour cells' biological activities, and there are some other specific peptides that can interact with immune cells to stimulate an antitumour immune response. One of such peptide is the formyl peptide receptors (FPRs), which are expressed abundantly and selectively on immune cells and in several metastatic tumour cells.[118] In the meantime, two specific ligands that target FPR1/2 (annexin-1, WKYMVm (m: d-Met), and fMLFK (f: formyl)), are reported as strong immunostimulators with the potential to switch on native immune cells and enhance immuno-mediated tumour treatment.[119] For instance, the prodrug cisplatin was conjugated to FPR1/2 formyl peptide to target FPR1/2-overexpressing cancer and immune cells and further instigate an antitumoral immune response for both immunotherapy and chemotherapy.[120] In short, there are promising potentials to achieve excellent therapeutic anticancer intervention with these immunostimulant peptides for the regulation of immune responses. Also, enhanced therapeutic efficacies could be harnessed when combined with other conventional therapeutic strategies.

15.13 TUMOUR-SPECIFIC PEPTIDES FOR CANCER THERANOSTICS

With the advent of combinatorial library technologies, an enormous number of structurally diverse peptides have been screened and identified. These methods allow for the discovery of structural peptides with high affinity and targeting potential.[121-123] These targeting peptides with varied sequences are broadly classified into three targeting groups based on their locations and functions: (i) tumour cell surface receptor targeting, (ii) tumour microenvironment targeting (tumour extracellular matrix (ECM) targeting, tumour vasculature targeting and tumour-associated cell targeting) and (iii) subcellular organelle targeting (cell nuclei, plasma membrane, mitochondrial, etc.). In this section, we present some of the interventions using therapeutic peptides for anticancer applications as summarized in Figure 15.2 and listed in Table 15.1.

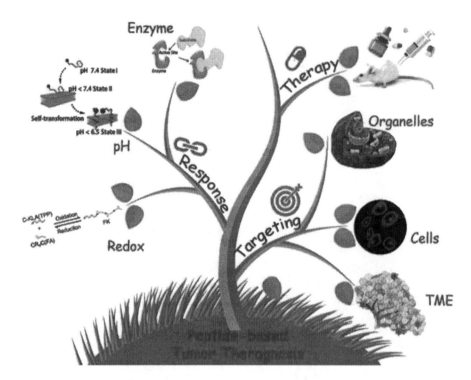

FIGURE 15.2 Schematic representation of diverse biological functions of peptides for tumour theragnosis, principally categorized into three functions, targeting, response and therapeutic applications, in the fabrication of peptide-based theranostic nanoplatforms. Figure reproduced with permission.[6] These functional peptides were employed to synthesize novel multifunctional biomaterial theranostic nanoplatforms with promising potential for tumour diagnosis and therapy.

15.13.1 Tumour Cell Surface Receptor Targeting

According to classification by the "hallmarks of cancer", there are six unique differences between tumour cells and normal cells. These differences are (i) prolonged and sustained proliferation signalling, (ii) the ability to evade growth suppressors, (iii) activating invasion and metastasis, (iv) enabling replicative immortality, (v) inducing angiogenesis and (vi) resisting cell death.[124, 125] These hallmarks give us far more enriching information about the devastating potential of tumour cells accrued through fast proliferation and mutation in comparison to normal cells. Also, tumour cells have the capability to overexpress enormous numbers of unique proteins resident on their cell membrane or secreted out of cells to facilitate the proliferation, invasion, migration and metastasis of tumour cells.[126] Interestingly, these overexpressed cell surface proteins allow for active interactions with prospective ligands for targeted tumour diagnosis and treatment.[127] It is possible to engineer peptides through the efficient screening provided by combinatorial library techniques to specifically bind and interact with the overexpressed cell surface proteins. For clinical applications of tumour targeting, these overexpressed proteins on the cancer cell surface have been explored in a number of investigations and include neuropilin-1 (NRP-1) receptors,[128] protein tyrosine phosphatase receptor type J (PTPRJ),[129] the epidermal growth factor receptor (EGFR),[130] the integrin αvβ6 receptor,[131] the insulin-like growth factor 1 receptor (IGF1R),[132] low-density lipoprotein receptor-related protein 1 (LRP1),[133] the transferrin receptor (Tf-R)[134] and the asialoglycoprotein receptor (ASGPR),[135] among others. Notably, specific receptors are overexpressed on different tumours, and

TABLE 15.1

Concise Summary of Representative Peptides Highlighting Their Different Types and Sequences, Respective Targeting/Responsive Sites, Unique Cellular Functions, Different Therapeutic Strategies and Distinct Imaging Methods

Type of Peptide	Peptide Sequences	Targeting/Responsive Sites	Cellular Functions	Therapeutic Strategies	Imaging Techniques	Ref.
Tumour-targeting peptides						
Tumour cell-targeting peptides	CDTFPYLGWWNPNEYRY	HER2	Epidermal growth factor receptor targeting	PDT	FL/PET[b]/CT[c]/MRI	148, 149
	iRGD	NRP-1, integrin $\alpha_v\beta_3/\alpha_v\beta_5$		Chemotherapy/PDT	FL	150, 151
	RPARPAR	Neuropilin-1	Neuropilin-1 targeting	NA	FL[a]	152
	TFFYGGSRGKRNNFKTEEY	LRP1	Low-density lipoprotein receptor-related protein 1 targeting	Chemotherapy/immunotherapy	MRI/FL/SERRS	153, 154
	NAVPNLRGDLQVLAQKVART			Immunotherapy	CT	155
	RGDLATLRQLAQEDGVVGVRK	Integrin $\alpha v\beta 6$	Integrin $\alpha_v\beta_6$ targeting	Surgery/PDT/immunotherapy	FL/PET/CT	156, 157
Tumour microenvironment-targeting peptides						
	CGLIIQKNEC/CNAGESSKNC			Chemotherapy	FL/CT	158, 159
	RGD	Integrin $\alpha_v\beta_3$, $\alpha_v\beta_5$	Tumour vasculature targeting	Chemotherapy/PDT$_{PTT}$e	FL/MRI	160–162
	CREKA	Fibronectin	Tumour extracellular matrix targeting	NA	MRI/FL	163, 164
	YEQDPWGVKWWY (M2pep)	M2 macrophages	Tumour-associated M2 macrophages targeting	Immunotherapy	FL	165, 166
	GPA	Fibroblast-activation protein-α (FAP-α)	Tumour-associated fibroblasts targeting	Chemotherapy	FL	167, 168
Subcellular organelle-targeting peptides						
	C_{16}-K(PpIX)RRKK-PEG$_8$	Cell plasma membrane	Plasma membrane targeting	PDT	FL	169–171
	AEQNPIYWARYADWLFTTPLLLLD-LALLVDADEGT (pHLIP)	Cell plasma membrane	Plasma membrane targeting	PDT	FL/CT	172
	C_{16}-K(R)-R$_4$	Cell plasma membrane	Plasma membrane targeting	Chemotherapy/PDT	FL	173
	C_{16}-K(TPE)-GGGH-GFLGK-PEG$_8$	Cell plasma membrane	Plasma membrane targeting	Chemotherapy	FL	174
	CGGGPKKKRKVGG	Cell nucleus	Nucleus targeting	NA	FL/MRI	175
	YGRKKRRQRRR (TAT)	Cell nucleus	Nucleus targeting	Chemotherapy/PDT	FL/MRI	176
	PKKKRKV	Cell nucleus	Nucleus targeting	PDT	FL	177
	FrFKFrFK	Mitochondria	Mitochondria targeting	NA	FL	105, 178

(Continued)

TABLE 15.1 (Continued)
Concise Summary of Representative Peptides Highlighting Their Different Types and Sequences, Respective Targeting/Responsive Sites, Unique Cellular Functions, Different Therapeutic Strategies and Distinct Imaging Methods

Type of Peptide	Peptide Sequences	Targeting/Responsive Sites	Cellular Functions	Therapeutic Strategies	Imaging Techniques	Ref.
Tumour microenvironment–responsive peptides						
Enzyme-responsive peptides	GGKGPLGLPG	MMP-9	Matrix metalloprotease responsive	NA	FL/MRI	179
	GPLGIAG	MMP-2	Matrix metalloprotease responsive	Chemotherapy	FL	180
	PLGLA	MMP-2	Matrix metalloprotease responsive	PDT/PTT/chemotherapy	FL/PAI[d]	181
	PLGVR	MMP-2	Matrix metalloprotease responsive	Chemotherapy/PDT/gene therapy	FL	182
	DEVD	Caspase-3	Caspase responsive	PDT/chemotherapy/immunotherapy	FL	183–185
	GRRGKGG	Cathepsin B	Cathepsin responsive	NA	FL	186
	KGRR	Cathepsin B	Cathepsin responsive	NA	FL	187
	GFLG	Cathepsin B	Cathepsin responsive	Chemotherapy/PDT	FL	188
	CKF	Cathepsin B	Cathepsin responsive	PDT	FL/PAI	189
	PTN	Legumain	Legumain responsive	NA	FL	190
	LNAAKKKK	Legumain	Legumain responsive	NA	FL/MRI	191
	AAN	Legumain	Legumain responsive	Chemotherapy	FL/MRI	192
pH-responsive peptides	VKVSVKVSVKVSVKVSE	Acid	Acid responsive	NA	NA	193
	pHLIP	Acid	Acid responsive	PDT/gene therapy	FL/CT	194
Redox-responsive peptides	R-S-S-R (disulphide, R refers to any groups)	GSH	GSH responsive	Chemotherapy/gene therapy	FL	195
	DYF, DFY, YFD, FYD, YDF, FDY	Tyrosinase	Enzymatic oxidation	NA	NA	196
Therapeutic peptides						
Immunostimulating peptides	Annexin-1, WKYMVm, and fMLFK	Formyl peptide receptors	Potent immunostimulators	Chemotherapy/immunotherapy	FL	118, 119
Autophagy-inducing peptide	CGTNVFNATFHIWHSGQFGT (beclin-1)	PI3Ks	Inducing autophagy	Chemotherapy	FL	115–117
Venom peptides	KLAKLAKKLAKLAK	Mitochondrial membranes	Inducing cell apoptosis	Chemotherapy/PDT	FL	197, 198
	GIGAVLKVLTTGLPALISWIKRKRQQ	Cell membranes	Disrupting cell membranes	Chemotherapy/immunotherapy	FL	111–114, 199

[a] Fluorescence (FL) imaging; [b] Positron emission tomography (PET) imaging; [c] X-ray computed tomography (CT) imaging; [d] Photoacoustic (PA) imaging; [e] Photothermal therapy (PTT), Not applicable (NA)

these allow for the development of diverse techniques that explore the application of peptide-based targeting for precision in tumour imaging and efficacious therapy.

In addition, it is possible for some tumour types to secrete other active receptors, which will give further clues to the discovery of more superior targeting ligands for clinical use or academic research.[121] An ideal strategy for the selection of tumour-targeting peptides for interaction with overexpressed proteins on cell surface is premised on the understanding of the abnormal biological activities these proteins perform in tumour cells such as fast proliferation, intense metabolic reactions and erratic invasion.[136] These cell surface proteins are mainly grouped as cell surface receptors, cell adhesion molecules and growth factor receptors. A concise summary of these tumour cell surface receptors, corresponding targeting peptides and cellular functions is captured in Table 15.1.

15.13.2 Tumour Microenvironment–Targeting Peptides

It is possible to design and fabricate different multifunctional nanoplatforms to precisely and efficiently deliver drugs or therapeutics to tumour sites through the integration of tumour-targeting peptides overexpressed on tumour cells.[137] A major fundamental factor that allowed for the targeting potential of tumour-targeting peptides is squarely based on the difference between the tissue microenvironments of tumour and normal cells.[138] Usually, what constitutes the whole microenvironment of tumour includes the tumour cells in addition to their surrounding cells, soluble factors and extracellular matrix.[139] These surrounding cells, often referred to as tumour-associated neighbouring cells, are made up of a number of other cellular molecules, including endothelial cells, fibroblasts, blood and lymphatic vasculatures, pericytes, muscle cells and diverse immune cells (e.g., macrophages, T, B an natural killer cells, neutrophils, dendritic cells and mast cells).[140] Notably, all of these tumour-associated cells can potentially interact with tumour cells either directly or indirectly through the release of cytokines, enzymes and growth factors to influence the growth of the tumour and regulate its microenvironment.[141] As such, in addition to tumour cells, the tumour microenvironment can be a promising and strategic target site to explore for tumour diagnosis and therapy. Thus, the unique tumour microenvironment allows tumour-targeting peptides to exhibit effective targeting potential and therapeutic efficacy. In brief, three peptide categories constitute the tumour-microenvironment–targeting peptides: tumour vasculature, tumour extracellular matrix and tumour associated cell-targeting peptides.

15.13.3 Subcellular Organelle–Targeting Peptides

It is evidently true that several diseases are caused from mutations or alterations at the subcellular level.[142] With the advent of precision medicine tools, tumour theranostics targets the subcellular organelles to optimize drug bioavailability, improve therapeutic efficacy and reduce the potential toxic side effects of conventional chemotherapeutics.[143] Conversely, there are natural subcellular obstacles inherent within most subcellular organelles and membranous structures that prevent easy access for these therapeutic actives to elicit their effects. Thus, it is imperative to develop an efficient subcellular organelle-targeting technique to precisely target the active sites within the intracellular organelles for therapeutic response. On a general note, vital subcellular organelles are crucial intracellular active sites responsible for substance metabolism and cellular energy. Notable among these subcellular organelles are the mitochondria, lysosomes, plasma membrane, nucleus and endoplasmic reticula.[144] It is noteworthy that both tumour and normal cells have varied and advanced membranous organelles with unique and specific functions to play in most intercellular activities. These organelles within both tumour and normal cells participate in essential cellular activities such as cell growth, cell proliferation, cell differentiation and cell death.[145] There are huge benefits for efficient therapeutic efficacy and precision tumour treatment when tumour cell organelles are specifically targeted. For instance, when the nucleus is targeted, a number of key metabolic processes are directly and intrinsically altered, such as control of DNA replication and transcription, genetic programming interference, regulation of cellular signalling transduction pathways and the fate of cells.[146] Using this approach of nuclear transporting and targeting, it is possible to treat and cure different types of diseases including degenerative diseases, infections, leukaemia, cancer and inflammatory conditions.[147] Summarily, subcellular targeting strategies are mainly tailored toward three subcellular organelles, the plasma membrane, nuclear and mitochondria.

15.14 NANOPARTICLE-BASED DRUG DELIVERY SYSTEMS FOR CANCER THERANOSTICS

15.14.1 Nanoparticulate Theranostic Systems for Tumour Targeting

Three important parameters of nanoparticles employed for medical treatments influence their efficiency as nano-drug delivery cargoes and in turn dictate their therapeutic efficacy: their specific sizes, unique shapes and surface structural architectures.[200] Generally, nanoparticles with sizes less than 100 nm are considered appropriate for cancer therapy because of their ability to effectively deliver their payloads and achieve the enhanced permeability and retention (EPR) effect.[201] It is possible for particles with smaller sizes, below 2 nm, to leak from the normal vasculature, thereby damaging the healthy cells, and those with diameter size less than 10 nm can readily be filtered through renal clearance.[202] Meanwhile, those with sizes beyond 100 nm are exposed to the clearance by phagocytosis from the systemic circulation.[203]

Nanoparticle shape also affects their cellular update and internalization. Within the tumour microenvironment, spherical nanoparticles have been reported to have enhanced cellular uptake.[204] Nanoparticles with rigid morphology and high aspect ratio are known to aggregate more steadily in macrophages compared to those possessing small and flexible morphology, which prolongs their half-life within the blood and ultimately decreases their clearance from the blood circulation.[9] Nanomaterials with the same chemical composition have been shown to exhibit different cellular toxicities based on the differences in their shapes. For instance, silica nanoparticles and nanowires behaved differently in cells. Notably, silica nanoparticles were shown to be less cytotoxic at higher concentrations, whereas silica nanowires were highly toxic at the same concentrations in an in vivo experiment using two different human epithelial cell lines.

In the meantime, nanoparticle surface architecture can influence their bioavailability and circulation time within the biological system. Employing the PEGylation chemistry on nanoparticles will reduce their opsonization and thereby escape clearance by the immune system.[205] As such, encapsulating nanoparticles with hydrophilic materials like polyethylene glycol (PEG), increases their in vivo circulation, improves their internalization and accumulation in tumours to deliver their payloads.[206] Put together, the therapeutic efficacies of nanoparticles for cancer treatment are largely influenced by their different characteristics. Different nanoparticle types for cancer nanomedicines are highlighted in Figure 15.3.

15.14.2 Exosomes as Nanoplatforms for Cancer Theranostics

Recently, the use of exosomes as potential and promising nanomaterials for cancer diagnosis and therapy has gained credible momentum. As endogenous bodies, these nanostructures are produced by diverse cells and taken in by recipient cells. Exosomes possess unique structural and compositional characteristics that confer on them low cytotoxicity and innate potential to escape immune monitoring and overcome biological hurdles. Notable among their excellent characteristic potential are their ability to stabilize their encapsulated agents, including proteins, nucleic acids or other therapeutic actives. They are able to navigate through cell membranes and deliver their payloads to the active site.[207] More importantly, residents upon the exosomal membrane are important molecular signatures such as diverse molecular proteins, nucleic acids, cytokines, proinflammatory factors and transcription factor receptors which enable exosomes to take part in several cellular activities.[208–210]

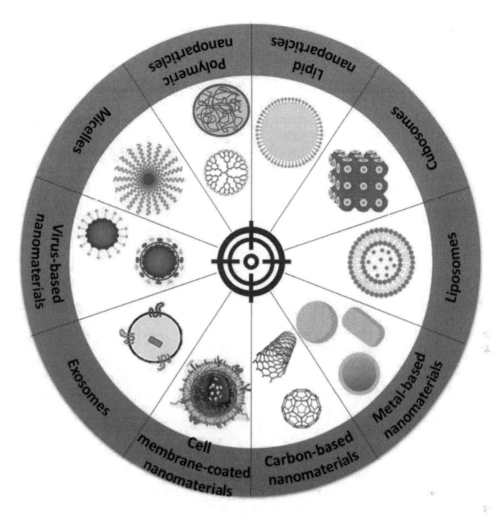

FIGURE 15.3 Schematic representation showing the different types of nanomaterials employed for the fabrication of peptide-based tumour-targeting applications. Figure re-printed with permission.[201]

Wang and co-workers (2021) explored the vast potential of nanostructures and produced a combined application of functionalized exosomes and chemo/gene/photothermal therapy.[208] In their experiment, exosomes were employed to encapsulate DOX and surface modified with magnetic nanoparticles which were subsequently linked with molecular signatures that have the potential to target miR-21 for responsive molecular imaging. These designed nanocarriers are tailored to the tumour site using an external magnetic field and trigger localized hyperthermia under NIR irradiation and deliver DOX as their payload. This multifunctional nanoplatform showed exceptional therapeutic efficacy on tumour size reduction up to 97.57%. This strategy is proposed to be the next paradigm shift in precision cancer nanomedicines.

Similar to this procedure, the work of Kwon and colleagues (2022) resulted in the fabrication of a novel nanoplatform for colorectal cancer therapy.[209] Within this system, isolated exosomes from the tumour cells were used to encapsulate DOX, functionalized with folate and coated with magnetic nanoparticles. The fabricated theranostic nanosystem exhibited enhanced apoptotic efficacy and improved inhibition of tumour growth for cancer therapy.

In an alternative investigation, Pei and co-workers (2021) employed a different strategy against colorectal cancer.[211] Their technique involved a combinatorial therapeutic procedure for the dual inhibition of FGL1 and TGF-β1 that allows for the simultaneous blockage of the immune checkpoint and modulate the tumour microenvironment. In this regard, they successfully fabricated a cRGD-modified exosome nanosystem ladened with high siFGL1 and siTGF-β1 loading efficiency, which enhanced the amount of infiltrated CD8+ T tumour cells while reducing the amount of immunosuppressive cells.

Conversely to the strategy of Pei and colleagues, Zhou and co-workers (2021) developed an exosome-based delivery nanoplatform to improve the immunotherapy of pancreatic ductal adenocarcinoma and reverse the tumour immunosuppression of M2-like TAMs by disrupting the galectin-9/dectin 1 axis.[212] By using exosomes derived from the bone marrow of mesenchymal stem cells, galectin-9 siRNA was encapsulated and coated with oxaliplatin prodrug. The synthesized nanoplatform showed enhanced tumour targeting efficiency and exhibited potent antitumoral efficacies through the suppression of tumour microphage polarization, the recruitment of cytotoxic T lymphocytes and regulatory T cell (Treg) downregulation.

Long non-coding RNA (lncRNA) MEG3 was encapsulated using exosome technology to target osteosarcoma in a more recent experiment by Huang and co-workers (2022).[213] In particular, c(RGDyK) peptide-modified exosomes were used to load MEG3 to effectively deliver their payload to target bone cancer cells in vitro and in vivo. As such, the fabricated nanosystem has viable therapeutic efficacy for osteosarcoma.

Furthermore, another innovative exosome-based nanoplatform was developed to deliver DOX to treat glioma.[213] A bioinspired neutrophil-exosome loaded with inflammatory chemotaxis with excellent penetrating potential against the blood brain barrier (BBB) was produced. This nanocarrier showed viable capability for the clinical management of glioma as well as other solid tumour or brain diseases.

The work of Zhu and colleagues (2021) synthesized novel theranostics nanovesicles made up of manganese carbonyl-loaded exosome as a radiosensitizer.[214] These nanoparticulate systems enhanced the efficient controlled release of carbon monoxide, followed by the generation of ROS under the influence of X-ray irradiation to facilitate the inhibition of tumour growth under very low-dose radiotherapy.

Last, ^{99m}Tc-radiolabel HER2-targeted exosomes were fabricated by Molavipordanjani and colleagues (2020) for tumour imaging.[215] This novel nanoplatform displayed increased affinity for SKOV-3 cells when compared to MCF-7, HT29, U87-MG or A549 cell lines. As such, the nanosystem demonstrated selective targeting for ovarian adenocarcinoma. Notably, there was increased accumulation of the irradiated nanoparticles at the active site and excellent visualization of the tumour in SKOV-3 tumour-bearing nude mouse models.

15.15 THERANOSTICS: THE PARADIGM SHIFT IN CANCER THERAPY

Theranostics nanoparticles can be defined as a set of multifunctional nanoplatforms, succinctly designed and well crafted for tailored and precision disease management through a synergistic combinatorial diagnostic and therapeutic potential together in one singular biodegradable and biocompatible nanovector.[216] An ideal, viable and potent theranostic nanoplatform must exhibit the following potentials: (i) rapid and selective accumulation at the desired active sites, (ii) ability to investigate and provide feedback on the morphologic and biochemical properties of diseases, (iii) precise and on-target delivery of a therapeutic dose of its payloads without any deleterious impact on healthy cells and (iv) rapid clearance from the living system or on-point biodegradation into non-toxic excretable by-products.[217] Despite the design and fabrication of several kinds of theranostic nanoparticulate systems (both organic and inorganic) over the last decade for cancer therapy, there none among them has met all these requirements until now.

A vast number of preclinical and ongoing clinical trials have demonstrated the potential capabilities of tumour active targeting using peptide-based or antibody-bound imaging nanoparticles and chemotherapeutics.[218, 219] Also, the use of active targeting has been shown to be valuable, particularly in the treatment of small metastases that are not well vascularized (<100 mm^3) to augment the application of enhanced permeability and retention (EPR) effect. Until now, most theranostic nanoparticles application have been based on passive targeting techniques with good potential but limited clinical outcomes. As such, the need to produce precise and effective biocompatible theranostic nanoplatforms for selective in vivo tumour targeting remains a high hurdle to cross. Within this section, we seek to explore tumour-targeted theranostics and highlight the present status, examine the challenges and offer future prospects of specific-active-tumour-theranostics nanoparticles.

On a general note, there are numerous techniques to engineer theranostic nanoparticles.[217] Such methods include (i) coupling or encapsulation of therapeutic actives like anticancer chemotherapeutics and photosensitizers to pre-formed imaging nanoparticles such as iron oxide nanoparticles, quantum dots or gold nanocages; (ii) using exiting therapeutic nanoparticles to tag imaging contrast agents such as optical or magnetic nanoparticles, fluorescent dyes or the addition of diverse radioisotopes; (iii) synergistic encapsulation of both therapeutic and imaging agents in one-pot nanoplatforms with biocompatible architecture such as ferritin peptide nanocages, polymeric- and porous silica- nanoparticles; and (iv) design and fabrication of novel nanoparticles such as [^{64}Cu] CuS, porphysomes and gold nanoshells/cages, among others with unique intrinsic and effective dual therapeutic and imaging characteristics that produce the desired outcomes. Additionally, surface PEGylation and conjugation with diverse targeting ligands will improve and prolong in vivo circulation half-life and enhance the precision tumour targeting potential of the theranostic nanoplatforms.[220]

While toxicity has been major limiting factor for the advancement of several nanoparticle-based imaging and therapeutic agents into clinical trials, progressive milestones have been achieved within the last decade, with the approval of over 35 imaging or therapeutic nanoparticles by the Food and Drug Administration (FDA) for clinical trials.[221] However, the use of theranostic nanoparticles remains in the onset stages of translational research despite the enormous efforts focused on preclinical investigations and no clinical trials to date. A viable alternative to worsen this slow momentum and fast-track the transition from preclinical to clinical applications is to engage the fabrication of theranostic nanoparticles using FDA-approved therapeutic or imaging nanoplatforms. A few such platforms, including biodegradable polymeric nanoparticles, gold nanoparticles of nanocages, iron oxide nanoparticles (presently in clinical application) and silica- and silica-gold nanoparticles, might have a greater possibility to fast-track the transitional process.

15.16 CONCLUSION AND FUTURE DIRECTIONS

Traditional strategies for tumour diagnosis and therapy have shown some remarkable outcomes, including effective tumour inhibition and broad clinical applications while also facing numerous limitations that hinder their robust efficiency, such as low drug bioavailability, unsatisfactory targeting potentials and deleterious side effects. Consequently, within the past decades, the advent of combinatorial peptide libraries has gained rapid momentum in the promotion of progressive discovery and developments of multifunctional peptide-based bio-platforms for enhanced tumour diagnosis and therapy due to the unique and excellent characteristics of bioactive peptides. It is possible to incorporate short bioactive peptides in traditional tumour diagnostic and therapeutic systems to elicit effective functions, including on-point tumour targeting and specific tumour response as well as improved tumour therapy. Within this chapter, we examined the application of peptide-based nanoplatforms for cancer theranostics. In short, therapeutic peptides, as minute constituents within multifunctional systems, contribute immensely to the actualization of various functions of the systems. They significantly influence the diagnostic and therapeutic efficacy with precise bioactivity. A number of nanomaterials, be they organic nanomaterials (natural polymers, amphiphilic polymer, dendritic polymers, semiconducting polymers, etc.); inorganic nanomaterials (metallic nanoparticles, quantum dots, mesoporous silica nanoparticles, etc.); or organic–inorganic hybrid nanomaterials, including polymer-coated inorganic nanoparticles and metal-organic frameworks, have been fabricated for application as tumour theranostic agents in recent times.

All these nanomaterials possess excellent characteristics and unique functions for tumour-targeted therapy. If properly harnessed in single multifunctional nanoplatforms, peptide-based nanosystems could have multiple functions, including specific targeting, enhanced selectivity, accurate drug release, effective tumour therapy and precise tumour diagnosis. More importantly, these all-in-one peptide-based nanosystems are capable of excellent application for tumour theranostic therapy. They would immensely advance research into antitumour therapies and accelerate translation from bench to bed. However, the transition from preclinical to clinical trials for these peptide-based nanoplatforms remain very slow, as only a few products have found application in clinical cancer health management. Unfortunately, despite the excellent progress achieved by functional peptide-based multifunctional platforms in in vitro and small animal experiments, their application in clinical tumour-related interventions is still stalled.

To conclude, peptide-based nanoplatforms for tumour theranostics represent a highly and thoroughly researched topic whose implementation in practice would represent a paradigm shift in treating cancer. If properly harnessed, it promises an effective treatment modality for the management of cancers and other related disease.

REFERENCES

1. Sung, H. et al. Global cancer statistics 2020: GLOBOCAN estimates of incidence and mortality worldwide for 36 cancers in 185 countries. *CA. Cancer J. Clin.* **71**, 209–249 (2021).
2. Hodson, R. Precision medicine. *Nature* **537**, S49–S49 (2016).
3. Rong, L., Lei, Q. & Zhang, X.-Z. Recent advances on peptide-based theranostic nanomaterials. *VIEW* **1**, 20200050 (2020).
4. Pattabiraman, V. R. & Bode, J. W. Rethinking amide bond synthesis. *Nature* **480**, 471–479 (2011).
5. Santos, R. et al. A comprehensive map of molecular drug targets. *Nat. Rev. Drug Discov.* **16**, 19–34 (2017).
6. Zhang, C. et al. Peptide-based multifunctional nanomaterials for tumor imaging and therapy. *Adv. Funct. Mater.* **28**, 1804492 (2018).
7. Sharma, R. et al. Functionalized peptide-based nanoparticles for targeted cancer nanotherapeutics: A state-of-the-art review. *ACS Omega* **7**, 36092–36107 (2022).
8. Moasses Ghafary, S. et al. Design and preparation of a theranostic peptideticle for targeted cancer therapy: Peptide-based codelivery of doxorubicin/curcumin and graphene quantum dots. *Nanomedicine Nanotechnol. Biol. Med.* **42**, 102544 (2022).
9. S. Adebowale, A., E. Choonara, Y., Kumar, P., C. du Toit, L. & Pillay, V. Functionalized nanocarriers for enhanced bioactive delivery to squamous cell carcinomas: Targeting approaches and related biopharmaceutical aspects. *Curr. Pharm. Des.* **21**, 3167–3180 (2015).
10. Ngema, L. M., Adeyemi, S. A., Marimuthu, T. & Choonara, Y. E. A review on engineered magnetic nanoparticles in non-small-cell lung carcinoma targeted therapy. *Int. J. Pharm.* **606**, 120870 (2021).
11. Adeyemi, S. A. et al. Folate-decorated, endostatin-loaded, nanoparticles for anti-proliferative chemotherapy in esophaegeal squamous cell carcinoma. *Biomed. Pharmacother.* **119**, 109450 (2019).
12. Siddique, S. & Chow, J. C. L. Recent advances in functionalized nanoparticles in cancer theranostics. *Nanomaterials* **12**, 2826 (2022).
13. Sun, X. et al. Effects of nanoparticle sizes, shapes, and permittivity on plasmonic imaging. *Opt. Express* **30**, 6051–6060 (2022).
14. Sim, S. & Wong, N. K. Nanotechnology and its use in imaging and drug delivery (Review). *Biomed. Rep.* **14**, 42 (2021).
15. Peng Wang, Y. et al. Multifunctional and multimodality theranostic nanomedicine for enhanced phototherapy. *J. Mater. Chem. B* (2023) doi:10.1039/D2TB02345H.
16. Lodhi, M. S. et al. A novel formulation of theranostic nanomedicine for targeting drug delivery to gastrointestinal tract cancer. *Cancer Nanotechnol.* **12**, 26 (2021).
17. Gou, Y. et al. Bio-inspired protein-based nanoformulations for cancer theranostics. *Front. Pharmacol.* **9**, (2018).

18. Ye, Y., Yu, J. & Gu, Z. Versatile protein nanogels prepared by in situ polymerization. *Macromol. Chem. Phys.* **217**, 333–343 (2016).
19. Ye, Y. *et al.* A melanin-mediated cancer immunotherapy patch. *Sci. Immunol.* **2**, eaan5692 (2017).
20. Ye, Y. *et al.* Synergistic transcutaneous immunotherapy enhances antitumor immune responses through delivery of checkpoint inhibitors. *ACS Nano* **10**, 8956–8963 (2016).
21. Karimi, M. *et al.* Albumin nanostructures as advanced drug delivery systems. *Expert Opin. Drug Deliv.* **13**, 1609–1623 (2016).
22. Maier, R., Fries, M. R., Buchholz, C., Zhang, F. & Schreiber, F. Human versus bovine serum albumin: A subtle difference in hydrophobicity leads to large differences in bulk and interface behavior. *Cryst. Growth Des.* **21**, 5451–5459 (2021).
23. Majorek, K. A. *et al.* Structural and immunologic characterization of bovine, horse, and rabbit serum albumins. *Mol. Immunol.* **52**, 174–182 (2012).
24. Sugio, S., Kashima, A., Mochizuki, S., Noda, M. & Kobayashi, K. Crystal structure of human serum albumin at 2.5 Å resolution. *Protein Eng. Des. Sel.* **12**, 439–446 (1999).
25. Chen, Q. *et al.* Near-infrared dye bound albumin with separated imaging and therapy wavelength channels for imaging-guided photothermal therapy. *Biomaterials* **35**, 8206–8214 (2014).
26. Chen, Q. *et al.* An albumin-based theranostic nano-agent for dual-modal imaging guided photothermal therapy to inhibit lymphatic metastasis of cancer post surgery. *Biomaterials* **35**, 9355–9362 (2014).
27. Han, H. *et al.* Enzyme-sensitive gemcitabine conjugated albumin nanoparticles as a versatile theranostic nanoplatform for pancreatic cancer treatment. *J. Colloid Interface Sci.* **507**, 217–224 (2017).
28. Rong, P. *et al.* Protein-based photothermal theranostics for imaging-guided cancer therapy. *Nanoscale* **7**, 16330–16336 (2015).
29. Lisitskiy, V. A. *et al.* Multifunctional human serum albumin-therapeutic nucleotide conjugate with redox and pH-sensitive drug release mechanism for cancer theranostics. *Bioorg. Med. Chem. Lett.* **27**, 3925–3930 (2017).
30. Peer, D. *et al.* Nanocarriers as an emerging platform for cancer therapy. *Nat. Nanotechnol.* **2**, 751–760 (2007).
31. Sim, C. *et al.* Photoacoustic-based nanomedicine for cancer diagnosis and therapy. *J. Controlled Release* **203**, 118–125 (2015).
32. Mo, Y., Barnett, M. E., Takemoto, D., Davidson, H. & Kompella, U. B. Human serum albumin nanoparticles for efficient delivery of Cu, Zn superoxide dismutase gene. *Mol. Vis.* **13**, 746–757 (2007).
33. Li, X. *et al.* Human transport protein carrier for controlled photoactivation of antitumor prodrug and real-time intracellular tumor imaging. *Bioconjug. Chem.* **26**, 955–961 (2015).
34. Fürst, W. & Banerjee, A. Release of glutaraldehyde from an albumin-glutaraldehyde tissue adhesive causes significant in vitro and in vivo toxicity. *Ann. Thorac. Surg.* **79**, 1522–1528 (2005).
35. Hu, D. *et al.* Activatable albumin-photosensitizer nanoassemblies for triple-modal imaging and thermal-modulated photodynamic therapy of cancer. *Biomaterials* **93**, 10–19 (2016).
36. Sheng, Z. *et al.* Smart human serum albumin-indocyanine green nanoparticles generated by programmed assembly for dual-modal imaging-guided cancer synergistic phototherapy. *ACS Nano* **8**, 12310–12322 (2014).
37. Chen, Q., Liang, C., Wang, C. & Liu, Z. An imageable and photothermal "Abraxane-like" nanodrug for combination cancer therapy to treat subcutaneous and metastatic breast tumors. *Adv. Mater.* **27**, 903–910 (2015).
38. Chen, Q. *et al.* Drug-induced self-assembly of modified albumins as nano-theranostics for tumor-targeted combination therapy. *ACS Nano* **9**, 5223–5233 (2015).
39. Jia, Q. *et al.* Biocompatible iron phthalocyanine—albumin assemblies as photoacoustic and thermal theranostics in living mice. *ACS Appl. Mater. Interfaces* **9**, 21124–21132 (2017).
40. Moon, H. *et al.* Multifunctional theranostic contrast agent for photoacoustics- and ultrasound-based tumor diagnosis and ultrasound-stimulated local tumor therapy. *J. Controlled Release* **218**, 63–71 (2015).
41. Wen, L., Yang, S., Zhong, J., Zhou, Q. & Xing, D. Thermoacoustic imaging and therapy guidance based on ultra-short pulsed microwave pumped thermoelastic effect induced with superparamagnetic iron oxide nanoparticles. *Theranostics* **7**, 1976–1989 (2017).
42. Gao, F.-P. *et al.* Supramolecular adducts of squaraine and protein for noninvasive tumor imaging and photothermal therapy in vivo. *Biomaterials* **35**, 1004–1014 (2014).
43. Chen, Q. & Liu, Z. Albumin carriers for cancer theranostics: A conventional platform with new promise. *Adv. Mater.* **28**, 10557–10566 (2016).

44. Lee, C. et al. An albumin nanocomplex-based endosomal pH-activatable on/off probe system. *Colloids Surf. B Biointerfaces* **144**, 327–334 (2016).
45. Pan, J. et al. Mimicking drug—substrate interaction: A smart bioinspired technology for the fabrication of theranostic nanoprobes. *Adv. Funct. Mater.* **27**, 1603440 (2017).
46. Zhou, L. et al. Size-Tunable Gd_2O_3@albumin nanoparticles conjugating chlorin e6 for magnetic resonance imaging-guided photo-induced therapy. *Theranostics* **7**, 764–774 (2017).
47. Sheng, Z. et al. Protein-assisted fabrication of nano-reduced graphene oxide for combined in vivo photoacoustic imaging and photothermal therapy. *Biomaterials* **34**, 5236–5243 (2013).
48. Liu, Y. et al. Artificially controlled degradable inorganic nanomaterial for cancer theranostics. *Biomaterials* **112**, 204–217 (2017).
49. Yang, W. et al. Albumin-bioinspired Gd: CuS nanotheranostic agent for in vivo photoacoustic/magnetic resonance imaging-guided tumor-targeted photothermal therapy. *ACS Nano* **10**, 10245–10257 (2016).
50. Chen, L. et al. Marriage of albumin—gadolinium complexes and mos2 nanoflakes as cancer theranostics for dual-modality magnetic resonance/photoacoustic imaging and photothermal therapy. *ACS Appl. Mater. Interfaces* **9**, 17786–17798 (2017).
51. Wei, K.-C. et al. 1,3-Bis(2-chloroethyl)-1-nitrosourea-loaded bovine serum albumin nanoparticles with dual magnetic resonance—fluorescence imaging for tracking of chemotherapeutic agents. *Int. J. Nanomedicine* **11**, 4065–4075 (2016).
52. Battogtokh, G. & Ko, Y. T. Graphene oxide-incorporated pH-responsive folate-albumin-photosensitizer nanocomplex as image-guided dual therapeutics. *J. Controlled Release* **234**, 10–20 (2016).
53. Chen, Q. et al. Protein modified upconversion nanoparticles for imaging-guided combined photothermal and photodynamic therapy. *Biomaterials* **35**, 2915–2923 (2014).
54. Wang, L. et al. GEM-loaded magnetic albumin nanospheres modified with cetuximab for simultaneous targeting, magnetic resonance imaging, and double-targeted thermochemotherapy of pancreatic cancer cells. *Int. J. Nanomedicine* **10**, 2507–2519 (2015).
55. Yu, J. et al. Multistimuli-regulated photochemothermal cancer therapy remotely controlled via Fe5C2 nanoparticles. *ACS Nano* **10**, 159–169 (2016).
56. Semkina, A. et al. Core—shell—corona doxorubicin-loaded superparamagnetic Fe3O4 nanoparticles for cancer theranostics. *Colloids Surf. B Biointerfaces* **136**, 1073–1080 (2015).
57. Dong, C. et al. A protein—polymer bioconjugate-coated upconversion nanosystem for simultaneous tumor cell imaging, photodynamic therapy, and chemotherapy. *ACS Appl. Mater. Interfaces* **8**, 32688–32698 (2016).
58. Sahu, A., Lee, J. H., Lee, H. G., Jeong, Y. Y. & Tae, G. Prussian blue/serum albumin/indocyanine green as a multifunctional nanotheranostic agent for bimodal imaging guided laser mediated combinatorial phototherapy. *J. Controlled Release* **236**, 90–99 (2016).
59. Huang, H. et al. pH-sensitive Au—BSA—DOX—FA nanocomposites for combined CT imaging and targeted drug delivery. *Int. J. Nanomedicine* **12**, 2829–2843 (2017).
60. An, F.-F. & Zhang, X.-H. Strategies for preparing albumin-based nanoparticles for multifunctional bioimaging and drug delivery. *Theranostics* **7**, 3667–3689 (2017).
61. Banyard, S. H., Stammers, D. K. & Harrison, P. M. Electron density map of apoferritin at 2.8-Å resolution. *Nature* **271**, 282–284 (1978).
62. Alkhateeb, A. A. & Connor, J. R. The significance of ferritin in cancer: Anti-oxidation, inflammation and tumorigenesis. *Biochim. Biophys. Acta BBA—Rev. Cancer* **1836**, 245–254 (2013).
63. Bellini, M. et al. Protein nanocages for self-triggered nuclear delivery of DNA-targeted chemotherapeutics in cancer cells. *J. Controlled Release* **196**, 184–196 (2014).
64. Truffi, M. et al. Ferritin nanocages: A biological platform for drug delivery, imaging and theranostics in cancer. *Pharmacol. Res.* **107**, 57–65 (2016).
65. Bhushan, B. et al. Ferritin nanocages: A novel platform for biomedical applications. *J. Biomed. Nanotechnol.* **10**, 2950–2976 (2014).
66. Zhen, Z. et al. Ferritin nanocages to encapsulate and deliver photosensitizers for efficient photodynamic therapy against cancer. *ACS Nano* **7**, 6988–6996 (2013).
67. Zhen, Z. et al. RGD-modified apoferritin nanoparticles for efficient drug delivery to tumors. *ACS Nano* **7**, 4830–4837 (2013).
68. Huang, P. et al. Dye-loaded ferritin nanocages for multimodal imaging and photothermal therapy. *Adv. Mater.* **26**, 6401–6408 (2014).
69. Crich, S. G. et al. Targeting ferritin receptors for the selective delivery of imaging and therapeutic agents to breast cancer cells. *Nanoscale* **7**, 6527–6533 (2015).

70. Conti, L. et al. L-Ferritin targets breast cancer stem cells and delivers therapeutic and imaging agents. *Oncotarget* **7**, 66713–66727 (2016).
71. Turino, L. N. et al. Ferritin decorated PLGA/paclitaxel loaded nanoparticles endowed with an enhanced toxicity toward mcf-7 breast tumor cells. *Bioconjug. Chem.* **28**, 1283–1290 (2017).
72. Elzoghby, A. O. Gelatin-based nanoparticles as drug and gene delivery systems: Reviewing three decades of research. *J. Controlled Release* **172**, 1075–1091 (2013).
73. Sahoo, N., Sahoo, R. Ku., Biswas, N., Guha, A. & Kuotsu, K. Recent advancement of gelatin nanoparticles in drug and vaccine delivery. *Int. J. Biol. Macromol.* **81**, 317–331 (2015).
74. Ninan, G., Jose, J. & Abubacker, Z. Preparation and characterization of gelatin extracted from the skins of rohu (*Labeo rohita*) and common carp (*Cyprinus carpio*). *J. Food Process. Preserv.* **35**, 143–162 (2011).
75. Foox, M. & Zilberman, M. Drug delivery from gelatin-based systems. *Expert Opin. Drug Deliv.* **12**, 1547–1563 (2015).
76. Nezhadi, S. H., Choong, P. F. M., Lotfipour, F. & Dass, C. R. Gelatin-based delivery systems for cancer gene therapy. *J. Drug Target.* **17**, 731–738 (2009).
77. Tran, T. T.-D., Tran, P. H.-L., Yoon, T.-J. & Lee, B.-J. Fattigation-platform theranostic nanoparticles for cancer therapy. *Mater. Sci. Eng. C* **75**, 1161–1167 (2017).
78. Li, W.-M., Chen, S.-Y. & Liu, D.-M. In situ doxorubicin—CaP shell formation on amphiphilic gelatin—iron oxide core as a multifunctional drug delivery system with improved cytocompatibility, pH-responsive drug release and MR imaging. *Acta Biomater.* **9**, 5360–5368 (2013).
79. Cheng, Z. et al. Gelatin-encapsulated iron oxide nanoparticles for platinum (IV) prodrug delivery, enzyme-stimulated release and MRI. *Biomaterials* **35**, 6359–6368 (2014).
80. Tran, T. T.-D., Tran, P. H.-L., Amin, H. H. & Lee, B.-J. Biodistribution and in vivo performance of fattigation-platform theranostic nanoparticles. *Mater. Sci. Eng. C* **79**, 671–678 (2017).
81. Tsai, L.-C., Hsieh, H.-Y., Lu, K.-Y., Wang, S.-Y. & Mi, F.-L. EGCG/gelatin-doxorubicin gold nanoparticles enhance therapeutic efficacy of doxorubicin for prostate cancer treatment. *Nanomed.* **11**, 9–30 (2016).
82. Suarasan, S. et al. Doxorubicin-incorporated nanotherapeutic delivery system based on gelatin-coated gold nanoparticles: Formulation, drug release, and multimodal imaging of cellular internalization. *ACS Appl. Mater. Interfaces* **8**, 22900–22913 (2016).
83. Hu, G., Chun, X., Wang, Y., He, Q. & Gao, H. Peptide mediated active targeting and intelligent particle size reduction-mediated enhanced penetrating of fabricated nanoparticles for triple-negative breast cancer treatment. *Oncotarget* **6**, 41258–41274 (2015).
84. Parkkinen, J., von Bonsdorff, L., Ebeling, F. & Sahlstedt, L. Function and therapeutic development of apo transferrin. *Vox Sang.* **83**, 321–326 (2002).
85. Gomme, P. T., McCann, K. B. & Bertolini, J. Transferrin: Structure, function and potential therapeutic actions. *Drug Discov. Today* **10**, 267–273 (2005).
86. Brandsma, M. E., Jevnikar, A. M. & Ma, S. Recombinant human transferrin: Beyond iron binding and transport. *Biotechnol. Adv.* **29**, 230–238 (2011).
87. Aisen, P., Leibman, A. & Zweier, J. Stoichiometric and site characteristics of the binding of iron to human transferrin. *J. Biol. Chem.* **253**, 1930–1937 (1978).
88. Tortorella, S. & Karagiannis, T. C. Transferrin receptor-mediated endocytosis: A useful target for cancer therapy. *J. Membr. Biol.* **247**, 291–307 (2014).
89. Dufès, C., Al Robaian, M. & Somani, S. Transferrin and the transferrin receptor for the targeted delivery of therapeutic agents to the brain and cancer cells. *Ther. Deliv.* **4**, 629–640 (2013).
90. Tros de Ilarduya, C. & Düzgüneş, N. Delivery of therapeutic nucleic acids via transferrin and transferrin receptors: Lipoplexes and other carriers. *Expert Opin. Drug Deliv.* **10**, 1583–1591 (2013).
91. Chen, M.-L., He, Y.-J., Chen, X.-W. & Wang, J.-H. Quantum-dot-conjugated graphene as a probe for simultaneous cancer-targeted fluorescent imaging, tracking, and monitoring drug delivery. *Bioconjug. Chem.* **24**, 387–397 (2013).
92. Wang, D., Li, Y., Tian, Z., Cao, R. & Yang, B. Transferrin-conjugated nanodiamond as an intracellular transporter of chemotherapeutic drug and targeting therapy for cancer cells. *Ther. Deliv.* **5**, 511–524 (2014).
93. Muthu, M. S., Kutty, R. V., Luo, Z., Xie, J. & Feng, S.-S. Theranostic vitamin E TPGS micelles of transferrin conjugation for targeted co-delivery of docetaxel and ultra bright gold nanoclusters. *Biomaterials* **39**, 234–248 (2015).
94. Sonali, R. P., et al. Transferrin liposomes of docetaxel for brain-targeted cancer applications: Formulation and brain theranostics. *Drug Deliv.* **23**, 1261–1271 (2016).

95. Jiao, Y., Sun, Y., Tang, X., Ren, Q. & Yang, W. Tumor-targeting multifunctional rattle-type theranostic nanoparticles for MRI/NIRF bimodal imaging and delivery of hydrophobic drugs. *Small* **11**, 1962–1974 (2015).
96. Peng, H. *et al*. Nuclear-targeted multifunctional magnetic nanoparticles for photothermal therapy. *Adv. Healthc. Mater.* **6**, 1601289 (2017).
97. Wang, D. *et al*. Transferrin-coated magnetic upconversion nanoparticles for efficient photodynamic therapy with near-infrared irradiation and luminescence bioimaging. *Nanoscale* **9**, 11214–11221 (2017).
98. Hou, L., Shan, X., Hao, L., Feng, Q. & Zhang, Z. Copper sulfide nanoparticle-based localized drug delivery system as an effective cancer synergistic treatment and theranostic platform. *Acta Biomater.* **54**, 307–320 (2017).
99. Wang, K. *et al*. Self-assembled IR780-loaded transferrin nanoparticles as an imaging, targeting and PDT/PTT agent for cancer therapy. *Sci. Rep.* **6**, 27421 (2016).
100. Kang, C. S., Ren, S., Sun, X. & Chong, H.-S. Theranostic polyaminocarboxylate—cyanine—transferrin conjugate for anticancer therapy and near-infrared optical imaging. *ChemMedChem* **11**, 2188–2193 (2016).
101. Zhu, M. *et al*. Indocyanine green-holo-transferrin nanoassemblies for tumor-targeted dual-modal imaging and photothermal therapy of glioma. *ACS Appl. Mater. Interfaces* **9**, 39249–39258 (2017).
102. Salvati, A. *et al*. Transferrin-functionalized nanoparticles lose their targeting capabilities when a biomolecule corona adsorbs on the surface. *Nat. Nanotechnol.* **8**, 137–143 (2013).
103. Fosgerau, K. & Hoffmann, T. Peptide therapeutics: Current status and future directions. *Drug Discov. Today* **20**, 122–128 (2015).
104. Lewis, R. J. & Garcia, M. L. Therapeutic potential of venom peptides. *Nat. Rev. Drug Discov.* **2**, 790–802 (2003).
105. Hamley, I. W. Small bioactive peptides for biomaterials design and therapeutics. *Chem. Rev.* **117**, 14015–14041 (2017).
106. Javadpour, M. M. *et al*. De novo antimicrobial peptides with low mammalian cell toxicity. *J. Med. Chem.* **39**, 3107–3113 (1996).
107. Dathe, M. *et al*. Hydrophobicity, hydrophobic moment and angle subtended by charged residues modulate antibacterial and haemolytic activity of amphipathic helical peptides. *FEBS Lett.* **403**, 208–212 (1997).
108. Adar, L., Shamay, Y., Journo, G. & David, A. Pro-apoptotic peptide-polymer conjugates to induce mitochondrial-dependent cell death. *Polym. Adv. Technol.* **22**, 199–208 (2011).
109. Han, K. *et al*. Dual-stage-light-guided tumor inhibition by mitochondria-targeted photodynamic therapy. *Adv. Funct. Mater.* **25**, 2961–2971 (2015).
110. Habermann, E. Bee and wasp venoms. *Science* **177**, 314–322 (1972).
111. Dempsey, C. E. The actions of melittin on membranes. *Biochim. Biophys. Acta* **1031**, 143–161 (1990).
112. Liu, S. *et al*. Melittin prevents liver cancer cell metastasis through inhibition of the Rac1-dependent pathway. *Hepatol. Baltim. Md* **47**, 1964–1973 (2008).
113. Huang, C. *et al*. Hybrid melittin cytolytic Peptide-driven ultrasmall lipid nanoparticles block melanoma growth in vivo. *ACS Nano* **7**, 5791–5800 (2013).
114. Jin, H. *et al*. Tumor ablation and therapeutic immunity induction by an injectable peptide hydrogel. *ACS Nano* **12**, 3295–3310 (2018).
115. Aita, V. M. *et al*. Cloning and genomic organization of beclin 1, a candidate tumor suppressor gene on chromosome 17q21. *Genomics* **59**, 59–65 (1999).
116. Pattingre, S. *et al*. Bcl-2 antiapoptotic proteins inhibit Beclin 1-dependent autophagy. *Cell* **122**, 927–939 (2005).
117. Wang, Y. *et al*. Self-assembled autophagy-inducing polymeric nanoparticles for breast cancer interference in-vivo. *Adv. Mater. Deerfield Beach Fla* **27**, 2627–2634 (2015).
118. Huang, J. *et al*. The G-protein-coupled formylpeptide receptor FPR confers a more invasive phenotype on human glioblastoma cells. *Br. J. Cancer* **102**, 1052–1060 (2010).
119. Kim, S. D. *et al*. A WKYMVm-containing combination elicits potent anti-tumor activity in heterotopic cancer animal model. *PloS One* **7**, e30522 (2012).
120. Wong, D. Y. Q., Yeo, C. H. F. & Ang, W. H. Immuno-chemotherapeutic platinum(IV) prodrugs of cisplatin as multimodal anticancer agents. *Angew. Chem.* **126**, 6870–6874 (2014).
121. Komin, A., Russell, L. M., Hristova, K. A. & Searson, P. C. Peptide-based strategies for enhanced cell uptake, transcellular transport, and circulation: Mechanisms and challenges. *Adv. Drug Deliv. Rev.* **110–111**, 52–64 (2017).

122. Eskandari, S., Guerin, T., Toth, I. & Stephenson, R. J. Recent advances in self-assembled peptides: Implications for targeted drug delivery and vaccine engineering. *Adv. Drug Deliv. Rev.* **110–111**, 169–187 (2017).
123. Sun, X. *et al*. Peptide-based imaging agents for cancer detection. *Adv. Drug Deliv. Rev.* **110–111**, 38–51 (2017).
124. Senga, S. S. & Grose, R. P. Hallmarks of cancer—the new testament. *Open Biol.* **11**, 200358 (2021).
125. Hanahan, D. & Weinberg, R. A. The hallmarks of cancer. *Cell* **100**, 57–70 (2000).
126. Shadidi, M. & Sioud, M. Selective targeting of cancer cells using synthetic peptides. *Drug Resist. Updat. Rev. Comment. Antimicrob. Anticancer Chemother.* **6**, 363–371 (2003).
127. Deckert, P. M. Current constructs and targets in clinical development for antibody-based cancer therapy. *Curr. Drug Targets* **10**, 158–175 (2009).
128. Jubb, A. M. *et al*. Neuropilin-1 expression in cancer and development. *J. Pathol.* **226**, 50–60 (2012).
129. Ostman, A., Hellberg, C. & Böhmer, F. D. Protein-tyrosine phosphatases and cancer. *Nat. Rev. Cancer* **6**, 307–320 (2006).
130. Yewale, C., Baradia, D., Vhora, I., Patil, S. & Misra, A. Epidermal growth factor receptor targeting in cancer: A review of trends and strategies. *Biomaterials* **34**, 8690–8707 (2013).
131. Desgrosellier, J. S. & Cheresh, D. A. Integrins in cancer: Biological implications and therapeutic opportunities. *Nat. Rev. Cancer* **10**, 9–22 (2010).
132. LeRoith, D. & Roberts, C. T. The insulin-like growth factor system and cancer. *Cancer Lett.* **195**, 127–137 (2003).
133. Maletínská, L. *et al*. Human glioblastoma cell lines: Levels of low-density lipoprotein receptor and low-density lipoprotein receptor-related protein. *Cancer Res.* **60**, 2300–2303 (2000).
134. Li, H. & Qian, Z. M. Transferrin/transferrin receptor-mediated drug delivery. *Med. Res. Rev.* **22**, 225–250 (2002).
135. Jain, K., Kesharwani, P., Gupta, U. & Jain, N. K. A review of glycosylated carriers for drug delivery. *Biomaterials* **33**, 4166–4186 (2012).
136. Zhao, J. *et al*. Multi-targeting peptides for gene carriers with high transfection efficiency. *J. Mater. Chem. B* **5**, 8035–8051 (2017).
137. Muntimadugu, E., Kommineni, N. & Khan, W. Exploring the potential of nanotherapeutics in targeting tumor microenvironment for cancer therapy. *Pharmacol. Res.* **126**, 109–122 (2017).
138. Ungefroren, H., Sebens, S., Seidl, D., Lehnert, H. & Hass, R. Interaction of tumor cells with the microenvironment. *Cell Commun. Signal. CCS* **9**, 18 (2011).
139. Joyce, J. A. Therapeutic targeting of the tumor microenvironment. *Cancer Cell* **7**, 513–520 (2005).
140. Turley, S. J., Cremasco, V. & Astarita, J. L. Immunological hallmarks of stromal cells in the tumour microenvironment. *Nat. Rev. Immunol.* **15**, 669–682 (2015).
141. Mbeunkui, F. & Johann, D. J. Cancer and the tumor microenvironment: A review of an essential relationship. *Cancer Chemother. Pharmacol.* **63**, 571–582 (2009).
142. Tuppen, H. A. L., Blakely, E. L., Turnbull, D. M. & Taylor, R. W. Mitochondrial DNA mutations and human disease. *Biochim. Biophys. Acta* **1797**, 113–128 (2010).
143. Qin, S.-Y., Cheng, Y.-J., Lei, Q., Zhang, A.-Q. & Zhang, X.-Z. Combinational strategy for high-performance cancer chemotherapy. *Biomaterials* **171**, 178–197 (2018).
144. Gweon, B. *et al*. Plasma effects on subcellular structures. *Appl. Phys. Lett.* **96**, 101501 (2010).
145. Evan, G. I. & Vousden, K. H. Proliferation, cell cycle and apoptosis in cancer. *Nature* **411**, 342–348 (2001).
146. Pouton, C. W., Wagstaff, K. M., Roth, D. M., Moseley, G. W. & Jans, D. A. Targeted delivery to the nucleus. *Adv. Drug Deliv. Rev.* **59**, 698–717 (2007).
147. Faustino, R. S., Nelson, T. J., Terzic, A. & Perez-Terzic, C. Nuclear transport: Target for therapy. *Clin. Pharmacol. Ther.* **81**, 880–886 (2007).
148. Geng, L. *et al*. HER2 targeting peptides screening and applications in tumor imaging and drug delivery. *Theranostics* **6**, 1261–1273 (2016).
149. Sörensen, J. *et al*. Measuring HER2-Receptor Expression In Metastatic Breast Cancer Using [68Ga] ABY-025 Affibody PET/CT. *Theranostics* **6**, 262–271 (2016).
150. Wang, Y. *et al*. Tumor-penetrating nanoparticles for enhanced anticancer activity of combined photodynamic and hypoxia-activated therapy. *ACS Nano* **11**, 2227–2238 (2017).
151. Xu, X. *et al*. ROS-responsive polyprodrug nanoparticles for triggered drug delivery and effective cancer therapy. *Adv. Mater.* **29**, 1700141 (2017).
152. Teesalu, T., Sugahara, K. N., Kotamraju, V. R. & Ruoslahti, E. C-end rule peptides mediate neuropilin-1-dependent cell, vascular, and tissue penetration. *Proc. Natl. Acad. Sci. U. S. A.* **106**, 16157–16162 (2009).

153. Gao, X. et al. Guiding brain-tumor surgery via blood—brain-barrier-permeable gold nanoprobes with acid-triggered MRI/SERRS signals. *Adv. Mater.* **29**, 1603917 (2017).
154. Qiao, C. et al. Traceable nanoparticles with dual targeting and ROS response for RNAi-based immunochemotherapy of intracranial glioblastoma treatment. *Adv. Mater. Deerfield Beach Fla* **30**, e1705054 (2018).
155. Hodgins, N. O. et al. Investigating in vitro and in vivo αvβ6 integrin receptor-targeting liposomal alendronate for combinatory γδ T cell immunotherapy. *J. Control. Release Off. J. Control. Release Soc.* **256**, 141–152 (2017).
156. Yu, X. et al. Inhibiting metastasis and preventing tumor relapse by triggering host immunity with tumor-targeted photodynamic therapy using photosensitizer-loaded functional nanographenes. *ACS Nano* **11**, 10147–10158 (2017).
157. Gao, D. et al. A near-infrared phthalocyanine dye-labeled agent for integrin αvβ6-targeted theranostics of pancreatic cancer. *Biomaterials* **53**, 229–238 (2015).
158. Pilch, J. et al. Peptides selected for binding to clotted plasma accumulate in tumor stroma and wounds. *Proc. Natl. Acad. Sci. U. S. A.* **103**, 2800–2804 (2006).
159. Zhang, B. et al. Targeting fibronectins of glioma extracellular matrix by CLT1 peptide-conjugated nanoparticles. *Biomaterials* **35**, 4088–4098 (2014).
160. Cai, Y. et al. Supramolecular 'trojan horse' for nuclear delivery of dual anticancer drugs. *J. Am. Chem. Soc.* **139**, 2876–2879 (2017).
161. Chen, J.-X., Xu, X.-D., Chen, W.-H. & Zhang, X.-Z. Multi-functional envelope-type nanoparticles assembled from amphiphilic peptidic prodrug with improved anti-tumor activity. *ACS Appl. Mater. Interfaces* **6**, 593–598 (2014).
162. Li, S.-Y. et al. A pH-responsive prodrug for real-time drug release monitoring and targeted cancer therapy. *Chem. Commun. Camb. Engl.* **50**, 11852–11855 (2014).
163. Simberg, D. et al. Biomimetic amplification of nanoparticle homing to tumors. *Proc. Natl. Acad. Sci. U. S. A.* **104**, 932–936 (2007).
164. Zhou, Z. et al. MRI detection of breast cancer micrometastases with a fibronectin-targeting contrast agent. *Nat. Commun.* **6**, 7984 (2015).
165. Ngambenjawong, C., Cieslewicz, M., Schellinger, J. G. & Pun, S. H. Synthesis and evaluation of multivalent M2pep peptides for targeting alternatively activated M2 macrophages. *J. Control. Release Off. J. Control. Release Soc.* **224**, 103–111 (2016).
166. Conde, J. et al. Dual targeted immunotherapy via in vivo delivery of biohybrid RNAi-peptide nanoparticles to tumour-associated macrophages and cancer cells. *Adv. Funct. Mater.* **25**, 4183–4194 (2015).
167. Aggarwal, S. et al. Fibroblast activation protein peptide substrates identified from human collagen I derived gelatin cleavage sites. *Biochemistry* **47**, 1076–1086 (2008).
168. Ji, T. et al. Transformable peptide nanocarriers for expeditious drug release and effective cancer therapy via cancer-associated fibroblast activation. *Angew. Chem.* **128**, 1062–1067 (2016).
169. Sato, A. K., Viswanathan, M., Kent, R. B. & Wood, C. R. Therapeutic peptides: Technological advances driving peptides into development. *Curr. Opin. Biotechnol.* **17**, 638–642 (2006).
170. Löwik, D. W. P. M., Leunissen, E. H. P., van den Heuvel, M., Hansen, M. B. & van Hest, J. C. M. Stimulus responsive peptide based materials. *Chem. Soc. Rev.* **39**, 3394–3412 (2010).
171. Qiu, W.-X. et al. A self-delivery membrane system for enhanced anti-tumor therapy. *Biomaterials* **161**, 81–94 (2018).
172. Luo, G.-F. et al. A self-transformable pH-driven membrane-anchoring photosensitizer for effective photodynamic therapy to inhibit tumor growth and metastasis. *Adv. Funct. Mater.* **27**, 1702122 (2017).
173. Li, S.-Y. et al. A versatile plasma membrane engineered cell vehicle for contact-cell-enhanced photodynamic therapy. *Adv. Funct. Mater.* **27**, 1604916 (2017).
174. Zhang, C. et al. A transformable chimeric peptide for cell encapsulation to overcome multidrug resistance. *Small* **14**, 1703321 (2018).
175. Cheng, F.-Y. et al. Stabilizer-free poly(lactide-co-glycolide) nanoparticles for multimodal biomedical probes. *Biomaterials* **29**, 2104–2112 (2008).
176. Han, S.-S. et al. Dual-pH sensitive charge-reversal polypeptide micelles for tumor-triggered targeting uptake and nuclear drug delivery. *Small Weinh. Bergstr. Ger.* **11**, 2543–2554 (2015).
177. Han, K. et al. Acidity-triggered tumor-targeted chimeric peptide for enhanced intra-nuclear photodynamic therapy. *Adv. Funct. Mater.* **26**, 4351–4361 (2016).
178. Horton, K. L., Stewart, K. M., Fonseca, S. B., Guo, Q. & Kelley, S. O. Mitochondria-penetrating peptides. *Chem. Biol.* **15**, 375–382 (2008).

179. Hou, Y. et al. Protease-activated ratiometric fluorescent probe for pH mapping of malignant tumors. *ACS Nano* **9**, 3199–3205 (2015).
180. Ji, T. et al. Designing liposomes to suppress extracellular matrix expression to enhance drug penetration and pancreatic tumor therapy. *ACS Nano* **11**, 8668–8678 (2017).
181. Li, S.-Y. et al. Protease-activable cell-penetrating peptide-protoporphyrin conjugate for targeted photodynamic therapy in vivo. *ACS Appl. Mater. Interfaces* **7**, 28319–28329 (2015).
182. Liu, Y. et al. A peptide-network weaved nanoplatform with tumor microenvironment responsiveness and deep tissue penetration capability for cancer therapy. *Adv. Mater. Deerfield Beach Fla* **27**, 5034–5042 (2015).
183. Li, S.-Y. et al. A ratiometric theranostic probe for tumor targeting therapy and self-therapeutic monitoring. *Biomaterials* **104**, 297–309 (2016).
184. Cheng, H. et al. Multi-Förster resonance energy transfer-based fluorescent probe for spatiotemporal matrix metalloproteinase-2 and caspase-3 imaging. *Anal. Chem.* **89**, 4349–4354 (2017).
185. Song, W. et al. Enhanced immunotherapy based on photodynamic therapy for both primary and lung metastasis tumor eradication. *ACS Nano* **12**, 1978–1989 (2018).
186. Ryu, J. H. et al. Non-invasive optical imaging of cathepsin B with activatable fluorogenic nanoprobes in various metastatic models. *Biomaterials* **35**, 2302–2311 (2014).
187. Shim, M. K. et al. Cathepsin B-specific metabolic precursor for in vivo tumor-specific fluorescence imaging. *Angew. Chem.* **128**, 14918–14923 (2016).
188. Yuan, Y. et al. Specific light-up bioprobe with aggregation-induced emission and activatable photoactivity for the targeted and image-guided photodynamic ablation of cancer cells. *Angew. Chem. Int. Ed Engl.* **54**, 1780–1786 (2015).
189. Ai, X. et al. In vivo covalent cross-linking of photon-converted rare-earth nanostructures for tumour localization and theranostics. *Nat. Commun.* **7**, 10432 (2016).
190. Wang, Y. et al. Protease-activatable hybrid nanoprobe for tumor imaging. *Adv. Funct. Mater.* **24**, 5443–5453 (2014).
191. Chen, Y.-J. et al. Peptide-based MRI contrast agent and near-infrared fluorescent probe for intratumoral legumain detection. *Biomaterials* **35**, 304–315 (2014).
192. Liu, Z. et al. Legumain protease-activated TAT-liposome cargo for targeting tumours and their microenvironment. *Nat. Commun.* **5**, 4280 (2014).
193. Murai, K., Higuchi, M., Kinoshita, T., Nagata, K. & Kato, K. Design of a nanocarrier with regulated drug release ability utilizing a reversible conformational transition of a peptide, responsive to slight changes in pH. *Phys. Chem. Chem. Phys. PCCP* **15**, 11454–11460 (2013).
194. Reshetnyak, Y. K., Andreev, O. A., Lehnert, U. & Engelman, D. M. Translocation of molecules into cells by pH-dependent insertion of a transmembrane helix. *Proc. Natl. Acad. Sci. U. S. A.* **103**, 6460–6465 (2006).
195. Chen, S. et al. A surface charge-switchable and folate modified system for co-delivery of proapoptosis peptide and p53 plasmid in cancer therapy. *Biomaterials* **77**, 149–163 (2016).
196. Lampel, A. et al. Polymeric peptide pigments with sequence-encoded properties. *Science* **356**, 1064–1068 (2017).
197. Han, K. et al. Dual-stage-light-guided tumor inhibition by mitochondria-targeted photodynamic therapy. *Adv. Funct. Mater.* **25**, 2961–2971 (2015).
198. Ellerby, H. M. et al. Anti-cancer activity of targeted pro-apoptotic peptides. *Nat. Med.* **5**, 1032–1038 (1999).
199. Hoshino, Y., Kodama, T., Okahata, Y. & Shea, K. J. Peptide imprinted polymer nanoparticles: A plastic antibody. *J. Am. Chem. Soc.* **130**, 15242–15243 (2008).
200. Bahrami, B. et al. Nanoparticles and targeted drug delivery in cancer therapy. *Immunol. Lett.* **190**, 64–83 (2017).
201. Niculescu, A.-G. & Grumezescu, A. M. Novel tumor-targeting nanoparticles for cancer treatment-a review. *Int. J. Mol. Sci.* **23**, 5253 (2022).
202. Venturoli, D. & Rippe, B. Ficoll and dextran vs. globular proteins as probes for testing glomerular permselectivity: Effects of molecular size, shape, charge, and deformability. *Am. J. Physiol.-Ren. Physiol.* **288**, F605–F613 (2005).
203. Decuzzi, P., Pasqualini, R., Arap, W. & Ferrari, M. Intravascular delivery of particulate systems: Does geometry really matter? *Pharm. Res.* **26**, 235–243 (2009).
204. Yao, Y. et al. Nanoparticle-based drug delivery in cancer therapy and its role in overcoming drug resistance. *Front. Mol. Biosci.* **7**, (2020).
205. Yang, Q. et al. Evading immune cell uptake and clearance requires PEG grafting at densities substantially exceeding the minimum for brush conformation. *Mol. Pharm.* **11**, 1250–1258 (2014).

206. Wong, C. *et al.* Multistage nanoparticle delivery system for deep penetration into tumor tissue. *Proc. Natl. Acad. Sci.* **108**, 2426–2431 (2011).
207. Bahardoust, M. & Bagheriâ Hosseinabadi, Z. Role of adipose-derived mesenchymal stem cells in the regeneration of cardiac tissue and improvement of cardiac function: A narrative review. *Biointerface Res. Appl. Chem.* **11**, 8446–8456 (2021).
208. Wang, J. *et al.* Designer exosomes enabling tumor targeted efficient chemo/gene/photothermal therapy. *Biomaterials* **276**, 121056 (2021).
209. Kwon, S.-H., Faruque, H. A., Kee, H., Kim, E. & Park, S. Exosome-based hybrid nanostructures for enhanced tumor targeting and hyperthermia therapy. *Colloids Surf. B Biointerfaces* **205**, 111915 (2021).
210. Zhang, F. *et al.* Mesenchymal stem cell-derived exosome: A tumor regulator and carrier for targeted tumor therapy. *Cancer Lett.* **526**, 29–40 (2022).
211. Pei, X. *et al.* Targeted exosomes for co-delivery of siFGL1 and siTGF-β1 trigger combined cancer immunotherapy by remodeling immunosuppressive tumor microenvironment. *Chem. Eng. J.* **421**, 129774 (2021).
212. Zhou, W. *et al.* Pancreatic cancer-targeting exosomes for enhancing immunotherapy and reprogramming tumor microenvironment. *Biomaterials* **268**, 120546 (2021).
213. Huang, X. *et al.* Engineered exosome as targeted lncRNA MEG3 delivery vehicles for osteosarcoma therapy. *J. Controlled Release* **343**, 107–117 (2022).
214. Zhu, D. *et al.* Delivery of manganese carbonyl to the tumor microenvironment using tumor-derived exosomes for cancer gas therapy and low dose radiotherapy. *Biomaterials* **274**, 120894 (2021).
215. Molavipordanjani, S. *et al.* 99mTc-radiolabeled HER2 targeted exosome for tumor imaging. *Eur. J. Pharm. Sci.* **148**, 105312 (2020).
216. Jokerst, J. V. & Gambhir, S. S. Molecular imaging with theranostic nanoparticles. *Acc. Chem. Res.* **44**, 1050–1060 (2011).
217. Chen, F., Ehlerding, E. B. & Cai, W. Theranostic nanoparticles. *J. Nucl. Med. Off. Publ. Soc. Nucl. Med.* **55**, 1919–1922 (2014).
218. Blackwell, K. L. *et al.* Overall survival benefit with lapatinib in combination with trastuzumab for patients with human epidermal growth factor receptor 2-positive metastatic breast cancer: Final results from the EGF104900 Study. *J. Clin. Oncol. Off. J. Am. Soc. Clin. Oncol.* **30**, 2585–2592 (2012).
219. Benezra, M. *et al.* Multimodal silica nanoparticles are effective cancer-targeted probes in a model of human melanoma. *J. Clin. Invest.* **121**, 2768–2780 (2011).
220. Wani, A. *et al.* Surface PEGylation of Mesoporous Silica Nanorods (MSNR): Effect on loading, release, and delivery of mitoxantrone in hypoxic cancer cells. *Sci. Rep.* **7**, 2274 (2017).
221. As, T. & Ss, G. Nanooncology: The future of cancer diagnosis and therapy. *CA. Cancer J. Clin.* **63**, (2013).

16 Antibody-Based Targeting of Nanoplatforms for Cancer Theranostics

Malvin Ofosu-Boateng, Seth Kwabena Amponsah and Benedicta Obenewaa Dankyi

CONTENTS

16.1 Introduction 275
16.2 Nanoparticle Types 276
16.3 Antibodies Used in Cancer Management 277
16.4 Antibody-Targeted Nanoparticles 278
    16.4.1 Conjugation by Adsorption 279
    16.4.2 Conjugation via a Linker (Covalent Conjugation) 279
    16.4.3 Conjugation via Adaptor Molecules 279
16.5 Theranostics of Antibody-Nanoparticle Conjugates in Cancer 281
16.6 Conclusion 282
References 283

16.1 INTRODUCTION

Cancer can be defined as the abnormal uncontrolled proliferation of cells beyond normal, with the potential of these cells invading or spreading to other organs. Cancer can be said to be one of the leading causes of death worldwide [1]. Most cancer deaths that were recorded in 2020 were mostly from breast, lung, colon, prostate, skin and stomach cancers [1]. There are several diagnostic and treatment options available for cancer. Often, chemotherapeutic agents remain the first line of treatment. There are, however, some challenges associated with the use of chemotherapeutic agents, some of which are poor accumulation of drugs in tumor cells and undesired off-target effects of drugs in healthy tissues [2].

Nanomedicine is the application of nanoparticles which have a size of between 1 and 100 nm in diameter in the diagnosis, monitoring and treatment of diseases [3]. The use of nanoparticles in medicines has several applications, some of which include drug delivery, medical imaging, gene delivery into individual cells and detection of proteins. One of the major applications of nanoparticles is drug delivery in the diagnosis and treatment of cancer. Nanoparticle use in cancer has been researched in an attempt to improve efficiency of treatment and overcome resistance [4]. Several nanoparticles are currently employed in the treatment of cancer, some of which are lipid-based nanoparticles, polymer-based nanoparticles, dendrimers and micelles.

The American Cancer Association has listed the common types of cancer treatment options to be surgery, chemotherapy, radiation therapy, targeted therapy, stem cell transplant, hormone therapy and immunotherapy. One of the options in cancer immunotherapy involves the use of monoclonal antibodies. The development of hybridoma technology by Köhler and Milstein [5] paved the way for the use of antibodies in cancer diagnosis and management. Monoclonal antibodies produced from hybridoma technology aid in the identification of surface antigens of cancer cells, which is useful in identification of cancer types. This technology also facilitates the use of antibodies as targeted vehicles which increase the specificity of particular agents [6].

16.2 NANOPARTICLE TYPES

Some of the very common nanoparticles that are available for use in the pharmaceutical industry are lipid-based nanoparticles, polymer-based nanoparticles, dendrimers and micelles. Lipid-based nanoparticles or liposomes consist of an aqueous compartment that is found in a closed lipid bilayer [7]. Most anticancer treatments have low water solubility; hence, liposomes offer a convenient drug delivery system that enhance the solubility of drugs.

Polymer-based or polymeric nanoparticles have surface adsorption properties that allows loading with active compounds of interest [8]. This imparts a controlled release attribute [9] to the nanoparticles and other advantages such as protection of the active drug and improved bioavailability.

Micelles, unlike other nanoparticles, have a hydrophobic core and a hydrophilic surface. They are considered surface surfactants. Drug moieties that have poor water solubility can be delivered in vivo with micelles. The small size of the micellar nanoparticles enhances the permeability and retention effect, which leads to increased drug concentration at tumor sites [10]. Micelles have also been employed in the delivery of small interfering ribonucleic acid (siRNA) to target sites [6].

There are also available inorganic nanoparticles that are used in cancer therapy. One example is iron oxide nanoparticles. A common biomedical application of iron oxide nanoparticles is to facilitate diagnosis of brain tumors using magnetic resonance imaging scans [11]. Even though the exact mechanism of their anticancer effect is unknown, it is postulated that iron oxide nanoparticles work by stimulating pro-inflammatory immune response [12]. This leads to T-cell infiltration into tumor cells, which inhibits tumor growth [13].

Mesoporous silica nanoparticles are also drug delivery nanocarriers, which can be loaded with drugs either through electrostatic adsorption, hydrophobic interactions or covalent binding [14]. Anticancer drugs like doxorubicin and paclitaxel have been loaded on mesoporous silica nanoparticles to target cancer cells [15]. There are other inorganic nanoparticles such as gold nanoparticles, carbon nanotubes and quantum dots that have been used in cancer diagnosis and drug delivery [16]. Table 16.1 shows some anticancer drugs currently on the market that are formulated as nanoparticles

TABLE 16.1
Examples of Approved Anticancer Drug Nanoparticles Available on the Market [17]

Drug	Nanoparticle Type	Indication	Approval Year
Oxaliplatin	Micelle	Lymphoma	2016
Docetaxel	Micelle	Head and neck cancer	2016
Magnablate	Iron nanoparticles	Prostate cancer	2016
Paclitaxel	Polymeric micelle	Breast cancer	2016
Vincristine	Liposome	Lymphoblastic leukemia	2012
Mifamurtide	Liposome	Osteosarcoma	2009
Leuprolide Acetate	Polymeric	Advanced prostate cancer	2002
Cytarabine	Liposome	Lymphomatous meningitis	1999
Neocarzinostatin	Polymeric	Hepatoma	1997
Daunorubicin	Liposome	HIV-associated Kaposi sarcoma	1996
Doxorubicin	Liposome	Breast cancer, ovarian cancer	1995

16.3 ANTIBODIES USED IN CANCER MANAGEMENT

One relevant characteristic of antibodies that makes them a favorable option in the treatment of cancer and other diseases is their ability to bind to particular epitopes of corresponding antigens [18]. Some monoclonal antibodies exert their anticancer effect through different mechanisms. Some bind to cell surface receptors and induce cell death either intrinsically or by releasing a conjugated drug. For instance, the drug gemtuzumab ozogamicin, which is used in the management of acute myeloid leukemia, normally binds to cells that express the CD33 antigen [19]. This is then followed by internalization of the conjugate that will cause death of cancer cells. Lexatumumab and conatumumab induce apoptosis in cancer cells via direct agonist activity on DR5 receptors [20]. The binding of these monoclonal antibodies to DR5 receptors on the cell surface leads to the formation of death-inducing signaling complex (DISC) [21], which activates several caspases. The caspases destroy different cellular substrates that eventually cause apoptotic cell death. Table 16.2 shows examples of tumor antigens that are targeted by monoclonal antibodies.

Monoclonal antibodies can also elicit anticancer activities by activating the immune system to kill cancer cells [23]. This happens through the recruitment of immune cells to the sites of the tumor.

TABLE 16.2
Examples of Tumor-Associated Antigens Targeted by Monoclonal Antibodies [22]

Antigen Type	Examples	Associated Tumors
Cluster of differentiation (CD) antigens	CD20	Non-Hodgkin's lymphoma
	CD30	Hodgkin's lymphoma
	CD33	Acute myelogenous leukemia
	CD38	Multiple myeloma
	CD52	Chronic lymphocytic leukemia
Growth factors	EGFR	Lung, breast, colon, head and neck cancer
	HER2	Breast, ovarian, prostate
	ErbB3	Breast, colon, prostate tumors
	c-met	Breast, ovarian, lung tumors
	IGF1R	Lung, thyroid, prostate, breast tumors
	RANKL	Prostate cancer
	TRAIL	Colon, lung cancer
Vascular targets	VEGF	Tumor vasculature
	VEGFR	Epithelium-derived solid tumors
	αVβ3	Tumor vasculature
	α5β1	Tumor vasculature
Glycoproteins	EpCAM	Breast, colon, lung cancer
	CEA	Breast, colon, lung cancer
	PSMA	Prostate cancer
	Folate binding protein	Ovarian tumor
	Carbonic anhydrase IX	Renal cell carcinoma
	gpA33	Colorectal carcinoma
Glycolipid gangliosides	GD2	Neuroectodermal tumors
	GD3	Neuroectodermal tumors
	GM2	Neuroectodermal tumors
Stromal and matrix antigens	FAP	Breast, colon, lung cancer
	Tenascin	Glioma, breast, prostate tumors

TABLE 16.3
Examples of Approved Monoclonal Antibodies Used in the Management of Cancer [25, 26]

Drug	Target	Indication	Approval Year
Isatuximab	CD38	Multiple myeloma	2020
Sacituzumab govitecan	TROP2	Triple-negative breast cancer	2020
Enfortumab vedotin	Nectin-4	Urothelial/bladder cancer	2019
Atezolizumab	PD-L1	Triple-negative breast cancer	2019
		Bladder	2016
Bevacizumab	VEGF	Ovarian cancer	2018
		Colorectal cancer	2004
Olaratumab	FDGFRα	Sarcoma	2016
Necitumumab	EGFR	Non-small cell lung cancer	2015
Ramucirumab	VEGFR2	Gastric cancer	2014
Bentuximab	CD30	Hodgkin's lymphoma	2011
Cetuximab	EGFR	Colorectal cancer	2006
Gemtuzumab ozogamicin	CD33	Acute myeloid leukemia	2000
Trastuzumab	HER2	Breast cancer	1998
Rituximab	CD20	B-cell lymphoma	1997

Trastuzumab, which is commonly used in the management of breast cancer, has several proposed mechanisms of action, one of which is antibody dependent cellular cytotoxicity. Arnould et al. in their study illustrated that the anti-tumor activity of trastuzumab is associated with increased natural killer cells numbers around tumor sites [24]. They also found an increase in the immunologic cytotoxic molecules Granzyme B and TiA1. Table 16.3 shows some approved antibodies employed in the management of cancer.

16.4 ANTIBODY-TARGETED NANOPARTICLES

A challenge commonly seen in cancer therapy is selectivity and specificity of chemotherapeutic agents to tumor cells. Over the last few years, a lot of work has been done and some of the strategies to improve specificity of these anticancer drugs include formulating them as nanoparticles. Nanoparticles can accumulate in tumor cells because of the permeable vasculature of these tissues and the enhanced permeation and retention effect of the nanoparticles. This principle is referred to as the passive targeted delivery. Monoclonal antibodies have also been utilized in cancer management because of their ability to bind to specific receptors and antigens most likely found on tumor cells. This specificity is conferred by their complementarity determining regions (CDRs). Antibody-targeted nanoparticles are used in the diagnosis and screening of cancer. In a study conducted by Chen et al., anti-human HER-2 antibody was conjugated to generation 5 poly (amidoamine) dendrimer-encapsulated gold nanoparticles to produce a contrast agent [27]. This antibody-nanoparticle demonstrated improved efficiency to target and image HER-2 positive tumors. Table 16.4 provides information on some approved antibody-nanoparticle formulations on the market.

One strategy that has further improved the overall function of chemotherapeutic agents is the use of monoclonal antibodies as ligands for targeting nanoparticles to specific site. This provides for more effective target recognition. The challenge that is associated with antibody targeting of nanoparticles is conjugation strategies that would preserve the functionality of the nanoparticles and the antibodies. The methods of conjugation used are adsorption, covalent binding and the use of adapters.

TABLE 16.4
Examples of Approved Antibody-Nanoparticle Formulations for Cancer Therapy [25]

Name	Indication	Approval Date
Saclituzumab govitecan	Triple negative breast cancer	2020
Trastuzumab deruxtecan	Breast cancer	2019
Enfurtumab vedotin	Bladder cancer	2019
Polatuzumab vedotin	B-cell lymphoma	2019
Moxetumomab pasudotox	Hairy-cell leukemia	2018
Inotuzumab ozogamicin	Acute lymphoblastic leukemia	2017
Trastuzumab emtansine	Breast cancer	2013
Brentuximab vedotin	Hodgkin's lymphoma	2011
Iodine tositumomab	Non-Hodgkin's lymphoma	2003
Ibritumomab tiuxetan	Non-Hodgkin's lymphoma	2002
Gemtuzumab ozogamicin	Acute myeloid leukemia	2000

16.4.1 Conjugation by Adsorption

Adsorption refers to the non-covalent binding of antibodies to nanoparticles. Here the antibody is attached to the nanoparticle through weak interactions such as Van der Waal's forces, hydrogen bonding, electrostatic and hydrophobic interactions [28]. Binding can also be between ionic bonds of opposite charges on the surface of the antibodies and the nanoparticles [29]. Trastuzumab (herceptin) has been conjugated to docetaxel polymeric nanoparticles in some studies, and it was found to induce higher cytotoxicity than the use of only the polymeric nanoparticle [30].

16.4.2 Conjugation via a Linker (Covalent Conjugation)

Antibodies can be linked covalently with nanoparticles using methods such as carbodiimide, maleimide chemistry or click chemistry reactions (copper catalyzed alkyne-azide cycloaddition reactions) [31]. Figure 16.1 illustrates the different covalent conjugation reactions between antibodies and nanoparticles.

Functional groups found on the amino acid residues of the antibodies are the selectable targets for bioconjugation [32]. In the carbodiimide method, the carboxyl groups on the surface of the nanoparticles are activated in order to form stable covalent bonds with the amine groups on the amino acids [33]. In the maleimide method, thiol groups found in the antigen binding fragments of the antibody interacts covalently with maleimide on the nanoparticle surface [34]. This conjugate can also be achieved with the aid of crosslinking agents such as Traut's reagent, succinimidyl acetate and sulfosuccinimidyl hexanoate, which attach the thiol groups to the nanoparticles [35]. The final method of covalent conjugation is the click chemistry reactions approach. Here, the azide group on an antibody forms covalent bonds with the alkyne groups on drug-containing nanoparticles [36]. This method has been successfully used in binding anti-CD63 antibody to polymeric gold nanoparticles that are used in the development of gold labels in biosensing techniques [37].

16.4.3 Conjugation via Adaptor Molecules

This method of conjugation of antibodies to nanoparticles is a non-covalent approach that utilizes the strong binding affinity between biotin (Figure 16.2) and a biotin binding analog such as avidin [38]. The use of adapter molecules helps to overcome the challenge of inappropriate positioning of the antibody on the nanoparticle [39]. In cancer management, biotin has been explored as a tumor

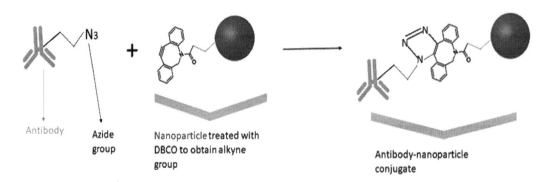

FIGURE 16.1 Covalent conjugation reactions between antibodies and nanoparticles.

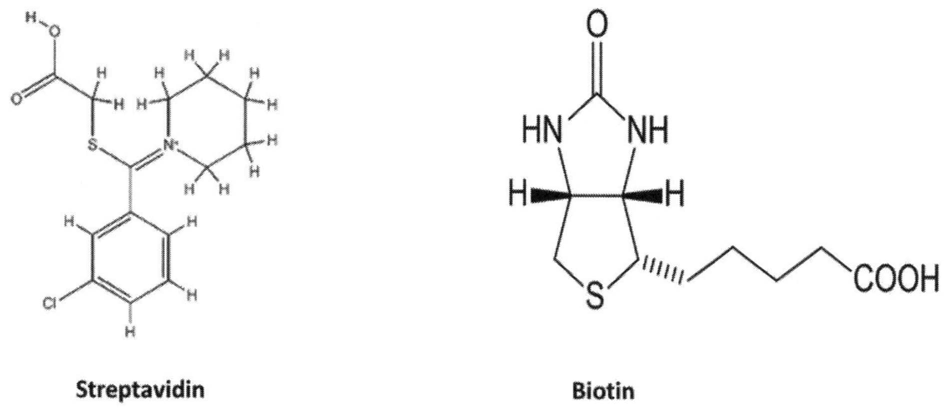

Streptavidin   **Biotin**

FIGURE 16.2 Chemical structures of streptavidin and biotin.

targeting moiety because a variety of cancer cells express biotin receptors. This makes it a favorable candidate for conjugates used in cancer therapy [39]. Neutravidin and streptavidin (Figure 16.2) are usually the avidin analogues [38] that are used to functionalize the nanoparticle because these agents can selectively bind to the biotinylated antibodies and prevents non-specific binding, especially to receptors found on sugar moieties [39].

16.5 THERANOSTICS OF ANTIBODY-NANOPARTICLE CONJUGATES IN CANCER

Theranostics is derived from the two words "therapy" and "diagnostics". Cancer theranostics can therefore be said to be approaches that are used in the diagnosis and treatment of cancer [40]. Despite several approaches to improve the diagnosis and treatment of cancer, antibody nanoplatforms have come to play a significant role in the current state in cancer theranostics. A number of imaging techniques for tumors have been developed. An example is optical imaging, which is a non-invasive method that makes use of fluorescent probes [41]. Initially nanoparticles that have been labeled with fluorochromes were used in imaging. Even though this technique is not obsolete, the use of antibodies has further advanced this procedure. Cy5.5-anti-CD20 nanoparticles have improved the resolution of images taken using this technique by reducing background signals, since the addition of the antibodies further localizes the binding of the fluorochrome-labeled nanoparticle to the tumor sites [42]. Similarly, the monoclonal antibody bevacizumab was conjugated with iron oxide nanoparticles to provide better images of cancers that express the vascular endothelial growth factor (VEGF) breast tumors [43]. This, therefore, serves as a starting point to explore the use of this combination both as an imaging contrast and a targeted treatment option. Iron oxide has been applied in several imaging methods for different cancers. Table 16.5 summarizes some of the antibody-nanoparticle conjugates that are used in cancer diagnostics.

Antibody-conjugated nanoparticles are currently being utilized in treating different cancers. There is active research to further understand the mechanism of action of antibody-conjugated nanoparticles that will enhance their therapeutic activities. Currently, antibodies that have been conjugated with nanoparticles of cytotoxic drugs, radioisotopes and nucleic acids are being investigated and developed [16] because of the improved intra-tumoral drug distribution, enhanced efficacy and controlled release of these agents. In the delivery of cytotoxic agents, most loaded nanoparticles are conjugated with HER2- or EGFR-related antibodies. Camptothecin-loaded silica nanoparticles conjugated with EGFR antibodies have been found to be effective in treating tumors like glioblastoma that over-express EGFR [44]. In another study, paclitaxel-loaded nanoparticles targeted with EGFR

TABLE 16.5
Examples of Antibody-Nanoparticle Conjugations Used in Cancer Diagnostics [42]

Nanoparticle	Antibody	Imaging Technique	Type of Cancer
Iron oxide nanoparticle	Anti-VEGF (bevacizumab)	Optical fluorescence imaging	Breast tumor
Cy5.5 nanoparticle	Anti CD20 antibody	Optical fluorescence imaging	Leukemia
Manganese oxide–mesoporous silica nanoparticles	Prostate-specific antigen antibody	T1-weighted MRI	Prostate cancer
Superparamagnetic iron oxide nanoparticles	Plectin-1 antibody	T2-weighted MRI	Pancreatic cancer
Superparamagnetic iron oxide nanoparticles	HER2 antibodies	T2-weighted MRI	Breast cancer
Superparamagnetic iron oxide nanoparticles	Epidermal growth factor receptor deletion mutant antibodies	T2-weighted MRI	Glioblastoma
2,3-dimercaptosuccinicacid modified superparamagnetic iron oxide nanoparticles	CD20 antibodies	T2-weighted MRI	Malignant lymphoma
Superparamagnetic iron oxide nanoparticles	AFP and GPC3 antibodies	T2-weighted MRI	Hepatocellular carcinoma

antibodies showed moderate efficacy in the management of lung tumors [45]. Pegylated liposomal doxorubicin has also shown high in-vivo cytotoxic action when bound with anti-nucleosome monoclonal antibodies in glioma tumors [46].

The principle of localization of delivery of antibody-targeted nanoparticles has also been applied in radiation treatment [16]. Cetuximab conjugated with superparamagnetic iron oxide nanoparticles has been found to increase the sensitivity of gliomas that have become resistant to radiation [16]. In a study conducted by Meissner et al., the antisense nucleic acid gene BCL2, an antiapoptotic gene that is normally overexpressed in acute leukemias, was formulated into a liposomal nanoparticle and targeted with CD20 monoclonal antibody rituximab [47]. This study showed a reduction of BCL2 proteins within the cancer cells suggesting suppression of the antiapoptotic gene.

Although antibody-conjugated nanoparticles are used either for their individual therapeutic potentials or diagnostic purposes, the theranostic potential of some of the antibody-conjugated nanoparticles is gradually gaining popularity. Over the last few years, superparamagnetic iron oxide nanoparticles (SPIONs) have seen a lot of clinical utility, including theranostic potential in different studies. Cetuximab bound to polyethylene glycol (PEG)-coated SPIONs has been successfully used to trigger apoptosis in human glioma cells and at the same time enhance contrast of SPIONs in T2-weighted magnetic resonance imaging (MRI) [48]. Superparamagnetic iron-platinum nanoparticles that were encapsulated with paclitaxel targeted to prostate-specific antigen antibodies have been shown to induce cytotoxicity to prostate cancer cells as well as functioning as an MRI contrast agent [49]. Iron oxide nanocrystals loaded with docetaxel and targeted with the single-chain variable fragment (scFv) of an antibody have also been shown to have theranostic properties, that is, enhancing contrast on T2-weighted MRI as well as killing prostate cancer cells [50]. Table 16.6 shows some available theranostic antibody nanoparticle combinations.

16.6 CONCLUSION

Antibodies and nanoparticles have both shown promising prospects in cancer management, where they have been extensively used in both diagnosis and treatment. Antibody-based nanoplatforms provide a means of enhancing diagnosis and treatment properties of anticancer agents by combining

TABLE 16.6
Antibody Nanoparticles Utilized for Their Theranostic Effects [16]

Nanoparticle	Antibody	Theranostic Indication
Superparamagnetic iron oxide nanoparticles	Anti-EGFR antibody (cetuximab)	Glioma tumors
Superparamagnetic iron-platinum nanoparticles encapsulated with paclitaxel	Anti-prostate membrane specific (PMSA) antigen antibody	Prostate cancer
Doxorubicin and iron oxide magnetic nanocrystal polymeric nanoparticles	Anti-HER2 antibody (trastuzumab)	Breast cancer
Docetaxel and iron oxide magnetic nanocrystal polymeric nanoparticles	Single-chain variable fragment against prostate stem cell antigen	Prostate cancer
Doxorubicin and indocyamine green polymeric nanoparticles	Anti-HER2 antibody	Breast cancer
Quantum dot lipid–based rapamycin nanoparticles	Anti-HER2 antibody (trastuzumab)	Breast cancer
Carbon nanotubes loaded with doxorubicin	Prostate stem cell–specific antigen antibodies	Prostate cancer

antibodies and nanoparticles for better outcomes. Although the theranostic use of antibody-based nanoplatforms in cancer management has gained popularity over the past few years, more studies are needed in this area to be able to understand and explore the full potential and capabilities.

REFERENCES

1. WHO. (2022). www.who.int/news-room/fact-sheets/detail/cancer (accessed November 25, 2022).
2. Gonzalez-Valdivieso, J., Girotti, A., Schneider, J., & Arias, F. J. (2021). Advanced nanomedicine and cancer: Challenges and opportunities in clinical translation. *International Journal of Pharmaceutics*, 599, 120438.
3. Kateb, B., Chiu, K., Black, K. L., Yamamoto, V., Khalsa, B., Ljubimova, J. Y., . . . Heiss, J. D. (2011). Nanoplatforms for constructing new approaches to cancer treatment, imaging, and drug delivery: What should be the policy?. *Neuroimage*, 54, S106–S124.
4. Markman, J. L., Rekechenetskiy, A., Holler, E., & Ljubimova, J. Y. (2013). Nanomedicine therapeutic approaches to overcome cancer drug resistance. *Advanced Drug Delivery Reviews*, 65(13–14), 1866–1879.
5. Köhler, G., & Milstein, C. (1975). Continuous cultures of fused cells secreting antibody of predefined specificity. *Nature*, 256(5517), 495–497.
6. Obeid, M. A., Aljabali, A. A., Alshaer, W., Charbe, N. B., Chellappan, D. K., Dua, K., . . . Tambuwala, M. M. (2021). Targeting siRNAs in cancer drug delivery. In *Advanced drug delivery systems in the management of cancer* (pp. 447–460). Academic Press.
7. Tenchov, R., Bird, R., Curtze, A. E., & Zhou, Q. (2021). Lipid nanoparticles—from liposomes to mRNA vaccine delivery, a landscape of research diversity and advancement. *ACS nano*, 15(11), 16982–17015.
8. Zielińska, A., Carreiró, F., Oliveira, A. M., Neves, A., Pires, B., Venkatesh, D. N., . . . Souto, E. B. (2020). Polymeric nanoparticles: Production, characterization, toxicology and ecotoxicology. *Molecules*, 25(16), 3731.
9. Schaffazick, S. R., Pohlmann, A. R., Dalla-Costa, T., & Guterres, S. S. (2003). Freeze-drying polymeric colloidal suspensions: Nanocapsules, nanospheres and nanodispersion. A comparative study. *European Journal of Pharmaceutics and Biopharmaceutics*, 56(3), 501–505.
10. Maeda, H., Wu, J., Sawa, T., Matsumura, Y., & Hori, K. (2000). Tumor vascular permeability and the EPR effect in macromolecular therapeutics: A review. *Journal of Controlled Release*, 65(1–2), 271–284.
11. Lakshmipriya, T., & Gopinath, S. C. (2021). Introduction to nanoparticles and analytical devices. In *Nanoparticles in analytical and medical devices* (pp. 1–29). Elsevier.
12. Soetaert, F., Korangath, P., Serantes, D., Fiering, S., & Ivkov, R. (2020). Cancer therapy with iron oxide nanoparticles: Agents of thermal and immune therapies. *Advanced Drug Delivery Reviews*, 163, 65–83.

13. Korangath, P., Barnett, J. D., Sharma, A., Henderson, E. T., Stewart, J., Yu, S. H., ... Ivkov, R. (2020). Nanoparticle interactions with immune cells dominate tumor retention and induce T cell—mediated tumor suppression in models of breast cancer. *Science Advances, 6*(13), eaay1601.
14. Tarn, D., Ashley, C. E., Xue, M. I. N., Carnes, E. C., Zink, J. I., & Brinker, C. J. (2013). Mesoporous silica nanoparticle nanocarriers: Biofunctionality and biocompatibility. *Accounts of Chemical Research, 46*(3), 792–801.
15. Alyassin, Y., Sayed, E. G., Mehta, P., Ruparelia, K., Arshad, M. S., Rasekh, M., ... Ahmad, Z. (2020). Application of mesoporous silica nanoparticles as drug delivery carriers for chemotherapeutic agents. *Drug Discovery Today, 25*(8), 1513–1520.
16. Carter, T., Mulholland, P., & Chester, K. (2016). Antibody-targeted nanoparticles for cancer treatment. *Immunotherapy, 8*(8), 941–958.
17. Anselmo, A. C., & Mitragotri, S. (2019). Nanoparticles in the clinic: An update. *Bioengineering & Translational Medicine, 4*(3), e10143.
18. Eras, A., Castillo, D., Suárez, M., Vispo, N. S., Albericio, F., & Rodriguez, H. (2022). Chemical conjugation in drug delivery systems. *Frontiers in Chemistry, 10*.
19. DeLeve, L. D. (2013). Cancer chemotherapy. In *Drug-induced liver disease* (pp. 541–567). Academic Press.
20. Ocker, M., & Höpfner, M. (2012). Apoptosis-modulating drugs for improved cancer therapy. *European Surgical Research, 48*(3), 111–120.
21. Wiezorek, J., Holland, P., & Graves, J. (2010). Death receptor agonists as a targeted therapy for cancer. *Clinical Cancer Research, 16*(6), 1701–1708.
22. Scott, A. M., Allison, J. P., & Wolchok, J. D. (2012). Monoclonal antibodies in cancer therapy. *Cancer Immunity, 12*(1).
23. Dastidar, D. G., Ghosh, D., & Das, A. (2022). Recent developments in nanocarriers for Cancer Chemotherapy. *OpenNano*, 100080.
24. Arnould, L. 1., Gelly, M., Penault-Llorca, F. A., Benoit, L., Bonnetain, F., Migeon, C., ... Coudert, B. (2006). Trastuzumab-based treatment of HER2-positive breast cancer: An antibody-dependent cellular cytotoxicity mechanism?. *British Journal of Cancer, 94*(2), 259–267.
25. Zahavi, D., & Weiner, L. (2020). Monoclonal antibodies in cancer therapy. *Antibodies, 9*(3), 34.
26. Tong, J. T., Harris, P. W., Brimble, M. A., & Kavianinia, I. (2021). An insight into FDA approved antibody-drug conjugates for cancer therapy. *Molecules, 26*(19), 5847.
27. Chen, J. S., Chen, J., Bhattacharjee, S., Cao, Z., Wang, H., Swanson, S. D., ... Wang, S. H. (2020). Functionalized nanoparticles with targeted antibody to enhance imaging of breast cancer in vivo. *Journal of Nanobiotechnology, 18*(1), 1–9.
28. Tallawi, M., Rosellini, E., Barbani, N., Cascone, M. G., Rai, R., Saint-Pierre, G., & Boccaccini, A. R. (2015). Strategies for the chemical and biological functionalization of scaffolds for cardiac tissue engineering: A review. *Journal of the Royal Society Interface, 12*(108), 20150254.
29. Goossens, J., Sein, H., Lu, S., Radwanska, M., Muyldermans, S., Sterckx, Y. G. J., & Magez, S. (2017). Functionalization of gold nanoparticles with nanobodies through physical adsorption. *Analytical Methods, 9*(23), 3430–3440.
30. Choi, J. S., Jang, W. S., & Park, J. S. (2018). Comparison of adsorption and conjugation of Herceptin on poly (lactic-co-glycolic acid) nanoparticles—Effect on cell internalization in breast cancer cells. *Materials Science and Engineering: C, 92*, 496–507.
31. Lee, N. K., Wang, C. P. J., Lim, J., Park, W., Kwon, H. K., Kim, S. N., ... Park, C. G. (2021). Impact of the conjugation of antibodies to the surfaces of polymer nanoparticles on the immune cell targeting abilities. *Nano Convergence, 8*(1), 1–11.
32. Talebzadeh, S., Queffélec, C., & Knight, D. A. (2019). Surface modification of plasmonic noble metal—metal oxide core—shell nanoparticles. *Nanoscale Advances, 1*(12), 4578–4591.
33. Lee, J. C., Donahue, N. D., Mao, A. S., Karim, A., Komarneni, M., Thomas, E. E., ... Wilhelm, S. (2020). Exploring maleimide-based nanoparticle surface engineering to control cellular interactions. *ACS Applied Nano Materials, 3*(3), 2421–2429.
34. Schmid, D., Park, C. G., Hartl, C. A., Subedi, N., Cartwright, A. N., Puerto, R. B., ... Goldberg, M. S. (2017). T cell-targeting nanoparticles focus delivery of immunotherapy to improve antitumor immunity. *Nature Communications, 8*(1), 1–12.
35. Wong, S. S. (1991). *Chemistry of protein conjugation and cross-linking*. CRC Press.
36. Vatansever, E. C., Kang, J., Tuley, A., Ward, E. S., & Liu, W. R. (2020). An optimal "Click" formulation strategy for antibody-drug conjugate synthesis. *Bioorganic & Medicinal Chemistry, 28*(24), 115808.

37. Finetti, C., Sola, L., Pezzullo, M., Prosperi, D., Colombo, M., Riva, B., . . . Chiari, M. (2016). Click chemistry immobilization of antibodies on polymer coated gold nanoparticles. *Langmuir*, *32*(29), 7435–7441.
38. Gascón, V., Márquez-Alvarez, C., Díaz, I., & Blanco, R. M. (2016). Hybrid ordered mesoporous materials as supports for permanent enzyme immobilization through non-covalent interactions. In: Abel M. Maharramov, Kamran T. Mahmudov, Maximilian N. Kopylovich, Armando J. L. Pombeiro, editors. *Non-covalent interactions in the synthesis and design of new compounds*. Wiley Online Library.
39. Juan, A., Cimas, F. J., Bravo, I., Pandiella, A., Ocaña, A., & Alonso-Moreno, C. (2020). An overview of antibody conjugated polymeric nanoparticles for breast cancer therapy. *Pharmaceutics*, *12*(9), 802.
40. Chircov, C., & Grumezescu, A. M. (2019). Basics in nanoarchitectonics. In *Nanoarchitectonics in biomedicine* (pp. 1–21). William Andrew Publishing.
41. Martelli, C., Dico, A. L., Diceglie, C., Lucignani, G., & Ottobrini, L. (2016). Optical imaging probes in oncology. *Oncotarget*, *7*(30), 48753.
42. Lin, X., O'Reilly Beringhs, A., & Lu, X. (2021). Applications of nanoparticle-antibody conjugates in immunoassays and tumor imaging. *The AAPS Journal*, *23*(2), 1–16.
43. Lin, R., Huang, J., Wang, L., Li, Y., Lipowska, M., Wu, H., . . . & Mao, H. (2018). Bevacizumab and near infrared probe conjugated iron oxide nanoparticles for vascular endothelial growth factor targeted MR and optical imaging. *Biomaterials Science*, *6*(6), 1517–1525.
44. Secret, E., Smith, K., Dubljevic, V., Moore, E., Macardle, P., Delalat, B., . . . Voelcker, N. H. (2013). Antibody-functionalized porous silicon nanoparticles for vectorization of hydrophobic drugs. *Advanced Healthcare Materials*, *2*(5), 718–727.
45. Karra, N., Nassar, T., Ripin, A. N., Schwob, O., Borlak, J., & Benita, S. (2013). Antibody conjugated PLGA nanoparticles for targeted delivery of paclitaxel palmitate: Efficacy and biofate in a lung cancer mouse model. *Small*, *9*(24), 4221–4236.
46. Gupta, B., & Torchilin, V. P. (2007). Monoclonal antibody 2C5-modified doxorubicin-loaded liposomes with significantly enhanced therapeutic activity against intracranial human brain U-87 MG tumor xenografts in nude mice. *Cancer Immunology, Immunotherapy*, *56*(8), 1215–1223.
47. Meissner, J. M., Toporkiewicz, M., Czogalla, A., Matusewicz, L., Kuliczkowski, K., & Sikorski, A. F. (2015). Novel antisense therapeutics delivery systems: In vitro and in vivo studies of liposomes targeted with anti-CD20 antibody. *Journal of Controlled Release*, *220*, 515–528.
48. Kaluzova, M., Bouras, A., Machaidze, R., & Hadjipanayis, C. G. (2015). Targeted therapy of glioblastoma stem-like cells and tumor non-stem cells using cetuximab-conjugated iron-oxide nanoparticles. *Oncotarget*, *6*(11), 8788.
49. Ling, Y., Wei, K., Luo, Y., Gao, X., & Zhong, S. (2011). Dual docetaxel/superparamagnetic iron oxide loaded nanoparticles for both targeting magnetic resonance imaging and cancer therapy. *Biomaterials*, *32*(29), 7139–7150.
50. Taylor, R. M., & Sillerud, L. O. (2012). Paclitaxel-loaded iron platinum stealth immunomicelles are potent MRI imaging agents that prevent prostate cancer growth in a PSMA-dependent manner. *International Journal of Nanomedicine*, *7*, 4341.

17 Carbohydrate-Based Targeting of Nanoplatforms for Cancer Theranostics

Emmanuel Boadi Amoafo, Malvin Ofosu-Boateng,
Kwasi Agyei Bugyei and Seth Kwabena Amponsah

CONTENTS

17.1 Introduction 286
17.2 Theranostics 287
17.3 Targeting of Nanoplatforms in Cancer 287
17.4 Carbohydrate-Based Targeting of Nanoplatforms 288
    17.4.1 Carbohydrates 288
    17.4.2 Design of Carbohydrate-Based Theranostic Nanoparticles 289
    17.4.3 Chitosan-Based Nanoparticles for Cancer Therapy 289
    17.4.4 Hyaluronic Acid–Based Theranostic Nanoparticles 291
References 292

17.1 INTRODUCTION

The term "cancer" refers to a collection of illnesses characterized by tissues that grow abnormally (neoplasm) and which lead to cells that lack normal function and/or morphology [1]. The abnormal cells usually have the potential to invade other organs, proliferate, and undergo angiogenesis, all of which can result in life-threatening cancers [2]. Cancer ranks as the second most common cause of death in the United States [3]. According to projections, by 2030, new cancer cases will be about 26 million, and cancer-related deaths will be about 17 million per year [3]. Early identification is essential for successful therapy, as significant mortality from cancers is a result of metastasis to other organs. Cancer treatment is challenging, particularly because tumor cells have become more resistant to available chemotherapeutic agents [4].

Usually, the choice of cancer treatment is dependent on the type of cancer, characteristics of cancer, location of cancer, patient's medical history, and previous treatment experiences. Chemotherapy still remains one of the most commonly used cancer treatment modalities. To effectively combat tumors, chemotherapy is usually administered either alone or in combination with surgery, radiotherapy, or some adjuvant therapy [5]. However, there are drawbacks to each technique, including invasiveness, systemic and local off-target adverse effects, nonspecific targeting, poor drug solubility, and multidrug resistance, among others [6]. The use of adjuvant therapies has become more popular, and this appears to improve treatment outcomes and reduce adverse effects, a few of which are summarized in Table 17.1.

In an attempt to improve cancer diagnostics and treatment, different nanoparticle systems have been developed. As a diagnostic and therapeutic tool for cancer, nanoplatforms hold great promise [14]. Generally, nanoparticles have complex structures and high surface area-to-volume ratios; hence, they can undergo chemical conjugation or physical encapsulation with several agents [15].

TABLE 17.1
Examples of Adjuvant Therapies in Cancer Treatment

Adjuvant Therapy	Type of Cancer	Advantages	Disadvantages	References
Hormone therapy	Breast cancer, ovarian cancer, endometrial cancer	There is an improved prognosis.	Cognitive impairment.	[7, 8]
Immunotherapy	Bladder cancer, Hodgkin's lymphoma, melanoma	High degree of accuracy and specificity. Prevents tumor recurrence and metastasis.	Expensive treatment. May lead to autoimmune diseases.	[9, 10]
Laser ablation	Breast cancer, bladder cancer, glioblastoma, prostate cancer	Negligibly invasive alternative to surgery. Predictable size of necrosis.	Not appropriate for large tumors.	[11]
Photodynamic therapy	Brain cancer, cervical cancer, gastric cancer, liver and lung cancer	Little invasiveness. Tolerated well by patients. Low cost compared to other therapies.	Complex scheduling. Not applicable when cancer has already metastasized. Photosensitivity after treatment.	[12, 13]

Nanoparticles may be used to target tumors passively or actively. They can also ferry an array of freights and target tumors. Furthermore, nanoparticles can be used to diagnose and treat cancer in a targeted manner, yielding high-quality images of tumor sites and enhancing therapeutic efficacy without damaging normal cells. This is one of the reasons therapeutic nanoparticles have been studied extensively as next-generation nanocarrier systems [16].

17.2 THERANOSTICS

The term "theranostics" combines the words "therapeutics" and "diagnostics". In theranostics, multiple techniques are used to provide comprehensive molecular images, diagnosis, and treatment tailored to the individual. Currently, efforts have been made to combine theranostics with nanotechnology, ultimately developing theranostic nanoplatforms and methodology [14]. In most clinical management, diagnosis precedes therapy. However, it is also possible to monitor early treatment and forecast treatment effectiveness by using therapy followed by diagnosis. Additionally, the co-development of therapeutics and diagnostics is a possibility [17]. In cancer therapy, a number of treatment approaches must be chosen based on the patient's features and the course of the disease because cancer is a highly varied and adaptable disease. Researchers working on cancer treatment alternatives anticipate that theranostic nanoparticles will give patients personalized therapy options and enhance their prognoses. Moreover, after treatments, theranostic nanoparticles can track therapy effectiveness, allowing clinicians to make tailored treatment decisions [14].

17.3 TARGETING OF NANOPLATFORMS IN CANCER

In technical terms, nanoparticles have an average size of less than 100 nm. They have distinctive properties that are usually not present in samples of the same material [18]. In terms of their structural composition, nanoparticles usually would have three layers: the surface, shell, and core (which is essentially the center of the nanoparticle) [19]. Nanoparticles have grown in significance in various disciplines due to their unique qualities, including high surface-to-volume ratios, dissimilarities, sub-micron sizes, and enhanced targeting systems [20].

There are a number of nanoparticle-based theranostic systems available, some of which include gold nanostructures, magnetic nanoparticles, polymeric nanoparticles, crystalline silica nanoparticles, and carbohydrate-based nanoparticles. The targetability and multifunctionality of nanoparticles make them ideal for theranostics [21]. Furthermore, nanoparticles can be used to deliver therapeutic agents at required concentrations to the targeted site, often based on external stimuli or molecular signals [22]. The merit of theranostic nanomedicine is the ability to track the release and delivery of drugs as well as their efficacy over time. Radionuclides, magnetic resonance imaging (MRI) contrast agents, or optical imaging agents can be chemically or physically affixed to nanoplatforms. In addition, anticancer medicines such as hydrophobic chemical medications, peptides, proteins, or genetic pharmaceuticals can be coupled to nanoparticles [14].

Nanoparticles can target the diseased site passively and/or actively. Theranostic nanoparticles can diagnose diseases by describing the disease's location, stage, and response to treatment once they have arrived at the site of the disease [23]. One of the well-investigated methods for delivering nanoparticles to cancerous tissues is passive targeting. This method makes use of enhanced permeability and retention effect [24]. Although this approach has a lot of potential, the physiological constraints and retention effect make it less successful. Additionally, the uptake of nanoparticles at the tumor site is significantly impacted by the physicochemical properties of the nanomedicine, such as shape, surface chemistry, and size [25, 26].

It is generally acknowledged that these constraints can be circumvented by optimizing nanoparticle design and using an active targeting mechanism. This has the tendency to improve accumulation of nanoparticles in tumors as opposed to unintended localization. One or more targeting moieties are typically functionalized on the surface of nanoparticles to facilitate active targeting. Consequently, they can recognize antigens or receptors which may be upregulated or highly expressed on tumor cells compared to healthy cells. Due to this, there have been several techniques that have been developed in an attempt to alter the structure of nanoparticles [27].

17.4 CARBOHYDRATE-BASED TARGETING OF NANOPLATFORMS

17.4.1 CARBOHYDRATES

The basic component of all carbohydrates is the monosaccharide. The majority of sugars found in nature are polysaccharides, and these are synthesized via a sequence of condensation reactions between monosaccharides [28]. Biopolymers that are most abundant on earth are carbohydrates, and they make up more than 80% of the biomass. Carbohydrates mostly perform structural and energy storage roles. Carbohydrate-carbohydrate and carbohydrate-protein interactions play an important role in cell differentiation, proliferation, adhesion, inflammation, and immunity [29, 30].

Carbohydrates are hydrophilic, biodegradable, stable, and nontoxic agents. Carbohydrates have carboxyl, hydroxyl, and amino groups in their structure, and this aids their interaction with biological tissues (Figure 17.1). Carbohydrates are also known for their bioadhesive properties, especially to mucosal surfaces, which allow them to stay in the body for a long time [31].

When considered as a whole, the properties of carbohydrates satisfy the criteria for a platform that is efficient for in vivo drug delivery and imaging [32]. Carbohydrates can be designed as biological instruments with specific shapes and functionalities thanks to breakthroughs in polymer science. For more than 20 years, researchers have studied the incorporation of carbohydrates, mostly polysaccharides, into hydrogels and nanofibers [28]. However, it may be difficult to analyze and reproduce such polysaccharides due to their polydispersive nature and branching structure. Due to this, it is relatively difficult to control for quality, reproducibility, and analysis. Currently, there are no detailed three-dimensional (3D) investigations of single chains, nor is there a definition of the structure-function correlation since chains are so flexible. Additionally, limited regioselectivity is a problem for chemical alterations that are used to adjust the polysaccharide characteristics, which makes the sample even more polydisperse [33].

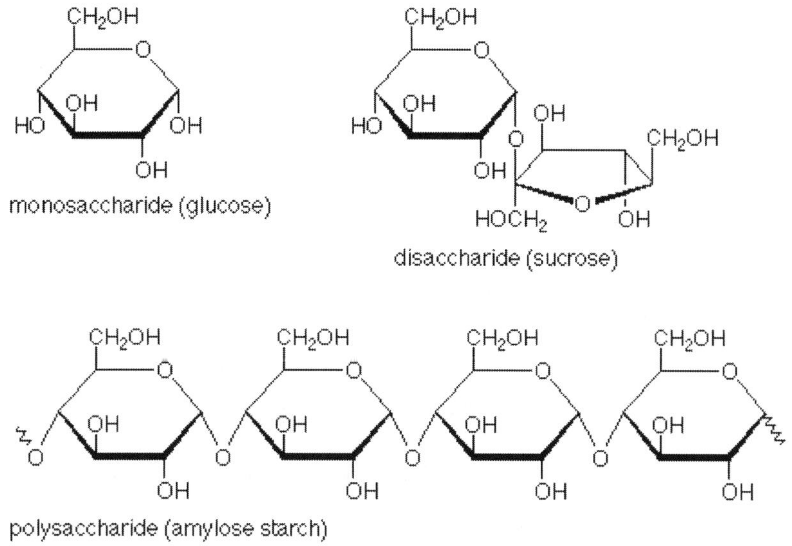

FIGURE 17.1 Basic structure of carbohydrates.

17.4.2 Design of Carbohydrate-Based Theranostic Nanoparticles

Using chemical modification and nanotechnology, carbohydrate-based nanoparticles can now be made to have a wide range of shapes and sizes for therapeutic and imaging purposes. The method of administration, type of imaging, and therapy must all be taken into account when designing carbohydrate-based nanoparticles for theranostic nanomedicine. These parameters may be influenced by the type of carbohydrate used as well as its salient features, size, and structure [34].

Due to their many functional groups, carbohydrates can be modified and used for a number of biomedical applications, some of which include tissue engineering, regenerative medicine, and drug delivery systems. For theranostics, both imaging and therapeutic compounds can be physically encapsulated within the carbohydrate-based nanoparticles or covalently bonded at the carbohydrate backbone. The carboxylic acid, aldehyde, hydroxyl, and amino groups are the most frequently employed functional groups that are available for chemical modification [35, 36].

Moieties capable of covalently binding to functional groups on carbohydrates include bile acids, fatty acids, cholesterol, polyester, polyanhydride, and some drugs such as paclitaxel [37]. Chemical interactions between carbohydrates and non-polar moieties, such as those that are enzyme-cleavable (ester bonds), pH-sensitive (hydrazone bonds), and reducible (disulfide bonds), may become labile in the presence of cancer cells. The resulting polymers, which are amphiphilic in nature, tend to self-assemble into nanoparticles in hydrous surroundings as a result of hydrophobic interactions [38]. Generally, non-covalent interactions are typically weak in comparison with covalent interactions. Among the most common non-covalent interactions used in making supramolecular carbohydrate nanoparticles for drug delivery are π–π stacking, electrostatic interaction, and host-guest recognition [39].

It is important to note that lipids, natural polymers, and synthetic polymers are the most common drug delivery vectors utilized today. Numerous nanocarriers for treating cancer have also been developed based on these materials, with promising results in the preclinical research [39].

17.4.3 Chitosan-Based Nanoparticles for Cancer Therapy

The positively charged, linear polysaccharide chitosan is made up of randomly distributed N-acetyl-D-glucosamine and β-(1–4)-linked D-glucosamine (Figure 17.2). In most cases, chitosan can be

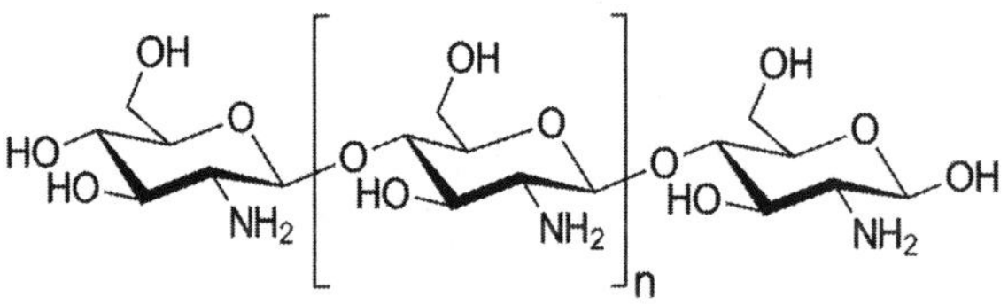

FIGURE 17.2 Basic structure of chitosan.

synthesized by deacetylating chitin, a major structural constituent found in the cell wall of fungi and also in crustaceans (e.g., shrimp and crabs) [40]. For chitosan to be distinguished from chitin, at least 60% of it needs to include glucosamine residues. Chitosan has similar biocompatibility and mucoadhesive properties as other polysaccharides. The deacetylation process can result in nontoxic amino sugars being absorbed into the body after biodegradation in the lysosomes [41–43]. Additionally, chitosan can be used to make nanoparticles, nanogels, films, and fibers through its ionic crosslinking and covalent bonding [44, 45]. Chitosan can combine with other biological molecules that are negatively charged and yield polyelectrolyte complexes. For instance, in gene therapy, polyplexes can be synthesized by combining nucleic acids with chitosan [46].

Primary amino groups also convey the other two crucial features of chitosan in biomedicine. First, due to its primary amine, chitosan usually has a pKa of 6.5 and would normally dissolve in acidic aqueous solutions [47]. Second, chitosan consists of a high density of positive charges, and this can promote cellular uptake. Nonetheless, it can also occasionally cause cellular toxicity [43]. Conversely, a chitosan-based drug delivery platform and its derivatives ought to exhibit excellent solubility and minimal toxicity under typical physiological circumstances. Such amphiphilic chitosan derivatives offer a promising delivery system for drugs in the treatment of cancer [39].

Chitosan can be used for imaging and therapy purposes when appropriately modified with nanoparticles. An effective method for delivering small molecules is chitosan–drug conjugates [48]. The construction of conjugates may involve two main covalent bonds. First, conjugates can be made where covalent bonds are formed between chitosan (chitosan derivatives) and drugs. The bonds can be hydrolyzed or degraded by enzymes under certain physiological conditions, such as cancer environments. The second type of conjugate is usually built with covalent bonds that are relatively stable: amide bonds [49].

Some clinical applications of chitosan in cancer theranostics include the use of glycol chitosan-based nanoparticles (GCNPs) and doxorubicin (DOX), which yields a hydrophobic chemotherapeutic agent (DOX-GCNP) or B-cell lymphoma 2 (BCL-2) small interfering ribonucleic acid (siRNA) encapsulated for gene therapy (siRNA-GCNP) [50]. Using this theranostic strategy, DOX-GCNPs and Bcl-2 siRNA-GCNPs are administered to cancer cells to suppress their anti-apoptotic defensive mechanisms while inducing apoptosis. As a cellular defensive mechanism, DOX dosing over time may cause overexpression of proteins that have anti-apoptotic properties such as BCL-2 that can reduce the effectiveness of chemotherapy. To synthesize DOX-GCNPs, 5-beta-cholanic acid is usually chemically coupled to the chitosan backbone, which is hydrophilic in nature. By dialysis in a dimethyl sulfoxide (DMSO)/water co-solvent environment, DOX is encapsulated to the inner cores of GCNPs. Alternatively, glycol chitosan that is thiolated can be used as the initiating material in the formulation of siRNA-GCNPs [51].

Although DOX and siRNA have different properties, both formulations of GCNPs share almost same physicochemical characteristics and would usually achieve uniform in vivo disposition in human prostate cancer (PC-3) tumor-bearing mice. In vivo near-infrared fluorescence imaging of Cy5.5-labeled GCNPs that are conjugated with DOX or loaded with siRNA have the tendency to accumulate in the tumor vasculature and cancer cells around the same time [50]. After 51 days post-treatment, the average weight of tumors in mice models was about nine times lower in the combination group compared with the free DOX or DOX-GCNP groups [50].

Another important treatment approach for cancer is photodynamic therapy, where photosensitizers produce reactive oxygen species (ROS), primarily hydroxyl radicals and singlet oxygen, when exposed to light of the right wavelength [52]. Various biological objects, including tumor cells, can be killed by ROS in the surrounding environment. Hence, ROS are promising candidates for treating tumors. When exposed to light, the photosensitizers amassed at the sites of the tumors can also provide a powerful fluorescence signal that can be used for diagnostic purposes [53]. However, as a result of their low water solubility, inadequate tumor-targeting properties, and phototoxicity, the use of photosensitizers as therapeutic agents is quite constrained. The aforementioned limitations can be circumvented by encapsulating photosensitizers in tumor-targeted drug delivery carriers made of chitosan [54].

Glycol chitosan, a water-soluble derivative of chitosan, is made by combining hydrophobic glycol molecules with chitosan. Glycol chitosan can self-assemble into nanoparticles for the delivery of imaging and therapeutic agents to the tumor site due to its amphiphilic character. Additionally, due to their functional groups, cationic nature, and biocompatibility, glycol chitosan nanoparticles loaded with photosensitizers and anti-tumor drugs are known to be promising candidates for cancer imaging and therapy [55].

17.4.4 Hyaluronic Acid–Based Theranostic Nanoparticles

Hyaluronic acid, an anionic and non-sulfated glycosaminoglycan, is made up of alternating D-glucuronic acid and D-N-acetylglucosamine disaccharide units that are connected by alternating 1,4 and 1,3 glycosidic linkages [56], as shown in Figure 17.3. Hyaluronic acid is an important component of human tissues such as epithelia, connective tissue, neural tissue, skin, vitreous, and synovial fluid. There are several functions that this polysaccharide plays in the body, such as the regulation of cell motility, proliferation, and adhesion between cells. Hyaluronic acid plays an important role in the development of embryos, healing of wounds, and inflammation. In an average 70-kg person, approximately 15 g of hyaluronic acid is present because it is an essential structural molecule [57].

The commercially available hyaluronic acid is produced by fermentation of *Streptococcus* species. Hyaluronic acid can be obtained from umbilical cords, synovial fluids, rooster combs, or the vitreous humor [58]. Hyaluronic acid offers good application prospects for use in biomedicine because of its widespread distribution in the body and inherent qualities such as non-immunogenicity, biodegradability, and biocompatibility. Additionally, non-immunogenic and very viscous hyaluronic acid solutions can be purchased commercially for use as an ophthalmic surgical aid. There is also application of hyaluronic acid in the management of osteoarthritis, where it is used as visco-supplementation for synovial fluids [59]. Hyaluronic acid is useful for drug delivery, particularly in cancer therapy where the overexpressed hyaluronic acid–specific cell receptors CD44 and CD88 that take part in interactions with tumor cell adhesion are suited [58, 60].

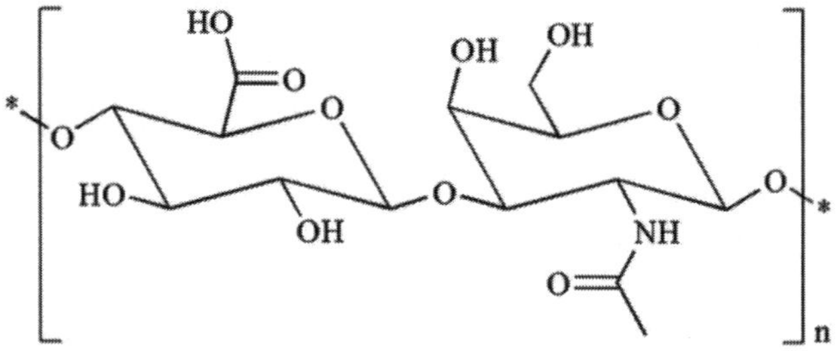

FIGURE 17.3 Basic structure of hyaluronic acid.

With the use of labile or stable covalent bonds, small drugs can be conjugated to hyaluronic acid to form prodrugs. Taxol and doxorubicin, model anti-cancer drugs, were conjugated directly to hyaluronic acid through ester bonds. These anticancer drugs were also conjugated by means of linkers ranging from small bifunctional molecules [61–63] to polymers modified with an enzyme-degradable peptide [64] or by a pH-sensitive linker (cis-aconityl) coupled with a redox disulfide bond [65]. By joining the 2'-OH of taxol via a succinate ester to adipic dihydrazide-modified hyaluronic acid, Prestwich et al. developed and synthesized a series of hyaluronic acid-taxol bioconjugates (HA-Taxol) [66]. Using carbodiimide chemistry and 1-ethyl-3-(3-dimethylaminopropyl) carbodiimide as a coupling agent, HA-Taxol was produced. Hyaluronic acid receptors are overexpressed in some human cancer cell lines such as colon, ovarian, and breast cancers; thus, HA-Taxol conjugates may exhibit selective toxicity. Prestwich et al. found minimal toxicity in mouse fibroblast cells treated parallel to cancer cells at the same concentrations [66].

In another study, a multifunctional nanocomposite was developed by loading copper sulfide (CuS) into Cy5.5-conjugated hyaluronic acid nanoparticles (HANP). This yielded an activatable Cy5.5–HANP/CuS (HANPC) nanocomposite. This was to aid tumor-targeted optical/photoacoustic image-guided photothermal therapy (PTT) [67]. It is important to note that theranostic design utilizes either imaging modality to identify the precise location of laser irradiation for PTT. The HANPC exhibits enhanced fluorescence signals when the nanoparticles degrade at the tumor site, so it can act as an imaging probe that can be activated and used for cancer imaging [28]. It is the proximity of the CuS to the nanoparticle core that quenches Cy5.5 fluorescence. Cy5.5 fluorescence is recovered in tumors once the hyaluronic acid backbone is degraded, resulting in significantly increased fluorescence signal intensity. Moreover, CuS serves as photoacoustic imaging as well as PTT due to its highly absorbing properties [28].

Unlike chitosan, which carries positive charges from the D-glucuronic acid unit, hyaluronic acid carries negative charges from the D-glucuronic acid unit, making it unsuitable for the direct transport of nucleic acids by electrostatic contact, since these are likewise negatively charged. Due to the usefulness of hyaluronic acid, methods were developed to overcome this flaw [39]. First, by generating a polyelectrolyte complex, hyaluronic acid can be added to an existing nucleic acid carrier to alter it and increase its effectiveness. The need for a nucleic acid carrier can subsequently be met by covalently conjugating hyaluronic acid to a positively charged unit [39].

Additionally, hyaluronic acid–based nanoparticles can be functionalized as a non-viral siRNA carrier with tumor imaging capabilities [68]. Using hyaluronic acid-grafted poly-dimethylaminoethyl methacrylate that can be chemically crosslink through disulfide bonds, Yoon et al. produced a stimuli-responsive, non-viral carrier of siRNA for gene silencing therapy [68]. Additionally, the siRNA delivery mechanism was created to react with two environmental triggers following nanoparticle uptake [28]. First, the disulfide linkages that crosslink the nanoparticle are reduced by glutathione, which is almost 1000 times more abundant inside the cell than outside. In order to break down the nanoparticle and release siRNA inside the cell, intracellular hyaluronidase first has to destroy hyaluronic acid [69].

Hyaluronic acid nanoparticles, as demonstrated by these examples, are not only useful disease-targeting carriers for imaging and therapy, but they can also be used in different theranostic strategies and nanoformulations.

REFERENCES

1. Kaur, C. and U. Garg, *Artificial intelligence techniques for cancer detection in medical image processing: A review.* Materials Today: Proceedings, 2023. 81(2): pp. 806–809.
2. Davatgaran-Taghipour, Y., et al., *Polyphenol nanoformulations for cancer therapy: Experimental evidence and clinical perspective.* International Journal of Nanomedicine, 2017. **12**: p. 2689–2702.
3. Thun, M.J., et al., *The global burden of cancer: Priorities for prevention.* Carcinogenesis (New York), 2009. **31**(1): p. 100–110.

4. Alavi, M. and A. Nokhodchi, *Micro- and nanoformulations of paclitaxel based on micelles, liposomes, cubosomes, and lipid nanoparticles: Recent advances and challenges*. Drug Discovery Today, 2022. **27**(2): p. 576–584.
5. Ito, K., et al., *Sequential therapy with crizotinib and alectinib in ALK-rearranged non—small cell lung cancer—a multicenter retrospective study*. Journal of Thoracic Oncology, 2017. **12**(2): p. 390–396.
6. Khan, H., et al., *Flavonoids nanoparticles in cancer: Treatment, prevention and clinical prospects*. Seminars in Cancer Biology, 2021. **69**: p. 200–211.
7. Carlson, M.J., K.W. Thiel, and K.K. Leslie, *Past, present, and future of hormonal therapy in recurrent endometrial cancer*. International Journal of Women's Health, 2014. **6**: p. 429–435.
8. Eeles, R.A., et al., *Adjuvant hormone therapy may improve survival in epithelial ovarian cancer: Results of the AHT randomized trial*. Obstetrical & Gynecological Survey, 2016. **71**(4): p. 223–224.
9. Tan, S., D. Li, and X. Zhu, *Cancer immunotherapy: Pros, cons and beyond*. Biomedicine & Pharmacotherapy, 2020. **124**: p. 109821 109821.
10. Hoteit, M., et al., *Cancer immunotherapy: A comprehensive appraisal of its modes of application (Review)*. Oncology Letters, 2021. **22**(3): p. 1.
11. Schena, E., P. Saccomandi, and Y. Fong, *Laser ablation for cancer: Past, present and future*. Journal of Functional Biomaterials, 2017. **8**(2): p. 19.
12. Cheng, G. and B. Li, *Nanoparticle-based photodynamic therapy: New trends in wound healing applications*. Materials Today Advances, 2020. **6**: p. 100049.
13. Calixto, G.M.F., et al., *Nanotechnology-based drug delivery systems for photodynamic therapy of cancer: A review*. Molecules (Basel, Switzerland), 2016. **21**(3): p. 342–342.
14. Choi, K.Y., et al., *Theranostic nanoplatforms for simultaneous cancer imaging and therapy: Current approaches and future perspectives*. Nanoscale, 2012. **4**(2): p. 33–342.
15. Koo, H., et al., *In vivo targeted delivery of nanoparticles for theranosis*. Accounts of Chemical Research, 2011. **44**(10): p. 1018–1028.
16. Ambrogio, M.W., et al., *Mechanized silica nanoparticles: A new frontier in theranostic nanomedicine*. Accounts of Chemical Research, 2011. **44**(10): p. 903–913.
17. Xiaoyuan, C. and S.T.C. Wong, *Chapter 1—cancer theranostics: An introduction*. Elsevier 2014. p. 3–8.
18. Boisseau, P. and B. Loubaton, *Nanomedicine, nanotechnology in medicine*. Comptes Rendus Physique, 2011. **12**(7): p. 620–636.
19. Tiwari, J.N., R.N. Tiwari, and K.S. Kim, *Zero-dimensional, one-dimensional, two-dimensional and three-dimensional nanostructured materials for advanced electrochemical energy devices*. Progress in Materials Science, 2012. **57**(4): p. 724–803.
20. Gavas, S., S. Quazi, and T.M. Karpiński, *Nanoparticles for cancer therapy: Current progress and challenges*. Nanoscale Research Letters, 2021. **16**(1): p. 173–173.
21. Xie, J., S. Lee, and X. Chen, *Nanoparticle-based theranostic agents*. Advanced Drug Delivery Reviews, 2010. **62**(11): p. 1064–1079.
22. Kirchdoerfer, R.N., et al., *Pre-fusion structure of a human coronavirus spike protein*. Nature (London), 2016. **531**(7592): p. 118–121.
23. Toy, R., et al., *Targeted nanotechnology for cancer imaging*. Advanced Drug Delivery Reviews, 2014. **76**: p. 79–97.
24. Bazak, R., et al., *Passive targeting of nanoparticles to cancer: A comprehensive review of the literature*. Molecular and Clinical Oncology, 2014. **2**(6): p. 904–908.
25. Pearce, A.K. and R.K. O'Reilly, *Insights into active targeting of nanoparticles in drug delivery: Advances in clinical studies and design considerations for cancer nanomedicine*. Bioconjugate Chemistry, 2019. **30**(9): p. 2300–2311.
26. Jahan, S.T., et al., *Targeted therapeutic nanoparticles: An immense promise to fight against cancer*. Journal of Drug Delivery, 2017: p. 9090324–9090325.
27. Bloise, N., et al., *Targeting the "sweet side" of tumor with glycan-binding molecules conjugated-nanoparticles: Implications in cancer therapy and diagnosis*. Nanomaterials (Basel, Switzerland), 2021. **11**(2): p. 289.
28. Swierczewska, M., et al., *Polysaccharide-based nanoparticles for theranostic nanomedicine*. Advanced Drug Delivery Reviews, 2016. **99**(Pt A): p. 70–84.
29. Rojo, J., J.C. Morales, and S. Penades, *ChemInform abstract: Carbohydrate-carbohydrate interactions in biological and model systems*. ChemInform, 2002. **33**(29).
30. Lee, Y.C. and R.T. Lee, *Carbohydrate-protein interactions: Basis of glycobiology*. Accounts of Chemical Research, 1995. **28**(8): p. 321–327.

31. Adachi, N., et al., *Cellular distribution of polymer particles bearing various densities of carbohydrate ligands.* Journal of Biomaterials Science. Polymer ed., 1994. **6**(5): p. 463–479.
32. Dankyi, B.O, et al., *Chitosan coated hydroxypropylmethyl cellulose microparticles of levodopa (and carbidopa): In vitro and rat model kinetic characteristics.* Current Therapeutic Research, 2020. **93**: p. 100612.
33. Gim, S., et al., *Carbohydrate-based nanomaterials for biomedical applications.* Wiley Interdisciplinary Reviews. Nanomedicine and Nanobiotechnology, 2019. **11**(5): p. e1558-n/a.
34. Ghosh, S.C., S. Neslihan Alpay, and J. Klostergaard, *CD44: A validated target for improved delivery of cancer therapeutics.* Expert Opinion on Therapeutic Targets, 2012. **16**(7): p. 635–650.
35. Lee, S.J., et al., *Tumor-homing photosensitizer-conjugated glycol chitosan nanoparticles for synchronous photodynamic imaging and therapy based on cellular on/off system.* Biomaterials, 2011. **32**(16): p. 4021–4029.
36. Choi, K.Y., et al., *Versatile RNA interference nanoplatform for systemic delivery of RNAs.* ACS Nano, 2014. **8**(5): p. 4559–4570.
37. Mizrahy, S. and D. Peer, *Polysaccharides as building blocks for nanotherapeutics.* Chemical society reviews, 2012. **41**: p. 2623–2640.
38. Zhang, W., et al., *ER stress potentiates insulin resistance through PERK-mediated FOXO phosphorylation.* Genes & Development, 2013. **27**(4): p. 441–449.
39. Liu, K., X. Jiang, and P. Hunziker, *Carbohydrate-based amphiphilic nano delivery systems for cancer therapy.* Nanoscale, 2016. **8**(36): p. 1691–16156.
40. Janes, K.A., P. Calvo, and M.J. Alonso, *Polysaccharide colloidal particles as delivery systems for macromolecules.* Advanced Drug Delivery Reviews, 2001. **47**(1): p. 83–97.
41. Tomihata, K. and Y. Ikada, *In vitro and in vivo degradation of films of chitin and its deacetylated derivatives.* Biomaterials, 1997. **18**(7): p. 567–575.
42. Lee, K.Y., W.S. Ha, and W.H. Park, *Blood compatibility and biodegradability of partially N-acylated chitosan derivatives.* Biomaterials, 1995. **16**(16): p. 1211–1216.
43. Kean, T. and M. Thanou, *Biodegradation, biodistribution and toxicity of chitosan.* Advanced Drug Delivery Reviews, 2010. **62**(1): p. 3–11.
44. Bhattarai, N., J. Gunn, and M. Zhang, *Chitosan-based hydrogels for controlled, localized drug delivery.* Advanced Drug Delivery Reviews, 2010. **62**(1): p. 83–99.
45. Park, J.H., et al., *Targeted delivery of low molecular drugs using chitosan and its derivatives.* Advanced Drug Delivery Reviews, 2010. **62**(1): p. 28–41.
46. Liu, Z., et al., *Polysaccharides-based nanoparticles as drug delivery systems.* Advanced Drug Delivery Reviews, 2008. **60**(15): p. 1650–1662.
47. Rudzinski, W.E. and T.M. Aminabhavi, *Chitosan as a carrier for targeted delivery of small interfering RNA.* International Journal of Pharmaceutics, 2010. **399**(1): p. 1–11.
48. Ringsdorf, H., *Structure and properties of pharmacologically active polymers.* Journal of Polymer Science. Polymer Symposia, 1975. **51**(1): p. 135–153.
49. Seymour, L.W., et al., *N-(2-hydroxypropyl)methacrylamide copolymers targeted to the hepatocyte galactose-receptor: Pharmacokinetics in DBA2 mice.* British Journal of Cancer, 1991. **63**(6): p. 859–866.
50. Yoon, H.Y., et al., *Glycol chitosan nanoparticles as specialized cancer therapeutic vehicles: Sequential delivery of doxorubicin and Bcl-2 siRNA.* Scientific Reports, 2014. **4**(1): p. 6878.
51. Yhee, J.Y., et al., *Cancer-targeted MDR-1 siRNA delivery using self-cross-linked glycol chitosan nanoparticles to overcome drug resistance.* Journal of Controlled Release, 2015. **198**: p. 1–9.
52. García Calavia, P., et al., *Photosensitiser-gold nanoparticle conjugates for photodynamic therapy of cancer.* Photochemical & Photobiological Sciences, 2018. **17**(11): p. 1534–1552.
53. Simões, J.C.S., et al., *Conjugated photosensitizers for imaging and PDT in cancer research.* Journal of Medicinal Chemistry, 2020. **63**(23): p. 14119–14150.
54. Soubhagya, A.S. and M. Prabaharan, *Chitosan-based theranostics for cancer therapy.* Springer International Publishing: Cham, 2021. p. 271–292.
55. Lee, S.J., et al., *Tumor specificity and therapeutic efficacy of photosensitizer-encapsulated glycol chitosan-based nanoparticles in tumor-bearing mice.* Biomaterials, 2009. **30**(15): p. 2929–2939.
56. Raemdonck, K., et al., *Polysaccharide-based nucleic acid nanoformulations.* Advanced Drug Delivery Reviews, 2013. **65**(9): p. 1123–1147.
57. Stern, R., *Hyaluronan catabolism: A new metabolic pathway.* European Journal of Cell Biology, 2004. **83**(7): p. 317–325.

58. Mizrahy, S. and D. Peer, *Polysaccharides as building blocks for nanotherapeutics.* Chemical Society Reviews, 2012. **41**(7): p. 2623–2640.
59. Yadav, A.K., P. Mishra, and G.P. Agrawal, *An insight on hyaluronic acid in drug targeting and drug delivery.* Journal of Drug Targeting, 2008. **16**(2): p. 91–107.
60. Platt, V.M. and F.C. Szoka, *Anticancer therapeutics: Targeting macromolecules and nanocarriers to hyaluronan or CD44, a hyaluronan receptor.* Molecular Pharmaceutics, 2008. **5**(4): p. 474–486.
61. Luo, Y., M.R. Ziebell, and G.D. Prestwich, *A hyaluronic acid–taxol antitumor bioconjugate targeted to cancer cells.* Biomacromolecules, 2000. **1**(2): p. 208–218.
62. Auzenne, E., et al., *Hyaluronic acid- paclitaxel: Antitumor efficacy against CD44(+) human ovarian carcinoma xenografts.* Neoplasia (New York, N.Y.), 2007. **9**(6): p. 479–486.
63. Leonelli, F., et al., *A New and simply available class of hydrosoluble bioconjugates by coupling paclitaxel to hyaluronic acid through a 4-hydroxybutanoic acid derived linker.* Helvetica Chimica Acta, 2005. **88**(1): p. 154–159.
64. Yi, L.U.O., et al., *Targeted delivery of doxorubicin by HPMA copolymer-hyaluronan bioconjugates.* Pharmaceutical Research, 2002. **19**(4): p. 396–402.
65. Lin, C.-J., et al., *Integrated self-assembling drug delivery system possessing dual responsive and active targeting for orthotopic ovarian cancer theranostics.* Biomaterials, 2016. **90**: p. 12–26.
66. Luo, Y. and G.D. Prestwich, *Synthesis and selective cytotoxicity of a hyaluronic acid–antitumor bioconjugate.* Bioconjugate Chemistry, 1999. **10**(5): p. 755–763.
67. Zhang, L., et al., *Activatable hyaluronic acid nanoparticle as a theranostic agent for optical/photoacoustic image-guided photothermal therapy.* ACS Nano, 2014. **8**(12): p. 12250–12258.
68. Yoon, H.Y., et al., *Bioreducible hyaluronic acid conjugates as siRNA carrier for tumor targeting.* Journal of Controlled Release, 2013. **172**(3): p. 653–661.
69. Vallet-Regí, M., et al., *Engineering mesoporous silica nanoparticles for drug delivery: Where are we after two decades?* Chemical Society reviews, 2022. **51**(13): p. 5365–5451.

18 Subcellular Targeting of Nanoparticles for Cancer Theranostics

Vivek Patel, Ajay J. Khopade and Jayvadan K. Patel

CONTENTS

18.1 Introduction 296
18.2 Nanoparticles 298
18.3 Nanoparticle Cellular Uptake through Material Design 298
    18.3.1 Nanoparticle Size and Shape 298
    18.3.2 Nanoparticle Charge 301
    18.3.3 Nanoparticle Elasticity 302
    18.3.4 Nanoparticle Surface Modifications with Targeting Ligands 302
18.4 Specific Subcellular Organelle Targeting 302
    18.4.1 Cytosolic Targeting 302
    18.4.2 Endo/Lysosomal Targeting 305
    18.4.3 Mitochondria Targeting 306
    18.4.4 Nucleus Targeting 307
    18.4.5 Golgi Apparatus Targeting 308
    18.4.6 Endoplasmic Reticulum Targeting 309
    18.4.7 Peroxisomal Targeting 309
    18.4.8 Proteasomal Targeting 310
18.5 Conclusion and Future Perspectives 310
References 310

18.1 INTRODUCTION

Traditional treatments of cancer such as radiation and chemotherapy therapy are frequently limited by side effects due to a lack of targeted delivery for carrying anticancer drugs to the neoplastic tissues and low drug concentration at the site. To overcome these limitations, subcellular organelle-targeted nanoformulations of anticancer drugs are required. Current developments in nanoformulation-based organelle-targeted drug delivery systems have demonstrated that they can transport drugs to pathological tissues via cell-membrane targeting and release drugs in the cytoplasm (1–3). To be effective, drug nanoparticles (NPs) must enter the cytoplasm of the targeted cell and gain access to a specific organelle such as endosome/lysosome, endoplasmic reticulum (ER), mitochondrion, nucleus, nucleolus, Golgi apparatus (GA), proteasomes and peroxisomes. (4). Organelle-specific delivery has emerged as a key strategy for targeted drug delivery research for cancer theranostics (5).

    Organelles are critical for cell function (6–15). Mitochondria are known as the powerhouse of the cells: they are in charge of the adenosine triphosphate (ATP) synthesis from food, calcium ion cycle and apoptosis regulation. The lysosome is involved in digestion, autophagy and cellular defense. Protein synthesis and transportation are facilitated by the endoplasm reticulum and Golgi apparatus.

Subcellular Targeting of Nanoparticles for Theranostics

The nucleus regulates gene expression as well as cell proliferation (16, 17). Taking advantage from the critical biological effects of various subcellular organelles, driving appropriate therapeutics into targeted organelles may open up a new avenue for cancer treatment (18–29). For example, cisplatin, a first-line chemotherapeutic drug, can crosslink with tumor DNA and reduce gene repression in a variety of cancer cells (30, 31). As a result, cisplatin driving into the nucleus has a higher therapeutic index than diffusing it into the cytoplasm of cancer cells. Therefore, driving into a specific organelle may reduce the side effects and prevent recurrence (32–34).

Organelle-targeted therapeutics have gained popularity in recent years as a means to improve efficiency and specificity while reducing toxicity (35–38). Previous research has shown that organelle-targeted nanoformulations can help overcome a variety of challenges, including drug resistance, drug premature leakage and high dosage requirements (39–49).

Günter Blobel of Rockefeller University was awarded the Nobel Prize in Physiology or Medicine in 1975 for his discovery that proteins have intrinsic signals that govern their transport and localization in the cell. The 'signal hypothesis' predicted that 'zip codes' were in charge of directing certain proteins to subcellular compartments such as the cytosol, cytoplasm, endosome/lysosome nucleus, nucleolus, mitochondria, endoplasmic reticulum, Golgi apparatus, proteasomes and peroxisomes (Figure 18.1) (50–52). The intracellular delivery of drug nanoformulations can significantly impact therapeutic efficacy. The biological efficacy of certain drugs is dependent on precise organelle targeting. As an example, drugs designed for gene therapy must eventually be directed to the cell nucleus in order for the therapeutic protein to be expressed. Certain drugs, such as RNAi, must target the cytosol in order to inhibit mRNA production in cells. Similarly, pro-apoptotic drugs can

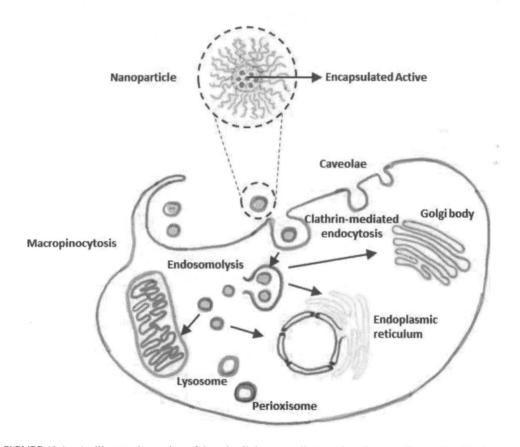

FIGURE 18.1 An illustrated overview of the subcellular organelle-targeting elements discussed in this chapter.

be precisely targeted to the mitochondria where they exert their therapeutic effects. Many proteins in signal transduction pathways are initially localized to a single organelle-targeted compartment, bind to protein partners in the cell (or are signaled by other proteins) and then change their position in the cell, contributing to a change in therapeutic function. Finally, we seek not only targeting of a specific organelle but also a higher level of enlightenment in which multiple signals can be used to target multiple organelles. Indeed, the ability to precisely target drugs to different organelles is altering cancer theranostics development (53).

This chapter focusses on therapeutic targeting to specific cellular organelles, as illustrated in Figure 18.1. In the case of methods to reach and target the organelle for therapy, each cellular organelle is concisely described. This chapter will also will also define and elucidate the various NP platforms currently available in pre-clinical, clinical and translational cancer theranostics as well as focusing on how nanoformulations physicochemical properties, such as size, shape, elasticity and surface modifications, affect cellular uptake. Finally, the future of cancer theranostics delivery will be discussed.

18.2 NANOPARTICLES

The Commission of the European Union defined NPs as "any naturally occurring or manufactured particulate material either unbound, as an aggregate or agglomerate, where at least half of the particles have one or more external dimensions between 1 and 100nm in size". Despite the fact that NPs are often defined by their size and generally defined by their properties not obtained by other particles with the same chemical composition, there is no formal consensus on the definition, for example, NP behavior and physical properties influenced by large surface area to volume ratio, high mobility in the free-state and the nanosized-dependent display of quantum effects, among other things (54–57).

NPs were prepared using various materials and techniques, resulting in NPs with versatile functional properties. The resulting diversity has translational potential for a variety of medical applications.

Table 18.1 summarizes the advantageous physical properties of the NP platforms discussed in this chapter. Given these characteristics, it is not surprising that NPs have received increasing attention over the last decade for their exciting translational potential in both cancer diagnosis and treatment. Many nanoformulations have been approved for clinical use in cancer chemotherapy, and several nanomedicines are currently undergoing clinical trials (see Table 18.2).

18.3 NANOPARTICLE CELLULAR UPTAKE THROUGH MATERIAL DESIGN

Nano-formulation researchers are focusing on controlling NPs in order to produce calculated nanoparticle–cell interactions by specifying physicochemical parameters such as size, shape, charge, elasticity and surface modifications with targeting ligands for cellular uptake among various types of cells.

18.3.1 Nanoparticle Size and Shape

Size and shape are critical physicochemical parameters that impact diffusivity, surface-to-cell membrane contact area, membrane adhesion and the strain energy required for cell plasma membrane movement (88).

Nanoparticle size is important for cellular interaction and uptake, but at the same time, other variables play equally important roles in cellular entry such as nanoparticle morphology, density, nanoparticle rate of sedimentation, cell phenotype and protein corona formation (89–94).

Larger NPs (≥250 nm) were internalized only via caveolae-mediated endocytosis, whereas smaller NPs (≤100 nm) were internalized through both caveolae- and clathrin-dependent endocytosis.

TABLE 18.1
NP Platforms and Favorable Physical Characteristics

Nanoparticle Illustration	Size (nm)	Favorable Physical Characteristics	Ref.
Natural polymers (albumin)	50–300	Biocompatible, biodegradable, non-toxic, non-immunogenic, straightforward to cross-link and chemically modify. Drugs readily incorporated into albumin polymer matrix.	(58–69)
Synthetic polymers (poly-lactic-co-glycolic acid (PLGA))	50–300	Synthetic material that is non-toxic and undergoes hydrolysis in vivo to produce biodegradable metabolites. Drugs easily incorporated into the matrix.	(59, 69)
Liposomes	20–1000	Well established as cancer therapeutics, can encapsulate hydrophilic and lipophilic drugs, soluble, low uptake by macrophages, favorable stability in vivo, and as colloids, protect drugs from breakdown.	(60, 69)
Micelles	10–100	Unique core-shell architecture; hydrophobic core acts as natural carrier environment, hydrophilic shell enables stability in aqueous solution, structural modifications can further augment tumor cell uptake.	(61, 69)
Dendrimers	5–20	Symmetrical branched polymeric macromolecules with a central core allow either encapsulation or conjugation of therapeutic agent. Self-assembling, polyvalent, chemically stable, non-toxic and soluble.	(62, 63, 69)
Iron-oxide nanoparticles (superparamagnetic iron-oxide nanoparticles (SPIONs))	10–100	Biocompatible and biodegradable, established clinical use as magnetic resonance imaging (MRI) contrast agents, controllable by an externally applied magnetic field; diverse formulations allow fine-tuning of physicochemical properties.	(64, 69)

TABLE 18.1 *(Continued)*
NP Platforms and Favorable Physical Characteristics

Nanoparticle Illustration	Size (nm)	Favorable Physical Characteristics	Ref.
Silica (porous silica nanoparticles (pSiNPs))			
	50–1000	Good chemical and thermal stability, large surface area and pore volume. Can encapsulate large amounts of bioactive molecules and promote controlled drug release. Also offer simple surface functionalization.	(65)
Gold nanoparticles (gold nanoparticles (AuNPs), gold nanorods)			
	1–100	Intense light absorption, high photothermal conversion rate and ease of synthesis in a variety of shapes and sizes. Good biocompatibility, colloidal stability and simple ligand conjugation chemistry.	(66, 69)
Carbon nanoparticles (carbon nanotubes (CNTs))			
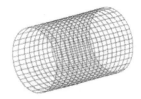	1–10	High carrying capacity and high propensity to traverse cell membranes Easily chemically modified or functionalized through formation of stable covalent bonds.	(67)
Quantum dots (QDs)			
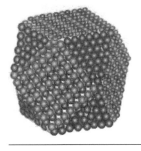	2–10	Broad absorption spectra and high fluorescence quantum yield, high photostability. Possibility of multiplexing (different colors of QDs used within one assay with a single excitation source). Can be combined with other NP platforms for cancer theranostics.	(68)

Similarly, when block copolymer NPs (micelles ≤50 nm in diameter and vesicles ≤100 nm in diameter) were synthesized with different sizes using different hydrophilic chain lengths, then distinct uptake patterns were seen due to the change in size. Smaller micelles were internalized more quickly than the larger vesicles, so nanoparticle size not only affects the underlying uptake kinetics but also the efficiency of nanoparticle delivery (95).

Besides nanoparticle size, both the shape anisotropy and orientation of the nanoparticle relative to the plasma membrane impact cellular uptake (96, 97). For example, when comparing cellular uptake between rod-like shapes and spheres, it was found that 50-nm spherical gold NPs had higher accumulation inside human prostate cancer cells compared to gold nanorods (98).

TABLE 18.2
Nanoparticles Clinically Approved or Under Clinical Trial

Product	Drug	Carrier components	Company	Stage	Ref.
Nab-paclitaxel (Abraxane)	Paclitaxel	Human serum albumin	Abraxis BioScience	FDA and EMA approved	(70)
Genexol-PM	Paclitaxel	Micelle: mPEG-PDLLA	Samyang Biopharm	Approved in Korea	(71)
Apealea	Paclitaxel	Micelle: two isoforms of N-retinoyl-L-cysteic acid Methyl ester sodium salt	Oasmia Pharmaceutical	EMA approved	(72)
Lipusu	Paclitaxel	Liposome: lecithin/cholesterol	Nanjing Luye Sike Pharmaceutical Co.	Phase IV	(73)
Taclantis/ Bevetex	Paclitaxel	Concentrate for nanodespersion: polyvinylpyrrolidone, cholesteryl sulfate and caprylic acid	Sun Pharma Advance Research Co. Ltd.	Approved in India Phase II in US	(74)
Doxil	Doxorubicin	Liposome: HSPC, cholesterol, mPEG-DSPE	Johnson & Johnson	FDA and EMA approved	**(75)**
Myocet	Doxorubicin	Liposome: phosphatidylcholine, cholesterol	Teva	EMA approved	(76)
ThermoDox	Doxorubicin	Thermosensitive liposomal doxorubicin	MedKoo Biosciences Inc.	Phase III completed	(77)
Nanoparticle generator	Doxorubicin	Porous silicon microparticle with polymeric doxorubicin	Houston Methodist Research Institute	Planning of Phase I	(78)
NC-6004	Cisplatin	Micelle: PEG-P(Glu)	Nano Carrier Co.	Phase I/II	(79)
Lipoplatin	Cisplatin	Liposome: SPC/cholesterol/ DPPG/mPEG DSPE	Regulon Inc.	Phase II/III	(80)
CRLX101	Camptothecin	PEG-modified β-cyclodextrin	Cerulean Pharma Inc.	Phase II	(81)
NKTR-102	Irinotecan	PEG (four-arm) conjugation	Nektar Therapeutics	Phase II	(82)
Onivyde	Irinotecan	Liposome: DSPC, cholesterol, mPEG-DSPE	Merrimack Pharmaceuticals	Phase II/III	(83)
DOTAP: Chol-TUSC2	TUSC2	DOTAP: Chol	Genprex, Inc.	Phase I/II	(84)
Mepact	Mifamurtide	Liposome: POPC, OOPS	Takeda Pharmaceutical	EMA approved	(85)
Marqibo	Vincristine sulfate	Liposome: sphingomyelin, cholesterol	Talon Therapeutics	FDA approved	(86)
Vyxeos	Cytarabine and daunorubicin	Liposome: DSPC, DSPG, cholesterol	Jazz Pharmaceuticals	FDA and EMA approved	**(87)**

Abbreviations: FDA, U.S. Food and Drug Administration; EME, European Medicines Agency; mPEG-PDLLA, Poly(D,L-lactide)-PEG-methyl ether; DSPE, 1,2-Distearoyl-sn-glycero-3-phosphoethanolamine; SPC, soybean phosphatidylcholine; DOTAP, 1,2-dioleoyl-3-trimethylammonium-propane; DSPG, 1,2-Distearoyl-sn-glycero-3-phosphoglycerol, sodium salt; POPC, 1-palmitoyl-2-oleoylphosphatidylcholine; OOPS, 1,2-dioleoylphosphatidylserine

18.3.2 Nanoparticle Charge

A synthetic nanoparticle has an either positive, negative or neutral surface charge, which is measured by zeta potential. Zeta potential of nanoformulations is quantified by electrophoretic mobility measurements of colloidal dispersions in aqueous medium. Therefore, a nanoparticle's zeta potential indicates the overall nanoparticle surface charge in the corresponding colloidal dispersion (99).

Surface charge is the dominant factor that influences internalization. Since the cell plasma membrane is negatively charged, based on Coulomb's law, positively charged NPs are more likely to be electrostatically attracted by the negatively charged plasma membrane, resulting in increased accumulation inside cells (100–104). However, several previous studies report that negatively charged NPs can overcome the negatively charged cell plasma membrane and accumulate within cells [146–150]. This indicates that nanoparticle cellular uptake is more complicated and goes well beyond the simplified notion of Coulomb-driven electrostatic interactions (105–110).

The generation of a zwitterionic surface means both positive and negative charges are present on a nanoparticle's surface and can be helpful in significant reduction of protein corona formation. This may alter and affect nanoparticle biodistribution and cellular interactions in comparison to only cationic or anionic nanoparticle designs (111). In summary, surface charge should be considered very carefully when designing NPs for medical and biological applications.

18.3.3 Nanoparticle Elasticity

Elasticity can impact targeting, bio-distribution and cellular uptake. Nanoparticle "softness" and "stiffness" can be quantified by Young's modulus, which defines the relationship between stress and strain for a material. Atomic force microscopy (AFM) is used to measure nanoparticle elasticity. It was found that "soft" NPs have higher cellular uptake rates than "stiff" NPs due to potential receptor diffusion and larger contact surface area with the cell plasma membrane (112–114).

18.3.4 Nanoparticle Surface Modifications with Targeting Ligands

The NP surface can be modified with targeting ligands to enable specific interaction and binding of NPs to cell surface receptors. This is a eminent concept in nanomedicine and referred to as "active targeting" (115–118). Targeting ligands that are generally used in nanomedicine include folate, transferrin, aptamers, peptides, small molecules, proteins, antibodies, antibody fragments and nucleic acids. To engineer active targeting NPs, a number of design parameters need to be considered and optimized for efficient targeting. These parameters include target ligand length, target ligand density, hydrophobicity and avidity (119–126). Table 18.3 summarizes examples of nanoparticle-targeting ligands that have been reported for in vitro and/or in vivo applications in nanomedicine.

NPs that do not exhibit specific surface targeting ligands are called passive targeting NPs. Passive targeting NP interactions with cells are non-specific. This type of non-specific interaction may facilitate NP uptake in healthy as well as diseased cells (143–146).

18.4 SPECIFIC SUBCELLULAR ORGANELLE TARGETING

18.4.1 Cytosolic Targeting

NPs that target specific cellular organelles must first be delivered to the cytosol. To reach the site of drug action in the cells, they have to cross a series of membrane barriers. There are three main barriers that must be overcome to facilitate cytosolic delivery:

- Escaping reticuloendothelial system (RES) detection
- Interaction with the cell membrane and internalization
- Intracellular trafficking and endosomal escape

18.4.1.1 Escaping RES Detection

Endothelial layers of the organs such as spleen, liver and bone marrow contain RES, which represents mononuclear macrophages that filter the blood for foreign pathogenic particles (147). When drug NPs are delivered into the bloodstream, they must avoid detection by RES and then only interact with the cell membrane to achieve cytosolic delivery.

Several approaches have been implemented to avoid macrophage detection and increase the circulatory time. A few examples used for cytosolic drug delivery include stealth liposomes and/

TABLE 18.3
Examples of Different Types of Nanoparticle-Targeting Ligands for Enhanced Nanoparticle Cellular Interaction with Targeted Cell Types

Targeting Ligand	Nanoparticle Core Material	Targeted Cell Types	Ref.
Peptides			
CLT1 peptide for fibronectins	PEG-PLA	C6 glioma cells	(127)
CREKA pentapeptide sequence for fibrin	DSPE-PEG2000	GL261 glioma cells	(128)
M2-macrophage targeting peptide	HPMA polymer	Tumor-associated macrophages	(129, 130)
RGD motif for integrin binding	Mesoporous silica	SCC-7 mouse squamous cell carcinoma, HT-29 human colon cancer cells	(131)
P160 targeting peptide found through phage display	Cadmium-selenide core zinc-sulfide shell quantum dots	MCF-7 human breast cancer cells	(132)
Chlorotoxin peptide for MMP2	Silver	U87MG glioblastoma	(133)
Glycoproteins			
Transferrin to cross blood–brain barrier	DSPC-cholesterol-POPG	U87MG glioblastoma, GL261 glioma cells	(134)
Antibodies and antibody fragments			
Anti-CD8a F(ab')2	PLGA-PEG	CD8+ T-cells	(135)
HuA33 monoclonal antibody	Poly(methacrylic) acid	LIM1889, LIM2405+, LIM2405- human colon cancer cells	(136)
HER2 monoclonal antibody	Gold plasmonic vesicles	SKBR-3 human breast cancers	(137)
Anti-epidermal growth factor receptor	Quantum dots	MDA-MB-231 mammary adenocarcinoma, BxPC-3 pancreatic adenocarcinoma	(138)
Nucleic acids			
Single-stranded oligonucleotide-based aptamers	Quantum dots	A549 lung adenocarcinoma	(139)
Anti-cMet DNA aptamer	Lipidated aptamer-based nanocarriers loaded with doxorubicin	H1838 non-small cell lung cancer	(140)
G- rich DNA aptamer	Zinc gallogermanate	4T1 mammary carcinoma	(141)
Small molecules			
Folic acid for folate receptors	Mesoporous silica	U20 osteosarcoma	(142)

Abbreviations: CTL1, fibronectin targeting peptide; CREKA, fibrin binding peptide cysteine, arginine, glutamic acid, lysine, aspartic acid; RGD, arginine, glycine aspartic acid, MMP2-Matrix metalloproteinase; PEG-PLA, poly(ethylene glycol)-poly(lactide); DSPE-PEG2000, 1,2-distearoyl-sn-glycero-3-phosphoethanolamine-N-[amino(polyethylene glycol)-2000]; HPMA, poly (N-(2-hydroxypropyl) methacrylamide); DSPC, 1,2-distearoyl-sn-glycero-3-phosphocholine; POPG, 1,2-distearoyl-sn-glycero-3-phosphocholine; PLGA-PEG, poly(lactic-co-glycolic acid) and polyethylene glycol

or targeted liposomes (for accumulation in target organs). Stealth liposomes can be developed by amphiphilic stabilizers (e.g., cholesterol) (148–150), phosphatidylinositol and gangliosides (151) or a hydrophilic surface by grafting with polyethylene glycol (152). The combination of stealth and targeted liposomes has proven very effective (153).

18.4.1.2 Cell Membrane Interaction and Internalization

Cell membranes are lipid bilayers made up of phospholipids, cholesterol, glycolipids and proteins. Proteins, particularly proteoglycans, are more important for cationic polymer-based drug delivery. Cationic polymer ions interact with negatively charged sulfated proteoglycans of cell membrane in an ionic cell membrane interaction (154–156). Conjugation with cell-penetrating peptides and proteins can also overcome the barrier of the cell membrane (157, 158). A protein transduction domain with 9–16 cationic amino acid residues could penetrate cell membrane and cross the nucleus (159–163).

Internalization can occur via endocytosis (initiated by electrostatic or hydrophobic interactions with the cell membrane or interaction with a cell-surface receptor), followed by endosomal escape, or via other mechanisms such as macropinocytosis or combinations of these (153). After entering the cytosol, the drug may either act or move to a subcellular compartment (e.g., nucleus, mitochondria and peroxisome). Table 18.4 depicts various cytosol-targeting agents.

TABLE 18.4
Cytosolic Targeting of Drug Therapeutics

Targeting Agents	Composition	Drug Delivered	Ref.
Liposomes	Lipid bilayer neutral lipids, DPPC and cholesterol	Amikacin (Arikace)	(164)
	Soybean oil and phospholipids	Cyclosporin	(165)
	MPEG–DSPE, HSPC and cholesterol	Doxorubicin (Doxil)	(166)
pH-sensitive liposomes	CHEMS and DOPE	Diptheria toxin A chain	(167)
	DOPE, N-succinyl–DOPE and PEG–ceramide	Gentamycin	(168)
Thermosensitive liposomes	DPPC, HSPC and cholesterol liposomes surface-modified with DSPE-PEG-2000:PNIPAMAAM17	Doxorubicin	(169)
Targeted thermal magnetic liposomes	DPPC, cholesterol, DSPE–PEG (2000) and DSPE–PEG (2000)–folate	Doxorubicin	(170, 171)
Polymeric micelles	PEG–poly(aspartic acid) block copolymer	Doxorubicin (Adriamycin)	(172)
pH-sensitive micelles	PEG-block-poly(aspartate-hydrazide) or PEG–p (Asp-Hyd) was modified using either levulinic acid or 4-acetyl benzoic acid attached via hydrazone bonds	Doxorubicin (Adriamycin)	(173)
pH-sensitive micelles with cell surface targeting	Amphiphilic block copolymers that self-assemble into spherical micelles, folate–PEG–poly (aspartate-hydrazone-adriamycin) with γ- carboxylic acid-activated folate	Doxorubicin (Adriamycin)	(174)
Thermosensitive micelles/polymers	Micellar cyclotriphosphazenes	Human growth hormone	(175)
	Biodegradable triblock copolymer of PLGA–PEG–PLGA (ReGel)	Paclitaxel	(176)
Cell-penetrating peptides	Doxorubicin bound to HPMA-based polymer with the cell-penetrating peptide Tat	Doxorubicin	(177)
Cationic polymers and cationic lipids	Polyethyleneimine	Genes (DNA)	(178, 179)
Virus mimetic	Hydrophobic polymer core (poly(L-histidine-co-phenylalanine) poly (His32-co-Phe6) and two layers of hydrophilic shell (one PEG end linked to core polymer, another end to BSA)	Doxorubicin	(180)
NES	NES (LQLPPLERLTL) encoded in a plasmid	Genes (DNA)	(181)
	NES (ALPPLERLTL) conjugated to DNA	Antisense oligonucleotide	(182)

Abbreviations: BSA, Bovine serum albumin; CHEMS, Cholesterylhemisuccinate; DOPE, Dioleoylphosphatidylethanolamine; DPPC, Dipalmitoylphosphatidylcholine; DSPE, Distearoylphosphatidylethanolamine; HPMA, N-(2-hydroxypropyl) methacrylamide; HSPC, Fully hydrogenated soy phosphatidylcholine; MPEG, Methoxypolyethylene glycol; NES, Nuclear export signal

18.4.2 Endo/Lysosomal Targeting

Endocytosis occurs when materials enter cells and are encapsulated by a small portion of the plasma membrane (called clathrin-coated pit formation), followed by a pinching off to form an endocytic vesicle. Similarly, caveolae are formed instead of clathrin in pinocytosis. These endomembrane pockets serve as primary entry points for NPs into the cell and selective conveyance to the cell's interior via the inner compartments known as lysosomes. Lysosomes are single-membrane vesicles with an acidic microenvironment of pH 4.5 to 5.0. Lysosomes contains over 60 hydrolytic enzymes for the digestion of phagocytosed materials, macromolecules (derived intra- or extra-cellularly) and biomolecules (such as proteins, lipids, carbohydrates and nucleic acids) for production of nutrients (183). Lysosome-targeted therapy makes use of the acidic lysosomal microenvironment (pH 4.5 to 5.0) for controlled drug release and activates a variety of therapeutic agents (184, 185). The secondary and tertiary amine groups of cationic polyethyleneimines are protonated in the acidic environment of the endosomes, resulting in osmotic swelling, endosomal rupture and endosomal escape (186, 187). Other methods of escaping the endosome have been used, such as pH-sensitive liposomes and polymers, membrane-disruptive polymers masked by PEG via disulfide groups, membrane-disruptive peptides and so on (188, 189). Table 18.5 shows examples of endosomal/lysosomal targeted nanoformulations for cancer.

TABLE 18.5
Endosomal/Lysosomal Targeted Nanoformulations for Cancer Theranostics

Composition	Size (nm)	Theranostics	Pathway	Outcome	Ref.
LDH-hydrazone-Do	90–120	pH	Proton sponge effect	Enhanced ROS generation	(190)
ORMOSIL	<100	Autophagy	Caveolae-mediated endocytosis	Endocytosis and subcellular localization	(191)
CALNN-capped gold NPs	5	–	Peptide cleavage by the protease cathepsin L	Biosensing, fluorescence Quenching	(192)
pRNA-3WJ	5–20	Immune	Endocytosis	Endosomal escape for siRNA delivery	(193)
FA-SLICS	–	pH	Receptor-mediated endosomal pathway	Controlling the release of loaded genes by the endosomal microenvironment	(194)
L-histidine-based polymeric micelles	–	pH	Endosomolytic activity	Ablating drug-sensitive ovarian cancer as well as drug-resistant counterpart cells	(195)
V-ATPase inhibition	–	Cisplatin	ERK/MEK pathway	Regulating autophagy that assists in chemoresistance in ovarian cancer	(196)
YPSMA-1-PEOz-PLA	–	pH	Degradative lysosomes Pathway	Efficient delivery of anticancer drugs for treating PSMA-positive prostate cancers	(197)
Nanosomes	–	Carbohydrate–lectin	LMP pathway	Facilitating targeted molecular delivery and intracellular traffic through an endocytic route without the influence of protein corona	(198)
T-UPSM	<50	pH	Lysosomal catabolism inhibition	KRAS mutant pancreatic cancer treatment through simultaneous lysosomal pH buffering and rapid drug release	(199)
F-Gly-MTX NPs	<50	Enzymatic, pH	Acid lysosomal compartment and protease	Release of the MTX via peptide bond cleavage in the presence of proteinase K	(200)

Abbreviations: LDHs, Layered double hydroxides; ROS, reactive oxygen species; LMP, lysosome membrane permeabilization; ORMOSIL NPs, Organically-modified silica nanoparticles; pRNA-3WJ, Three-way junction motif of packaging RNA molecules; FA-SLICS, Folic acid functionalized Schiff-base linked imidazole chitosan;/extracellular signal—regulated kinase (ERK)/mitogen-activated protein kinase kinase (MEK) pathway; YPSMA-1-PEOz-PLA, Anti-(prostate specific membrane antigen) antibody-poly(2-ethyl-2-oxazoline)-poly(D, L-lactide); T-UPSM, Triptolide prodrug-loaded ultra-pH-sensitive micelles; F-Gly-MTX NPs, Methotrexate-conjugated glycine-coated magnetic nanoparticles

18.4.3 MITOCHONDRIA TARGETING

The mitochondrial membrane is a two-layer phospholipid membrane with embedded proteins that participate in the respiratory chain complex, ATP synthase electron transport, calcium metabolism, ROS generation, immunity regulation and protein import machinery. As a result, it is known as the "powerhouse" of eukaryotic cells (201–203). Mitochondrion-targeted delivery of theranostics including paclitaxel, etoposide, betulinic acid and ceramide, which efficiently regulate mitochondrial function, may hold promise for efficient cancer treatment (204–207). Mitochondrial targeting is accomplished through the use of delocalized lipophilic cations (DLCs) such as triphenylphosphonium. DLCs reduced the activation energy connected to deionization prior to its transport through the hydrophobic inner mitochondrial membrane (208, 209). DLCs are distributed over the large surface area of inner mitochondrial membrane, increasing the ionic strength and reducing interaction with surrounding water molecules. Cations move across the inner mitochondrial membrane with the more negative strength, increasing the concentration of mitochondria-targeted conjugates in the matrix (210). A variety of triphenylphosphonium-decorated NPs such as nanoprobe, Fe3O4 NPs, carbon dots, mesoporous silica NPs, gold NPs and polymer NPs (211–221) have been investigated for mitochondria-specific theranostics.

Mitochondrial-targeted liposomes made up of positively charged amphiphilic molecules could also be used as potential mitochondrial transporters. The most common example is DQAsomes, which are produced by the dicationic mitochondriotropic compound dequalinium (222). They can transport drugs to mitochondria due to their high mitochondrial affinity (223, 224). It was reported that the DQA80s nanosomes consisting of dequalinium/1,2-dioleoyl-3-trimethylammonium-propane (DOTAP)/1,2-dioleoyl-sn-glycero-3-phosphoethanolamine (DOPE) demonstrated higher cellular uptake and more cytotoxicity than DQAsomes (225).

Mitochondrial targeting signal (MTS) peptides can also be used for mitochondrial targeting to some extent (226). Because mitochondria are 10°C hotter than other organelles, thermo-responsive drug delivery to mitochondria is possible (227, 228). Table 18.6 shows examples of nanomaterials for mitochondria-targeted cancer therapy.

TABLE 18.6
Mitochondria-Targeted Nanomaterials for Cancer Theranostics

Nanomaterials	Therapy	Targeting Ligands	Pathway	Ref.
Nanodiamonds	Chemotherapy	FA, MLS peptide	Receptor- and protein-mediated targeting	(229)
Liposome		TPP	Mitochondrial transmembrane potential mediated accumulation	(230)
PLGA		TPP	GSH-triggered TPP exposure	(231)
UCNPs@TiO$_2$	Photodynamic therapy	TPP	Mitochondrial transmembrane potential mediated accumulation	(232)
PS-peptide NPs		MLS	ROS-accelerated cellular internalization, cationic hydrophobic peptide-mediated targeting	(233)
AIE dots		FA, TPP	Receptor-mediated targeting, mitochondrial transmembrane potential mediated accumulation	(234)
Fe$_3$O$_4$	Photothermal therapy	TPP	Mitochondrial transmembrane potential mediated accumulation	(235)
F127 NPs		Biotin, TPP	Receptor-mediated targeting, mitochondrial transmembrane potential mediated accumulation	(236)
TiO$_2$–Au	Radiotherapy	TPP	Mitochondrial transmembrane potential mediated accumulation	(237)
MOFs	Combined therapy	Ru(bpy)$_3^{2+}$	Mitochondrial transmembrane potential mediated accumulation	(238)

Abbreviations: TPP, Triphenylphosphonium; FA, Folate; MLS, Mitochondria localizing sequence; UCNPs, Upconversion nanoparticles

18.4.4 NUCLEUS TARGETING

The nucleus is the central organelle, contains the majority of the genetic material and plays an important role in cell growth, differentiation, proliferation, apoptosis and so on. The nucleus is known as the cell's control center because it regulates various cell activities by managing gene expressions.

The nucleus is the double lipid bilayer of the nuclear envelope. This pair of peripheral bilayers is perforated by nuclear pore complexes (NPCs). NPCs are large proteinaceous structures that serve as selective portals for macromolecule nucleoplasmic transport. Nucleocytoplasmic transport occurs through the pores formed by NPCs embedded in the nuclear envelope (239). An illustrated overview of the nuclear structure and nuclear entry of the drug-loaded NPs through the nuclear pore complexes is depicted in Figure 18.2.

The domains of unstructured phenylalanine-glycine (FG) repeats in the inner NPC channel act as a steric permeability barrier to larger macromolecules in the form of a disordered filamentous cloud or hydrogel (240–242). The NPC channel is around 39 nm in diameter, though the previously described low-density unstructured domains limit the opening to around 10 nm (240, 243). Therefore, ions and small molecules can freely transport to the nuclear compartment via passive diffusion, whereas macromolecules larger than ~40 kDa require active transport (241, 244).

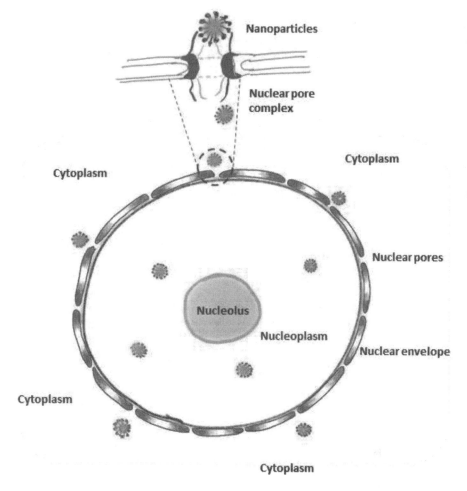

FIGURE 18.2 An illustrated overview of nuclear entry of the drug-loaded NPs through the nuclear pore complexes.

TABLE 18.7
Nucleus-Targeted Nanoformulations for Cancer Theranostics

Composition	Size (nm)	Pathway	Outcome	Ref.
DOX@MSN@TAT/RGD	<50	Vasculature-/membrane-to nucleus sequential drug delivery strategy exploiting RGD and TAT dual peptides	Enhanced in vivo chemotherapeutic efficacy in clinics	(264)
IONPs-TAT/PEG	20	Enhancing the therapeutic effect by nuclear targeting	Simultaneous cancer targeting, imaging and PTT	(265)
UCNPs@TiO2-Ce6-TAT	25	Generation of multiple ROS ($\cdot$OH, O_2^- and $^1O_2$) using the dual photosensitizer	Enhancing PDT can be achieved against MDR cancer	(266)
Chitosan nanoparticles-NLS	25	Study on impaired nuclear import and aberrant nanoparticle intracellular trafficking in glioma	Effect of nanoparticle size and NLS density on nuclear targeting in cancer and normal cells	(267)
AuNPs@PEG@RGD/NLS	35	Mechanical stiffness of the nucleus and stimulate the overexpression of lamin A/C located around the nuclear membrane	Increasing nuclear stiffness and slowing cancer cell migration and invasion	(268)
NLS/RGD-AgNPs	35	ROS generation and effect on cell cycle and cell division	DNA damage and apoptosis	(269)
Acridin-9-methanol nanoparticles	60	Nucleus targeted intracellular controlled drug release	Fluorescent imagine and killing the cancer cells	(270)

Abbreviations: AgNPs, Silver nanoparticles; AuNPs, Gold nanoparticles; MSN, Mesoporous silica nanoparticles; NLS, Nuclear localization sequence; TAT, a HIV peptide; RGD, RGDRGDRGDRGDPGC peptide; MDR, Multidrug resistance; IONP-TAT, Iron oxide nanoparticles-YGRKKRRQRRR peptide

This active transport mechanism is aided by nuclear transport receptors, specifically importin-β/karyopherin-β class receptors, which bind specific macromolecules to be transported across the nuclear envelope frequently with the assistance of adaptor proteins (241). Transport receptors have specific binding of the oligopeptide sequences on the cargo to the receptors known as nuclear localization signals (NLSs), and they can transport the vehicles via the NPCs.

A number of functional NPs, including micelles, MSNPs, QDs, gold NPs, peptide assemblies and drug conjugates, have been shown to target the nucleus via NLS conjugation to the NP surface (245–260). Among all the NLSs, the TAT peptide has been shown to be an efficient moiety for delivering NPs to cell nuclei via binding receptors importin-α and -β (261, 262). The proteins that do not have an identified NLSs have been observed to localize to the nucleolus through interactions with other molecules, such as nucleolin, a protein that may localize to the nucleolus partially due to RNA binding. Recent research has also suggested that nucleolar "hub proteins" play an important role in nucleolar localization (263). Therefore, nucleolar targeting can be achieved through NLSs by being dragged via NLSs containing protein or by binding to a nucleolar hub protein. Table 18.7 shows examples of nanomaterials for nucleus-targeted cancer therapy.

18.4.5 Golgi Apparatus Targeting

The Golgi apparatus is important in post-translational modifications, tumor growth, drug resistance, cancer metastasis and immune evasion (271). Specific therapeutic drug accumulation in the Golgi apparatus may impair protein folding and result in stress signal, leading to cancer cell apoptosis. The GA also transports macromolecules to the plasma membrane, lysosomes and the cell's exterior in secretory vesicles. The GA transports in the inverse direction of drug delivery.

Overexpression of Golgi-associated proteins, pharmacological agents and certain pathological changes can all cause profound changes in the GA which can lead to a variety of neurodegenerative disorders (272–274).

Therapeutic agents targeting the Golgi network have been used in cancer diagnosis and treatment (275, 276). One effective Golgi targeting strategy is to target rapamycin mammalian target (mTOR). mTOR is a phosphatidylinositol 3-kinase (PI3K)-related kinase that responds to cellular and environmental nutrition and energy conditions. The mTOR signal can be activated by a variety of stimuli, including nutrients, ATP and stress signals (277). mTOR is strongly linked to cancer, and activation of mTOR in certain tumor cells inhibits PI3K. Inactivation of mTOR, on the other hand, causes an increase in PI3K activity and as a result a decrease of the antiproliferative effect of mTOR inhibition, killing the tumor (278). As a result, mTOR is a promising therapeutic target in the Golgi network.

Because the Golgi apparatus has been described to have a mechanism for stress-induced initiation of apoptosis (272), this validates future efforts to target therapeutics to the Golgi as a mechanism for specific induction of apoptosis in malignant cells.

18.4.6 Endoplasmic Reticulum Targeting

The ER membrane, which accounts for half of the total membrane of an average cell, is formed by a series of lamellar and tubular cavities. The ER is continuous with the nuclear envelope and extends to the cell periphery, enlacing most cellular organelles, such as mitochondria, the Golgi apparatus and peroxisomes (279). Ribosomes were coated to the rough ER, where proteins are transported into the ER via cotranslational process. One end of the protein is translocated into the ER, while the other end of the protein is assembled in the ribosome during cotranslational transport. The ER is known for producing proteins and lipids that are intended for intracellular organelles and the cell surface (280). The ER also regulates Ca^{2+} signaling via its Ca^{2+}-binding proteins and homeostasis (281).

ER stress has been identified as an important signal pathway that causes cancer death, making the ER an important target for improved cancer treatment. (282–285). For highly effective cancer therapy, several biocompatible metal complexes can be used to target ER stress (286). ER-targeted oxovanadium (IV) vitamin-B6 complex, [VO(HL)(acdppz)]Cl, is being developed for cancer PDT. ER-targeted comb-like polymer PAsp-g-(PEG-ICG) was recently developed for enhanced chemotherapy. Paclitaxel (PTX)-loaded nano-micelles (PTX@PAsp-g-(PEG-ICG)) demonstrated a significant synergistic effect for improved in vivo cancer treatment due to the ER-targeted ROS production under laser irradiation (2 W/cm^2, 5 min), demonstrating the potential application of ER targeting for improved cancer treatment.

ER-targeted nanoformulations have received less attention than nucleus- and mitochondria-targeted theranostics, mostly because the targeting mechanisms are still unknown and development of related nanomaterials is difficult.

18.4.7 Peroxisomal Targeting

Peroxisomes are single-layer multifunctional spherical vesicles found throughout the cytoplasm (287). The term 'peroxisome' was coined because hydrogen peroxide is formed and degraded in the organelle. Peroxisomes are responsible for numerous biochemical and metabolic pathways. Peroxisome dysfunction is linked to aging and several other diseases, making it an intriguing pharmaceutical target.

Peroxisomal disorders are caused by either peroxin mutations or deficiencies in peroxisomal enzymes that result in peroxisomal biogenesis disorder (PBD). PBDs are classified into the Zellweger spectrum of disease (Zellweger syndrome, neonatal adrenoleukodystrophy) and rhizomelic chondrodysplasia punctata (RCDP) type I disease or infantile Refsum's disease based on clinical severity. Single peroxisomal enzyme deficiencies, in addition to peroxin mutations that affect enzyme delivery to the peroxisome, have been linked to a number of human diseases, including adult Refsum's disease, X-linked adrenoleukodystrophy, RCDP type II and III, primary hyperoxaluria type I and acatalasemia.

Peroxisome targeting is extremely useful in restoring, altering or supplementing peroxisomal function. It could be used to treat single-enzyme deficiencies, PBD and diseases associated with altered ROS levels (288).

18.4.8 Proteasomal Targeting

This is an ATP-dependent protease that degrades abnormal proteins found in the cytosol, nucleus or endoplasmic reticulum. The 20S proteasomal core is capped by two 19S regulatory subunits on each end of the total proteasome (53).

Proteasome inhibition has been used therapeutically in the treatment of inflammatory diseases and cancer therapy (289). While proteasome activation can be used for treatment of neurodegenerative diseases and cancer therapy (290), as a result, it has become a popular target for drug therapy.

Bortezomib is a proteasome inhibitor that targets the 20S subunit of the ubiquitin-protease system. It is the first small molecule that has been approved for treatment of multiple myeloma (291, 292). Many other drugs are currently being developed to target other proteasome components (293, 294) and other components of the proteasomal degradation pathway such as ubiquitin ligases and ubiquitin-activating and -conjugating enzymes (295).

18.5 CONCLUSION AND FUTURE PERSPECTIVES

The ultimate goal of nanoparticle-based targeted drug delivery is to deliver a drug to a specific location within the cell while minimizing side effects and maximizing therapeutic efficiency. The most widely targeted and studied organelles are the cytoplasm, nucleus and mitochondria. Most drugs require cytosolic targeting before they can reach any organelle within the cell. Drug carriers are still being developed to overcome barriers to cytosolic delivery and endosomal escape. Capturing and delivering oncogenic proteins to the proteasome is a promising cancer treatment strategy. Manipulation of current therapeutics with specific targeting signals ensures accumulation in specific cell compartments.

The majority of NP drugs currently approved for clinical trials are simple formulations of liposomes or polymeric NPs delivered via passive targeting (296). There are few active cellular-targeted NPs-drugs constructs in the ClinicalTrials.gov pipeline (e.g., Trial# NCT03774680, NCT02979392 etc.) for treating/diagnosing cancer. However, no active subcellular organelle-targeting NP drugs have been approved as of yet. It clear that more understanding and control over preparations as well as navigation of NP drug constructs to the organelles of disease-affected tissue are desperately needed. Furthermore, newer technologies (for example, protein switch and virus-like delivery systems, among others) mimic nature and enable a more sophisticated approach to active subcellular organelle-targeting drug delivery.

REFERENCES

1. Guo X, Wei X, Chen Z, et al. Multifunctional nanoplatforms for subcellular delivery of drugs in cancer therapy. Prog. Mater. Sci. 2020; 107:10059.

2. Wilhelm S, Tavares AJ, Dai Q, Ohta S, Audet J, Dvorak HF, et al. Analysis of nanoparticle delivery to tumours. Nat. Rev. Mater. 2016; 1:16014.
3. Shi J, Kantoff PW, Wooster R, Farokhzad OC. Cancer nanomedicine: Progress, challenges and opportunities. Nat. Rev. Cancer. 2017; 17:20.
4. Torchilin VP. Recent approaches to intracellular delivery of drugs and DNA and organelle targeting. Annu. Rev. Biomed. Eng. 2006; 8:343–375.
5. Torchilin VP, Khaw B-A, Weissig V. Intracellular targets for DNA delivery: Nuclei and mitochondria. Somat. Cell Mol. Genet 2002; 27:49–64.
6. Mills EL, Kelly B, Logan A, et al. Succinate dehydrogenase supports metabolic repurposing of mitochondria to drive inflammatory macrophages. Cell. 2016; 167(2):457–470.
7. Shri Y, Massague J. Mechanisms of TGF-beta signaling from cell membrane to the nucleus. Cell. 2033; 113(6):685–700.
8. Vyas S, Zaganjor E, Haigis MC. Mitochondria and cancer. Cell. 2016; 166(3):555–566.
9. Weinberg SE, Sena LA, Chandel NS. Mitochondria in the regulation of innate and adaptive immunity. Immunity. 2015; 42(3):4006–4017.
10. Zong WX, Rabinowitz JD, White E. Mitochondria and cancer. Mol. Cell.2016; 61(5):667–676.
11. Weinberg SE, Chandel NS. Targeting mitochondria metabolism for cancer therapy. Nat. Chem. Biol. 2015; 11(1):9–15.
12. Ashley CE, Carnes EC, Philips GK, et al. The targeted delivery of multicomponent cargos to cancer cell by nanoporous particle-supported lipid bilayers. Nat. Mater. 2011; 10(5):389–397.
13. Wang M, Kaufman RJ. The impact of the endoplasmic reticulum protein folding environment on cancer development. Nat. Rev. Cancer. 2014; 14(9):581–597.
14. Philips MJ, Voelz GK. Structure and function of ER membrane contact sites with other organelles. Nat. Rev. Mol. Cell Biol. 2016; 17(2):69–82.
15. Yu L, McPhee CK, Zheng L, et al. Termination of autophagy and reformation of lysosomes regulated by mTOR. Nature. 2010; 465(7300):942–946.
16. Zhu H, Fan J, Du J, et al. Fluorescent probes for sensing and imaging within specific cellular organelles. Acc. Chem. Res. 2016; 49(10):2115–2126.
17. Wang Y, Wei G, Zhang X, et al. Multistage targeting strategy using magnetic composite nanoparticles for synergism of photothermal therapy and chemotherapy. Small. 2017:1702994–1703003.
18. Feng G, Liu J, Zhang CJ, et al. Artemisinin and AIEgen conjugate for mitochondria-targeted and image-guided chemo-and photodynamic cancer cell ablation. ACS Appl. Mater. Interfaces. 2018; 10(14):11546–11553.
19. Qiu K, Wang J, Song C, et al. Crossfire for two-photon photodynamic therapy with fluorinated ruthenium (II) photosensitizers. ACS Appl. Mater. Interfaces. 2017; 9(22):18482–18492.
20. Wu C, Wang L, Tian Y, et al. "Triple-Punch" Anticancer strategy mediated by near-infrared photosensitizer/CpG oligonucleotides dual-dressed and mitochondria targeted nanographene. ACS Appl. Mater. Interfaces. 2018; 10(8):6942–6955.
21. Tu Z, Qiao H, Yan Y, et al. Directed graphene based nanoplatforms for hyperthermia: Overcoming multiple drug resistance. Angew Chem. Int. Ed. 2018; 57(35):11198–11202.
22. Wang Z, Chen Y, Zhang H, et al. Mitochondria targeting polydopamine nanocomposites as chemophotothermal therapeutics for cancer. Bioconjugate Chem. 2018; 29(7):2415–2425.
23. Liu Y, Li H, Xie J, et al. Facile construction of mitochondria targeting nanoparticles for enhanced phototherapeutic effects. Biomater Sci. 2017; 5(5):1022–1031.
24. Liu j, Liang H, Li M, et al. Tumor acidity activating multifunctional nanoplatform for NIR-mediated multiple enhanced photodynamic and photothermal tumor therapy. Biomaterials. 2018; 157:107–124.
25. Zhou W, Wang X, Hu M, et al. A mitochondrion-targeting copper complex exhibits potent cytotoxicity against cisplatin-resistant tumor cells through multiple mechanisms of action. Chem. Sci. 2014; 5(7):2761–2770.
26. He H, Wang J, Wang H, et al. Enzymatic cleavage of branched peptides for targeting mitochondria. J. Am. Chem. Soc. 2018; 140(4):1215–1218.
27. Zheng Y, Ji X, Yu B, et al. Enrichment-triggered prodrug activation demonstrated through mitochondria-targeted delivery of doxorubicin and carbon monoxide. Nat. Chem. 2018; 10(7):787–794.
28. Jeena MT, Palanikumar L, Go EM, et al. Mitochondria localization induced self-assembly of peptide amphiphiles for cellular dysfunction. Nat. Commun. 2017; 8(1):26–35.
29. Wang H, Gao Z, Liu X, et al. Targeted production of reactive oxygen species in mitochondria to overcome cancer drug resistance. Nat. Commun. 2018; 9(1):562–577.

30. Liang XJ, Meng H, Wang Y, et al. Metallofullerene Nanoparticles circumvent tumor resistance to cisplatin by reactivating endocytosis. Proc. Natl. Acad. Sci. USA. 2010; 107(16):7449–7454.
31. Zhang Y, Wang F, Li M, et al. Self stabilized hyaluronate nanogel for intracellular codelivery of doxorubicin and cisplatin to osteosarcoma. Adv. Sci. 2018:1700821–1700832.
32. Chen WH, Luo GF, Zhang XZ, et al. Recent advances in subcellular targeted cancer therapy based on functional materials. Adv. Mater. 2018:1802725–1802763.
33. Gao W, Cao W, Zhang H, et al. Targeting lysosomal membrane permeabilization to induce and image apoptosis in cancer cells by multifunctional Au-ZnO hybrid nanoparticles. Chem. Commun. 2014; 50(60):8117–8120.
34. Pan L, Liu J, Shi J, et al. Cancer cell Nucleus-targeting nanocomposites for advanced tumor therapeutics. Chem. Soc. Rev. 2018; 47(18):6930–6946.
35. Biswas S, Torchilin VP, Nanopreparations for organelle-specific delivery in cancer. Adv. Drug. Delivery Rev. 2014; 66:26–41.
36. Ma X, Gong N, Zhong L, et al. Future of nanotherapeutics: Targeting the cellular sub-organelles. Biomaterials. 2016; 97:10–21.
37. Qiu K, Chen Y, Rees T, et al. Organelle-targeting metal complexes: From molecular design to bioapplications. Coord. Chem. Rev. 2019; 378:66–86.
38. Milane L, Trivedi M, Singh A, et al. Mitochondrial biology, targets and drug delivery. J. Controlled Release. 2015; 207:40–58.
39. Mo R, Sun Q, Xue J, et al. Multistage pH-responsive liposomes for mitochondrial-targeted anticancer drug delivery. Adv. Mater. 2012; 24(27):3659–3665.
40. Noh I, Lee D, Kim H, et al. Enhanced photodynamic cancer treatment by mitochondria-targeting and brominated near-infrared fluorophores. Adv. Sci. 2018; 5(3):1700481–1700491.
41. Zhang L, Wang D, Yang K, et al. Mitochondria-targeted artificial "Nano-RBCs' for amplified synergistic cancer phototherapy by a single NIR irradiation. Adv. Sci. 2018:1800049–1800063.
42. Ling D, Bae BC, Park W, et al. Photodynamic efficacy of photosensitizers under an attenuated light dose via lipid Nano-carrier-mediated nuclear targeting. Biomaterials. 2012; 33(21):5478–5486.
43. Jin E, Zhang B, Sun X, et al. Acid-active cell-penetrating peptides for in-vivo tumor-targeting drug delivery. J. Am. Chem. Soc. 2013; 135(2):933–940.
44. Pan L, He Q, Liu J, et al. Nuclear-targeted drug delivery of TAT peptide-conjugated Monodisperse mesoporous silica nanoparticles. J. Am. Chem. Soc. 2012; 134(13):5722–5725.
45. Yuan H, Fales AM, Vo-Dinh T, et al. TAT peptide-functionalized gold nanostars: Enhanced intracellular delivery and efficient NIR photothermal therapy using ultralow irradiance. J. Am. Chem. Soc. 2012; 134(28):11358–11361.
46. Biswas S, Doswadkar NS. Deshpande PP, et al. Liposomes loaded with paclitazel and modified with novel triphenylphosphonium-PEG-PE conjugate possess low toxicity, Target Mitochondria and demonstrate enhanced antitumor effects in-vitro and in-vivo. J. Controlled Release. 2012; 159(3): 393–402.
47. Yang G, Xu L, Xu J, et al. Smart nanoreactors for pH-responsive tumor homing, mitochondria-targeting and enhanced photodynamic-Immunotherapy of cancer. Nano Lett. 2018; 18(4):2475–2484.
48. Zhou F, Wu S, Wu B, et al. Mitochondria-targeting single-walled carbon nanotubes for cancer photothermal therapy. Small. 2011; 7(19):2727–2735.
49. Gao P, Pan W, Li N, et al. Boosting cancer therapy with organelle-targeted nanomaterials. ACS Appl. Mater. Interfaces. 2019:1–85.
50. Heemels MT. Medicine Nobel goes to pioneer of protein guidance mechanisms. Nature. 1999; 401(6754):625.
51. Hagmen M. Protein zip codes make Nobel journey. Science. 1999; 286(5440):666.
52. Shields D. Gunter Blobel—still passionate after all these years. Trends Cell Biol. 2001; 11(8):349–350.
53. Mossalam M, Dixon AS, Lim C. Controlling subcellular delivery to optimize therapeutic effect. Ther. Deliv. 2010; 1(1):169–193.
54. Bleeker EJ, De Jong WH, Geertsma RE, et al. Considerations on the EU definition of a nanomaterial: Science to support policy making. Regul. Toxicol. Pharmacol. 2013; 65(1):119–125.
55. Kreyling WG, Semmler-Behnke M, Chaudhry Q. A complementary definition of nanomaterial. Nano Today.2010; 5(3):165–168.
56. Auffan M, Rose J, Bottero J-Y, Lowry GV, et al. Towards a definition of inorganic nanoparticles from an environmental, health and safety perspective. Nat. Nanotechnol. 2009; 4(10):634–641.
57. Dobson P, Jarvie, H., King, S. Nanoparticle. Encyclopedia Britannicawww.britannica.com/science/nanoparticle. 2015.

58. Elzoghby AO, Samy WM, Elgindy NA. Albumin-based nanoparticles as potential controlled release drug delivery systems. J. Control. Release. 2012; 157(2):168–182.
59. Kumari A, Yadav SK, Yadav SC. Biodegradable polymeric nanoparticles based drug delivery systems. Colloids Surf., 2010; B 75(1):1–18.
60. Garg T, K Goyal A. Liposomes: Targeted and controlled delivery system. Drug Deliv. Lett. 2014; 4(1):62–71.
61. Mohamed S, Parayath NN, Taurin S, et al. Polymeric nano-micelles: Versatile platform for targeted delivery in cancer. Ther. Deliv. 2014; 5(10):1101–1121.
62. Abbasi E, Aval SF, Akbarzadeh A, et al. Dendrimers: Synthesis, applications, and properties. Nanoscale Res. Lett. 2014; 9(1):1–10.
63. Kaminskas LM, Boyd BJ, Porter CJ. Dendrimer pharmacokinetics: The effect of size, structure and surface characteristics on ADME properties. Nanomedicine. 2011; 6(6):1063–1084.
64. Mok H, Zhang M. Superparamagnetic iron oxide nanoparticle-based delivery systems for biotherapeutics. Expert Opin. Drug Deliv. 2013; 10(1):73–87.
65. Rosenholm JM, Mamaeva V, Sahlgren C, et al. Nanoparticles in targeted cancer therapy: Mesoporous silica nanoparticles entering preclinical development stage. Nanomedicine. 2012; 7(1):111–120.
66. Cabral RM, Baptista PV. Anti-cancer precision theranostics: A focus on multifunctional gold nanoparticles. Expert Rev. Mol. Diagn. 2014; 14(8):1041–1052.
67. Rastogi V, Yadav P, Bhattacharya SS, et al. Carbon nanotubes: An emerging drug carrier for targeting cancer cells. J. Drug Deliv. 2014; 670815:1–23.
68. Valizadeh A, Mikaeili H, Samiei M, et al. Quantum dots: Synthesis, bioapplications, and toxicity. Nanoscale Res. Lett. 2012; 7(1):1–14.
69. Sun T, Zhang YS, Pang B, et al. Engineered nanoparticles for drug delivery in cancer therapy. Angew. Chem. Int. Ed. 2014; 53(46):12320–12364.
70. U.S. FDA. ABRAXANE® for Injectable Suspension (paclitaxel protein-bound particles for injectable suspension) (albumin-bound). www.accessdata.fda.gov/drugsatfda_docs/label/2013/021660s037lbl.pdf. 2013.
71. Lee S, Kim Y, Cho CH, et al. An open-label, randomized, parallel, phase II trial to evaluate the efficacy and safety of a cremophor-free polymeric micelle formulation of paclitaxel as first-line treatment for ovarian cancer: A Korean gynecologic oncology group study (KGOG-3021). Cancer Res. Treat. 2018; 50:195–203.
72. Borgå O, Henriksson R, Bjermo H, et al. Maximum tolerated dose and pharmacokinetics of paclitaxel micellar in patients with recurrent malignant solid tumours: A dose-escalation study. Adv. Ther. 2019; 36:1150–1163.
73. Bernabeu E, Cagel M, Lagomarsino E, et al. Paclitaxel: What has been done and the challenges remain ahead. Int. J. Pharm. 2017; 526:474–495.
74. Jain M, Gupte S, Patil S, et al. Paclitaxel injection concentrate for nanodispersion versus nab-paclitaxel in women with metastatic breast cancer: A multicenter, randomized, comparative phase II/III study. Breast Cancer Res. Treat. 2016; 156:125–134.
75. Silverman L, Barenholz Y. In vitro experiments showing enhanced release of doxorubicin from Doxil1 in the presence of ammonia may explain drug release at tumor site. Nanomedicine. 2015; 11:1841–1850.
76. Swenson C, Perkins W, Roberts P, et al. Liposome technology and the development of Myocet™ (liposomal doxorubicin citrate). Breast. 2001; 10:1–7.
77. Dou Y, Hynynen K, Allen C. To heat or not to heat: Challenges with clinical translation of thermosensitive liposomes. J. Control Release. 2017; 249:63–73.
78. Xu R, Zhang G, Mai J, et al. An injectable nanoparticle generator enhances delivery of cancer therapeutics. Nat. Biotechnol. 2016; 34:414–418.
79. Deshmukh AS, Chauhan PN, Noolvi MN, et al. Polymeric micelles: Basic research to clinical practice. Int. J. Pharm. 2017; 532:249–268.
80. Stathopoulos G, Antoniou D, Dimitroulis J, et al. Comparison of liposomal cisplatin versus cisplatin in non-squamous cell non-small-cell lung cancer. Cancer Chemother Pharmacol. 2011; 68:945–950.
81. Clark AJ, Wiley DT, Zuckerman JE, et al. CRLX101 nanoparticles localize in human tumors and not in adjacent, nonneoplastic tissue after intravenous dosing. Proc. Natl. Acad. Sci USA. 2016; 113: 3850–3854.
82. Neal JW, Wakelee H, Padda SK, et al. PS01.04: A Phase II study of etirinotecan pegol (NKTR-102) in patients with refractory brain metastases and advanced lung cancer. J. Thorac. Oncol. 2016; 11:271–272.
83. Adiseshaiah PP, Crist RM, Hook SS, et al. Nanomedicine strategies to overcome the pathophysiological barriers of pancreatic cancer. Nat. Rev. Clin. Oncol. 2016; 13:750–765.

84. Lu C, Stewart DJ, Lee JJ, et al. Phase I clinical trial of systemically administered TUSC2 (FUS1)-nanoparticles mediating functional gene transfer in humans. PLoS One. 2012; 7:34833.
85. Morgan S, Mackay A. Summary of Product Characteristics. Springer; 2010.
86. Bedikian AY, Vardeleon A, Smith T, et al. Pharmacokinetics and urinary excretion of vincristine sulfate liposomes injection in metastatic melanoma patients. J. Clin. Pharmacol. 2006; 46:727–737.
87. Pillai G. Applications Targeted Nano Drugs Delivery Systems. Elsevier; 2019.
88. Agarwal R, Singh V, Jurney P, et al. Mammalian cells preferentially internalize hydrogel nanodiscs over nanorods and use shape specific uptake mechanisms. Proc. Natl. Acad. Sci. 2013; 110:17247–17252.
89. Chithrani BD, Ghazani, AA, Chan WC. Determining the size and shape dependence of gold nanoparticle uptake into mammalian cells. Nano Lett. 2006; 6:662–668.
90. Jiang W, Kim BYS, Rutka JT, et al. Nanoparticle-mediated cellular response is size-dependent. Nat. Nanotechnol. 2008; 3:145–150.
91. Cho EC, Zhang Q, Xia Y. The effect of sedimentation and diffusion on cellular uptake of gold nanoparticles. Nat. Nanotechnol. 2011; 6:385–391.
92. MacParland SA, Tsoi KM, Ouyang B, et al. Phenotype determines nanoparticle uptake by human macrophages from liver and blood. ACS Nano. 2017; 11:2428–2443.
93. Santos TD, Varela J, Lynch I, et al. Quantitative assessment of the comparative nanoparticle-uptake efficiency of a range of cell lines. Small. 2011; 7:3341–3349.
94. Cheng X, Tian X, Wu A, et al. Protein corona influences cellular uptake of gold nanoparticles by phagocytic and nonphagocytic cells in a size-dependent manner. ACS Appl. Mater. Interfaces. 2015; 7:20568–20575.
95. Chang T, Lord MS, Bergmann B, et al. Size effects of self-assembled block copolymer spherical micelles and vesicles on cellular uptake in human colon carcinoma cells. J. Mater. Chem. B. 2014; 2:2883–2891.
96. Yang K, Ma YQ, Computer simulation of the translocation of nanoparticles with different shapes across a lipid bilayer. Nat. Nanotechnol. 2010; 5:579–583.
97. Dasgupta S, Auth T, Gompper G, Shape and orientation matter for the cellular uptake of nonspherical particles. Nano Lett. 2014; 14:687–693.
98. Arnida A, Ghandehari Malugin H, Cellular uptake and toxicity of gold nanoparticles in prostate cancer cells: A comparative study of rods and spheres. J. Appl. Toxicol. 2010; 30:212–217.
99. Hunter RJ, Zeta Potential in Colloid Science: Principles and Applications. Elsevier; 1988.
100. Jiang Y, Huo S, Mizuhara T, et al. The interplay of size and surface functionality on the cellular uptake of Sub-10 nm gold nanoparticles. ACS Nano. 2015; 9:9986–9993.
101. Chertok B, David AE, Yang VC, Polyethyleneimine-modified iron oxide nanoparticles for brain tumor drug delivery using magnetic targeting and intra-carotid administration. Biomaterials. 2010; 31:6317–6324.
102. Li X, Chen Y, Wang M, et al. Amesoporous silica nanoparticle–PEI–fusogenic peptide system for siRNA delivery in cancer therapy. Biomaterials. 2013; 34:1391–1401.
103. Ngamcherdtrakul W, Morry J, Gu S, et al. Cationic polymer modified mesoporous silica nanoparticles for targeted siRNA delivery to HER2+ breast cancer. Adv. Funct. Mater. 2015; 25:2646–2659.
104. Vigderman L, Manna P, Zubarev ER, et al. Quantitative replacement of cetyl trimethylammonium bromide by cationic thiol ligands on the surface of gold nanorods and their extremely large uptake by cancer cells. Angew. Chem. Int. Ed. 2012; 51:636–641.
105. Ayala V, Herrera AP, Latorre-Esteves M, et al. Effect of surface charge on the colloidal stability and in vitro uptake of arboxymethyl dextran-coated iron oxide nanoparticles. J. Nanopart. Res. 2013; 15:1874.
106. Zhou Y, Shi L, Li Q, et al. Imaging and inhibition of multi-drug resistance in cancer cells via specific association with negatively charged CdTe quantum dots. Biomaterials. 2010; 31:4958–4963.
107. Lee JS, Ankone M, Pieters E, et al. Circulation kinetics and biodistribution of dual-labeled polymersomes with modulated surface charge in tumor-bearing mice: Comparison with stealth liposomes. J. Control. Release. 2011; 155:282–288.
108. Xie Y, Qiao H, Su Z, et al. PEGylated carboxymethyl chitosan/calcium phosphate hybrid anionic nanoparticles mediated hTERT siRNA delivery for anticancer therapy. Biomaterials. 2014; 35:7978–7991.
109. Secret E, Maynadier M, Gallud A, et al. Anionic porphyrin-grafted porous silicon nanoparticles for photodynamic therapy. Chem. Commun. 2013; 49:4202–4204.
110. Chithrani BD, Chan WCW. Elucidating the mechanism of cellular uptake and removal of protein-coated gold nanoparticles of different sizes and shapes. Nano Lett. 2007; 7(6):1542–1550.
111. Moyano DF, Saha K, Prakash G, et al. Fabrication of corona free nanoparticles with tunable hydrophobicity. ACS Nano. 2014; 8:6748–6755.

112. Anselmo AC, Zhang M, Kumar S, et al. Elasticity of nanoparticles influences their blood circulation, phagocytosis, endocytosis, and targeting. ACS Nano. 2015; 9:3169–3177.
113. Guo P, Liu D, Subramanyam K, et al. Nanoparticle elasticity directs tumor uptake. Nat. Commun. 2018; 9:1–9.
114. Yi X, Gao H. Kinetics of Receptor-Mediated Endocytosis of Elastic Nanoparticles; Nanoscale. 2017; 9:454–463.
115. Cho K, Wang X, Nie S, et al. Therapeutic nanoparticles for drug delivery in cancer. Clin. Cancer Res. 2008; 14:1310–1316.
116. Bertrand N, Wu J, Xu X, et al. Cancer nanotechnology: The impact of passive and active targeting in the era of modern cancer biology. Adv. Drug Deliv. Rev. 2013; 66:2–25.
117. Byrne JD, Betancourt T, Brannon-Peppas L. Active targeting schemes for nanoparticle systems in cancer therapeutics. Adv. Drug Deliv. Rev. 2008; 60:1615–1626.
118. Belfiore L, Saunders DN, Ranson M, et al. Towards clinical translation of ligand-functionalized liposomes in targeted cancer therapy: Challenges and opportunities. J. Control. Release. 2018; 277:1–13.
119. Hak S, Helgesen E, Hektoen HH, et al. The effect of nanoparticle polyethylene glycol surface density on ligand-directed tumor targeting studied in vivo by dual-modality imaging. ACS Nano. 2012; 6:5648–5658.
120. Zern BJ, Chacko AM, Liu J, et al. Reduction of nanoparticle avidity enhances the selectivity of vascular targeting and PET detection of pulmonary inflammation. ACS Nano. 2013; 7:461–2469.
121. Qhattal HSS, Hye T, Alali A, et al. Hyaluronan polymer length, grafting density, and surface poly(ethylene glycol) coating influence in vivo circulation and tumor targeting of hyaluronan-grafted liposomes. ACS Nano. 2014; 8:5423–5440.
122. Pozzi D, Colapicchioni V, Caracciolo G, et al. Effect of polyethyleneglycol (PEG chain length on the bio-nano- interactions between PEGylated lipid nanoparticles and biological fluids: From nanostructure to uptake in cancer cells. Nanoscale. 2014; 6:2782–2792.
123. Mou Q, Ma Y, Zhu X, et al. A small-molecule nanodrug consisting of amphiphilic targeting ligand-chemotherapy drug conjugate for targeted cancer therapy. J. Control. Release. 2016; 230:34–44.
124. Ding HM, Ma YQ. Role of physicochemical properties of coating ligands in receptor-mediated endocytosis of nanoparticles. Biomaterials. 2012; 33:5798–5802.
125. Ding H, Tian W, Ma Y. Designing nanoparticle translocation through membranes by computer simulations. ACS Nano. 2012; 6:1230–1238.
126. Colombo M, Fiandra L, Alessio G, et al. Tumour homing and therapeutic effect of colloidal nanoparticles depend on the number of attached antibodies. Nat. Commun. 2016; 7.
127. Zhang B, Shen S, Liao Z, et al. Targeting fibronectins of glioma extracellular matrix by CLT1peptide-conjugated nanoparticles. Biomaterials. 2014; 35:4088–4098.
128. Chung EJ, Cheng Y, Morshed R, et al. Fibrin-binding, peptide amphiphile micelles for targeting glioblastoma. Biomaterials. 2014; 35:1249–1256.
129. Cieslewicz M, Tang J, Yu JL, et al. Targeted delivery of proapoptotic peptides to tumor-associated macrophages improves survival. Proc. Natl. Acad. Sci. 2013; 110:15919–15924.
130. Ngambenjawong C, Pun SH. Multivalent polymers displaying M2 macrophage targeting peptides improve target binding avidity and serum stability. ACS Biomater. Sci. Eng. 2017; 3:2050–2053.
131. Zhang J, Yuan ZF, Wang Y, et al. Multifunctional envelope-type mesoporous silica nanoparticles for tumor-triggered targeting drug delivery. J. Am. Chem. Soc. 2013; 135:5068–5073.
132. Zhang MZ, Yu Y, Yu RN, et al. Tracking the downregulation of folate receptor-α in cancer cells through target specific delivery of quantum dots coupled with antisense oligonucleotide and targeted peptide. Small. 2013; 94183–94193.
133. Locatelli E, Naddaka M, Uboldi C, et al. Targeted delivery of silver nanoparticles and alisertib: In vitro and in vivo synergistic effect against glioblastoma. Nanomedicine. 2014; 9:839–849.
134. Lam FC, Morton SW, Wyckoff J, et al. Enhanced efficacy of combined temozolomide and bromodomain inhibitor therapy for gliomas using targeted nanoparticles. Nat. Commun. 2018; 9.
135. Schmid D, Park CG, Hartl CA, et al. T cell targeting nanoparticles focus delivery of immunotherapy to improve antitumor immunity. Nat. Commun. 2017; 8:1–11.
136. Dai Q, Yan Y, Ang CS, et al. Monoclonal antibody-functionalized multilayered particles: Targeting cancer cells in the presence of protein coronas. ACS Nano. 2015; 9:2876–2885.
137. Song J, Zhou J, Duan H. Self-assembled plasmonic vesicles of SERS-encoded amphiphilic gold nanoparticles for cancer cell targeting and traceable intracellular drug delivery. J. Am. Chem. Soc. 2012; 134:13458–13469.
138. Kotagiri N, Li Z, Xu X, et al. Antibody quantum dot conjugates developed via copper-free click chemistry for rapid analysis of biological samples using a microfluidic microsphere array system. Bioconjug. Chem. 2014; 25:1272–1281.

139. Engelberg S, Modrejewski J, Walter JG, et al. Cancer cell-selective, clathrin-mediated endocytosis of aptamer-decorated nanoparticles, Onco. Target. 2018; 9:20993–21006.
140. Prusty DK, Adam V, Zadegan RM, et al. Supramolecular aptamer nano-constructs for receptor-mediated targeting and light-triggered release of chemotherapeutics into cancer cells. Nat. Commun. 2018; 9.
141. Wang J, Ma Q, Hu XX, et al. Autofluorescence-free targeted tumor imaging based on luminous nanoparticles with composition-dependent size and persistent luminescence. ACS Nano. 2017; 11:8010–8017.
142. Porta F, Lamers GEM, Morrhayim J, et al. Folic acid-modified mesoporous silica nanoparticles for cellular and nuclear targeted drug delivery. Adv. Healthc. Mater. 2013; 2:281–286.
143. Shi J, Kantoff PW, Wooster R, et al. Cancer nanomedicine: Progress, challenges and opportunities. Nat. Rev. Cancer. 2016; 17:20–37.
144. Wilhelm S, Tavares AJ, Chan WCW. Reply to "Evaluation of nanomedicines: Stick to the basics". Nat. Rev. Mater. 2016; 1:16074.
145. Salvati A, Pitek AS, Monopoli MP, et al. Transferrin-functionalized nanoparticles lose their targeting capabilities when a biomolecule corona adsorbs on the surface. Nat. Nanotechnol. 2013; 8:137–143.
146. Tonigold M, Simon J. Pre-adsorption of antibodies enables targeting of nanocarriers despite a biomolecular corona. Nat. Nanotechnol. 2018; 13:862–869.
147. Yaseen MA, Yu J, Jung B, et al. Biodistribution of encapsulated indocyanine green in healthy mice. Mol Pharm. 2009; 6(5):1321–1332.
148. Liu D, Huang L. Role of cholesterol in the stability of pH-sensitive, large unilamellar liposomes prepared by the detergent—dialysis method. Biochim Biophys Acta. 1989; 981(2):254–260.
149. Liu D, Huang L. pH-sensitive, plasma-table liposomes with relatively prolonged residence in circulation. Biochim Biophys Acta. 1990; 1022(3):348–354.
150. Collins D, Litzinger DC, Huang L. Structural and functional comparisons of pH-sensitive liposomes composed of phosphatidylethanolamine and three different diacylsuccinylglycerols. Biochim Biophys Acta. 1990; 1025(2):234–242.
151. Liu D, Mori A, Huang L. Large liposomes containing ganglioside GM1 accumulate effectively in spleen. Biochim Biophys Acta. 1991; 1066(2):159–165.
152. Kono K, Igawa T, Takagishi T. Cytoplasmic delivery of calcein mediated by liposomes modified with a pH-sensitive poly(ethylene glycol) derivative. Biochim Biophys Acta. 1997; 1325(2):143–154.
153. Torchilin VP. Recent approaches to intracellular delivery of drugs and DNA and organelle targeting. Annu. Rev. Biomed. Eng. 2006; 8:343–375.
154. Alberts B, Johnson A, Lewis J, et al. Molecular Biology of the Cell. 5. Garland Science; 2008.
155. Guillem VM, Alino SF. Transfection pathways of nonspecific and targeted PEI-polyplexes. Gene. Ther. Mol. Biol. 2004; 8:369–384.
156. Poon GM, Gariepy J. Cell-surface proteoglycans as molecular portals for cationic peptide and polymer entry into cells. Biochem. Soc. Trans. 2007; 35(Pt 4):788–793.
157. Nitin N, Santangelo PJ, Kim G, et al. Peptide-linked molecular beacons for efficient delivery and rapid mRNA detection in living cells. Nucleic Acids. Res. 2004; 32(6):e58.
158. Ziegler A, Nervi P, Durrenberger M, et al. The cationic cell-penetrating peptide CPP(TAT) derived from the HIV-1 protein TAT is rapidly transported into living fibroblasts: Optical, biophysical, and metabolic evidence. Biochemistry. 2005; 44(1):138–148.
159. Nori A, Jensen KD, Tijerina M, et al. Tat-conjugated synthetic macromolecules facilitate cytoplasmic drug delivery to human ovarian carcinoma cells. Bioconjug Chem. 2003; 14(1):44–50.
160. Sethuraman VA, Bae YH. Tat peptide-based micelle system for potential active targeting of anticancer agents to acidic solid tumors. J. Control Release 2007; 118(2):216–224.
161. Wadia JS, Stan RV, Dowdy SF. Transducible TAT-HA fusogenic peptide enhances escape of TAT fusion proteins after lipid raft macropinocytosis. Nat. Med. 2004; 10(3):310–315.
162. Richard JP, Melikov K, Vives E, et al. Cell-penetrating peptides. A re-evaluation of the mechanism of cellular uptake. J. Biol. Chem. 2003; 278(1):585–590.
163. Jarver P, Langel U. The use of cell-penetrating peptides as a tool for gene regulation. Drug Discov Today. 2004; 9(9):395–402.
164. Li Z, Zhang Y, Wurtz W, et al. Characterization of nebulized liposomal amikacin (arikace) as a function of droplet size. J. Aerosol. Med. Pulm. Drug Deliv. 2008; 21(3):245–254.
165. Venkataram S, Awni WM, Jordan K, et al. Pharmacokinetics of two alternative dosage forms for cyclosporine: Liposomes and intralipid. J. Pharm. Sci 1990; 79(3):216–219.
166. James ND, Coker RJ, Tomlinson D, et al. Liposomal doxorubicin (Doxil): An effective new treatment for Kaposi's sarcoma in aids. Clin. Oncol. (R Coll Radiol) 1994; 6(5):294–296.

167. Chu CJ, Dijkstra J, Lai MZ, et al. Efficiency of cytoplasmic delivery by pH-sensitive liposomes to cells in culture. Pharm Res. 1990; 7(8):824–834.
168. James ND, Coker RJ, Tomlinson D, et al. Liposomal doxorubicin (Doxil): An effective new treatment for Kaposi's sarcoma in aids. Clin. Oncol. (R Coll Radiol) 1994; 6(5):294–296.
169. Lutwyche P, Cordeiro C, Wiseman DJ, et al. Intracellular delivery and antibacterial activity of gentamicin encapsulated in pH-sensitive liposomes. Antimicrob Agents Chemother. 1998; 42(10):2511–2520.
170. Han HD, Shin BC, Choi HS. Doxorubicin-encapsulated thermosensitive liposomes modified with poly(n-isopropylacrylamide-co-acrylamide): Drug release behavior and stability in the presence of serum. Eur. J. Pharm Biopharm. 2006; 62(1):110–116.
171. Pradhan P, Banerjee R, Bahadur D, et al. Targeted magnetic liposomes loaded with doxorubicin. Methods Mol. Biol. 2010; 605:279–293.
172. Pradhan P, Giri J, Rieken F, et al. Targeted temperature sensitive magnetic liposomes for thermochemotherapy. J. Control. Release. 2009; 142(1):108–121.
173. Yokoyama M, Miyauchi M, Yamada N, et al. Characterization and anticancer activity of the micelle-forming polymeric anticancer drug adriamycin-conjugated poly(ethylene glycol)-poly(aspartic acid) block copolymer. Cancer Res. 1990; 50(6):1693–1700.
174. Alani AW, Bae Y, Rao DA, et al. Polymeric micelles for the pH-dependent controlled, continuous low dose release of paclitaxel. Biomaterials. 2010; 31(7):1765–1772.
175. Toti US, Moon SH, Kim HY, et al. Thermosensitive and biocompatible cyclotriphosphazene micelles. J. Control. Release. 2007; 119(1):34–40.
176. Masaki T, Rathi R, Zentner G, et al. Inhibition of neointimal hyperplasia in vascular grafts by sustained perivascular delivery of paclitaxel. Kidney Int. 2004; 66(5):2061–2069.
177. Nori A, Jensen KD, Tijerina M, et al. Subcellular trafficking of HPMA copolymer—Tat conjugates in human ovarian carcinoma cells. J. Control. Release. 2003; 91(1–2):53–59.
178. Boussif O, Lezoualc'h F, Zanta MA, et al. A versatile vector for gene and oligonucleotide transfer into cells in culture and in vivo: Polyethylenimine. Proc. Natl. Acad. Sci. USA. 1995; 92(16):7297–7301.
179. Jeong JH, Song SH, Lim DW, et al. DNA transfection using linear poly(ethylenimine) prepared by controlled acid hydrolysis of poly(2-ethyl-2-oxazoline). J. Control. Release. 2001; 73(2–3):391–399.
180. Lee ES, Kim D, Youn YS, et al. A virus-mimetic nanogel vehicle. Angew Chem. Int. Ed. Engl. 2008; 47(13):2418–2421.
181. Kakar M, Cadwallader AB, Davis JR, et al. Signal sequences for targeting of gene therapy products to subcellular compartments: The role of CRM1 in nucleocytoplasmic shuttling of the protein switch. Pharm Res. 2007; 24(11):2146–2155.
182. Meunier L, Mayer R, Monsigny M, et al. The nuclear export signal-dependent localization of oligonucleopeptides enhances the inhibition of the protein expression from a gene transcribed in cytosol. Nucleic. Acids. Res. 1999; 27(13):2730–2736.
183. Alberts B, Johnson A, Lewis J, et al. Molecular Biology of the Cell. Garland Science; 2008.
184. Shen Y, Sun Y, Yan R, et al. Rational engineering of semiconductor QDs Enabling remarkable O2 production for Tumor-targeted photodynamic therapy. Biomaterials. 2017; 148:31–40.
185. Wang T, Wang D, Yu H, et al. Intracellulary acidiswitchable multifunctional micelles for combinational photo/chemotherapy of the drug-resistant tumor. ACS Nano. 2016; 10(3):3496–3508.
186. Boussif O, Lezoualc'h F, Zanta MA, et al. A versatile vector for gene and oligonucleotide transfer into cells in culture and in vivo: Polyethylenimine. Proc. Natl. Acad. Sci. USA. 1995; 92(16):7297–7301.
187. Akinc A, Thomas M, Klibanov AM, et al. Exploring polyethylenimine-mediated DNA transfection and the proton sponge hypothesis. J. Gene. Med. 2005; 7(5):657–663.
188. Duvall CL, Convertine A, Benoit DS, et al. Intracellular delivery of a proapoptotic peptide via conjugation to a raft synthesized endosomolytic polymer. Mol. Pharm. 2009; 7(2):468–476.
189. Murthy N, Campbell J, Fausto N, et al. Bioinspired pH-responsive polymers for the intracellular delivery of biomolecular drugs. Bioconjug Chem. 2003; 14(2):412–419.
190. Liu CG, Kankala RK, Liao HY, et al. Engineered pH-responsive hydrazone-carboxylate complexes-encapsulated 2D matrices for cathepsin-mediated apoptosis in cancer. J. Biomed. Mater. Res. A. 2019; 107A(6):1184–1194.
191. Wu CY, Wu YF, Jin Y, et al. Endosomal/lysosomal location of organically modified silica nanoparticles following caveolae-mediated endocytosis. RSC Adv. 2019; 9(24):13855–13862.
192. Sée V, Free P, Cesbron Y, et al. Cathepsin L digestion of nanobioconjugates upon endocytosis. ACS Nano. 2009; 3(9):2461–2468.
193. Xu CC, Haque F, Jasinski DL, et al. Favorable biodistribution, specific targeting and conditional endosomal escape of RNA nanoparticles in cancer therapy. Cancer Lett. 2018; 414:57–70.

194. Shi BY, Zhang H, Bi JX, et al. Endosomal pH responsive polymers for efficient cancer targeted gene therapy. Colloid Surface B. 2014; 119:55–65.
195. Kim D, Lee ES, Oh KT, et al. Doxorubicin-loaded polymeric micelle overcomes multidrug resistance of cancer by double-targeting folate receptor and early endosomal pH. Small. 2008; 4(11):2043–2050.
196. Kulshrestha A, Katara GK, Ibrahim SA, et al. Targeting V-ATPase isoform restores cisplatin activity in resistant ovarian cancer: Inhibition of autophagy, endosome function, and ERK/MEK pathway. J. Oncol. 2019; 2343876.
197. Gao Y, Li Y, Li Y, et al. PSMA-mediated endosome escape-accelerating polymeric micelles for targeted therapy of prostate cancer and the real time tracing of their intracellular trafficking. Nanoscale. 2015; 7(2):597–612.
198. Koide R, Nishimura SI. Antiadhesive nanosomes facilitate targeting of the lysosomal GlcNAc salvage pathway through derailed cancer endocytosis. Angew Chem. Int. Ed. Engl. 2019; 58(41):14513–14518.
199. Kong C, Li Y, Liu ZS, et al. Targeting the oncogene KRAS mutant pancreatic cancer by synergistic blocking of lysosomal acidification and rapid drug release. ACS Nano. 2019; 13(4):4049–4063.
200. Nosrati H, Mojtahedi A, Danafar H, et al. Enzymatic stimuli-responsive methotrexate-conjugated magnetic nanoparticles for target delivery to breast cancer cells and release study in lysosomal condition. J. Biomed Mater. Res. A. 2018; 106(6):1646–1654.
201. Hollenbeck PJ, Saxton WM. The axonal transport of mitochondria. J. Cell Sci. 2005; 118:5411–5419.
202. Zamzami N, Kroemer G. The mitochondrion in apoptosis: How Pandora's box opens. Nat. Rev. Mol. Cell Bio. 2001; 2:67–71.
203. Mcbride HM, Neuspiel M, Wasiak S. Mitochondria: More than just a powerhouse. Curr Biol. 2006; 16:R551–R560.
204. Costantini P, Jacotot E, Decaudin D, et al. Mitochondrion as a novel target of anticancer chemotherapy. J. Natl. Cancer I. 2000(92):1042–1053.
205. Robertson JD, Gogvadze V, Zhivotovsky B, et al. Distinct pathways for stimulation of cytochrome release by etoposide. J. Biol. Chem. 2000; 275:32438–32443.
206. André N, Braguer D, Brasseur G, et al. Paclitaxel induces release of cytochrome c from mitochondria isolated from human neuroblastoma cells. Cancer Res. 2000; 60:5349–5353.
207. Kidd JF, Pilkington MF, Schell MJ, et al. Paclitaxel affects cytosolic calcium signals by opening the mitochondrial permeability transition pore. J. Biol. Chem. 2002; 277:6504–6510.
208. Kelley SO, Stewart KM, Mourtada R. Development of novel peptides for mitochondrial drug delivery: Amino acids featuring delocalized lipophilic cations. Pharm Res. 2011; 28:2808–2819.
209. Ross MF, Kelso GF, Blaikie FH, et al. Lipophilic triphenylphosphonium cations as tools in mitochondrial bioenergetics and free radical biology. Biochemistry (Moscow). 2005; 70:222–230.
210. Yousif LF, Stewart KM, Kelley SO. ChemInform abstract: Targeting mitochondria with organelle-specific compounds: Strategies and applications. ChemBioChem. 2009; 10:1939–1950.
211. Dong YC, Cho H, Kwon K, et al. Triphenylphosphonium-conjugated poly(ε-caprolactone)-based self-assembled nanostructures as nanosized drugs and drug delivery carriers for mitochondria-targeting synergistic anticancer drug delivery. Adv. Funct. Mater. 2015; 25:5479–54791.
212. Jung HS, Han J, Lee J, et al. Enhanced NIR radiation-triggered hyperthermia by mitochondrial targeting. J. Am. Chem Soc. 2015; 137:3017–3023.
213. Hu Q, Gao M, Feng G, et al. Mitochondria-targeted cancer therapy using a light-up probe with aggregation-induced-emission characteristics. Angew Chem. Inter. Ed. 2014; 53:14225–14229.
214. Yin C, Zhu H, Xie C, et al. Organic nanoprobe cocktails for multilocal and multicolor fluorescence imaging of reactive oxygen species. Adv. Funct. Mater. 2017; 27:1700493.
215. Wang B, Wang Y, Wu H, et al. A mitochondria-targeted fluorescent probe based on TPP-conjugated carbon dots for both one- and two-photon fluorescence cell imaging. RSC Adv. 2014; 4:49960–49963.
216. Wu X, Sun S, Wang Y, et al. A fluorescent carbon-dots-based mitochondria-targetable nanoprobe for peroxynitrite sensing in living cells. Biosens Bioelectron. 2017; 90:501–507.
217. Zhuang Q, Jia H, Du L, et al. Targeted surface-functionalized gold nanoclusters for mitochondrial imaging. Biosens Bioelectron. 2014; 55:76–82.
218. Chen S, Lei Q, Qiu WX, et al. Mitochondria-targeting "nanoheater" for enhanced photothermal/chemotherapy. Biomaterials. 2017; 117:92–104.
219. López V, Villegas MR, Rodríguez V, et al. Janus mesoporous silica nanoparticles for dual targeting of tumor cells and mitochondria. ACS Appl. Mater. Int. 2017; 9:26697–26706.
220. Choi YS, Kwon K, Yoon K, et al. Photosensitizer-mediated mitochondria-targeting nanosized drug carriers: Subcellular targeting, therapeutic, and imaging potentials. Int. J. Pharm. 2017; 520:195–206.

221. Xing L, Lyu J Y, Yang Y, et al. pH-Responsive de-PEGylated nanoparticles based on triphenylphosphine-quercetin self-assemblies for mitochondria-targeted cancer therapy. Chem. Commun. 2017; 53:8790–8793.
222. Weissig V, Lizano C, Torchilin VP. Micellar delivery system for dequalinium—a lipophilic cationic drug with anticarcinoma activity. J. Liposome Res. 2008; 8:391–400.
223. Weissig V, Lasch J, Erdos G, et al. DQAsomes: A novel potential drug and gene delivery system made from dequalinium™. Pharm Res. 1998; 15:334–337.
224. D'Souza GGM, Rammohan R, Cheng SM, et al. DQAsome-mediated delivery of plasmid DNA toward mitochondria in living cells. J. Control Release. 2003; 92:189–197.
225. Bae Y, Jung MK, Lee S, et al. Dequalinium-based functional nanosomes show increased mitochondria targeting and anticancer effect. Eur. J. Pharm. Biopha 2018; 124:104–115.
226. Abe Y, Shodai T, Muto T, et al. Structural basis of presequence recognition by the mitochondrial protein import receptor Tom20. Cell 2000; 100:551–560.
227. Chretien D, Benit P, Ha HH, et al. Mitochondria are physiologically maintained at close to 50 °C. PLoS Biol. 2018; 16:e2003992.
228. Wang D, Huang H, Zhou M, et al. A thermoresponsive nanocarrier for mitochondria-targeted drug delivery. Chem Commun. 2019; 55:4051–4054.
229. Chan MS, Liu LS, Leung HM, et al. Cancer-cell specific mitochondria targeted drug delivery by dual ligand functionalized nanodiamonds circumvent drug resistance. ACS Appl. Mater. Interfaces. 2017; 9(13):11780–11789.
230. Zhou J, Zhao WY, Ma X, et al. Ther anticancer efficacy of paclitaxel liposomes modified with mitochondrial targeting conjugate in resistant lung cancer. Biomaterials. 2013; 34(14):3626–3638.
231. Zhou W, Yu H, Zhang LJ, et al. Redox-triggered activation of nanocarriers for mitochondria targeting cancer chemotherapy. Nanoscale. 2017; 9(43):17044–17053.
232. Yu Z, Sun Q, Pan W, et al. A near-infrared triggered nanophotosensitizer inducing domino effect on mitochondrial reactive oxygen species burst for cancer therapy. ACS Nano. 2015; 9(11): 11064–11074.
233. Han K, Lei Q, Wang SB, et al. Dual stage light guided tumor inhibition by mitochondria targeted photodynamic therapy. Adv. Funct. Mater. 2015; 25(20)2961–2971.
234. Feng G, Qin W, Hu Q, et al. Cellular and mitochondrial dual targeted organic dots with aggregation induced emission characteristics for image guided photodynamic therapy. Adv. Healthcare. Mater.2015; 4(17):2667–2676.
235. Guo R, Peng H, Tian Y, et al. Mitochondria targeting magnetic composite nanoparticles for enhanced phototherapy of cancer. Small. 2016; 12(33):4541–4552.
236. Wang H, Chang J, Shi M, et al. A dual targeted organic photothermal agent for enhanced photothermal therapy. Angew. Chem. Int. Ed. 2019; 58(4):1609–1673.
237. Li N, Yu L, Wang J, et al. A mitochondria targeted nanoradiosensitizer activating reactive oxygen species burst for enhanced radiation therapy. Chem. Sci. 2018; 9(12):3159–3164.
238. Ni K, Lan G, Veroneau S, et al. Nanoscale metal organic frameworks for mitochondria targeted radiotherapy radiodynamic therapy. Nat. Commun, 2018; 9(1)4321–4333.
239. Feldherr CM. The nuclear annuli as pathways for nucleocytoplasmic exchanges. J. Cell Biol. 1962; 14:65–72.
240. Alber F, Dokudovskaya S, Veenhoff L, et al. The molecular architecture of the nuclear pore complex. Nature. 2007; 450:7170.
241. Terry L, Shows E, Wente S. Crossing the nuclear envelope: Hierarchical regulation of nucleocytoplasmic transport. Science. 2007; 318(5855):1412–1416.
242. Lim R, Aebi U, Fahrenkrog B, et al Towards reconciling structure and function in the nuclear pore complex. Histochem. Cell Biol. 2008; 129(2):105–116.
243. Pante N, Kann M. Nuclear pore complex is able to transport macromolecules with diameters of about 39 nm. Mol. Biol. Cell. 2002; 13(2):425–434.
244. Lange A, Mills R, Lange C, et al. Classical nuclear localization signals: Definition, function, and interaction with importin alpha. J. Biol. Chem. 2007; 282(8):5101–5105.
245. Sun B, Luo C, Zhang X, et al. Probing the impact of sulfur/selenium/carbon linkages on prodrug nanoassemblies for cancer therapy. Nat. Commun. 2019; 10:3211.
246. Pan L, He Q, Liu J, et al. Nuclear-targeted drug delivery of TAT peptide-conjugated monodisperse mesoporous silica nanoparticles. J. Am. Chem. Soc. 2012; 134:5722–5725.
247. Pan L, Liu J, Shi J. Intranuclear photosensitizer delivery and photosensitization for enhanced photodynamic therapy with ultralow irradiance. Adv. Funct. Mater. 2014; 24:7318–7327.

248. Erlei J, Zhang B, Sun X, et al. Acid-active cell-penetrating peptides for in vivo tumor-targeted drug delivery. J. Am. Chem. Soc. 2013; 135:933–940.
249. Yu J, Xie X, Xu X, et al. Development of dual ligand-targeted polymeric micelles as drug carriers for cancer therapy in vitro and in vivo. J. Mater Chem B. 2014; 2:2114–2126.
250. Wang GH, Cai YY, Du JK, et al. TAT-conjugated chitosan cationic micelle for nuclear-targeted drug and gene co-delivery. Colloid Surface B. 2018; 162:326–334.
251. Jing Y, Xiong X, Ming Y, et al. A multifunctional micellar nanoplatform with pH-triggered cell penetration and nuclear targeting for effective cancer therapy and inhibition to lung metastasis. Adv. Healthc Mater. 2018; 7:1700974.
252. Tang PS, Sathiamoorthy S, Lustig LC, et al. The role of ligand density and size in mediating quantum dot nuclear transport. Small. 2014; 10:4182–4192.
253. Yang C, Uertz J, Yohan D, et al. Peptide modified gold nanoparticles for improved cellular uptake, nuclear transport, and intracellular retention. Nanoscale. 2014; 6:12026–12033.
254. Hsiangkuo Y, Fales AM, Tuan VD. TAT peptide-functionalized gold nanostars: Enhanced intracellular delivery and efficient NIR photothermal therapy using ultralow irradiance. J. Am. Chem. Soc. 2012; 134:11358–11361.
255. Desplancq D, Groysbeck N, Chiper M, et al. Cytosolic diffusion and peptide-assisted nuclear shuttling of peptide-substituted circa 102 gold atom nanoclusters in living cells. ACS Appl. Nano. Mater. 2018; 1:4236–4246.
256. Narayanan K, Yen SK, Dou Q, et al. Mimicking cellular transport mechanism in stem cells through endosomal escape of new peptide-coated quantum dots. Sci. Rep. 2013; 3:2184.
257. Maity AR, Stepensky D. Nuclear and perinuclear targeting efficiency of quantum dots depends on density of peptidic targeting residues on their surface. J. Control Release. 2017; 257:32–39.
258. Wlodarczyk MT, Dragulska SA, Camacho-Vanegas O, et al. Platinum (II) complex-nuclear localization sequence peptide hybrid for overcoming platinum resistance in cancer therapy. ACS Biomater Sci. Eng. 2018; 9:463–467.
259. Tomizaki KY, Kishioka K, Kataoka S, et al. Non-covalent loading of anti-cancer doxorubicin by modularizable peptide self-assemblies for a nanoscale drug carrier. Molecules. 2017; 22:1916.
260. Hao X, Li Q, Guo J, et al. Multifunctional gene carriers with enhanced specific penetration and nucleus accumulation to promote neovascularization of HUVECs in vivo. ACS Appl. Mater. Inter. 2017; 9:35613–35627.
261. Patel SS, Belmont BJ, Sante JM, et al. Natively unfolded nucleoporins gate protein diffusion across the nuclear pore complex. Cell. 2007; 129:83–96.
262. Alber F, Dokudovskaya S, Veenhoff LM, et al. The molecular architecture of the nuclear pore complex. Nature. 2007; 450:695–701.
263. Boyne JR, Whitehouse A. Nucleolar trafficking is essential for nuclear export of intronless herpesvirus mRNA. Proc. Natl. Acad. Sci. USA. 2006; 103(41):15190–15195.
264. Pan L, Liu J, He Q, et al. MSN-mediated sequential vascular-to cell nuclear-targeted drug delivery for efficient tumor regression. Adv. Mater. 2014; 26(39):6742–6748.
265. Peng H, Tang J, Zheng R, et al. Nuclear-targeted multifunctional magnetic nanoparticles for photothermal therapy. Adv. Healthcare Mater. 2017; 6(7):1601289.
266. Yu Z, Pan W, Li N, et al. A nuclear targeted dual-photosensitizer for drug-resistant cancer therapy with NIR activated multiple ROS. Chem. Sci. 2016; 7:7.
267. Tammam SN, Azzazy HME, Lamprecht A. The effect of nanoparticle size and NLS density on nuclear targeting in cancer and normal cells; Impaired nuclear import and aberrant nanoparticle intracellular trafficking in glioma. J. Control Release. 2017; 253:30–36.
268. Ali MR, Wu Y, Ghosh D, et al. Nuclear membrane-targeted gold nanoparticles inhibit cancer cell migration and invasion. ACS Nano. 2017; 11(4):3716.
269. Austin LA, Kang B, Yen CW, et al. Nuclear targeted silver nanospheres perturb the cancer cell cycle differently than those of nanogold. Bioconjug Chem. 2011; 22(11):2324–2331.
270. Jana A, Saha B, Banerjee DR, et al. Photocontrolled nuclear-targeted drug delivery by single component photoresponsive fluorescent organic nanoparticles of acridin-9-methanol. Bioconjug Chem. 2013; 24(11):1828.
271. Moenner M, Pluquet O Bouchecareilh M, et al. Integrated endoplasmic reticulum stress responses in cancer. Cancer Res. 2007; 67(22):10631–10634.
272. Vella F. Molecular biology of the cell (third edition): By B Alberts, D Bray, J Lewis, M Raff, K Roberts and J D Watson. pp 1361. Garland Publishing, New York and London. Mol Biol Cell New York Garland 1994; 22:600–651.

273. Aridor M, Hannan LA. Traffic jam: A compendium of human diseases that affect intracellular transport processes. Traffic. 2000; 1:836–851.
274. Yoshida H. ER stress and diseases. Febs. J. 2007; 274:630–658.
275. Tarragó-Trani MT, Storrie B. Alternate routes for drug delivery to the cell interior: Pathways to the Golgi apparatus and endoplasmic reticulum. Adv. Drug Deliver. Rev. 2007; 59:782–797.
276. Rajendran L, Knölker HJ, Kai S. Subcellular targeting strategies for drug design and delivery. Nat. Rev. Drug Discov. 2010; 9:29–42.
277. Helena P, José Manuel L, Paula S. The mTOR signalling pathway in human cancer. Int. J. Mol. Sci. 2012; 13:1886–1918.
278. Hanahan D, Weinberg R. Hallmarks of cancer: The next generation. Cell 2011; 144:646–674.
279. Palade G. Intracellular aspects of the process of protein synthesis. Science 1975; 189(4200):347–358.
280. Stanley KK, Howell KE. Tgn38/41. A molecule on the move. Trends Cell Biol. 1993; 3(8):252–255.
281. Molloy SS, Anderson ED, Jean F, et al. Bi-cycling the furin pathway. From TGN localization to pathogen activation and embryogenesis. Trends Cell Biol. 1999; 9(1):28–35.
282. Wang Y, Luo S, Zhang C, et al. An NIR-Fluorophore-based therapeutic endoplasmic reticulum stress inducer. Adv. Mater. 2018; 30:1800475–180483.
283. Yu L, Wang Q, Yeung KW, et al. A biotinylated and endoplasmic reticulum-targeted glutathione responsive Zinc(II) Phthalocyanine for targeted photodynamic therapy. Chem. Asian J. 2018; 13(22):2509–2517.
284. Zhou Y, Cheung YK, Ma C, et al. Endoplasmic reticulum localized two photon absorbing boron dipyrromethenes as advanced photosensitizers for photodynamic therapy. J. Med. Chem. 2018; 61(9):3952–3961.
285. Dabrowski JM, Arnaut LG. Photodynamic therapy (PDT) of cancer: From local to systemic treatment. Photochem. Photobiol. Sci. 2015; 14(10):1765–1780.
286. Xu CY, Bailly-Maitre B, Reed JC. Endoplasmic reticulum stress: Cell life and death decisions. J. Clin. Invest. 2005; 115(10):2656–2664.
287. De Duve C, Baudhuin P. Peroxisomes (microbodies and related particles). Physiol. Rev. 1966; 46(2):323–357.
288. Terlecky SR, Koepke JI. Drug delivery to peroxisomes: Employing unique trafficking mechanisms to target protein therapeutics. Adv. Drug Deliv. Rev. 2007; 59(8):739–747.
289. Nalepa G, Rolfe M, Harper JW. Drug discovery in the ubiquitin-proteasome system. Nat. Rev. Drug Discov. 2006; 5(7):596–613.
290. Dahlmann B. Role of proteasomes in disease. BMC Biochem. 2007; 8(1):S3.
291. Richardson PG, Barlogie B, Berenson J, et al. A Phase 2 study of bortezomib in relapsed, refractory myeloma. N. Engl. J. Med. 2003; 348(26):2609–2617.
292. Kyle RA, Rajkumar SV. Multiple myeloma. N. Engl. J. Med. 2004; 351(18):1860–1873.
293. Adams J. The development of proteasome inhibitors as anticancer drugs. Cancer Cell. 2004; 5(5):417–421.
294. Burger AM, Seth AK. The ubiquitin-mediated protein degradation pathway in cancer: Therapeutic implications. Eur. J. Cancer. 2004; 40(15):2217–2229.
295. Robinson PA, Ardley HC. Ubiquitin-protein ligases—novel therapeutic targets? Curr. Protein. Pept. Sci. 2004; 5(3):163–176.
296. Bobo D, Robinson KJ, Islam J, et al. Nanoparticle-based medicines: A review of FDA-Approved materials and clinical trials to date. Pharm. Res. 2016; 33:2373–2387.

19 Characterization Techniques for Stimuli-Responsive Delivery Nanoplatforms in Cancer Treatment

Deepak Kulkarni, Dipak Gadade, Harshad Kapare, Namdev L. Dhas and Mayuri Ban

CONTENTS

19.1 Introduction 322
19.2 General Characteristics of Nanoplatforms 324
    19.2.1 Fourier Transform Infrared 324
    19.2.2 Differential Scanning Colorimetry 325
    19.2.3 X-ray Diffraction 325
    19.2.4 Transmission Electron Microscopy 325
    19.2.5 Drug Release 326
    19.2.6 Drug Entrapment Efficiency 327
    19.2.7 Particle Size and Polydispersity Index 328
    19.2.8 Zeta Potential/Surface Charge Determination 328
19.3 Characterization of Stimuli-Responsive Nanosystems 330
    19.3.1 pH-Responsive Nanosystems 330
    19.3.2 Redox-Responsive Nanosystems 331
    19.3.3 Enzyme-Responsive Nanosystems 331
    19.3.4 Light-Responsive Systems 333
    19.3.5 Ultrasound-Responsive Systems 334
References 335

19.1 INTRODUCTION

Nanodrug delivery is one of the main efficient paradigms in the field of biomedical science. Biocompatible and biodegradable nanosystems are well employed for drug delivery (1). Targeted delivery and efficient biodistribution are major advantages of nanodrug delivery systems used for cancer therapy. The targeting of overexpressed biomolecules in the tumor microenvironment is an important aspect of the execution of nanocarrier-based drug delivery systems (2). Stimuli-responsive systems are an emerging and promising area in nanotechnology-based delivery of drugs. Endogenous and exogenous stimuli govern the delivery of drugs from nanocarriers. pH, enzymes and redox systems are general endogenous stimuli, while temperature, light, ultrasound and electric and magnetic fields are exogenous stimuli. Figure 19.1 illustrates all the stimuli-responsive nanosystems for targeting the tumor micro-environment. Multi-stimuli–responsive nanodrug delivery systems are also well reported in the literature. In multi-stimuli–based systems, drug release depends on two or more stimuli simultaneously, for example, pH–redox or pH–temperature (3–7).

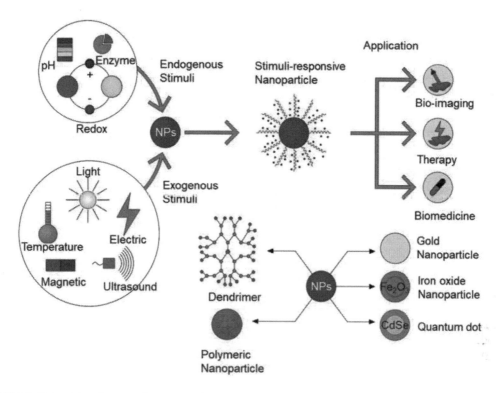

FIGURE 19.1 Stimuli-responsive nanosystems for targeting the tumor micro-environment (7).

The characterization of these stimuli-responsive nanosystems is one important aspect after fabrication. The shape, size, morphology, surface charge, drug loading/entrapment and drug release are general important characteristics which govern the in vivo performance of fabricated nanosystems (8, 9). General characteristics like Fourier transform infrared (FTIR) spectroscopy demonstrate the chemical nature and functionality of drugs and the material used for fabrication of nanoplatforms and characterize the chemical modifications done for the functionalization of nanoplatforms (10). Differential scanning colorimetry (DSC) is a useful tool to characterize the compatibility of formulation components and drugs used in nanosystems (11). X-ray diffraction (XRD) is a tool which characterizes the successful fabrication of nanosystems for drug delivery (12). Drugs with a crystalline nature show a sharp peak with better intensity in X-ray diffraction. The sharp peak disappears in X-ray diffraction of nanosystems when the drug is encapsulated within nanocarriers indicating the conversion of drug from crystalline form to amorphous form (13). Particle size and morphology also play important roles in biodistribution and accumulation at the target site. Spherical nanosized particles have greater surface area and provide better scope for in vivo absorption (14). The polydispersity index (PDI) provides information about uniformity in particle size of these nanocarriers. The uniformity of particle size plays an important role in the absorption of nanoplatforms (15). The zeta potential is an important characteristic which indicates the surface charge and associated stability of the nanosystem. The modulation of the surface can be characterized by determining the zeta potential. The surface modification for targeting a specific protein, receptor or gene can change the zeta potential to an anionic or cationic pole (16). Drug loading is an important process for the therapeutic potentiation of nanocarriers. Drug loading in nanocarriers is always challenging. The drug loading capacity or entrapment efficiency of nanoplatforms is an important characteristic for nanosystems used for drug delivery. The percentage of entrapment efficiency is calculated for nanosystems to analyze drug loading efficiency (17, 18). The surface morphology of nanoplatforms

TABLE 19.1
General Characterization Techniques for Nanoplatforms. Modified from Reference (8)

Characterization Technique	Interpretation
X-ray diffraction	Crystallinity and crystal structure
X-ray absorption spectroscopy (XAS)	X-ray absorption coefficient determination and identification of non-crystalline nanoparticles
Small-angle X-ray scattering (SAXS)	Particle size and particle size distribution of nanoparticles
X-ray photoelectron spectroscopy (XPS)	Elemental composition and electronic structure
Fourier transform infrared	Surface chemistry and composition and ligand binding on surface of nanoparticles
Nuclear magnetic resonance (NMR)	Structure, shape and size of nanoparticles. Ligand binding and related characterization
Brunauer-Emmett-Teller (BET)	Surface area of nanoparticles
Thermal gravimetric analysis (TGA)	Mass analysis
UV-visible spectroscopy	Size of nanoparticles, optical properties and concentration
Dynamic light scattering (DLS)	Hydrodynamic diameter
Nanoparticle tracking analysis (NTA)	Particle size and particle size distribution of nanoparticles
Finite mixture of regression (FMR)	Shape, particle size and particle size distribution of nanoparticles
Transmission electron microscopy	Shape, particle size and surface morphology of nanoparticles
Scanning electron microscopy	Surface morphology and dispersion of nanoparticles in cellular system
Atomic force microscopy (AFM)	Particle size in three dimensions, surface morphology and dispersion of nanoparticles in cellular system

is characterized by emission spectroscopy. Transmission electron microscopy (TEM) and scanning electron microscopy (SEM) are techniques which are useful to analyze surface morphology as well as particle size of hollow and solid nanosystems, respectively (19, 20). The general characterization techniques for nanoplatforms are summarized in Table 19.1.

Stimuli-responsive systems require specialized characterization to determine drug release in response to stimuli. In vitro simulation is the methodology used to characterize the stimuli-triggered response of these stimuli-responsive nanosystems. Drug release from nanosystems in response to endogenous as well as exogenous stimuli is studied by simulation. In vivo techniques to estimate drug release behavior using animal models are also reported in the literature (21, 22).

19.2 GENERAL CHARACTERISTICS OF NANOPLATFORMS

This chapter summarizes general as well as distinct characterization techniques for stimuli-responsive nanosystems with reference to the available literature and reported research and will be of interest for researchers working in the area of stimuli-responsive nanosystems for targeting the tumor microenvironment.

19.2.1 Fourier Transform Infrared

Fourier transform infrared spectroscopy measures absorption of electromagnetic radiation in the mid-infrared range wavelengths (4000–400 cm^{-1}). As a molecule absorbs IR light, its dipole moment is altered and becomes IR active. Spectra give information on functional groups, their interactions and their molecular structures (23). This technique is applicable for evaluation of nanostructure carriers in terms of pre-formulation for compatibility studies and formulation characterization for molecular dispersion, encapsulation and stability. Advancements in FTIR techniques like the in situ attenuated total reflectance-FTIR method in combination with differential electrochemical mass

Characterization Techniques for Stimuli-Responsive Delivery

spectrometry methods have been applied by researchers to investigate the effects of nanostructures on oxidation of ethanol, which confirms the effects of breakdown (24). Shukla et al. investigated the surfactant bonding to FePt NPs formulated with oleic acid and oleylamine using FTIR for investigation of the bonding mechanism (8, 25).

The FTIR technique is also applied for monitoring molecular changes in microorganisms in applications of nanoparticles, specifically in classification of bacteria composition. Biochemical bonds like stretching, bending, scissoring and twisting can be detected based on their molecular rotational degree and form of movement (26).

19.2.2 Differential Scanning Colorimetry

DSC is a thermoanalytical technique used to calculate the difference between the amount of heat needed to raise a sample's temperature and that of a reference. This technique is also applicable for characterization of nanostructure carriers for compatibility, encapsulation and stability studies. Various researchers routinely use this technique in nanoparticle characterization to identify phase transitions studies.

The DSC technique is applied in nanotechnology for quantitative analysis for the determination of heat of fusion and the extent of crystallization for crystalline materials. This technique is also used for determination of glass transition temperature Tg and understanding various behaviors in chemical reactions. This is also applicable for determination of interactions between drugs and excipients. DSC is also applicable in measures of crystallization and plasticization effects in nanocomposites (27).

19.2.3 X-ray Diffraction

X-ray diffraction is one of the most popular techniques for characterizing NPs. XRD typically reveals the lattice parameters, phase, crystalline grain size and crystalline structure. The latter parameter is computed using the Scherrer equation and the broadening of the strongest peak of an XRD measurement for a specific sample. Utilizing XRD methods has the benefit of generating statistically representative, volume-averaged results. These methods are routinely applied to powdered samples, usually following the drying of the relevant colloidal solutions (8, 28).

For particles less than 3 nm, the XRD peaks are too broad, making it inappropriate for amorphous materials. Upadhyay et al. found that the average crystallite size of magnetite NPs ranged from 9 to 53 nm using X-ray line broadening. In addition to instrumental broadening, the main causes of the widening of the XRD peaks were particle/crystallite size and lattice tension (28). The versatile method of X-ray diffraction is frequently used in the field of nanotechnology to characterize and gather precise data on the chemical makeup, crystal structure and crystalline grain size of nanoparticles. Beginning with the identification of the sample material's phase, XRD is used to characterize nanoparticles. The crystal type in the sample is established by a search-match procedure, which is carried out in the areas where the peaks of high intensities are detected.

For XRD analysis, thin films of nanoparticles are created by drop-casting the nanoparticle onto a suitable low-background substrate. An appropriate goniometer is used to measure multiple reflections from nanoparticle samples that have the desired orientation. A sample with random orientation, however, could just need a short scan through a 2-diffraction (29).

19.2.4 Transmission Electron Microscopy

The interaction between a thin sample and an electron beam with a consistent current density and energy typically between 60 and 150 keV is used in transmission electron microscopy. When the electron beam strikes the sample, a fraction of the electrons are transferred, while the remainder are either inelastically or elastomatically scattered (30). Size, sample density and elemental composition are a few of the factors that influence the strength of the interaction. The final image is built using the data gathered from the delivered electrons.

Structural and analytical characterization at the nanometer level has become very important in recent years for all kinds of materials. A transmission electron microscope is the instrument of choice for this purpose. Parameters such as grain size, grain size, lattice type, morphological information, crystallographic details, chemical composition, phase type and distribution can be obtained from transmission electron micrographs. Electron diffraction patterns of nanomaterials are also used to obtain quantitative information, including size, phase identification, orientation relationships and crystal defects in lattice structures (31).

As revealed in the previous section, the unique collection of physical properties of NPs, including their optical, magnetic, electronic and catalytic capabilities, as well as their interactions with biological systems, are defined by size and morphology. TEM is the most commonly used method to analyze the size and shape of nanoparticles because it directly observes the sample and provides the most accurate estimate of nanoparticle homogeneity. However, some limitations should be considered when applying this method, such as the challenge of measuring inaccurate images caused by a large number of particles or orientation effects (32, 33).

19.2.5 Drug Release

In vitro drug release from nanoparticles is a pharmaceutically crucial quality control parameter, as it is generally correlated with in vivo performance. In vivo drugs provide a clearer picture of the in vivo performance of nanoparticulate systems. There is plentiful evidence indicating the role of the drug release characteristics of nanoparticulate systems in the biopharmaceutical fate of drugs (34).

The polymers providing extended drug release to design smart nanoparticles provides the physiological and clinical advantages. The physiological benefits which can be attained with extended-release smart nanoparticles include decreased renal clearance, potential to reduce cytotoxicity and higher therapeutic indices, while their clinical advantages include improved efficacy and reduced toxicity (35).

The endogenous or exogenous triggers either increase or accelerate from the smart nanoparticles (36). The endogenous stimuli upon which triggered nanosystems work include changes in the pH of microenvironment of nanoparticles, enzymatic catalytic triggers, redox-responsive microchemical changes in smart nanoparticles, endogenous thermal response, tumor hypoxia and glucose-responsive release. Nanosystems are also reported with exogenous triggers for drug release, including temperature alterations, exposure to light of varied wavelengths with time and intensity modulation, ultrasound thermal effects and/or mechanical changes through cavitation or radiation and controlled electrical impulses and alteration in magnetic field moment and intensity. There are few reports revealing dual- and multi-responsive drug release (37).

19.2.5.1 Smart Nanoparticle Drug Release Studies

The in vitro drug release study of smart nanoparticles gives an estimate of the percentage of drug release over a period of time. A sample of smart nanoparticles is dispersed in a suitable buffer or dissolution media to obtain the desired concentration of the drug. Then the dispersion is placed in a dialysis bag or tubing (MWCO 3500–5000 Da). It is later immersed in 10 to 100 ml of buffer solution at $37 \pm 0.5°C$ using the agitation or stirring method. Alternatively, the smart nanoparticulate formulation can be directly added to the dissolution media. The centrifugation method can be used to estimate drug release in the presence of the trigger (38). The drug is released when the nanosystem is subjected to a specific trigger like pH (39), magnetic field (40), light (41) or enzyme (42). Then 0.5 to 1.0 ml of sample is withdrawn at regular predetermined internals and replaced with fresh buffer or dissolution media. Samples can be analyzed by a validated analytical method to determine drug release. Ultraviolet spectroscopy or fluorescence spectroscopy alone or coupled with chromatographic methods can be utilized for the purpose of analysis. The drug release rate from the smart nanoparticle system is determined. Further, mathematical modeling can be performed for deeper understanding of the drug release mechanism (43, 44).

Smart "on–off"-responsive nanoflowers were reported providing controlled release. These nanoflowers release the drug at the desired target site upon stimulus by change in pH or subjecting nanosystems either to an alternating magnetic field or near-infrared light irradiation. The nanosystem was intelligently designed to bind doxorubicin inside with amino-functionalized mesoporous silica-gold nanoflowers. The system switches on drug release of the doxorubicin only at pH 5 and turns it off it at pH 7.4. The amino group works as a molecular switch for drug release, while β-CD assists the nanosystem in controlling drug release by creating stearic hindrance. This nanosystem has theranostic advantages for imaging and targeted drug delivery to the tumor (45).

Smart-size transforming nanoparticles were reported for drug release into deeper cancerous tissues. A polymeric prodrug was prepared by combining camptothesin with a positively charged three-armed star and charge reversal polymer which forms a self-assembled neutral polymeric drug nanocomplex. In response to the acidic microenvironment of the tumor, smaller nanoparticles are released. Drug release studies performed under four different conditions by variation of pH from 6.8 and 7.4 and in the presence of glutathione at concentrations of 2 μm or 10 mM were reported. Higher drug release was observed in at pH 6.8 for 2 hr followed by incubation at pH 7.4 with 10 mM glutathione. The nanosystem was reported to possess theranostic value for significant tumor inhibition in in vivo as well as in vitro conditions (46).

The selection of an appropriate type of the nanoformulation is crucial for the desired in vitro drug release and ultimate fate of the drug in vivo. The differences in the drug release pattern of lipid nanocarriers clearly reveal the importance of the selection of a suitable type of carrier for drug delivery. Drug release from solid lipid nanoparticles (SLNs) and nanostructured lipid carriers (NLCs) of levofloxacin showed a linear pattern and biphasic initial burst drug release followed by a slow release pattern. SLN released almost 80% of the drug in first 4 hr, whereas NLC released less than 60% of the drug. The drug release pattern for both formulations was similar after 12 hr, but almost all drug content was released at 24 hr from SLN and 48 hr from NLC (47).

19.2.6 Drug Entrapment Efficiency

Drug entrapment efficiency is the quantity of active pharmaceutical or biological agent incorporated in the nanoparticulate system. It can be analyzed by dispersing or dissolving the drug-loaded smart nanoparticles in an organic solvent such as methanol. The solution is centrifuged and a supernatant is collected for the analysis and estimation of drug entrapped in the nanoparticles. Variables including centrifugation speed and time are optimized prior to analysis. The entrapment efficiency is computed using the following equation (47):

Entrapment Efficiency = {(Total amount of drug − Free drug)/(Total amount of drug)} × 100

The entrapment efficiency of nanosystems can be modified with the functionalization of nanoparticles. It was revealed in the literature that amino functionalization of a mesoporous silica nanosystem could achieve encapsulation efficiency as high as 97.06 ± 1.94% when doxorubicin hydrochloride was loaded, unaffected by the presence of nanogold (45).

The optimal amount of drug to use for higher drug loading and drug entrapment can be determined by changes in the drug/polymer ratio. Higher entrapment efficiency was reported with a 1/1 indomethacin/polymer ratio at 10 gm/l polymer concentration with particle size 225.6 ± 25.07 nm. Authors report that at the lower drug/polymer ratio 0.33/1, an insufficient amount of drug could not satisfy the hydrogen bonding required for the formation of micelles with pullulan-g-poly(N-isopropylacrylamide) polymer (48).

In order to optimize the entrapment efficiency of nanoparticles, in situ synthetic conditions should be optimized. During preparation of the solid lipid nanoparticles of levofloxacin, it was revealed that the drug migrates to the aqueous phase containing pluronic instead of getting trapped in the lipid phase. Adjusting reaction conditions to pH 7.4 where levofloxacin improved the entrapment efficiency to 12.5 by almost twice (i.e. increased from 5.2 to 12.5) (47). Levofloxacin has pKa values 5.7 and 7.9. Therefore, it remains in zwitterionic form between pH 5.7 to 7.9, converting it to hydrophobic nature and allowing entrapment in the lipid phase (49). Further addition of the acetone to reaction vessel improved the entrapment efficiency to 20.1% while converting solid lipid nanoparticles to nanostructured lipid carriers by addition of 3 wt% Crodamol GTCC-LQ oil increased entrapment efficiency further to 55.9%. The oil disrupted the crystal packing of solid lipid nanoparticles (50).

19.2.7 Particle Size and Polydispersity Index

Any material with at least one dimension smaller than 100 nm is referred as a nanoparticle. The particle size is an important characteristic of the nanoparticles included in the study of smart nanoparticle powers (51). The polydispersity index or heterogeneity index is an parameter that is used in characterization of the particle size distribution of smart nanoparticles. It is a dimensionless parameter showing particle size distribution in the given sample. In the case of polymers and polymeric nanoparticles, it is used as an measure of molecular weight distribution (49). Table 19.2 summarizes the polydispersity index–based interpretation of nanoparticles.

The polydispersity index was evaluated to check the effect of salt concentration on the stability and particle size distribution of the mesoporous silica nanoparticles of doxorubicin. A spontaneous increase in the polydispersity index was reported above salt concentration 0.6 mol/L, as depicted in Figure 19.2. The change in the polydispersity index of mesoporous nanoparticles was attributed to the destabilization of nanoparticles at the higher concentration. The mesoporous silica nanoparticles are a stable normal saline solution, making them suitable for injectable preparation (52).

Polydispersity index can be used an tool for the optimization of pullulan-g-poly(N-isopropylacrylamide)–based smart nanoparticles. These nanoparticles were developed by either the dialysis or nanoprecipitation method using solvent dimethyl formamide or dimethyl sulfoxide. The polydispersity index suggests nanoprecipitation in the presence of dimethyl formamide was suitable for the development of indomethacin polymeric nanoparticles. The effect of the drug/polymer ratio was also optimized by using the polydispersity index, and the optimized drug polymer ratio was 0.5/1, forming nanoparticles with a narrower size distribution. An increased drug/polymer ratio led to agglomeration of nanoparticles (48). Figure 19.2 illustrates the effect of sodium chloride concentration on PDI.

19.2.8 Zeta Potential/Surface Charge Determination

Zeta potential is an analytical parameter for the determination of surface characteristics of smart nanoparticles. It is based on the electric potential of smart nanoparticles at the shear plane of

TABLE 19.2
Polydispersity Index of Nanoparticles (49)

Polydispersity Index	Interpretation
Zero	Perfectly uniform sample
≤0.05	Highly monodispersed sample
≤0.20	Acceptable particle size distribution for polymeric nanoparticles
≤0.30	Acceptable particle size distribution for lipid nanoparticles
≥0.70	Very broad particle size distribution
1.00	Polydispersed sample with multiple particle sizes

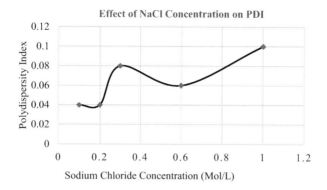

FIGURE 19.2 Effect of sodium chloride concentration on the polydispersity index of mesoporous nanoparticles.

FIGURE 19.3 Stability behavior of colloidal smart nanoparticles based on zeta potential value.

particles. Nanoparticles generally have excess surface charge at the surface, which attracts counter ions, forming a stern layer. A shear plane or diffused layer is present adjacent to the stern layer. The electric potential present at the boundary of the electrical double layer is referred as zeta potential, and the typical range of the zeta potential is –100 to +100 mV. It can be used for the optimization of smart nanoparticles based on the prediction of surface interactions and chemical stability of pharmaceutical formulations. Understanding of the zeta potential is crucial for the formulator and can reduce time spent in development (53). Figure 19.3 illustrates the stability behavior of colloidal smart nanoparticles based on zeta potential value.

Zeta potential is used a tool for the characterization and optimization of pH-responsive starch nanoparticles. In the absence of a surfactant, starch nanoparticles tend to coagulate, while the addition of cetyltrimethylammonium bromide, sodium dodecylbenzenesulfonate and polyethylene glycol improves stability. The increasing concentration of these surfactants increased the zeta potential in an almost linear order (54).

Zeta potential can also be used as an indicator of crosslinking of multifunctional fluorescent and magnetic chitosan nanosystem pH-responsive drug release. Fe_3O_4 nanoparticles and CdTe quantum dots were mixed with the acidic chitosan solution. The superparamagnetic properties of the former provide the magnetic nature, while the latter leads to the fluorescent characteristics of nanoparticles.

Positive surface charge ranging from +5 to +35 mV at neutral pH is used as an indicator for the preparation of composite particles before crosslinking. The crosslinking of chitosan nanoparticles showed a decrease in zeta potential over the period of time owing to crosslinking free amino functional groups by glutaraldehyde (55). The cationic 2-aminoethyldiisopropyl group decorated on the bolla pattern polymer having dual-end PEGylation of poly (β-amino ester) were reported. The alteration in critical micellar concentration and zeta potential shows the protonation of the polymer, which releases the drug in the acidic tumor microenvironment around pH 5. The zeta potential value of these macrophage membrane coated was around 2 mv at pH 7.4, while it was around 7 mv at pH 5 (56).

Zeta potential was used as characterization tool for the smart polymeric prodrug camptothesin. The smart polymeric prodrug nanosystem rapidly changed zeta potential from −6.8 to 5.2 mV when incubated at pH 6.8 and compared against an insensitive nanosystem, which is not able to change zeta potential in similar conditions (46).

19.3 CHARACTERIZATION OF STIMULI-RESPONSIVE NANOSYSTEMS

All the general characteristics necessary for nanocarriers are summarized previously, while some specific simulation based in vitro and animal model based in vivo characterizations are carried out for stimuli-responsive nanoplatforms for drug release triggered by exogenous or endogenous stimuli.

19.3.1 pH-Responsive Nanosystems

pH-dependent drug release from nanoplatforms needs to be characterized to confirm the efficiency of drug release in the tumor microenvironment. Various previous studies reported the characterization of pH-based drug release from nanosystems in the tumor microenvironment.

Surya et al. reported the fabrication and characterization of a pH-responsive mucoadhesive drug delivery system for controlled and targeted release of anticancer drugs. Various general examinations like FTIR and XRD were carried out to confirm the chemical and physical characteristics of the formulation. Swelling behavior at a specific pH was one of the characterizations performed. Swelling behavior analysis was studied at pH. The drug loading and encapsulation efficiency was also characterized, and it was found that the encapsulation efficiency was 72.5%, while total drug loading was 42%. SEM, TEM and DLS analysis of a pH-responsive formulation system was done to analyze surface morphology and particle size. TGA demonstrated the thermal stability of the formulation. An in vitro simulation study was carried out to investigate pH-dependent drug release behavior. A drug release study was carried out at gastric pH (pH 1.2) for 2 hr, and further study was done at intestinal pH (pH 7.4). The results demonstrated that only 15.6% of the drug was released at gastric pH (pH 1.2), while 82.5% of the drug release was at intestinal pH (pH 7.4) in controlled manner up to 6 hr (57).

Jeshvaghani et al. demonstrated the fabrication and characterization of pH-responsive nanocarriers for sustained release of anticancer drugs. The nanocarrier was fabricated from polyethylene glycol (PEG)/graphene oxide (GO)/silk fibroin (SF). The general FTIR and XRD analysis confirmed the presence of all the formulation components and their physical and chemical compatibility. The encapsulation efficiency of the nanocarrier was found to be 87.75%, while drug loading efficiency was about 46%. DLS demonstrated the particle size (293.7 nm), while the highly negative zeta potential (−102.9mV) showed the presence of hydroxyl groups in formulation components. SEM analysis demonstrated the spherical nature of the nanocarrier. The in vitro simulation–based drug release analysis revealed that the release of drug (doxorubicin) was retarded at pH 1.2. Dissolution- and diffusion-controlled drug release was observed at pH 7.4 and 5.4, respectively. A cytotoxicity study by 3-(4,5-dimethylthiazol-2-yl)-2,5-diphenyltetrazolium bromide (MTT) assay revealed an anticancer potential (58).

19.3.2 REDOX-RESPONSIVE NANOSYSTEMS

Redox reactions play a role as endogenous stimuli which trigger drug release from the nanosystem in the tumor microenvironment to achieve targeted drug delivery. Researchers have reported multiple case studies of fabrication and characterization of nanoplatforms.

Yang et al. reported the fabrication of polymeric nanoparticles with a backbone of low molecular weight heparin (LMWH). LMWH was conjugated with chlorin e6 and alpha-tocopherol succinate using cystamine. The cystamine acted as a linker which sensed the redox reaction. The resulting nanoparticles were amphiphilic in nature and provided scope for self-assembly in water. The anticancer drug paclitaxel was loaded in nanoparticles. The increased near-infrared intensity and generation of reactive oxygenated species of chlorin e6 was observed after the redox reaction. The redox reaction triggered drug release from the nanoparticles. DLS was performed for particle size analysis, and TEM was performed to observe the surface morphology before, during and after drug release. Confocal laser scanning microscopy (CLSM) was performed to study the cellular uptake of the drug. Flow cytometric analysis was found useful to study internalization. An in vitro cytotoxicity study was carried out using MTT assay and MCF-7 cells. Apoptosis was studied using an annexin V-FITC/PI double-staining assay kit. Wound healing assay was also performed to assess the migration of MCF-7 cells (59).

Ren et al. demonstrated the preparation of copolymer-based nanomicelles for redox-responsive delivery of anticancer drugs (docetaxel and indocyanine green). The copolymer was fabricated using arginine/glycine/aspartic acid as an active targeting agent functionalized with polyethylene glycol (PEG) and polycaprolactone (PCL), which serve as polar and non-polar ends, respectively. FTIR confirmed the presence of all the formulation ingredients with identification of characteristic bonds. The surface morphology of the fabricated nanomicelles was studied using TEM. The particle size distribution and zeta potential were analyzed using a laser particle analyzer. MTT and in vivo antitumor assays were performed to study anticancer activity (60).

19.3.3 ENZYME-RESPONSIVE NANOSYSTEMS

Targeted and controlled delivery of drugs in the tumor microenvironment can be achieved using stimuli-responsive nanosystems. Various in vivo enzymes present in the tumor microenvironment trigger the release of drugs from nanoplatforms. Multiple characterization of these nanosystems are illustrated ahead with reference to relevant literatures.

Guo and coworkers developed enzyme-responsive nanoparticles for the delivery of curcumin, fabricated from triblock polymer (polycaprolactone–peptide-polyethylene glycol). The prepared nanoparticles were implemented for targeting the microenvironment of a lung tumor. Matrix-metalloproteinase, the overexpressing protein in lung cancer, was the trigger factor for drug release in the tumor microenvironment. The triblock polymer was characterized by FTIR and proton NMR to confirm the presence of the formulation components. The polydispersity index was determined to confirm particle size uniformity. The morphology of nanoparticles was studied using TEM. The simulated drug release was studied at pH 7.4 and 6.5 (61).

Tang et al. developed an enzyme-responsive lipid-polymer–based nanosystem for targeted delivery of anticancer drugs in the tumor microenvironment. Polyethyleneimine-lecithin and PEG were used to fabricate the lipid-polymer based nanosystem. The dichloroacetate (DCA) and indocyanine green (ICG), – both the drugs were conjugated to polyethyleneimine through the formation of dense amide bonds. Overexpressive amidase in the tumor microenvironment was the trigger enzyme for the in vivo release of drugs. Particle size analysis and PDI was performed using DLS. The morphology of the fabricated nanosystem was analyzed using TEM. The components of the formulations were analyzed and confirmed by obtaining the UV-visible spectra of drugs and the lipid-polymer system. The fluorescence spectra were also obtained to characterize the ICG and lipid-polymer system. Drug loading was characterized using proton NMR and UV-visible spectroscopic analysis.

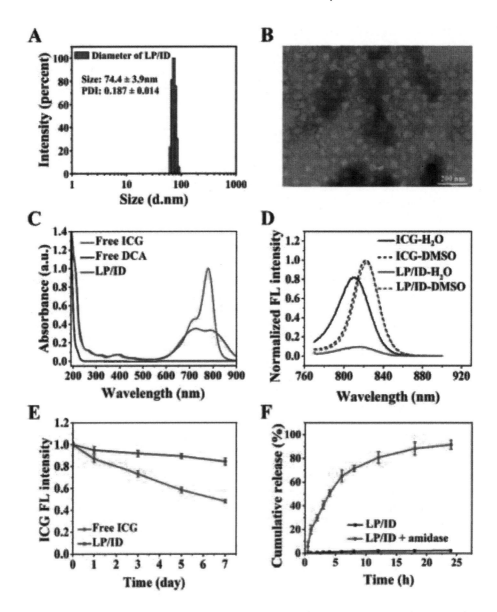

FIGURE 19.4 Characterization of lipid-polymer based nanoplatforms. Reproduced with permission from reference (62). Copyright Royal Society of Chemistry 2022.

Cellular uptake and cytotoxicity study showed the anticancer potential of the proposed nanosystem (62). Figure 19.4 illustrates the characterization of lipid-polymer based nanoplatforms.

19.3.3.1 Temperature-Responsive Nanosystems

Thermo-responsive drug delivery is a promising approach for targeted delivery in the tumor microenvironment. The rise in temperature (hyperthermia) is the important governing factor for drug release. The characterization of thermo-responsive nanosystems is illustrated in the following, with reference to available literature.

Xu et al. developed a magnetic thermo-responsive nanosystem for controlled release of an anticancer drug (doxorubicin). The nanosystem was fabricated from aminated magnetic mesoporous silica conjugated with carboxyl-modified DNA20. The prepared nanosystem demonstrated the controlled release of doxorubicin. Various general and specific characterizations were done to confirm the appropriate fabrication and functional efficiency of the nanosystem. Spherical surface morphology was characterized from SEM and TEM. The Brunauer-Emmett-Teller analyzer was used to determine the surface area. The nanosystem was having the grafting of amino groups and the weight loss in TG (thermogravimetric) analysis confirm the amino group grafting on nanosystem. The zeta potential demonstrated the surface charge of the nanocarrier. The change in zeta potential from −13.7 to 16.4 mV showed the grafting of amino groups on the nanosystem. FTIR also confirmed amino group grafting. Temperature-based drug release was observed from the nanosystem. The nanosystem showed lower drug release at 37°C and better drug release at 43°C. The cellular uptake and cytotoxicity revealed the anticancer efficiency of the fabricated nanosystem (63).

Maia et al. reported the development of thermo-responsive liposomes for delivery of anticancer drugs (gadodiamide). Gadolinium-based complexes (gadodiamide) have potential anticancer activity but are limited by inadequate cellular internalization. Liposomal drug delivery is advantageous to overcome this limitation. DLS, DSC and small angle X-ray scattering techniques were used for characterization of liposomes. TEM demonstrated the morphology of the fabricated liposomes. Entrapment efficiency and drug loading were determined. In vitro drug release analysis demonstrated the thermosensitive response of the liposome and resulting delivery of the anticancer drug (gadodiamide). The cytotoxicity of the fabricated nanosystem was studied by MTT assay, which revealed the potential activity of the fabricated liposomes (64).

19.3.4 Light-Responsive Systems

Light is an exogenous stimulus which triggers responsive drug release from nanocarriers. Specifically, near-infrared (NIR) lasers are finding applications in biomedicine in light-responsive drug delivery.

Tang et al. fabricated NIR laser–stimulated polymeric vesicles (polymerosomes) for delivery of anticancer drugs. Poly (propylene sulfide) 20-bl-poly (ethylene glycol) 12 (PPS20-b-PEG12) block copolymer was used for fabrication of polymerosomes. Zinc phthalocyanine loaded in the shell of polymerosomes was used as a photosensitizer, and the anticancer drug doxorubicin·HCl was loaded in the inner aqueous core. In characterization, proton NMR was used to confirm the structure of the copolymer block. Gel permeation chromatography (GPC) was used to determine the molecular weight and its distribution (65). Figure 19.5 illustrates the characterization of NIR-laser stimulated polymeric vesicles.

The DLS technique was used to determine the hydrodynamic diameter of the polymerosomes. The diameter was found to increase from 130 to 150 nm after encapsulation of the drug and photosensitizer. The morphology, shape and size of polymerosomes was determined using cryo-TEM. The cellular uptake of prepared polymerosomes was determined using A375 cell lines with the help of confocal microscopy, and cytotoxicity was studied by MTT assay (65).

Wang et al. reported the fabrication of redox- and light-responsive bioinspired MnO_2 hybrid (BMH) hydrogel for treatment of melanoma and also for treatment of wounds due to multidrug resistant bacterial infection. With excellent shear thinning and adhesive properties, the hydrogel demonstrated the light- and redox-responsive release of an anticancer drug (doxorubicin) in the tumor microenvironment. Nanosheets of MnO_2 were characterized for topography by TEM and AFM. The light-responsive drug release was also characterized for MnO_2 nanosheets. Caffeic acid and catechol were conjugated with chitosan, and this conjugation was characterized by FTIR and solid-state NMR. XRD and X-ray photoelectron spectroscopy were used to confirm the integration of MnO_2 sheets in hydrogel. The porosity of the hydrogel was determined using SEM. Elemental analysis of the hydrogel was performed using an energy-dispersive spectrometer (EDS). Anticancer activity was studied in mice by inducing A-375 cells for the development of a xenograft model (66).

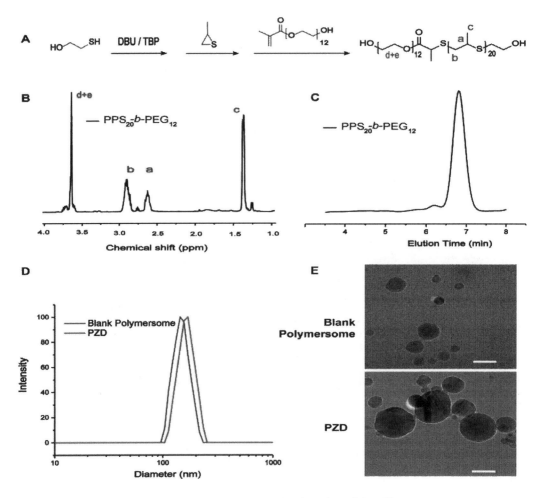

FIGURE 19.5 Characterization of NIR-laser stimulated polymeric vesicles (65).

19.3.5 ULTRASOUND-RESPONSIVE SYSTEMS

Ultrasound-assisted drug delivery of anticancer drugs is a prime approach to target cancerous tumors with better efficiency. The general and specific characterization of ultrasound responsive systems is reported by various researchers.

Li et al. reported the fabrication of ultrasound-responsive microbubbles for delivery of indocyanine green and folic acid (FA). The microbubbles were covalently conjugated with ICG and FA. Optical and green fluorescence microscopy was used to analyze the spherical morphology of the microbubbles. The existence of ICG and FA was confirmed by UV spectrophotometry. The particle size and zeta potential were determined for the microbubbles. The drug release was studied to analyze the ultrasound-responsive release behavior of the microbubbles. The cellular uptake and in vitro anticancer study was reported by the researchers. The in vivo anticancer study was performed in a cancer-induced mouse model (67). Figure 19.6 illustrates the characterization of ultrasound-responsive microbubbles.

de Matos and coworkers fabricated ultrasound-responsive liposomes for delivery of an anticancer drug (mistletoe lectin-1). The liposomes were characterized for size and zeta potential by zetasizer. The entrapment efficiency and loading capacity were determined for the prepared liposomes. The liposomes were loaded with a nanoemulsion, and it was confirmed by TEM. The stability study was

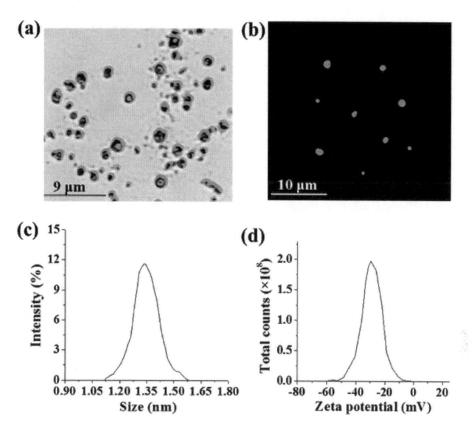

FIGURE 19.6 Characterization of ultrasound-responsive microbubbles. Reproduced with permission from (67). Copyright Royal Society of Chemistry 2018.

performed by determining the drug retention capacity. Further, the cellular uptake and cytotoxicity were studied for the prepared formulation (68).

REFERENCES

1. McNamara K, Tofail SAM. Nanoparticles in biomedical applications. Advances in Physics: X. 2017 Jan 2;2(1):54–88.
2. Baudino TA. Targeted cancer therapy: The next generation of cancer treatment. Curr Drug Discov Technol. 2015;12(1):3–20.
3. Polymer-based stimuli-responsive nanosystems for biomedical applications—Joglekar—2013—Biotechnology Journal—Wiley Online Library [Internet]. [cited 2023 Feb 8]. Available from: https://onlinelibrary.wiley.com/doi/abs/10.1002/biot.201300073.
4. Handa M, Singh A, Flora SJS, Shukla R. Stimuli-responsive polymeric nano systems for therapeutic applications. Curr Pharm Des. 2022;28(11):910–921.
5. Kumar P, Salve R, Gajbhiye KR, Gajbhiye V. Chapter 1—An overview of stimuli-responsive nanocarriers: State of the art. In: Gajbhiye V, Gajbhiye KR, Hong S, editors. Stimuli-Responsive Nanocarriers. Academic Press; 2022. p. 1–27. [Internet] [cited 2023 Feb 8]. Available from: www.sciencedirect.com/science/article/pii/B9780128244562000047.
6. Blum AP, Kammeyer JK, Rush AM, Callmann CE, Hahn ME, Gianneschi NC. Stimuli-responsive nanomaterials for biomedical applications. J Am Chem Soc. 2015 Feb 18;137(6):2140–2154.
7. Pham SH, Choi Y, Choi J. Stimuli-responsive nanomaterials for application in antitumor therapy and drug delivery. Pharmaceutics. 2020 Jul;12(7):630.

8. Mourdikoudis S, Pallares RM, Thanh NTK. Characterization techniques for nanoparticles: Comparison and complementarity upon studying nanoparticle properties. Nanoscale. 2018 Jul 13;10(27):12871–12934.
9. Ikhmayies SJ. Characterization of nanomaterials. JOM. 2014 Jan 1;66(1):28–29.
10. Kiefer J, Grabow J, Kurland HD, Müller FA. Characterization of nanoparticles by solvent infrared spectroscopy. Anal Chem. 2015 Dec 15;87(24):12313–12317.
11. Okoko J, Alonge A, Ngoddy P. High quality-cassava flour (HQCF) composites: Their thermal characteristics in retrospect. IOP Conf Ser: Earth Environ Sci. 2020 Feb;445(1):012043.
12. Holder CF, Schaak RE. Tutorial on powder X-ray diffraction for characterizing nanoscale materials. ACS Nano. 2019 Jul 23;13(7):7359–7365.
13. Whitfield P, Mitchell L. X-ray diffraction analysis of nanoparticles: Recent developments, potential problems and some solutions. Int J Nanosci. 2004 Dec;3(6):757–763.
14. Rasmussen MK, Pedersen JN, Marie R. Size and surface charge characterization of nanoparticles with a salt gradient. Nat Commun. 2020 May 11;11(1):2337.
15. Clayton KN, Salameh JW, Wereley ST, Kinzer-Ursem TL. Physical characterization of nanoparticle size and surface modification using particle scattering diffusometry. Biomicrofluidics. 2016 Sep 21;10(5):054107.
16. Clogston JD, Patri AK. Zeta Potential Measurement. In: McNeil SE, editor. Characterization of Nanoparticles Intended for Drug Delivery. Totowa, NJ: Humana Press; 2011. p. 63–70. [Internet] [cited 2023 Feb 8] (Methods in Molecular Biology). Available from: https://doi.org/10.1007/978-1-60327-198-1_6.
17. Karavelidis V, Karavas E, Giliopoulos D, Papadimitriou S, Bikiaris D. Evaluating the effects of crystallinity in new biocompatible polyester nanocarriers on drug release behavior. IJN. 2011 Nov 24;6:3021–3032.
18. Sadat T, Kashi J, Eskandarion S, Esfandyari-Manesh M, Mahmoud S, Marashi A, et al. Improved drug loading and antibacterial activity of minocycline-loaded PLGA nanoparticles prepared by solid/oil/water ion pairing method. IJN. 2012 Jan 10;7:221–234.
19. Smith DJ. Characterization of Nanomaterials Using Transmission Electron Microscopy. 2015 Aug 10 [cited 2023 Feb 8]; Available from: https://books.rsc.org/books/edited-volume/1940/chapter/2569325/Characterization-of-Nanomaterials-Using.
20. Buhr E, Senftleben N, Klein T, Bergmann D, Gnieser D, Frase CG, et al. Characterization of nanoparticles by scanning electron microscopy in transmission mode. Meas Sci Technol. 2009 Aug 1;20(8):084025.
21. Thomas RG, Surendran SP, Jeong YY. Tumor microenvironment-stimuli responsive nanoparticles for anticancer therapy. Front Mol Biosci. 2020 Dec 18;7. [Internet] [cited 2023 Feb 8] Available from: www.frontiersin.org/articles/10.3389/fmolb.2020.610533/full.
22. Stimuli-Responsive Nanoparticles for Controlled Drug Delivery in Synergistic Cancer Immunotherapy—Zhang—2022—Advanced Science—Wiley Online Library [Internet]. [cited 2023 Feb 8]. Available from: https://onlinelibrary.wiley.com/doi/full/10.1002/advs.202103444.
23. Blanco Andujar C. Sodium carbonate mediated synthesis of iron oxide nanoparticles to improve magnetic hyperthermia efficiency and induce apoptosis [Doctoral]. Doctoral thesis, UCL (University College London). UCL (University College London); 2014 [Internet] [cited 2023 Feb 10]. Available from: https://discovery.ucl.ac.uk/id/eprint/1430360/.
24. Cleavage of the C—C Bond in the ethanol oxidation reaction on platinum. Insight from experiments and calculations | The Journal of Physical Chemistry C [Internet] [cited 2023 Feb 10]. Available from: https://pubs.acs.org/doi/10.1021/acs.jpcc.6b03117.
25. Shukla N, Liu C, Jones PM, Weller D. FTIR study of surfactant bonding to FePt nanoparticles. J Magn Magn Mater. 2003 Oct 1;266(1):178–184.
26. Eid MM. Characterization of nanoparticles by FTIR and FTIR-microscopy. In: Mallakpour S, Hussain CM, editors. Handbook of Consumer Nanoproducts. Singapore: Springer; 2021. p. 1–30. [Internet] [cited 2023 Feb 10] Available from: https://doi.org/10.1007/978-981-15-6453-6_89-1.
27. Koshy O, Subramanian L, Thomas S. Chapter 5—differential scanning calorimetry in nanoscience and nanotechnology. In: Thomas S, Thomas R, Zachariah AK, Mishra RK, editors. Thermal and Rheological Measurement Techniques for Nanomaterials Characterization. Elsevier; 2017. p. 109–122. [Internet] [cited 2023 Feb 10] (Micro and Nano Technologies). Available from: www.sciencedirect.com/science/article/pii/B9780323461399000050.
28. Upadhyay S, Parekh K, Pandey B. Influence of crystallite size on the magnetic properties of Fe_3O_4 nanoparticles. J Alloys Compd. 2016 Sep 5;678:478–485.
29. Coatings | Free Full-Text | Introduction to Advanced X-ray Diffraction Techniques for Polymeric Thin Films. [Internet] [cited 2023 Feb 10]. Available from: www.mdpi.com/2079-6412/6/4/54.

30. Reimer L. Introduction. In: Reimer L, editor. Energy-Filtering Transmission Electron Microscopy. Berlin, Heidelberg: Springer; 1995. p. 1–42. [Internet] [cited 2023 Feb 10] (Springer Series in Optical Sciences). Available from: https://doi.org/10.1007/978-3-540-48995-5_1.
31. Eskandari MJ, Gostariani R, Asadabad MA, Eskandari MJ, Gostariani R, Asadabad MA. Transmission Electron Microscopy of Nanomaterials. Electron Crystallography. IntechOpen; 2020 [Internet] [cited 2023 Feb 10]. Available from: www.intechopen.com/chapters/72080.
32. Jun YW, Seo JW, Cheon J. Nanoscaling laws of magnetic nanoparticles and their applicabilities in biomedical sciences. Acc Chem Res. 2008 Feb;41(2):179–189.
33. Pan Y, Neuss S, Leifert A, Fischler M, Wen F, Simon U, et al. Size-dependent cytotoxicity of gold nanoparticles. Small. 2007 Nov;3(11):1941–1949.
34. Ragelle H, Danhier F, Préat V, Langer R, Anderson DG. Nanoparticle-based drug delivery systems: A commercial and regulatory outlook as the field matures. Expert Opin Drug Deliv. 2017 Jul;14(7):851–864.
35. Kalaydina RV, Bajwa K, Qorri B, Decarlo A, Szewczuk MR. Recent advances in "smart" delivery systems for extended drug release in cancer therapy. IJN. 2018 Aug 20;13:4727–4745.
36. Li F, Qin Y, Lee J, Liao H, Wang N, Davis TP, et al. Stimuli-responsive nano-assemblies for remotely controlled drug delivery. J Control Release. 2020 Jun 10;322:566–592.
37. El-Sawy HS, Al-Abd AM, Ahmed TA, El-Say KM, Torchilin VP. Stimuli-responsive nano-architecture drug-delivery systems to solid tumor micromilieu: Past, Present, and future perspectives. ACS Nano. 2018 Nov 27;12(11):10636–10664.
38. Smart nanoplatform for sequential drug release and enhanced chemo-thermal effect of dual drug loaded gold nanorod vesicles for cancer therapy | J Nanobiotechnol | Full Text [Internet]. [cited 2023 Feb 10]. Available from: https://jnanobiotechnology.biomedcentral.com/articles/10.1186/s12951-019-0473-3.
39. Huang TW, Lu HT, Ho YC, Lu KY, Wang P, Mi FL. A smart and active film with tunable drug release and color change abilities for detection and inhibition of bacterial growth. Mater Sci Eng C Mater Biol Appl. 2021 Jan;118:111396.
40. Zamora-Mora V, Fernández-Gutiérrez M, González-Gómez Á, Sanz B, Román JS, Goya GF, et al. Chitosan nanoparticles for combined drug delivery and magnetic hyperthermia: From preparation to in vitro studies. Carbohydr Polym. 2017 Feb 10;157:361–370.
41. Dong Y, Li S, Li X, Wang X. Smart MXene/agarose hydrogel with photothermal property for controlled drug release. Int J Biol Macromol. 2021 Nov 1;190:693–699.
42. Liu R, Hu C, Yang Y, Zhang J, Gao H. Theranostic nanoparticles with tumor-specific enzyme-triggered size reduction and drug release to perform photothermal therapy for breast cancer treatment. Acta Pharmaceutica Sinica B. 2019 Mar 1;9(2):410–420.
43. Tığlı Aydın RS, Pulat M. 5-Fluorouracil encapsulated chitosan nanoparticles for pH-stimulated drug delivery: Evaluation of controlled release kinetics. J Nanomaterials. 2012 May 29;2012:e313961.
44. Sang Y, Miao P, Chen T, Zhao Y, Chen L, Tian Y, et al. Fabrication and evaluation of graphene oxide/hydroxypropyl cellulose/chitosan hybrid aerogel for 5-fluorouracil release. Gels. 2022 Oct;8(10):649.
45. Liu F, Huang P, Huang D, Liu S, Cao Q, Dong X, et al. Smart "on-off" responsive drug delivery nanosystems for potential imaging diagnosis and targeted tumor therapy. Chem Eng J. 2019 Jun 1;365:358–368.
46. Hao Q, Wang Z, Zhao W, Wen L, Wang W, Lu S, et al. Dual-responsive polyprodrug nanoparticles with cascade-enhanced magnetic resonance signals for deep-penetration drug release in tumor therapy. ACS Appl Mater Interfaces. 2020 Nov 4;12(44):49489–49501.
47. Islan GA, Tornello PC, Abraham GA, Duran N, Castro GR. Smart lipid nanoparticles containing levofloxacin and DNase for lung delivery. Design and characterization. Colloids Surf B Biointerfaces. 2016 Jul 1;143:168–176.
48. Constantin M, Bucătariu S, Stoica I, Fundueanu G. Smart nanoparticles based on pullulan-g-poly(N-isopropylacrylamide) for controlled delivery of indomethacin. Int J Biol Macromol. 2017 Jan;94(Pt A):698–708.
49. Danaei M, Dehghankhold M, Ataei S, Hasanzadeh Davarani F, Javanmard R, Dokhani A, et al. Impact of particle size and polydispersity index on the clinical applications of lipidic nanocarrier systems. Pharmaceutics. 2018 May 18;10(2):57.
50. Smith T, Affram K, Nottingham EL, Han B, Amissah F, Krishnan S, et al. Application of smart solid lipid nanoparticles to enhance the efficacy of 5-fluorouracil in the treatment of colorectal cancer. Sci Rep. 2020 Oct 12;10(1):16989.
51. Powers KW, Palazuelos M, Moudgil BM, Roberts SM. Characterization of the size, shape, and state of dispersion of nanoparticles for toxicological studies. Nanotoxicology. 2007 Jan 1;1(1):42–51.

52. Mishra AK, Pandey H, Agarwal V, Ramteke PW, Pandey AC. Nanoengineered mesoporous silica nanoparticles for smart delivery of doxorubicin. J Nanopart Res. 2014 Jul 3;16(8):2515.
53. Kumar A, Dixit CK. 3—Methods for characterization of nanoparticles. In: Nimesh S, Chandra R, Gupta N, editors. Advances in Nanomedicine for the Delivery of Therapeutic Nucleic Acids. Woodhead Publishing; 2017. p. 43–58. [Internet] [cited 2023 Feb 10] Available from: www.sciencedirect.com/science/article/pii/B9780081005576000031.
54. Surfactant effects on the particle size, zeta potential, and stability of starch nanoparticles and their use in a pH-responsive manner | springerprofessional.de [Internet]. [cited 2023 Feb 10]. Available from: www.springerprofessional.de/en/surfactant-effects-on-the-particle-size-zeta-potential-and-stabi/13352928.
55. Magnetic and fluorescent multifunctional chitosan nanoparticles. |INIS [Internet]. [cited 2023 Feb 10]. Available from: https://inis.iaea.org/search/search.aspx?orig_q=RN:39040561.
56. Zhang Y, Cai K, Li C, Guo Q, Chen Q, He X, et al. Macrophage-membrane-coated nanoparticles for tumor-targeted chemotherapy. Nano Lett. 2018 Mar 14;18(3):1908–1915.
57. Surya R, Mullassery MD, Fernandez NB, Thomas D, Jayaram PS. Synthesis and characterization of a pH responsive and mucoadhesive drug delivery system for the controlled release application of anti-cancerous drug. Arab J Chem. 2020 May 1;13(5):5262–5276.
58. Jeshvaghani PA, Pourmadadi M, Yazdian F, Rashedi H, Khoshmaram K, Nigjeh MN. Synthesis and characterization of a novel, pH-responsive sustained release nanocarrier using polyethylene glycol, graphene oxide, and natural silk fibroin protein by a green nano emulsification method to enhance cancer treatment. Int J Biol Macromol. 2023 Jan 31;226:1100–1115.
59. Full article: Development of redox-responsive theranostic nanoparticles for near-infrared fluorescence imaging-guided photodynamic/chemotherapy of tumor [Internet]. [cited 2023 Feb 14]. Available from: www.tandfonline.com/doi/full/10.1080/10717544.2018.1451571.
60. Ren L, Nie J, Wei J, Li Y, Yin J, Yang X, et al. RGD-targeted redox responsive nano micelle: Co-loading docetaxel and indocyanine green to treat the tumor. Drug Delivery. 2021 Jan 1;28(1):2024–2032.
61. Guo F, Wu J, Wu W, Huang D, Yan Q, Yang Q, et al. PEGylated self-assembled enzyme-responsive nanoparticles for effective targeted therapy against lung tumors. J Nanobiotechnol. 2018 Jul 16;16(1):57.
62. Tang L, Xiao Q, Yin Y, Mei Y, Li J, Xu L, et al. An enzyme-responsive and NIR-triggered lipid—polymer hybrid nanoplatform for synergistic photothermal/chemo cancer therapy. Biomater Sci. 2022 May 4;10(9):2370–2383.
63. Xu Y, Zhu Y, Kaskel S. A smart magnetic nanosystem with controllable drug release and hyperthermia for potential cancer therapy. RSC Adv. 2015 Nov 20;5(121):99875–99883.
64. Maia ALC, e Silva ATM, César ALA, Giuberti CS, Evangelista FCG, Lemos J de A, et al. Preparation and characterization of gadolinium-based thermosensitive liposomes: A potential nanosystem for selective drug delivery to cancer cells. J Drug Deliv Sci Technol. 2021 Oct 1;65:102686.
65. Tang Q, Hu P, Peng H, Zhang N, Zheng Q, He Y. Near-infrared laser-triggered, self-immolative smart polymersomes for in vivo cancer therapy. IJN. 2020 Jan 10;15:137–149.
66. Wang S, Zheng H, Zhou L, Cheng F, Liu Z, Zhang H, et al. Injectable redox and light responsive MnO2 hybrid hydrogel for simultaneous melanoma therapy and multidrug-resistant bacteria-infected wound healing. Biomaterials. 2020 Nov 1;260:120314.
67. Li Y, Huang W, Li C, Huang X. Indocyanine green conjugated lipid microbubbles as an ultrasound-responsive drug delivery system for dual-imaging guided tumor-targeted therapy. RSC Adv. 2018 Sep 24;8(58):33198–33207.
68. de Matos MBC, Deckers R, van Elburg B, Lajoinie G, de Miranda BS, Versluis M, et al. Ultrasound-sensitive liposomes for triggered macromolecular drug delivery: Formulation and in vitro characterization. Front Pharmacol. 2019 Dec 5;10:1463.

20 Toxicological Aspects of Tumour Microenvironment-Responsive Nanoplatforms

Ofosua Adi-Dako, Benoit Banga N'guessan, Joseph Adusei-Sarkodie, Nii Hutton Mills and Doris Kumadoh

CONTENTS

20.1 Introduction 339
20.2 Characterization of Tumour Microenvironment 340
    20.2.1 Overview of Tumour Microenvironment..340
    20.2.2 Key Components of Tumour Microenvironment..341
    20.2.3 Role of Tumour Microenvironment in Drug Resistance 341
20.3 Development of Tumour Microenvironment-Responsive Nanoplatforms...................342
    20.3.1 Overview of Tumour Microenvironment-Responsive Nanoplatforms 342
    20.3.2 Design Strategies for Tumour Microenvironment-Responsive Nanoplatforms 342
    20.3.3 Key Features of Tumour Microenvironment-Responsive Nanoplatforms.....342
20.4 Toxicological Evaluation of Tumour Microenvironment-Responsive Nanoplatforms 343
    20.4.1 Overview of Toxicological Evaluation..343
    20.4.2 Key Factors Affecting Toxicity of Nanoplatforms 343
    20.4.3 Methods for Evaluating Toxicity of Nanoplatforms 344
20.5 Case Studies of Tumour Microenvironment-Responsive Nanoplatforms 344
    20.5.1 Overview of Case Studies..344
    20.5.2 Clinical Trials of Tumour Microenvironment-Responsive Nanoplatforms..............345
    20.5.3 Comparison of Efficacy and Toxicity of Different Nanoplatforms 345
20.6 Conclusion 346
    20.6.1 Summary of Key Findings...346
    20.6.2 Future Directions in the Development of Tumour Microenvironment-Responsive Nanoplatforms 346
    20.6.3 Importance of Toxicological Evaluation in the Development of Nanoplatforms for Cancer Therapy 347
References..347

20.1 INTRODUCTION

The therapeutic effectiveness of cancer drugs can be enhanced using nanoplatforms (Yang et al., 2021). These are designed to precisely target cancer cells while mitigating damage to healthy cells, addressing the limitations associated with traditional chemotherapy and radiation therapy (Murciano-Goroff et al., 2020). The method involves enclosing therapeutic drugs in nanoscale delivery agents like dendrimers, liposomes, and nanoparticles. These can be modified to selectively engage with the tumour microenvironment (Mitchell et al., 2021).

The desire to boost therapeutic potency and reduce the adverse impacts of traditional chemotherapy and radiation therapy has sparked interest in developing nanoplatforms for cancer therapy (Murciano-Goroff et al., 2020). Notably, nanoparticle-based delivery systems can extend the time therapeutic drugs circulate and gather in tumours, thereby improving their effectiveness. Moreover, nanoplatforms can circumvent drug resistance mechanisms that compromise the effectiveness of conventional chemotherapy (Su et al., 2021). Therefore, nanoplatforms have been developed to specifically target cancer cells, enhancing the therapeutic efficiency of anti-cancer medications while minimizing their toxicity to healthy cells.

Despite their promise, nanoplatforms for cancer therapy face several hurdles before their full potential can be tapped (Gavas et al., 2021). Creating stable nanoplatforms that effectively target cancer cells without eliciting an immune response remains challenging. Thorough toxicological testing of nanoplatforms is crucial to ensure their safety and effectiveness in a clinical setting (Fadeel et al., 2018). Nonetheless, to improve the outcome of cancer treatments, the development of nanoplatforms for cancer therapy must persist.

The tumour microenvironment plays a pivotal role in cancer progression. It consists of various cellular and non-cellular components such as immune cells, blood vessels, extracellular matrix, and cytokines, which interact with cancer cells and affect their response to treatment (Qin et al., 2020). The tumour microenvironment can also influence the resistance of cancer cells to chemotherapy and radiation therapy (Zou, 2022), making it an essential target in devising efficient cancer treatments.

Studies have shown that targeting specific components of the tumour microenvironment can enhance cancer therapy effectiveness (Bejarano et al., 2021; Jin & Jin, 2020; Tang et al., 2021). For example, focusing on the angiogenesis process can suppress tumour growth and facilitate drug delivery (Lugano et al., 2020; Zhao & Adjei, 2015). Likewise, targeting immune cells within the tumour microenvironment can enhance immunotherapy effectiveness and amplify the immune response to cancer (Murciano-Goroff et al., 2020).

The tumour microenvironment can also affect the development of resistance to cancer treatments (Seebacher et al., 2021). For example, tumour-associated macrophages can produce cytokines that promote cancer cell survival and treatment resistance (Pittet et al., 2022). Thus, the tumour microenvironment must be considered when devising new cancer treatments, and potential effects on it must be assessed.

The aim of this book chapter is to impart a comprehensive understanding of the toxicity of nanoplatforms used in cancer therapy. It will emphasize the importance of considering the toxicological characteristics of these nanoplatforms and the significant factors affecting their toxicity. It will also provide an overview of recent studies and examples of tumour microenvironment-responsive nanoplatforms, along with a discussion of the challenges and future prospects of this scientific field. Ultimately, this knowledge will contribute to the development of safer and more effective nanoplatforms for cancer therapy.

20.2 CHARACTERIZATION OF TUMOUR MICROENVIRONMENT

20.2.1 Overview of Tumour Microenvironment

Cancer cells are not isolated entities; they interact intimately with their surrounding tissues and cells. This neighbouring environment, filled with an intricate network of diverse cell types such as immune cells, fibroblasts, blood vessels, and extracellular matrix, plays a critical role in tumour evolution and progression (Hanahan & Weinberg, 2011).

These neighbouring cells closely communicate with the cancerous cells, managing numerous biological processes such as new blood vessel growth (angiogenesis), eluding the immune system, and resisting therapeutic treatments. The interplay of these components forms what we call the 'tumour microenvironment.'

Research has highlighted the tumour microenvironment's importance as a potential therapeutic target (Bejarano et al., 2021). Limiting the supply of oxygen and nutrients to the tumour, for instance, by constraining angiogenesis, could potentially inhibit tumour expansion and metastasis (Zhao & Adjei, 2015). Moreover, modulating the immune system has been proven beneficial in cancer treatment. This is because immune cells within the tumour microenvironment are capable of identifying and destroying cancer cells (Rosenberg & Restifo, 2015).

To sum up, the tumour microenvironment is a dynamic and multifaceted system that is crucial for the understanding of tumour biology and growth. Future investigations of the interactions between tumour cells and their microenvironment may lead to the development of new and improved cancer treatments (Hanahan & Weinberg, 2011).

20.2.2 Key Components of Tumour Microenvironment

The tumour milieu encompasses a plethora of cellular and molecular constituents crucial for the evolution, infiltration, and metastasis of the tumour. Cancer cells, stromal cells, extracellular matrix, cytokines, growth factors, angiogenesis inducers, and immune cells represent some key components of the tumour environment.

Stromal cells, such as fibroblasts and immune cells, impart structural reinforcement, govern angiogenesis, and tailor the immune response, thereby occupying a pivotal position within the tumour's ecological framework. Extracellular matrix substances discharged by fibroblasts are acknowledged as instrumental in the formation of the tumour stroma and the tumour's proliferation (Anderson & Simon, 2020). Immune cells, including T cells and macrophages, may either impede or expedite tumour growth, contingent on the balance of pro- and anti-tumorigenic cues they perceive (Wang et al., 2023).

Additional elements inhabiting the tumour environment comprise cytokines and growth factors like vascular endothelial growth factor (VEGF) and interleukin-6 (IL-6) (Dong, 2021). Their roles in advancing angiogenesis, enhancing tumour cell longevity, and amplifying tumour cell dispersion and invasion have been well delineated (Lima et al., 2021). By crafting a conducive habitat for tumour cells, these factors are indispensable in the inception and progression of tumours.

20.2.3 Role of Tumour Microenvironment in Drug Resistance

The tumour microenvironment, which includes the extracellular matrix, soluble substances, and benign cells in contact with malignant cells, plays a crucial role in how cancer cells respond to treatment. Drug resistance is greatly impacted by this microenvironment, with drug-resistance genes frequently being expressed within it (Seebacher et al., 2021). Various environmental factors within the tumour microenvironment, like hypoxia, oxidative stress, and growth factors, can modify this gene expression (Gatenby & Gillies, 2004; Petrova et al., 2018).

The tumour microenvironment can enhance cancer cell survival and growth, leading to treatment resistance. This is due to the production of cytokines and growth factors. For instance, studies have linked elevated levels of vascular endothelial growth factor (VEGF) in the microenvironment with increased drug resistance (Ghosh et al., 2008; Vasan et al., 2019). Additionally, the tumour microenvironment can create an immunosuppressive environment that limits the immune system's ability to target cancer cells (Zou, 2022). This reduces the effectiveness of immunotherapies which leverage the immune system to combat cancer (Yuan et al., 2016).

The microenvironment also boosts drug resistance by nurturing the growth of tumour-associated stromal cells, such as fibroblasts, which create a protective layer around the cancer cells. These stromal cells can produce various substances, like matrix metalloproteinases, that alter the extracellular matrix and hinder the dispersal of drugs within the tumour, thereby diminishing their efficacy (Y. Ni et al., 2021). Moreover, these cells can form a thick extracellular matrix network that slows drug diffusion and impairs drug delivery to the cancer cells, further contributing to treatment resistance (McMillin et al., 2013; Y. Ni et al., 2021).

20.3 DEVELOPMENT OF TUMOUR MICROENVIRONMENT-RESPONSIVE NANOPLATFORMS

20.3.1 Overview of Tumour Microenvironment-Responsive Nanoplatforms

The innovative medication delivery system known as cancer-responsive nanoplatforms, has been designed to target cancer cells more accurately. These nanoplatforms are built to react to modifications in the cancerous environment, such as diverse cell types, pH shifts, and temperature fluctuations (Gong et al., 2020). Such responsiveness allows them to specifically aim at cancer cells and transport medications directly to the tumour (Zhang et al., 2022). These nanoplatforms can be programmed to hold various drugs like chemotherapy, immunotherapy, and targeted therapies, and can be made from different materials, such as polymers, liposomes, and nanoparticles (Jain & Stylianopoulos, 2010; Mitchell et al., 2021).

One major strength of these cancer-responsive nanoplatforms is their potential to bypass the limitations of traditional drug delivery systems. Regular drug delivery methods often struggle to target cancer cells accurately, resulting in the dispersion of medication to healthy cells and tissues. This can lead to undesired side effects and decrease the treatment's efficiency (Liu et al., 2020). Cancer-responsive nanoplatforms can overcome these restrictions by delivering medications straight to the tumour while specifically aiming at cancer cells, thereby reducing the chances of negative effects (Peng et al., 2022).

Another advantage of these cancer-responsive nanoplatforms is their ability to enhance the behaviour and effects of drugs. For example, these nanoplatforms can extend the time medication stays in the body, guard against drug breakdown in the bloodstream, and boost drug concentration at the tumour site (Uthaman et al., 2018). This can result in improved treatment outcomes and fewer toxicities and side effects.

In summary, cancer-responsive nanoplatforms offer a novel and hopeful approach to cancer treatment. By bypassing the shortcomings of traditional drug delivery systems and enhancing the behaviour and effects of drugs, they can improve treatment results while minimizing toxicity and unwanted effects.

20.3.2 Design Strategies for Tumour Microenvironment-Responsive Nanoplatforms

A crucial aspect of cancer treatment involves devising techniques for designing nanoplatforms sensitive to the tumour's surrounding environment. Cancer therapy delivery systems have become more accurate and potent due to advancements in nanotechnology. The low oxygen and high acidity environment surrounding the tumour significantly influences the effectiveness of cancer therapies.

A common strategy involves using pH-responsive polymers like poly(2-(diisopropylamino) ethyl methacrylate) (pDPA) and poly(ethylene glycol) (PEG) to build nanoplatforms reactive to the tumour's surroundings (Yusa, 2022; Zhang et al., 2022). These polymers are capable of responding to the tumour's acidic milieu, triggering the release of the therapeutic substance. A different design method employs hypoxia-responsive polymers like poly(2-nitroimidazole) (PNI) or poly(N-isopropylacrylamide) (PNIPAM) to enhance drug delivery to oxygen-deficient tumour regions (Li et al., 2020; Zhang et al., 2022).

Moreover, the efficiency of nanoplatforms in targeting cancer cells can be improved by integrating custom ligands like antibodies or peptides (Wang et al., 2021). This could enhance the effectiveness of treatment while minimizing potential damage to normal cells. Employing dual-responsive nanoplatforms, responsive to both pH and hypoxia, could further enhance the potency of cancer therapies. In conclusion, crafting effective cancer treatments relies heavily on the design of nanoplatforms that are responsive to the conditions of the tumour's microenvironment.

20.3.3 Key Features of Tumour Microenvironment-Responsive Nanoplatforms

Enhancing the potency of cancer treatments is possible through the use of specially designed nanoplatforms that react to the distinct features of a tumour's environment, such as acidity levels, oxidative stress, and low oxygen levels. The purpose of these nanodevices is to deliver treatment substances to cancerous regions, steering clear of healthy tissue. This results in fewer side effects and improved treatment effectiveness (Liu et al., 2020).

A key trait of these cancer-responsive nanoplatforms is their capability to home in on tumour-specific antigens (TAAs) or tumour-related markers, also known as TAMs. By attaching TAA- or TAM-targeting peptides or antibodies to nanoparticles, they can bind to cancer cells with high precision, promoting the build-up of therapeutic drugs at the tumour location (Cheng et al., 2021; Liu et al., 2020). This ability to target enhances the precision and potency of cancer treatments.

These tumour-responsive nanoplatforms are also equipped with the crucial ability to conquer cancer-related obstacles, such as the tumour blood vascular permeability and retention (BTB/BBR) effect and the expulsion of therapeutic drugs by cancer cells (Chang et al., 2021; Gong et al., 2020). To tackle these challenges, nanoplatforms can be given unique features, like a high drug-holding capacity, compact size, and a 'stealth' surface that enables evasion of the immune system and longer circulation time (Liu et al., 2020). These characteristics empower better treatment outcomes by amplifying the penetration and retention of therapeutic drugs within tumour tissues.

20.4 TOXICOLOGICAL EVALUATION OF TUMOUR MICROENVIRONMENT-RESPONSIVE NANOPLATFORMS

20.4.1 Overview of Toxicological Evaluation

Thorough toxicity assessments are crucial when developing and applying nanoplatforms responsive to the tumour microenvironment in cancer therapy. These nanoplatforms aim to target and regulate therapeutic drug delivery to tumour sites, thereby reducing the harmful effects and side effects associated with standard chemotherapy. Yet, it's also essential to consider the potential toxicity of the nanoplatforms themselves (Ajdary et al., 2018). They might interfere with normal bodily functions and have detrimental outcomes (Damasco et al., 2020; Patel & Shah, 2017).

We can evaluate the toxicity of these cancer-specific nanoplatforms through different methods, like *in vitro* tests, animal models, and clinical trials. *In vitro* tests provide a quick and affordable approach, but they may not accurately predict the human response to these substances (Kumar et al., 2017; Lama et al., 2020). Animal models offer a more precise toxicity measurement of nanoplatforms (N'Guessan, Amponsah et al., 2021). Still, differences in toxicity and pharmacokinetics across species limit their applicability. Clinical trials are the gold standard for assessing nanoplatforms' toxicity, albeit being costly and time-intensive (Younis et al., 2022; Zhang et al., 2020).

Despite these challenges, it's crucial to carry out toxicity evaluations of nanoplatforms responsive to the tumour microenvironment. This ensures their safe and effective application in cancer therapy. Due to advances in nanotechnology, newly designed nanoplatforms show improved biocompatibility and reduced toxicity. However, more research is needed to fully comprehend their toxicological profiles (Li et al., 2012). In order to progress in nanomedicine and ensure the safe and effective utilization of nanoplatforms in cancer therapy, innovative techniques for assessing nanoplatforms toxicity, such as in silico models and *in vitro-in vivo* correlation, have been developed (Kumar et al., 2017; Li et al., 2016; Stillman et al., 2020).

20.4.2 Key Factors Affecting Toxicity of Nanoplatforms

While nanotechnology has the capacity to revolutionize healthcare, concerns about toxicity remain. Elements such as the size and structure of the nanomaterials significantly influence their toxicity levels. Studies have revealed that smaller nanomaterials exhibit greater toxicity as they can easily penetrate cells and tissues. Similarly, nanomaterials with more intricate structures pose greater risks due to their larger surface areas that facilitate interactions with cells and tissues (Karlsson et al., 2009; Kumar et al., 2012; Sukhanova et al., 2018).

The chemical composition of nanomaterials also plays a crucial role in toxicity. For example, nanoparticles made up of specific materials like silver and gold, are known to induce the production of reactive oxygen species (ROS), leading to increased toxicity (Harper et al., 2008; Sharifi et al., 2012). Conversely, nanoparticles composed of biodegradable polymers, such as poly(lactide-co-glycolide) (PLGA), have shown reduced toxicity levels (Gagliardi et al., 2021). Thus, the toxicity levels of these nanoplatforms are directly related to the materials used in their construction.

Surface functionalization of nanomaterials also influences toxicity. Surface modifications, such as PEGylation, can reduce nanoparticle toxicity by minimizing their interaction with cells (Harper et al., 2008). However, surface modifications that enhance cellular interactions may also increase toxicity. Therefore, a thorough analysis of the surface functionalization process is necessary when designing nanomaterials to ensure minimized toxicity (Buchman et al., 2019).

20.4.3 Methods for Evaluating Toxicity of Nanoplatforms

In the field of *in vitro* cell culture studies, we can evaluate the toxicity of nanoplatforms. Here, we grow human cells in a laboratory setting and expose them to varying levels of nanoplatforms to observe their impact. By looking at the remaining live cells after exposure and any changes in cell structure, we can measure their toxicity (Krewski et al., 2010; N'Guessan, Asiamah et al., 2021). Different kinds of cells such as cancer cells, immune cells, and healthy cells can be put to the test for nanoplatforms toxicity in these *in vitro* cell culture experiments.

In vivo animal testing offers another method for investigating nanoplatforms toxicity. Here, live animals are administered the nanoplatforms while we monitor the influence on their behaviour and overall health condition. To determine toxicity, we note any changes in the animals' weight, behaviour, and organ functionality (Huang et al., 2021). Research into nanoplatforms toxicity in different organs, like the liver, lungs, and heart, can also be carried out on live animals (Kalangi & Bhosale, 2022; N'Guessan, Amponsah et al., 2021).

Another way to study nanoplatforms toxicity is through computational methods. By understanding the nanoplatforms chemical makeup and physical attributes, computer simulations and algorithms can predict their toxicity (Raies & Bajic, 2016). These computational strategies can forecast the potential toxicity of nanoplatforms in various organs, such as the liver, lungs, and heart (Forest, 2022). These techniques help reduce the need for animal testing and provide insights into the molecular pathways linked to toxicity (Huang et al., 2021).

20.5 CASE STUDIES OF TUMOUR MICROENVIRONMENT-RESPONSIVE NANOPLATFORMS

20.5.1 Overview of Case Studies

Nanoplatforms that react to the tumour microenvironment have shown promise in cancer treatment because of their ability to precisely target cancer cells and provide unique therapeutic actions. These nanoplatforms, constructed from a variety of materials like polymeric nanoparticles, liposomes, and inorganic nanoparticles, are designed to respond to the specific conditions found within tumour tissues. Their aim is to improve upon traditional chemotherapy and radiotherapy by providing targeted, controlled drug delivery, overcoming issues like non-specific targeting, poor biodistribution, and high toxicity.

A wealth of case studies has illustrated the potential benefits of these responsive nanoplatforms in cancer treatment. For example, polymeric nanoparticles with pH-responsive properties have been used to deliver cancer drugs to tumour tissues (Jin et al., 2018; Palanikumar et al., 2020). These nanoparticles can release their drug payload in response to the acidic conditions found in tumour tissues, which enhances drug delivery effectiveness. Similarly, liposomal nanoplatforms that target tumour blood vessels have shown promising results (Mitchell et al., 2021). These nanoplatforms are

Toxicological Aspects of TME-Responsive Nanoplatforms

able to actively seek out and penetrate the blood vessels within tumours, leading to a higher concentration of the therapeutic agent in the targeted area.

Inorganic nanoplatforms like gold nanorods and iron oxide nanoparticles have also been explored as responsive platforms for cancer therapy (Iriarte-Mesa et al., 2020). By attaching specific targeting elements such as peptides and antibodies, these inorganic nanoplatforms can accurately target cancer cells. Moreover, they can be engineered to deliver thermal or photothermal therapies, which may enhance the efficacy of traditional cancer treatments by inducing localised heating in tumour tissues (Mitchell et al., 2021). These case studies reflect the potential of responsive nanoplatforms as a valuable approach to cancer therapy.

20.5.2 CLINICAL TRIALS OF TUMOUR MICROENVIRONMENT-RESPONSIVE NANOPLATFORMS

In the realm of innovative healthcare solutions, clinical trials are pivotal for the development of new medical treatments and technologies. This holds true even for the microenvironment-responsive nanoplatforms designed to combat cancer. The unique features of the tumour microenvironment are leveraged by these nanoplatforms to target and annihilate cancer cells. The goal of these trials is to evaluate the safety, efficacy, and applicability of these nanoplatforms for cancer treatment.

The development of stimuli-responsive nanomaterials has marked substantial progress in cancer treatment, and their effectiveness has been confirmed in preclinical studies (Gong et al., 2020; Peng et al., 2022; Thomas et al., 2020; Uthaman et al., 2018). Ongoing research into cancer treatment explores tumour microenvironment-responsive nanoplatforms for both diagnostic and therapeutic purposes (Gong et al., 2020; Peng et al., 2022; Uthaman et al., 2018). Despite numerous studies focusing on tumour-targeted nanomedicine, no clinical trials involving these nanoplatforms in humans have been documented yet (Adityan et al., 2020; Gong et al., 2020; Peng et al., 2022; Thomas et al., 2020; Uthaman et al., 2018).

At the moment, no actively targeted nanoparticles have received FDA approval for cancer treatment (Adityan et al., 2020). Although a few nanoparticles are in clinical trials, they need further investigation before receiving approval for human use (Adityan et al., 2020; Thomas et al., 2020). The development process of these nanoplatforms is intricate and ongoing, but their potential benefits are substantial. They could enhance cancer therapy efficacy by delivering therapeutic agents directly to the tumour site while minimizing adverse effects on healthy tissues (Gong et al., 2020; Thomas et al., 2020).

In summary, while the development of stimuli-responsive nanomaterials and tumour microenvironment-responsive nanoplatforms has shown promising results in preclinical studies, there is still a long way to go before they can be tested in humans. The potential benefits of these nanoplatforms are significant, as they could enhance the efficacy of cancer therapy by delivering drugs or other therapeutic agents directly to the tumour site while minimizing side effects on healthy tissues. Further research and clinical trials are needed to determine the safety and efficacy of these nanoplatforms in humans.

20.5.3 COMPARISON OF EFFICACY AND TOXICITY OF DIFFERENT NANOPLATFORMS

Nanoplatforms are a novel means of medication administration designed for enhanced effectiveness and reduced toxicity in targeted regions. The efficacy of nanoplatforms depends largely on aspects like their size, surface charge, and targeting approach. For therapeutic applications, it is also important to consider the targeting and accumulation of the nanoplatform in the desired tissue or organ. The nanoplatform must be able to selectively target and accumulate in the desired location to maximize its therapeutic efficacy (Yusuf et al., 2023).

Factors like immune system detectability, biodegradability, and biocompatibility determine nanoplatforms' toxicity. To reduce toxicity, biodegradable nanoplatforms based on polycaprolactone with reduced toxicity in both *in vitro* and *in vivo* scenarios could be utilized (Patra et al., 2018). The nature of the drug delivered also critically affects nanoplatforms' efficacy and toxicity. One study by Meng et al. (2021) discusses the different types of nanoplatforms used for mRNA delivery, including lipid-based and polymeric nanoplatforms. The study also compares the efficacy and toxicity of different nanoplatforms and highlights the importance of considering the nature of the drug delivered when selecting a nanoplatform for drug delivery. (Meng et al., 2021).

In conclusion, the efficacy and toxicity of nanoplatforms depend on various factors such as size, surface charge, targeting approach, immune system detectability, biodegradability, and biocompatibility. The use of ideal nano-drug delivery systems is decided primarily based on the biophysical and biochemical properties of the targeted drugs being selected. The safety assessment is vital for use in living beings to recognize potential risks and create preventive measures. Nanoplatforms have the potential to revolutionize cancer therapy by increasing treatment efficacy and minimizing side effects (Guo et al., 2020; Patra et al., 2018; Sharma et al., 2021; Siafaka et al., 2021; Yetisgin et al., 2020; Yusuf et al., 2023).

20.6 CONCLUSION

20.6.1 Summary of Key Findings

The book chapter we're discussing here delves into the potential toxicology aspects related to nanoplatforms specially devised to interact with tumour environments. We aimed to enhance the success of cancer treatments by providing a comprehensive review of the ongoing studies focusing on the design and development of these unique nanoplatforms. It's crucial to fully grasp the possible toxic effects of these nanoplatforms, as it directly affects their safety and usefulness in a clinical setup.

A key point we emphasized is the need to meticulously assess the biocompatibility and potential toxicity of nanoplatforms. This is crucial to limit any unwanted impact on patients. We delved into the potential cytotoxicity, genotoxicity, and immunotoxicity that these nanoplatforms may possess. In addition, we outlined various *in vitro* and *in vivo* techniques, such as cell-based tests, animal model studies, and human clinical trials, which are standard procedures to gauge the toxicity levels of these nanoplatforms.

We concluded the chapter by stressing the need for further investigations into the toxicological aspects of cancer microenvironment-responsive nanoplatforms. The pursuit of developing safe and effective cancer treatments necessitates a deeper understanding of toxicity mechanisms and factors influencing the biocompatibility of these nanoplatforms. By providing a comprehensive overview of the current research landscape regarding the toxic effects of nanoplatforms, this chapter underscores the importance of continued research in this field.

20.6.2 Future Directions in the Development of Tumour Microenvironment-Responsive Nanoplatforms

In the field of cancer treatment, the development of tumour microenvironment-reactive nanoplatforms has been on the rise. Recent advances in nanotechnology have enabled us to design smart and efficient nanocarriers that deliver therapeutic drugs accurately to cancer cells, minimizing collateral damage to healthy tissues. However, the challenge remains in creating nanoplatforms capable of adeptly adapting to the complex and ever-changing tumour microenvironment. Moving forward, we must devise and refine nanoplatforms that can actively respond to the varied physiological and pathological conditions within the cancer microenvironment.

One potential avenue for progress is incorporating stimuli-responsive materials into the design of these nanoplatforms. These materials have the capacity to change their physical and chemical properties in response to specific signals in the cancer microenvironment. For example, we can construct pH-sensitive nanocarriers that release drugs in response to the acidic conditions found within tumours or undergo pH-induced structural transformations (Z. Li et al., 2021). Similarly, we can use the high concentration of reducing agents like glutathione in tumour cells to build redox-responsive nanocarriers (Chen et al., 2020). Temperature-sensitive nanocarriers, which release drugs or change phase in response to the elevated temperatures of tumours, are also a possibility (Amin et al., 2022).

Another promising frontier is the development of multifunctional nanoplatforms. These could integrate various therapeutic methods and imaging probes for customized cancer treatment. Such nanoplatforms can be designed to react to specific tumour biomarkers and can be loaded with a broad range of therapeutic agents including chemotherapy drugs, immunotherapy substances, and gene therapies (Mi et al., 2016). Moreover, these nanoplatforms can house imaging probes like magnetic resonance imaging (MRI) or near-infrared fluorescence (NIRF) to monitor drug administration and tumour response in real time. By incorporating multiple therapeutic approaches into a single nanoplatform, we can achieve personalized cancer treatment that delivers medication accurately to cancer cells while limiting the unintended effects on healthy tissues.

20.6.3 Importance of Toxicological Evaluation in the Development of Nanoplatforms for Cancer Therapy

Nanoplatforms represent an encouraging advancement in cancer treatment delivery, offering benefits such as enhanced bioavailability, precision targeting, and regulated medication dispersal. A thorough toxicological evaluation is crucial to ensure the safety and efficacy of these evolving nanoplatforms used for cancer care. Factors including the physicochemical attributes of the nanoparticles and their interaction with biological systems are pivotal in assessing nanomaterial toxicity. The paramount goal is to maintain safety, mitigate any potential adverse reactions, and augment the therapeutic impact of these nanoplatforms.

Toxicological scrutiny of nanoplatforms extends to investigating any deleterious effects of the nanomaterial on human health and the environment. This indispensable evaluation uncovers potential risks, thereby fortifying the safety of nanoplatforms during various stages—development, manufacturing, and utilization. Toxicological studies of nanoplatforms should encompass components like particle dimensions, form, surface chemistry, and dosage levels. The examination should also focus on potential effects of nanoplatforms on diverse cells and organs, such as the immune system, liver, and kidneys. Ensuring nanoplatforms toxicity assessment is integral to guaranteeing the safety and efficacy of these innovative medical interventions.

To summarize, developing nanoplatforms for cancer treatment is a fascinating research area with the potential to revolutionize cancer care. However, rigorous toxicological testing is obligatory to affirm the safety and effectiveness of these interventions. Regular assessment of these novel therapeutics for safety and efficacy should be an ongoing practice. By integrating comprehensive toxicological evaluations, nanoplatforms for cancer treatment can be fine-tuned for optimum therapeutic benefits while minimizing any potential adverse impact on human health and the environment.

REFERENCES

Adityan, S., Tran, M., Bhavsar, C., & Wu, S. Y. (2020). Nano-therapeutics for modulating the tumour microenvironment: Design, development, and clinical translation. *Journal of Controlled Release*, *327*, 512–532. https://doi.org/10.1016/j.jconrel.2020.08.016

Ajdary, M., Moosavi, M. A., Rahmati, M., Falahati, M., Mahboubi, M., Mandegary, A., Jangjoo, S., Mohammadinejad, R., & Varma, R. S. (2018). Health concerns of various nanoparticles: A review of their in vitro and in vivo toxicity. *Nanomaterials (Basel)*, *8*(9). https://doi.org/10.3390/nano8090634

Amin, M., Lammers, T., & Ten Hagen, T. L. M. (2022). Temperature-sensitive polymers to promote heat-triggered drug release from liposomes: Towards bypassing EPR. *Advanced Drug Delivery Reviews*, *189*, 114503. https://doi.org/10.1016/j.addr.2022.114503

Anderson, N. M., & Simon, M. C. (2020). The tumor microenvironment. *Current Biology*, *30*(16), R921–R925. https://doi.org/10.1016/j.cub.2020.06.081

Bejarano, L., Jordão, M. J. C., & Joyce, J. A. (2021). Therapeutic targeting of the tumor microenvironment. *Cancer Discovery*, *11*(4), 933–959. https://doi.org/10.1158/2159-8290.Cd-20-1808

Buchman, J. T., Hudson-Smith, N. V., Landy, K. M., & Haynes, C. L. (2019). Understanding nanoparticle toxicity mechanisms to inform redesign strategies to reduce environmental impact. *Accounts of Chemical Research, 52*(6), 1632–1642. https://doi.org/10.1021/acs.accounts.9b00053

Chang, D., Ma, Y., Xu, X., Xie, J., & Ju, S. (2021). Stimuli-responsive polymeric nanoplatforms for cancer therapy [review]. *Frontiers in Bioengineering and Biotechnology, 9*. https://doi.org/10.3389/fbioe.2021.707319

Chen, M., Hu, J., Wang, L., Li, Y., Zhu, C., Chen, C., Shi, M., Ju, Z., Cao, X., & Zhang, Z. (2020). Targeted and redox-responsive drug delivery systems based on carbonic anhydrase IX-decorated mesoporous silica nanoparticles for cancer therapy. *Scientific Reports, 10*(1), 14447.

Cheng, Z., Li, M., Dey, R., & Chen, Y. (2021). Nanomaterials for cancer therapy: Current progress and perspectives. *Journal of Hematology & Oncology, 14*(1), 85. https://doi.org/10.1186/s13045-021-01096-0

Damasco, J. A., Ravi, S., Perez, J. D., Hagaman, D. E., & Melancon, M. P. (2020). Understanding nanoparticle toxicity to direct a safe-by-design approach in cancer nanomedicine. *Nanomaterials (Basel), 10*(11). https://doi.org/10.3390/nano10112186

Dong, C. (2021). Cytokine regulation and function in T cells. *Annual Review of Immunology, 39*, 51–76. https://doi.org/10.1146/annurev-immunol-061020-053702

Fadeel, B., Farcal, L., Hardy, B., Vázquez-Campos, S., Hristozov, D., Marcomini, A., Lynch, I., Valsami-Jones, E., Alenius, H., & Savolainen, K. (2018). Advanced tools for the safety assessment of nanomaterials. *Nature Nanotechnology, 13*(7), 537–543. https://doi.org/10.1038/s41565-018-0185-0

Forest, V. (2022). Experimental and computational nanotoxicology-complementary approaches for nanomaterial hazard assessment. *Nanomaterials (Basel), 12*(8). https://doi.org/10.3390/nano12081346

Gagliardi, A., Giuliano, E., Venkateswararao, E., Fresta, M., Bulotta, S., Awasthi, V., & Cosco, D. (2021). Biodegradable polymeric nanoparticles for drug delivery to solid tumors [review]. *Frontiers in Pharmacology, 12*. https://doi.org/10.3389/fphar.2021.601626

Gatenby, R. A., & Gillies, R. J. (2004). Why do cancers have high aerobic glycolysis? *Nature Reviews Cancer, 4*(11), 891–899. https://doi.org/10.1038/nrc1478

Gavas, S., Quazi, S., & Karpiński, T. M. (2021). Nanoparticles for cancer therapy: Current progress and challenges. *Nanoscale Research Letters, 16*(1), 173. https://doi.org/10.1186/s11671-021-03628-6

Ghosh, S., Sullivan, C. A. W., Zerkowski, M. P., Molinaro, A. M., Rimm, D. L., Camp, R. L., & Chung, G. G. (2008). High levels of vascular endothelial growth factor and its receptors (VEGFR-1, VEGFR-2, neuropilin-1) are associated with worse outcome in breast cancer. *Human Pathology, 39*(12), 1835–1843. https://doi.org/10.1016/j.humpath.2008.06.004

Gong, F., Yang, N., Wang, X., Zhao, Q., Chen, Q., Liu, Z., & Cheng, L. (2020). Tumor microenvironment-responsive intelligent nanoplatforms for cancer theranostics. *Nano Today, 32*, 100851. https://doi.org/10.1016/j.nantod.2020.100851

Guo, X., Wei, X., Chen, Z., Zhang, X., Yang, G., & Zhou, S. (2020). Multifunctional nanoplatforms for subcellular delivery of drugs in cancer therapy. *Progress in Materials Science, 107*, 100599. https://doi.org/10.1016/j.pmatsci.2019.100599

Hanahan, D., & Weinberg, R. A. (2011). Hallmarks of cancer: The next generation. *Cell, 144*(5), 646–674. https://doi.org/10.1016/j.cell.2011.02.013

Harper, S., Usenko, C., Hutchison, J. E., Maddux, B. L. S., & Tanguay, R. L. (2008). In vivo biodistribution and toxicity depends on nanomaterial composition, size, surface functionalisation and route of exposure. *Journal of Experimental Nanoscience, 3*(3), 195–206. https://doi.org/10.1080/17458080802378953

Huang, H. J., Lee, Y. H., Hsu, Y. H., Liao, C. T., Lin, Y. F., & Chiu, H. W. (2021). Current strategies in assessment of nanotoxicity: Alternatives to in vivo animal testing. *International Journal of Molecular Sciences, 22*(8). https://doi.org/10.3390/ijms22084216

Iriarte-Mesa, C., López, Y. C., Matos-Peralta, Y., de la Vega-Hernández, K., & Antuch, M. (2020). Gold, silver and iron oxide nanoparticles: Synthesis and bionanoconjugation strategies aimed at electrochemical applications. *Topics in Current Chemistry (Cham), 378*(1), 12. https://doi.org/10.1007/s41061-019-0275-y

Jain, R. K., & Stylianopoulos, T. (2010). Delivering nanomedicine to solid tumors. *Nature Reviews Clinical Oncology, 7*(11), 653–664. https://doi.org/10.1038/nrclinonc.2010.139

Jin, M., Jin, G., Kang, L., Chen, L., Gao, Z., & Huang, W. (2018). Smart polymeric nanoparticles with pH-responsive and PEG-detachable properties for co-delivering paclitaxel and survivin siRNA to enhance antitumor outcomes. *International Journal of Nanomedicine, 13*, 2405–2426. https://doi.org/10.2147/ijn.S161426

Jin, M.-Z., & Jin, W.-L. (2020). The updated landscape of tumor microenvironment and drug repurposing. *Signal Transduction and Targeted Therapy, 5*(1), 166. https://doi.org/10.1038/s41392-020-00280-x

Kalangi, S. K., & Bhosale, R. (2022). Editorial: New approaches in toxicity testing of nanotherapeutics [editorial]. *Frontiers in Pharmacology, 13*. https://doi.org/10.3389/fphar.2022.922551

Karlsson, H. L., Gustafsson, J., Cronholm, P., & Möller, L. (2009). Size-dependent toxicity of metal oxide particles—a comparison between nano- and micrometer size. *Toxicology Letters, 188*(2), 112–118. https://doi.org/10.1016/j.toxlet.2009.03.014

Krewski, D., Acosta, D., Jr., Andersen, M., Anderson, H., Bailar, J. C., 3rd, Boekelheide, K., Brent, R., Charnley, G., Cheung, V. G., Green, S., Jr., Kelsey, K. T., Kerkvliet, N. I., Li, A. A., McCray, L., Meyer, O., Patterson, R. D., Pennie, W., Scala, R. A., Solomon, G. M., . . . Zeise, L. (2010). Toxicity testing in the 21st century: A vision and a strategy. *Journal of Toxicology and Environmental Health - Part B: Critical Reviews, 13*(2–4), 51–138. https://doi.org/10.1080/10937404.2010.483176

Kumar, V., Kumari, A., Guleria, P., & Yadav, S. K. (2012). Evaluating the toxicity of selected types of nanochemicals. *Reviews of Environmental Contamination and Toxicology, 215*, 39–121. https://doi.org/10.1007/978-1-4614-1463-6_2

Kumar, V., Sharma, N., & Maitra, S. S. (2017). In vitro and in vivo toxicity assessment of nanoparticles. *International Nano Letters, 7*(4), 243–256. https://doi.org/10.1007/s40089-017-0221-3

Lama, S., Merlin-Zhang, O., & Yang, C. (2020). In vitro and in vivo models for evaluating the oral toxicity of nanomedicines. *Nanomaterials (Basel), 10*(11). https://doi.org/10.3390/nano10112177

Li, X., Wang, L., Fan, Y., Feng, Q., & Cui, F.-Z. (2012). Biocompatibility and toxicity of nanoparticles and nanotubes. *Journal of Nanomaterials, 2012*, 548389. https://doi.org/10.1155/2012/548389

Li, X., Wang, Z., Zhao, T., Yu, B., Fan, Y., Feng, Q., Cui, F. Z., & Watari, F. (2016). A novel method to in vitro evaluate biocompatibility of nanoscaled scaffolds. *Journal of Biomedical Materials Research Part A, 104*(9), 2117–2125. https://doi.org/10.1002/jbm.a.35743

Li, Y., Jeon, J., & Park, J. H. (2020). Hypoxia-responsive nanoparticles for tumor-targeted drug delivery. *Cancer Letters, 490*, 31–43. https://doi.org/10.1016/j.canlet.2020.05.032

Li, Z., Huang, J., & Wu, J. (2021). pH-Sensitive nanogels for drug delivery in cancer therapy. *Biomaterials Science, 9*(3), 574–589.

Lima, L. G., Ham, S., Shin, H., Chai, E. P. Z., Lek, E. S. H., Lobb, R. J., Müller, A. F., Mathivanan, S., Yeo, B., Choi, Y., Parker, B. S., & Möller, A. (2021). Tumor microenvironmental cytokines bound to cancer exosomes determine uptake by cytokine receptor-expressing cells and biodistribution. *Nature Communications, 12*(1), 3543. https://doi.org/10.1038/s41467-021-23946-8

Liu, H., Yao, J., Guo, H., Cai, X., Jiang, Y., Lin, M., Jiang, X., Leung, W., & Xu, C. (2020). Tumor microenvironment-responsive nanomaterials as targeted delivery carriers for photodynamic anticancer therapy. *Frontiers in Chemistry, 8*, 758. https://doi.org/10.3389/fchem.2020.00758

Lugano, R., Ramachandran, M., & Dimberg, A. (2020). Tumor angiogenesis: Causes, consequences, challenges and opportunities. *Cellular and Molecular Life Sciences, 77*(9), 1745–1770. https://doi.org/10.1007/s00018-019-03351-7

McMillin, D. W., Negri, J. M., & Mitsiades, C. S. (2013). The role of tumour–stromal interactions in modifying drug response: Challenges and opportunities. *Nature Reviews Drug Discovery, 12*(3), 217–228. https://doi.org/10.1038/nrd3870

Meng, C., Chen, Z., Li, G., Welte, T., & Shen, H. (2021). Nanoplatforms for mRNA therapeutics. *Advanced Therapeutics, 4*(1), 2000099. https://doi.org/10.1002/adtp.202000099

Mi, Y., Shao, Z., Vang, J. et al. (2016). Application of nanotechnology to cancer radiotherapy. *Cancer Nano, 7*, 11. https://doi.org/10.1186/s12645-016-0024-7

Mitchell, M. J., Billingsley, M. M., Haley, R. M., Wechsler, M. E., Peppas, N. A., & Langer, R. (2021). Engineering precision nanoparticles for drug delivery. *Nature Reviews Drug Discovery, 20*(2), 101–124. https://doi.org/10.1038/s41573-020-0090-8

Murciano-Goroff, Y. R., Warner, A. B., & Wolchok, J. D. (2020). The future of cancer immunotherapy: Microenvironment-targeting combinations. *Cell Research, 30*(6), 507–519. https://doi.org/10.1038/s41422-020-0337-2

N'Guessan B, B., Amponsah, S. K., Iheagwara, I. B., Seidu, M. A., Frimpong-Manso, S., Ofori-Attah, E., Bekoe, E. O., Sarkodie, J. A., Appiah-Opong, R., & Asiedu-Gyekye, I. J. (2021). Toxicity, mutagenicity and trace metal constituent of Termitomyces schimperi (Pat.) R. Heim (Lyophyllaceae) and kaolin, a recipe used traditionally in cancer management in Cote d'Ivoire. *Journal of Ethnopharmacology, 276*, 114147. https://doi.org/10.1016/j.jep.2021.114147

N'Guessan, B. B., Asiamah, A. D., Arthur, N. K., Frimpong-Manso, S., Amoateng, P., Amponsah, S. K., Kukuia, K. E., Sarkodie, J. A., Opuni, K. F., Asiedu-Gyekye, I. J., & Appiah-Opong, R. (2021). Ethanolic extract of Nymphaea lotus L. (Nymphaeaceae) leaves exhibits in vitro antioxidant, in vivo anti-inflammatory and cytotoxic activities on Jurkat and MCF-7 cancer cell lines. *BMC Complementary Medicine and Therapies, 21*(1), 22. https://doi.org/10.1186/s12906-020-03195-w

Ni, Y., Zhou, X., Yang, J., Shi, H., Li, H., Zhao, X., & Ma, X. (2021). The role of tumor-stroma interactions in drug resistance within tumor microenvironment [review]. *Frontiers in Cell and Developmental Biology, 9*. https://doi.org/10.3389/fcell.2021.637675

Palanikumar, L., Al-Hosani, S., Kalmouni, M., Nguyen, V. P., Ali, L., Pasricha, R., Barrera, F. N., & Magzoub, M. (2020). pH-responsive high stability polymeric nanoparticles for targeted delivery of anticancer therapeutics. *Communications Biology, 3*(1), 95. https://doi.org/10.1038/s42003-020-0817-4

Patel, P., & Shah, J. (2017). Safety and toxicological considerations of nanomedicines: The future directions. *Current Clinical Pharmacology, 12*(2), 73–82. https://doi.org/10.2174/1574884712666170509161252

Patra, J. K., Das, G., Fraceto, L. F., Campos, E. V. R., Rodriguez-Torres, M. D. P., Acosta-Torres, L. S., Diaz-Torres, L. A., Grillo, R., Swamy, M. K., Sharma, S., Habtemariam, S., & Shin, H.-S. (2018). Nano based drug delivery systems: Recent developments and future prospects. *Journal of Nanobiotechnology, 16*(1), 71. https://doi.org/10.1186/s12951-018-0392-8

Peng, S., Xiao, F., Chen, M., & Gao, H. (2022). Tumor-microenvironment-responsive nanomedicine for enhanced cancer immunotherapy. *Advanced Science (Weinh), 9*(1), e2103836. https://doi.org/10.1002/advs.202103836

Petrova, V., Annicchiarico-Petruzzelli, M., Melino, G., & Amelio, I. (2018). The hypoxic tumour microenvironment. *Oncogenesis, 7*(1), 10. https://doi.org/10.1038/s41389-017-0011-9

Pittet, M. J., Michielin, O., & Migliorini, D. (2022). Clinical relevance of tumour-associated macrophages. *Nature Reviews Clinical Oncology, 19*(6), 402–421. https://doi.org/10.1038/s41571-022-00620-6

Qin, X., Li, T., Li, S., Yang, H., Wu, C., Zheng, C., You, F., & Liu, Y. (2020). The tumor biochemical and biophysical microenvironments synergistically contribute to cancer cell malignancy. *Cellular & Molecular Immunology, 17*(11), 1186–1187. https://doi.org/10.1038/s41423-019-0282-5

Raies, A. B., & Bajic, V. B. (2016). In silico toxicology: Computational methods for the prediction of chemical toxicity. *Wiley Interdisciplinary Reviews: Computational Molecular Science, 6*(2), 147–172. https://doi.org/10.1002/wcms.1240

Rosenberg, S. A., & Restifo, N. P. (2015). Adoptive cell transfer as personalized immunotherapy for human cancer. *Science, 348*(6230), 62–68. https://doi.org/10.1126/science.aaa4967

Seebacher, N. A., Krchniakova, M., Stacy, A. E., Skoda, J., & Jansson, P. J. (2021). Tumour microenvironment stress promotes the development of drug resistance. *Antioxidants (Basel), 10*(11). https://doi.org/10.3390/antiox10111801

Sharifi, S., Behzadi, S., Laurent, S., Forrest, M. L., Stroeve, P., & Mahmoudi, M. (2012). Toxicity of nanomaterials. *Chemical Society Reviews, 41*(6), 2323–2343. https://doi.org/10.1039/c1cs15188f

Sharma, S., Parveen, R., & Chatterji, B. P. (2021). Toxicology of nanoparticles in drug delivery. *Current Pathobiology Reports, 9*(4), 133–144. https://doi.org/10.1007/s40139-021-00227-z

Siafaka, P. I., Okur, N. Ü., Karantas, I. D., Okur, M. E., & Gündoğdu, E. A. (2021). Current update on nanoplatforms as therapeutic and diagnostic tools: A review for the materials used as nanotheranostics and imaging modalities. *Asian Journal of Pharmaceutical Sciences, 16*(1), 24–46. https://doi.org/10.1016/j.ajps.2020.03.003

Stillman, N. R., Kovacevic, M., Balaz, I., & Hauert, S. (2020). In silico modelling of cancer nanomedicine, across scales and transport barriers. *NPJ Computational Materials, 6*(1), 92. https://doi.org/10.1038/s41524-020-00366-8

Su, Z., Dong, S., Zhao, S.-C., Liu, K., Tan, Y., Jiang, X., Assaraf, Y. G., Qin, B., Chen, Z.-S., & Zou, C. (2021). Novel nanomedicines to overcome cancer multidrug resistance. *Drug Resistance Updates, 58*, 100777. https://doi.org/10.1016/j.drup.2021.100777

Sukhanova, A., Bozrova, S., Sokolov, P., Berestovoy, M., Karaulov, A., & Nabiev, I. (2018). Dependence of nanoparticle toxicity on their physical and chemical properties. *Nanoscale Research Letters, 13*(1), 44. https://doi.org/10.1186/s11671-018-2457-x

Tang, T., Huang, X., Zhang, G., Hong, Z., Bai, X., & Liang, T. (2021). Advantages of targeting the tumor immune microenvironment over blocking immune checkpoint in cancer immunotherapy. *Signal Transduction and Targeted Therapy, 6*(1), 72. https://doi.org/10.1038/s41392-020-00449-4

Thomas, R. G., Surendran, S. P., & Jeong, Y. Y. (2020). Tumor microenvironment-stimuli responsive nanoparticles for anticancer therapy [review]. *Frontiers in Molecular Biosciences, 7*. https://doi.org/10.3389/fmolb.2020.610533

Uthaman, S., Huh, K. M., & Park, I. K. (2018). Tumor microenvironment-responsive nanoparticles for cancer theragnostic applications. *Biomaterials Research, 22*, 22. https://doi.org/10.1186/s40824-018-0132-z

Vasan, N., Baselga, J., & Hyman, D. M. (2019). A view on drug resistance in cancer. *Nature, 575*(7782), 299–309. https://doi.org/10.1038/s41586-019-1730-1

Wang, J., Li, Y., & Nie, G. (2021). Multifunctional biomolecule nanostructures for cancer therapy. *Nature Reviews Materials*, 6(9), 766–783. https://doi.org/10.1038/s41578-021-00315-x

Wang, Q., Shao, X., Zhang, Y., Zhu, M., Wang, F. X. C., Mu, J., Li, J., Yao, H., & Chen, K. (2023). Role of tumor microenvironment in cancer progression and therapeutic strategy. *Cancer Medicine*. https://doi.org/10.1002/cam4.5698

Yang, M., Li, J., Gu, P., & Fan, X. (2021). The application of nanoparticles in cancer immunotherapy: Targeting tumor microenvironment. *Bioactive Materials*, 6(7), 1973–1987. https://doi.org/10.1016/j.bioactmat.2020.12.010

Yetisgin, A. A., Cetinel, S., Zuvin, M., Kosar, A., & Kutlu, O. (2020). Therapeutic nanoparticles and their targeted delivery applications. *Molecules*, 25(9). https://doi.org/10.3390/molecules25092193

Younis, M. A., Tawfeek, H. M., Abdellatif, A. A. H., Abdel-Aleem, J. A., & Harashima, H. (2022). Clinical translation of nanomedicines: Challenges, opportunities, and keys. *Advanced Drug Delivery Reviews*, 181, 114083. https://doi.org/10.1016/j.addr.2021.114083

Yuan, Y., Jiang, Y. C., Sun, C. K., & Chen, Q. M. (2016). Role of the tumor microenvironment in tumor progression and the clinical applications [review]. *Oncology Reports*, 35(5), 2499–2515. https://doi.org/10.3892/or.2016.4660

Yusa, S.-I. (2022). Development and application of pH-responsive polymers. *Polymer Journal*, 54(3), 235–242. https://doi.org/10.1038/s41428-021-00576-x

Yusuf, A., Almotairy, A. R. Z., Henidi, H., Alshehri, O. Y., & Aldughaim, M. S. (2023). Nanoparticles as drug delivery systems: A review of the implication of nanoparticles' physicochemical properties on responses in biological systems. *Polymers*, 15(7), 1596. https://www.mdpi.com/2073-4360/15/7/1596

Zhang, C., Yan, L., Wang, X., Zhu, S., Chen, C., Gu, Z., & Zhao, Y. (2020). Progress, challenges, and future of nanomedicine. *Nano Today*, 35, 101008. https://doi.org/10.1016/j.nantod.2020.101008

Zhang, R., Liu, T., Li, W., Ma, Z., Pei, P., Zhang, W., Yang, K., & Tao, Y. (2022). Tumor microenvironment-responsive BSA nanocarriers for combined chemo/chemodynamic cancer therapy. *Journal of Nanobiotechnology*, 20(1), 223. https://doi.org/10.1186/s12951-022-01442-5

Zhao, Y., & Adjei, A. A. (2015). Targeting angiogenesis in cancer therapy: Moving beyond vascular endothelial growth factor. *Oncologist*, 20(6), 660–673. https://doi.org/10.1634/theoncologist.2014-0465

Zou, W. (2022). Immune regulation in the tumor microenvironment and its relevance in cancer therapy. *Cellular & Molecular Immunology*, 19(1), 1–2. https://doi.org/10.1038/s41423-021-00738-0

21 Commercialization Challenges of Tumor Microenvironment– Responsive Nanoplatforms

*Snigdha Das Mandal, Devanshu J Patel,
Surjyanarayan Mandal and Jayvadan K. Patel*

CONTENTS

21.1 Introduction 352
21.2 Scale-Up Challenges of Nanomedicines: Preparation Methods and Their Challenges 353
21.3 Regulatory Challenges of Nanomedicines: Generics or Nanosimilars 355
21.4 Clinical Challenges of Nanomedicines 356
21.5 Conclusion 358
References 359

21.1 INTRODUCTION

Nanomedicines are multifaceted engineered nanoscale structures and are extensively used in the treatment and targeting of diseases because of their numerous therapeutic benefits (1). Since the last decade, nanotechnology has been considered a revolutionary technology due to its applicability in multiple fields, including medicine. Due to numerous anticipated beneficial effects, nanomedicines have attracted attention from scientists around the globe. These benefits include protection of the active entity from degradation, enhanced solubility and bioavailability, superior pharmacokinetics, reduced toxicity, enhanced therapeutic efficacy, decreased drug immunogenicity and, most importantly, capability of delivering the drug at the target site (2). Due to exhaustive pharmaceutical research, an increasing number of products containing nanomaterials with different applications, especially with tumors, have surfaced. The use of nanotechnology in the development of new medicines is now part of research around the globe which is capable of providing new and innovative medical solutions to address unmet medical needs (3). However, the definition of nanomaterials has been controversial among various scientific and international regulatory corporations (4). Several opinions come out to agree on a consensual definition of nanomedicines which in turn based on the fact that nanomaterials possess different physicochemical properties than those of their conventional equivalents, due to their small size, charge and stability (5). These physicochemical properties provide a set of opportunities in the development of drug products; however, their safety issues are still major concerns. The physicochemical properties of the nanomedicines which usually alters the pharmacokinetics of drug moiety like absorption (potential to cross biological barriers), effective distribution, elimination and metabolism and their toxic properties which may help in establishing the concerns over the application of the nanoformulations (6). Several nanomedicine products have been approved by the European Medicines Agency (EMA) and the US Food and Drug Administration (FDA) for various therapeutic indications (7). In general, high therapeutic efficacy and safety are the two primary factors for a product in order to qualify as commercially successful. But most nanomedicines fail to meet these requirements; hence their successful commercialization is hampered to a large extent. The limited clinical success of nanomedicines is mainly due to challenges in formulation development and optimization, batch-to-batch consistency for the stability indicating parameters, robust and reproducible manufacturing and subsequent scale-up, characterization and screening analytical methods, regulatory barriers and instability in biological environments (8, 9). Specific properties of nanomedicines that are responsible for exerting benefits may also generate some safety concerns and impose challenges on their scale-up and clinical usefulness. However, in reality, there are several challenges that are overshadowing the beneficial effects. The application of nanotechnology in medicine has led to significant progress in providing better health outcomes. Successful translation of nanomedicines implies upgrading of industrial infrastructures, which includes setting up an equipped analytical lab for characterization, good manufacturing and pilot manufacturing/scale-up facilities according to regulatory standards. It is now crucial to impose quality control checks on every step of nanomedicine manufacturing, which will maintain the quality and efficacy of batches (10). Additionally, sensitive and robust analytical techniques that are able to detect minute variations and adequately characterize the product should be established. There are three major challenges associated with these nanomedicines: scale-up challenges of formulations, regulatory challenges and clinical challenges.

21.2 SCALE-UP CHALLENGES OF NANOMEDICINES: PREPARATION METHODS AND THEIR CHALLENGES

The scale-up of nanomedicines includes the proper development and optimization of manufacturing methods along with transfer of the technology for large-scale industrial manufacturing. Both analytical and manufacturing process limitations of nanomedicines in small-scale preparation may lead to failure to translate their preparation methods from the lab scale to the commercial scale (11). The small-scale preparation procedure shortfall may lead to failure to translate any method from the laboratory scale to industrial scale. In general terms, the production of nanomedicines is a challenging task in terms of reproducibility and quality of the product. Moreover, a well-designed scale-up procedure will ensure the quality of the nanomedicine, cost-effectiveness and a timely product launch.

Literature data reveals that there are only a few reported manufacturing processes of nanomedicines, especially using emulsion-based and nanoprecipitation methods, which are supported by scale-up aspects for industrial manufacturing. An adequately designed scale-up procedure will not only guarantee the quality of nanomedicines but also cost-effectiveness and a timely launch of the product to the market. Controlled drug/gene delivery to treat tumors and cell-based diagnosis are featured in the applications of nanomedicines (12). Only a few reported laboratory-scale manufacturing processes of nanomedicines are available as scale-up methods, especially those such as emulsion-based and nanoprecipitation methods, as well as using high-pressure homogenization techniques or spray drying (13). The limitations of preparation methods and subsequent hurdles in the scale-up of nanomedicines are considered the foremost challenges to their commercialization and large-scale production.

First, scale-up has mainly been observed to affect the characteristics of nanoparticles, such as particle size, drug encapsulation, process residual materials, colloidal stability and surface morphology, although some authors reported that nearly multifold scale-up of nanomedicine using an emulsion-based method from the laboratory batch size did not alter the encapsulation percentage of drug (14). Moreover, an increase in speed and time of agitation reduces particle sizes, which in turn decreases the polymer concentration in the final composition. Furthermore, particle size reduction and to obtain the narrow particle size distribution, polymer concentration is to be reduced which in turn reported to reduce the entrapment efficiency.

Moreover, it is frequently observed that the scale-up process normally reduces the drug loading of nanoformulations. In connection with the challenges in cancer therapy, the use of nanomaterials along with well-designed nanodrug delivery systems are able to minimize local and systemic toxicity by modulating the physiological processes and targeting drug moieties to the site of action. It is also anticipated that nanomedicines will revolutionize the entire health-care system due to their ability to target the affected area and hence minimize the toxicity of the active in the treatment of cancer. However, the design of efficient cancer nanotherapeutics remains a great challenge, and only a few nanoformulations or nanomedicines have entered clinical trials (15).

The physicochemical properties of nanomedicines play a vital role in the biocompatibility as well as expected toxicity in the biological systems (16). Therefore, the physicochemical parameters of nanomedicines (as depicted in) as a drug delivery system need to be carefully characterized to avoid their potential toxicity to healthy cells (17). Additionally, since these nanocarriers tend to interact with biomolecules and may tend to aggregate, forming a protein corona, it can be anticipated that the regular function of nanomedicines may be disturbed, rendering them ineffective in cancer treatment (18). In addition to the described physicochemical properties, the storage of nanomaterials and their subsequent stability may also have a direct impact on their pharmacological performance (19). Another major challenge in drug delivery is safety for human health, so issues may be associated with nanomaterials that may not have an immediate impact or not be noticeable at the initial phase of the treatment, but as they are reported to alter physiological processes, subsequent non-compatibility may arise. The use of nanocarriers in the treatment of cancer may result in unwanted toxicity through unfavorable interactions with biological entities (19, 20). Several studies have revealed the detrimental properties of nanocarriers due to their toxicity; hence toxicity assessment is in great demand for filing to any regulated market.

In addition to all these facts, a significant obstruction in commercialization of developed nanomedicines is the clinical study and which in turn because of the lack of in-depth understanding of nanomedicine-bio interfacial interactions. Furthermore, the difference in the drug release in vitro to that in vivo allows establishing in vitro–in vivo correlation (IVIVC) and is emerging as a major issue. Furthermore, the manufacturing of nanomedicines for commercialization is a key obstacle, as large scale-production is technically challenging (21).

Generally, only small quantities of nanomedicine are used for preclinical and clinical trial studies relating to cancer treatment. The large-scale production of nanoformulations, however, is quite challenging, as their physicochemical properties may vary from batch to batch. Moreover, the involvement of the complex multi-stage manufacturing processes of nanomedicines and the high cost of raw materials renders these sophisticated nanotherapeutics costly. As a well-planned and well-designed manufacturing process is critically important to scale up these therapeutics, huge cost is anticipated, but the noticeable clinical benefits can justify the manufacturing costs (22). Few scientists have also investigated the possibility of scale-up for the nanoprecipitation method is through a membrane contactor. In this method, a non-aqueous organic phase containing hydrophobic materials was introduced through the membrane pores of a filter in order to generate small droplets that react with the aqueous phase (containing surfactant), and ultimately nanoparticles are produced (23). The nanoprecipitation method showed high flux that supports an industrial scale-up method for the manufacture of nanomedicines. Second, the stability of materials used in the method and toxic solvent usage (e.g., chloroform or dichloromethane as organic phase) further limits the process. Owing to the toxic potential of organic solvents, pharmaceutical industries are now mostly avoiding their usage and are shifting to processes based on the use of aqueous systems. Hence, novel methodology to be introduced where aqueous solvents or other solvents with low toxicity will be used for the production of nanomedicines and can be scaled up the pharmaceutical industries. Basically, a shorter manufacturing time is always preferred for a single pilot batch. Galindo-Rodriguez et al. reported that the nanoprecipitation method had taken less time as compared with the emulsion-based method.

Another key issue is the challenge of regulatory approval of nanomedicines, as there are no specific or stringent FDA guidelines for products with nanomaterials. The regulatory findings on nanoformulated drugs are based on individual assessments or evaluations, which are time consuming and cause setbacks in commercialization. It is also evident that the difficulties in approval will tend to increase due to the development of multifunctional nanoformulation technologies. Thus, to solve the problems associated with nanomedicines for cancer treatment, a validated development strategy needs to be incorporated before the scale-up and subsequently used in medicine for better treatment to human beings. The preclinical pharmacokinetic protocols for nanomedicines should be designed in such a way that they can compensate for the complications involved in cancer cell treatment due to the altered physiology and the tumor microenvironment (24). Furthermore, case-by-case investigations are required to be evaluated in order to conclude nanomedicine tremendous potential of cancer treatment. Along with this, a comprehensive set of regulatory guidelines for approval becomes a must to expedite the scale-up of cancer nanotherapeutics (24, 25).

21.3 REGULATORY CHALLENGES OF NANOMEDICINES: GENERICS OR NANOSIMILARS

With the introduction of nanotechnology and the respective evolution of nanomedicines, a lot of scientific areas have benefited extensively, and this is especially noteworthy in the area of new drug product development. However there are several controversies associated with basic concepts related to these nanosystems and their commercialization. Due to the properties conferred by nanomedicines, the challenges of huddles for nanotechnology implementation, specifically in the pharmaceutical development of new drug products and respective regulatory issues are critically presented. A new area is emerging as a promising approach, termed theranostics, which combines diagnostics or imaging with therapy. In clinical practice, these nanomedicines are able to compensate for unmet medical needs by drug targeting, controlled and site-specific release of high-toxicity molecules like anticancer agents, exploiting multiple mechanisms of action, maximizing efficacy with reduced dose and toxicity, improving transport across biological barriers or favoring preferential distribution within the body (e.g., in areas with cancer lesions) (26, 27). Due to their small size, nanomedicines have a high specific surface area and hence particle surface energy is increased, making nanomedicines much more reactive as that of conventional medicines. Nanomaterials have a tendency to adsorb several biomolecules like proteins and lipids when they come in contact with biological fluids, and subsequently a biomolecule adsorption layer, known as a "corona", is formed on the surface of colloidal nanoparticles (28). However, the composition of the corona is a factor in the route of entry into the body and the particular fluid that nanomedicines come across, for example, gastro-intestinal fluid, blood or lung fluid. Furthermore, these dynamic changes can influence the corona constitution as the nanomedicines cross from one biological compartment to another (29).

In the last two decades, nanomedicines have been successfully introduced into clinical practice after continuous development in pharmaceutical research. In Europe, the nanomedicine market is composed of not only nanoparticles but also liposomes, nanocrystals, nanoemulsions, polymeric-protein conjugates and nanocomplexes (30). In the process of approval, nanomedicines were initially introduced on the basis of benefit to the risk involved. Another related challenge is the development of a framework for the evaluation of follow-on nanomedicines at the time of the reference medicine's patent expiration (31). Nanomedicines may be developed from actives sourced either from biological or non-biological sources or both. Biological nanomedicines are obtained from biological sources, while non-biological nanomedicines are developed using synthetic active ingredients and hence are classified as non-biological complex drug (NBCD) products (32, 33). In order to introduce a generic medicine in the pharmaceutical regulatory or rest of the world countries market, several parameters need to be optimized and demonstrated, as described in this chapter. For biological and non-biological nanomedicines, a more complete in vitro analysis as well as in vivo evaluation by a validated bio-analytical method is required to establish a perfect IVIVC. A comparison of quality, safety, efficacy and bioequivalence should be made with the reference drug product, which in turn leads to therapeutic equivalence of the developed nanomedicines (34). For regulatory purposes, the European Medicines Agency (EMA) has set the framework, especially for biological nanomedicines. This framework is basically a regulatory approach that includes the recommendations for comparative quality attributes non-clinical as well as clinical studies (35). The regulatory approach for approval of non-biological complex drug products is still ongoing. Since biological complex drug (BCD) products and NBCDs have some common features, sometimes the NBCD framework set by the EMA can be applied as the base for regulation of BCDs. However, the structure of BCDs cannot be fully characterized, and the in vivo results are influenced by the manufacturing process itself. Moreover, the draft regulatory roadmaps for some NBCD groups like nanoparticles, iron carbohydrate complexes and liposomes also help regulatory bodies create a final framework for NBCDs. The EMA has already released some guidance for developers of both new nanomedicines and nanosimilars in the preparation of applications for marketing authorization (36).

Published data on the therapeutic non-equivalence of nanoparticulate NBCDs and other similar products has created awareness of the regulatory gap to evaluate with reference complex drug products, which in turn defines the level of similarity required in order to get regulatory approval and subsequently practical use by healthcare experts. This needs to be resolute to which extent such complex formulations can be safely used as therapeutic alternatives or as a substitute or interchangeable medicinal products in the already established chronic therapy. This was a topic upon approval of generic and well-characterized small molecules with demonstrated bioequivalence to the reference drug product. A special challenge was produced by the nanoparticulate based NBCD products are quality issues and with potentially different fate in the biodistribution, efficacy followed by toxicity and also relating to a wide scientific gap of understanding the biointeractions (32, 37). The EMA has drafted several guidelines related to liposomal formulations (38) and iron nanomedicines for intravenous use (39), clearly stating that the current generic hypothesis with the established equivalence testing is not valid for these complex drugs. Because NBCDs are not biologicals by definition, a biosimilar approach cannot be applied. However, BCDs with complex structures, pharmacology and toxicology, along with their immunogenicity profiles, have also led to a similar kind of approach where evidence has to be provided on the comparability with the reference drug product for their subsequent use.

The FDA also published guidance for nanoparticle characterization and a genuine proposal for bioequivalence testing of colloidal iron sucrose intravenous injections (40). In on hand the EMA proposes for quality, nonclinical and clinical bioequivalence parameters to assess safety and efficacy for NBCDs while the US-FDA briefs on the quality including nanosizing and clinical bioequivalence testing parameters (Table 21.1). The FDA follows a complete evidence-based approach to be adapted in order to compare biosimilar nanomedicine products with the respective reference biological products.

Both BCDs and NBCDs are large molecular structures obtained through complex manufacturing processes. Hence, the products cannot be fully characterized by physicochemical means, and their in vivo performance cannot be established in a classical bioequivalence approach.

Therefore, in order to characterize their profile and extended efficacy quality, nonclinical and clinical tests are required to evaluate follow-on similars. Being not the same but similar, the extent of similarity has to be defined in a totality-of-evidence approach to define their therapeutic equivalence. The absence of unknown clinically relevant differences has to be granted for safety reasons. In addition, biological and nonbiological (synthetic) complex drugs may have nanoparticular characteristics and their follow-on are nanosimilars. For evidence development, different kinds of evidence are integrated and are necessary to address residual uncertainty. Different standards for different product categories and not a "one size fits all" format are used (Table 21.1). A globally harmonized approach is thus lacking for this new category of synthetic drug products with high complexity, including nanomedicine characteristics.

21.4 CLINICAL CHALLENGES OF NANOMEDICINES

The clinical translation of nanomedicines is an expensive and time-consuming process. The technologies for developing nanomedicines are usually more complex than those for conventional formulations (41, 42). Key factors related to the clinical development of nanomedicines include

TABLE 21.1
Differences in Requirements for the Evaluation of Intravenous Iron Preparations between the EMA and the US FDA

Evaluation Requirement	EMA	US FDA
Quality	Qualitative and quantitative composition Close/identical to the reference product Physicochemical characterization Particle size distribution Morphology Complex stability Labile iron release in plasma	Qualitative and quantitative composition Close/identical to the reference product Physicochemical characterization Particle size distribution Morphology
Nonclinical	In a relevant animal model (distribution, metabolism excretion). Emphasis on targeting, accumulation and retention at least in: • Plasma • Reticular endothelial system (RES) • Target tissues/organs	No specific recommendations
Clinical bioequivalence	Single-dose parallel or crossover study Analytes (serum): • Total iron • Transferrin-bound iron	Single-dose, randomized, parallel trial analytes (serum): • Total iron • Transferrin-bound iron
Clinical safety/efficacy	Therapeutic equivalence study: • Not necessary when totality of data (quality, nonclinical data, human PK study) demonstrate similarity Therapeutic equivalence study: • May be necessary in a relevant patient population when minor differences observed	No specific recommendations

large-scale manufacturing, biological challenges, biocompatibility and safety, intellectual property (IP) and government regulations as well as overall cost-effectiveness in comparison to current available therapies. These factors impose significant hurdles, limiting the commercialization of nanomedicines based on their safety and therapeutic effectiveness (43, 44). The interaction of nanotechnology and biology, that is, the influence of disease pathophysiology on the accumulation, distribution, retention and efficacy of the nanomedicines, as well as the pharmacokinetic and toxicity correlation between in vivo behavior in animals and humans, are deciding parameters for their successful translation. Therefore, applying a disease-driven approach on developing and optimizing nanomedicines in order to exploit the pathophysiological changes in disease biology has been taken into consideration to accelerate the clinical translation (45). Furthermore, few points to be evaluated like the relationship between the pathophysiology and the heterogeneity of the human disease and the mechanism of enhanced permeation and retention (EPR) with reduced accumulation of drug moiety in non-target organs. As little research has been done in order to comprehensively understand the correlations between nanomedicine behavior and patient biology in specific clinical applications or disease heterogeneity in patients, which are likely the foremost reasons for the failure in the translation of well-designed nanomedicines in clinical trials (46). These biological challenges can be a significant restriction for pharmaceutical industry investment into nanomedicines. In order to reduce investment risk for nanomedicines, the preclinical data needs to be evaluated carefully for therapeutic efficacy and safety relevant to human disease.

Furthermore, the evaluation of the pharmacokinetic and dynamic profile of nanomedicines should be done in multiple preclinical animal models in order to find their suitability for clinical disease, that is, reproducible results for the specific disease versus specific animal models. Additionally, animal models that reflect only a narrow spectrum of the clinical disease may provide useful data that can predict their suitability for treating a specific patient sub-group (45). The anatomy and physiology of the animal species used for the preclinical studies should be compared to those of humans and taken into account based on the routes of administration. Preclinical studies on nanomedicines should also be conducted under appropriate randomization to reduce bias, as well as being evaluated against proper controls, including the gold standard treatment and not just free drug solution (47). These factors are not emphasized in many currently published studies; hence it becomes difficult to assess clinical applicability of nanomedicines and their interpretation. Several other parameters include designing preclinical studies to optimize nanoformulation performance in vivo, dosing schedules and treatment regimens based on the specific clinical disease, as well as understanding the influence of disease progression and severity on nanomedicine performance, especially to combat cancer. This will conclude whether specific subjects (patients) may respond more positively to nanomedicine-based treatment. Interestingly, the majority of the publications on nanomedicine formulations focus on development and clinical trials, especially in cancer targeting (48). Despite the large number of publications, the ratio of translation of published studies to clinical applications has been very disappointing. Cancer targeting of nanomedicines has generally been based on the EPR effect, despite the fact that EPR-mediated accumulation has only been reported for some cancer types (49). Generally cancer is highly heterogeneous and can show inter- and intra-patient variability as the disease progresses. Hence the approach for designing nanomedicine-based treatment is unlikely to result in clinically beneficial outcomes. However, the EPR effect exploited for nanomedicine involving inflammatory diseases like rheumatoid arthritis, atherosclerosis and inflammatory bowel disease that causes leakiness of inflamed blood vessels (50). While the physicochemical properties of nanoformulations are expected to play a significant role on pharmacokinetics, drug distribution and accumulation in tumors, there are numerous inconveniences in comparing pre-clinical studies. In particular, differences in cell line and tumor size, effective dose, lack of acceptable pharmacokinetic data and differences in the meta-analysis limit further progress in this field, especially for cancer treatment. Similarly, physiochemical properties of the delivery system such as size, surface properties (modulation through pegylation), surface charge for stability, targeting efficiency through ligands and stability in blood or serum at physiological temperature are not uniformly documented. It should be further noted that not all diseases can be accessed with nanomedicines due to several biological barriers and that the EPR effect is also unlikely to be present in all clinical implications. Therefore, EPR is not referred to as the only determinant of efficacy of nanomedicines but rather influenced by the extent of cellular uptake and more importantly kinetics of drug release within target tissues (45). Furthermore, the advantages of ligand-gated nanomedicines in the clinical research phase have so far been negligible; despite enhanced accumulation in target sites being more predicted in preclinical studies. Potential reasons for this inconsistency have previously been cited as factors including physiological barriers, target accessibility and expression and formulation stability. Detailed analysis of the extent of nanomedicine accumulation, cellular internalization and intracellular functionality followed by intracellular degradation will also be important factors for clinical validation and translation. Therefore, by taking a disease-driven approach to nanomedicine development, it will be possible to build comprehensive preclinical data that best predicts efficacy for patients and supports translatable clinical development.

21.5 CONCLUSION

The potential applications of nanoformulations or nanomedicines for the diagnosis, prevention and treatment of deadly diseases like cancer are currently very extensive and broad. Practical application of nanomedicines therefore requires a degree of creativity and visionary power, well-defined approaches and systematic development, allowing them to reach new frontiers.

In this chapter, scientists around the globe provided an overview not only of some fascinating developments but also their applications. Since the science is currently expanding at a great pace, this chapter cannot include all aspects of present nanomedicine in detail; however, several challenges and their remedies are well documented. Nanotechnology has already had an significant impact on its clinical applications, which are expected to grow exponentially in the years to come. Located at the intersection of a number of fundamental disciplines, nanomedicine relies on chemical knowledge to provide required modifications of nanoformulations in order to enable the encapsulation of drug and diagnostic agent, provide detailed understanding of the pathophysiology of cancer to enable efficient targeting and therapy and finely engineer and manipulate the design of new nanodrug delivery systems. The main efforts and the majority of nanomedicine clinical applications are currently focused on treatment and efficient diagnostics of cancer. Novel drug delivery systems provide new opportunities to overcome the limitations associated with traditional drug therapy for cancer and to achieve both therapeutic and diagnostic functions in the same platform. The efficiency of drug delivery to a particular tumor site is dependent on the physicochemical properties, which in fact help to modify the design constraints, including clearance by the mononuclear phagocyte system or extravasation from circulation at the tumor site by the EPR effect. However, the lack of uniformity in pre-clinical trials of nanomedicines has raised questions about the systematic comparison of these studies. From the large number of pre-clinical trials, it is surprising to notice that little quantitative data has been found useful in designing nanomedicines for the treatment of cancer. The inadequate experimental design and variability of experimental conditions also contribute to slow development, followed by low efficiency on clinical impact. Finally, in this chapter, we attempt to summarize the current challenges in nanotherapeutics and provide an outlook on the future of this important field.

REFERENCES

1. Peer D, Karp JM, Hong S, et al. Nanocarriers as an emerging platform for cancer therapy. Nat. Nanotechnol. 2007; 2(12): 751–760.
2. Anselmo AC, Mitragotri S. Nanoparticles in the clinic. Bioeng. Transl. Med. 2016; 1(1): 10–29.
3. Bleeker EA, de Jong WH, Geertsma RE, et al. Considerations on the EU definition of a nanomaterial: Science to support policy making. Regul. Toxicol. Pharmacol. 2013; 65: 119–125.
4. Tinkle S, McNeil SE, Mühlebach S, et al. Nanomedicines: Addressing the scientific and regulatory gap. Ann. N. Y. Acad. Sci. 2014; 1313: 35–56.
5. Sara S, Joao S, Alberto P, et al. Nanomedicine: Principles, properties, and regulatory issues. Front. Chem. 2018; 6: 1–15.
6. Agrahari V, Hiremath P. Challenges associated and approaches for successful translation of nanomedicines into commercial products. Nanomedicine. 2017; 12: 819–823.
7. Troiano G, Nolan J, Parsons D, et al. A quality by design approach to developing and manufacturing polymeric nanoparticle drug products. AAPSJ. 2016; 18(6): 1354–1365.
8. Meng J, Agrahari V, Youm I. Advances in targeted drug delivery approaches for the central nervous system tumors: The inspiration of nanobiotechnology. J. Neuroimmune. Pharmacol. 2016; 12(1): 84–98.
9. Dobrovolskaia MA. Pre-clinical immunotoxicity studies of nanotechnology-formulated drugs: Challenges, considerations and strategy. J. Control. Release. 2015; 220: 571–583.
10. Desai N. Challenges in development of nanoparticle-based therapeutics. AAPSJ. 2012; 14(2): 282–295.
11. Rishi P, Jayachandra Babu R, Srinath P. Nanomedicine scale-up technologies: Feasibilities and challenges. AAPS PharmSci Tech. 2014; 15(6): 1527–1534.
12. Akgol S, Ulucan-Karnak F, Kuru CI, et al. The usage of composite nanomaterials in biomedical engineering applications. Biotech. Bioeng. 2021; 118: 2906–2922.
13. Galindo-Rodriguez SA, Puel F, Briançon S, et al. Comparative scale-up of three methods for producing ibuprofen-loaded nanoparticles. Eur. J. Pharm. Sci. 2005; 25(4): 357–367.
14. Saraf S. Process optimization for the production of nanoparticles for drug delivery applications. Expert Opin. Drug Deliv. 2009; 6: 187–196.
15. Mahmood R, Malik A, Waquar S, et al. New challenges in the use of nanomedicine in cancer therapy. Bioengineered. 2022; 13(1): 759–773.

16. Wu H, Ramanathan RK, Zamboni BA, et al. Population pharmacokinetics of pegylated liposomal CKD-602 (S-CKD602) in patients with advanced malignancies. J. Clin. Pharmacol. 2012; 52(2): 180–194.
17. Choi CHJ, Alabi CA, Webster P, et al. Mechanism of active targeting in solid tumors with transferrin-containing gold nanoparticles. Proc. Nat. Acad. Sci. 2010; 107(3): 1235–1240.
18. de Mendoza AE-H, Campanero MA, Mollinedo F, et al. Lipid nanomedicines for anticancer drug therapy. J. Biomed. Nanotechnol. 2009; 5(4): 323–343.
19. Gabizon A, Isacson R, Rosengarten O, et al. An open-label study to evaluate dose and cycle dependence of the pharmacokinetics of pegylated liposomal doxorubicin. Cancer Chemother Pharmacol. 2008; 61(4): 695–702.
20. Solomon R, Gabizon AA. Clinical pharmacology of liposomal anthracyclines: Focus on pegylated liposomal doxorubicin. Clin. Lymphoma and Myeloma. 2008; 8(1): 21–32.
21. Jain P, Pawar RS, Pandey RS, et al. In-vitro in-vivo correlation (IVIVC) in nanomedicine: Is protein corona the missing link? Biotechnol. Adv. 2017; 35: 889–904.
22. Betker JL, Gomez J, Anchordoquy TJ. The effects of lipoplex formulation variables on the protein corona and comparisons with in vitro transfection efficiency. J. Control Release. 2013; 171(3): 261–268.
23. Catherine C, Hatem F. Preparation of nanoparticles with a membrane contactor. JMSR. 2005; 266(1): 115–120.
24. Dawidczyk CM, Russell LM, Searson PC. Nanomedicines for cancer therapy: State-of-the-art and limitations to pre-clinical studies that hinder future developments. Front. Chem. 2014; 69(2): 1–13.
25. Ando M, Yonemori K, Katsumata N, et al. Phase I and pharmacokinetic study of nab-paclitaxel, nanoparticle albumin-bound paclitaxel, administered weekly to Japanese patients with solid tumors and metastatic breast cancer. Cancer Chemother. Pharmacol. 2012; 69: 457–465.
26. Chan VS. Nanomedicine: An unresolved regulatory issue. Regul. Toxicol. Pharmacol. 2006; 46: 218–224.
27. Moghimi SM, Hunter A, Murray J. Long-circulating and target specific nanoparticles: Theory to practice. Pharmacol. Rev. 2001; 53: 283–318.
28. Zhao Y, Chen C. Nano on reflection. Nat. Nanotechnol. 2016; 11: 828–834.
29. Pino P, Pelaz B, Zhang Q, et al. Protein corona formation around nanoparticles—from the past to the future. Mater. Horiz. 2014; 1: 301–313.
30. Agrahari V, Hiremath P. Challenges associated and approaches for successful translation of nanomedicines into commercial products. Nanomedicine. 2017; 12: 819–823.
31. Bleeker EA, de Jong WH, Geertsma RE, et al. Considerations on the EU definition of a nanomaterial: Science to support policy making. Regul. Toxicol. Pharmacol. 2013; 65: 119–125.
32. Tinkle S, McNeil SE, Mühlebach S, et al. Nanomedicines: Addressing the scientific and regulatory gap. Ann. N. Y. Acad. Sci. 2014; 1313: 35–56.
33. Hussaarts L, Muhlebach S, Shah VP, et al. Equivalence of complex drug products: Advances in and challenges for current regulatory frameworks. Ann. N. Y. Acad. Sci. 2017; 1407: 39–49.
34. Astier A, Barton Pai A, Bissig M, et al. How to select a nanosimilar. Ann. N. Y. Acad. Sci. 2017; 1407: 50–62.
35. Muhlebach S, Borchard G, Yildiz S. Regulatory challenges and approaches to characterize nanomedicines and their follow-on similars. Nanomedicine. 2015; 10: 659–674.
36. Schellekens H, Stegemann S, Weinstein V, et al. How to regulate nonbiological complex drugs (NBCD) and their follow-on versions: Points to consider. AAPSJ. 2014; 16: 15–21.
37. Ehmann F, Sakai-Kato K, Duncan R, et al. Next-generation nanomedicines and nanosimilars: EU regulators' initiatives relating to the development and evaluation of nanomedicines. Nanomedicine. 2013; 8: 849–856.
38. EMA. Reflection paper on the data requirements for intravenous liposomal products developed with reference to an innovator liposomal product. www.ema.europa.eu.
39. EMA CHMP/SWP. Draft reflection paper on the data requirements for intravenous iron-based nanocolloidal products developed with reference to an innovator medicinal product. www.ema.europa.eu.
40. FDA. Draft guidance for industry on bioequivalence recommendations for iron sucrose injection. http://google2.fda.gov/.
41. Teli MK, Mutalik S, Rajanikant GK. Nanotechnology and nanomedicine: Going small means aiming big. Curr. Pharm. Des. 2010; 16: 1882–1892.
42. Sainz V, Conniot J, Matos AI, et al. Regulatory aspects on nanomedicines. Biochem. Biophys. Res. Commun. 2015; 468: 504–510.
43. Allen TM, Cullis PR. Drug delivery systems: Entering the mainstream. Science. 2004; 303: 1818–1822.

44. Zhang L, Gu FX, Chan J M, et al. Nanoparticles in medicine: Therapeutic applications and developments. Clin. Pharmacol. Ther. 2008; 83: 761–769.
45. Hare JI, Lammers T, Ashford MB, et al. Challenges and strategies in anti-cancer nanomedicine development: An industry perspective. Adv. Drug. Deliv. Rev. 2017; 108: 25–38.
46. Hua S, Wu SY. The use of lipid-based nanocarriers for targeted pain therapies. Front. Pharmacol. 2013; 4: 143–147.
47. Danhier F, Feron O, Preat V. To exploit the tumor microenvironment: Passive and active tumor targeting of nanocarriers for anti-cancer drug delivery. J. Control Release. 2010; 148: 135–146.
48. Park K. The drug delivery field at the inflection point: Time to fight its way out of the egg. J. Control Release. 2017; 267: 2–14.
49. Puri A, Loomis K, Smith B, et al. Lipid-based nanoparticles as pharmaceutical drug carriers: From concepts to clinic. Crit. Rev. Ther. Drug Carrier Syst. 2009; 26: 523–580.
50. Min Y, Caster JM, Eblan MJ. Clinical translation of nanomedicine. Chem. Rev. 2015; 115: 11147–11190.

Index

2-(diisopropylamino) ethyl methacrylate (DPA), 43
4-(prop-2-ynyloxy) benzaldehyde, 43

A

acquired immune deficiency syndrome (AIDS), 230
acrylic acid (AA), 127
active targeting, 302
acute lymphoblastic leukaemia, 229
acute myeloid leukaemia, 229
adenocarcinoma, 218
adenosine triphosphate (ATP), 121
aggregation-induced emission (AIE), 79
alanine-alanine-asparagine-leucine (AANL) peptide, 31
alkyne-terminated surfactant, 32
amphipathic α-helical peptide, 257
amphiphilic polymers, 42
amphiphilic stabilizers, 303
angiogenesis, 258
antibody drug conjugates, 230
antimicrobial effect, 220
anti-proliferation efficacy, 254
antisense nucleic acid gene BCL2, 282
anti-tumour immune response (ACIR), 134
anxelekto receptor tyrosine kinases (AXL-RTK), 239
aptamers, 226
artesunate (AS), 255
asialoglycoprotein receptor (ASGPR), 258
atomic force microscopy (AFM), 302
ATP-dependent protease, 310
ATP synthase electron transport, 306
automated SELEX, 228
automation technology, 228
autophagosome formation, 257
autophagy regulation, 257
avidin, 279
azobenzene (azo), 123, 159

B

BALB/c nude mice, 43
beclin-1, 257
berberine (BBR), 128
β cyclodextrin (βCD), 86
β-galactosidase (β-gal), 123
biocompatibility, 42, 354
biodegradability, 42
bio-inspired nanoparticles, 38
biological complex drug (BCD), 356
biopolymers, 288
biopsy, 40
biotin, 279
bismuth selenide (Bi 2Se3), 83
bis(triethoxysilyl)ethane (BTSE), 193
black phosphorus (BP), 109
bortezomib (BTZ), 129
bovine serum albumin (BSA), 238, 247
BP quantum dots (BPQDs), 109

breast cancer resistant protein (BCRP), 40
BT-474 cells, 41

C

calcium phosphate (CaP), 253
camptothecin (CPT), 85, 131
cancer-associated Fibroblasts (CAFs), 5
cancer cell membrane (CM), 90
cancer deaths, 40
cancer nanotherapeutics, 354
capecitabine (cap), 91
carbodiimide chemistry, 279
carbon nanotubes (CNT), 161
carboxymethyl chitosan (CMCS), 129
caveolae, 305
Caveolae-mediated endocytosis, 298
CD44-targeting, 48
cell SELEX, 228
cellulose nano fiber (CNF), 128
chemo dynamic therapy (CDT), 88
chemotherapeutics, 261
chemotherapy, and radiotherapy (RT), 147
chitosan (CS), 126, 290
clathrin-dependent endocytosis, 298
click chemistry reactions, 279
clinical proteomics, 40
clustered regularly interspaced short palindromic repeats (CRISPR), 234
collagen, 7
collagen-altering enzymes, 8
combinational photothermal therapy, 210
complementarity determining regions (CDRs), 278
computed tomography (CT), 122, 146, 196, 251
confocal laser scanning microscopy (CLSM), 236
controlled drug/gene delivery, 353
core shell nanoparticles, 42
co-translational process, 309
Critical micellar concentration (CMC), 196
customized or primer-free capillary electrophoresis SELEX, 228
cyclodextrins (CDs), 104
cyclooxygenase – 2 (COX-2), 8
cytochrome C (Cyt C), 235
cytokines, 261
cytosol, 302
cytosol-targeting agents, 304

D

dark- field microscopy (DFM), 197
death inducing signalling complex (DISC), 277
dendrimers, 44
dendritic cells (DCs), 83
developed docetaxel (DTX), 134
D-glucuronic acid, 292
D hydroxyl radicals (•OH), 98
dichloroacetate (DCA), 331

363

differential scanning colorimetry (DSC), 323
dimercaptosuccinic acid (DMSA), 163
dioxygenase (IDO), 134
disulfide crosslinked Nanoparticles, 218
DNA methyl transferase (DNMT), 40
doxorubicin (DOX), 41, 108, 123, 149
drug delivery systems (DDSs), 221
drug retention capacity, 335
DT-diaphorase (DTD), 117

E

egg white (ovalbumin), 247
endocytosis, 255
endocytosis, 305
endomembrane, 305
endoplasmic reticulum (ER), 296
enhanced permeability and retention (EPR), 2, 136
entrapment efficiency (EE), 79, 327
epidermal growth factor (EGF), 10
epidermal growth factor receptor (EGFR), 258
epigallocatechin gallate (EGCG), 254
epigenetic profiling, 40
epithelial cell adhesion molecule (EpCAM), 239
epithelial-to-mesenchymal Transition (EMT), 6
erratic invasion, 261
ε-caprolactone (PCL), 84
ethyl dimethyl aminopropyl carbodiimide (EDC), 238
European Medicines Agency (EMA), 352
exosomes, 262
expression cassette SELEX, 228
extracellular matrix (ECM), 2, 186

F

ferritin, 252
ferroxidase, 252
fibroblast-activation protein-a (FAP-a), 22
fibroblast activation protein (FAP), 5
fibroblast growth factor, 4
fibronectin, 7
fluorescence imaging (FLI), 121
fluorescence imaging-guided chemotherapy, 254
fluorescence resonance energy transfer, 235
fluorescence spectroscopy, 326
fluorescent reporter (Rho-TP), 123
fluorochromes, 281
folic acid, 41
folic acid (FA), 41, 86, 127, 334
formyl peptide receptors (FPRs), 257

G

gadolinium (Gd), 123, 250
gangliosides, 303
gelatin, 253
gel permeation chromatography (GPC), 333
glioblastoma, 41
GLOBOCAN, 40
glucocorticoid suppressible hyperaldosteronism (GSH), 236
glutathione (GSH), 23, 131
glycol chitosan, 291
gold nanoparticles (AuNPs), 41, 126, 196

gold nanorods (AuNRs), 41, 123
golgi apparatus (GA), 296
G-quadruplex, 227
granzyme B, 278
graphene oxides (GOs), 41, 238

H

HeLa cells, 41
hematoporphyrin monomethyl ether (HMME), 236
heparin (Hep), 127
heterogeneity index, 328
high intensity focused ultrasound (HIFU), 190
high-pressure homogenization techniques, 353
honokiol (HK), 127
human epidermal growth factor receptor 2 (HER2), 199
human serum albumin (HAS), 247
human umbilical cord-derived vascular endothelial cells (HUVEC), 13
hyaluronic acid (HA), 133, 291
hybridisation chain reaction (HCR), 231
hydrogen peroxide (H2O2), 98, 157
hydroxyl radical (·OH), 87
hydroxyl radicals (•OH), superoxide anions (•O2), 156
hyperthermia, 332
hypochlorous acid (HOCL), 98
hypoxia-induced transcription factors (HIF), 114
hypoxia inducible factor-1-alpha (HIF-1-alpha), 9
hypoxia-inducible factor-1 (HIF-1), 22
hypoxia-inducible factor-1 (HIF-1) alpha, 115
hypoxia inducible factor (HIF), 91
hypoxia-responsive polymers, 342

I

immunogenic cell death (ICD), 134
immunogenicity, 42, 247
immuno-stimulators, 257
immunosuppression, 9
immunotherapy, 41
immunotherapy, photothermal therapy (PTT), 147
including singlet oxygen (1O2), 156
indocyanine green (ICG), 84, 104, 123, 331
indocyanine (ICG), 248
indoximod (IND), 135
inductively coupled plasma atomic emission spectroscopy (ICP-AES), 149
inorganic nanoplatforms, 345
insulin-like growth factor 1 receptor (IGF1R), 258
integrin αvβ6 receptor, 258
intellectual property (IP), 357
interleukin-6 (IL-6), 341
internalization, 304
interstitial fluid pressure (IFP), 186
in vitro–in vivo correlation (IVIVC), 343, 354
ionizing radiation, 225
iron (II) phthalocyanine (FePc), 249
isoelectric point (pI), 247

L

laminin, 7
large unilamellar vesicles (LUVs), 200
L-ferritin, 252

Index

lipid polymer hybrid nanoparticles (LPHNs), 86
local field enhancements (LSPR), 197
localization, 288
low-density lipoprotein receptor related protein 1 (LRP1), 258
low frequency ultrasound (LFUS), 199
low-intensity focused ultrasound (LIFU), 199
low molecular weight heparin (LMWH), 331
lysosomes, 305
lysyl oxidase (LOX), 8

M

macrophage, 5
magnetic drug targeting (MDT), 177
magnetic hydrogel (MH), 127
magnetic hyperthermia (MH), 147
magnetic iron oxide nanoparticles (MIONPs), 214
magnetic nanoparticle (MNP), 122, 176
magnetic resonance guided focused ultrasound surgery (MRgFUS), 217
magnetic resonance imaging (MRI), 82, 122, 146, 196, 282
magnetic resonance (MR), 149, 214
magnetofection (MF), 176
maleimide chemistry, 279
mannose-6-phosphate glycopolypeptide (M6P GP), 131
matrix metalloproteinase (MMP), 7, 23, 131
MCF-7 cells, 41
melittin, 257
mercaptopropyltrimethoxysilane (MPTMS), 79
mesoporous PDA (mPDA), 129
mesoporous silica-coated gold nanorods (MMSGNRs), 219
meso-porous silica (MS), 161
mesoporous silica nanoparticles (MSNs), 43, 79
metal organic frameworks (MOFs), 24
metastasis, 41
methoxy poly (ethylene glycol)-g-poly (aspartic acid)-g-tyrosine (CPPT), 128
methyl amino ethyl methacrylate (DMAMEA), 132
Michigan Cancer Foundation 7 (MCF 7), 236
microbubbles (MBs), 185
micrometastases, 41
micropinocytosis, 304
mitochondrial targeting signal (MTS), 306
molecular dynamic therapy (MDT), 93
molybdenum disulphide (MoS2), 238
monoclonal antibody (mAb), 83
mononuclear phagocyte system (MPS), 136, 186
monosaccharide, 288
MR diagnostics, 251
mucin 1 protein (MUC1), 235
multi drug resistance (MDR), 45, 130, 286
multifunctionality, 288
multilamellar vesicles (MLVs), 200

N

nanocarriers (NCs), 174
nano emulsions (NEs), 185
nanomaterials, 38
nanoparticles (NPs), 118
nanoprecipitation, 354
natural killer lymphocytes (NK Cells), 232
N-doped mesoporous carbon (NMCS), 130

near infrared fluorescent imaging (NIRFI), 42, 218
near infrared II, 214
near-infrared (NIR), 149
near infrared photothermal induced therapy (NIR-PIT), 5
near-infrared region (NIR), 122
negatively charged phospholipids, 42
neuropilin-1 (NRP-1) receptors, 258
neutravidin, 281
N-hydroxy succinimide (NHS), 238
NIR fluorescence imaging, 249
nitric oxide (NO), 24
non-biological complex drug (NBCD), 355
non-equilibrium capillary electrophoresis of equilibrium mixtures (NECEEM), 228
non-small cell Lung Cancer (NSCLC), 12
nuclear localization signals (NLSs), 308
nuclear pore complexes (NPCs), 307

O

obatoclax (OBX), 122
oleylamine, 253
oligo (ethylene glycol) methyl ether methacrylate (OEGMA), 43
olycaprolactone (PCL), 200
O-nitro benzyl (ONB), 133
optical imaging, 281
organelles, 296
organic drug-polymer bond linkers, 25
origami, 235
ovarian cancer cells (OCs), 124
oxaliplatin (OXA), 126

P

paclitaxel (PTX), 87, 135, 165
palladium nanoparticles (Pd NPs), 134
pancreatic cancer, 40
pancreatic ductal adeno carcinoma (PDAC), 5
particle replication in non-wetting template (PRINT), 13
PDT (photodynamic treatment), 164
PEG-b- poly[2-(diisopropylamino)ethyl methacrylate]-(PDPA) copolymer, 25
PEGylation chemistry, 262
peptides, 246
perfluoro hexane (PFH), 195
pericytes, 261
Peroxisomal biogenesis disorder (PBD), 310
peroxisomes, 309
phagocytosis, 262
pharmacokinetic, 357
phenylalanine-glycine (FG), 307
phosphatidylinositol, 303
phosphatidylinositol 3-kinase (PI3K), 309
photoacoustic imaging (PAI), 122
photoacoustic (PA), 248
photoacoustic (PA), 78
photochemical internalization (PCI), 165
photodynamic (PD), 249
photodynamic therapy, 41
photodynamic treatment (PDT), 33, 41, 104, 147, 157, 175
photolabile protecting groups (PPGs), 157
photo SELEX, 228
photosensitizer (PS), 164, 175, 257

photosensitizer Sino porphyrin sodium (SPS), 93
photothermal therapy (PTT), 41, 83, 126, 149, 175, 248, 249
phototoxicity, 291
pH-responsive polymerosomes, 42
pH-responsive polymers, 342
pH-sensitive hydrogel, 45
pH-sensitive nanocarriers, 346
pinocytosis, 305
plasmid DNA (pDNA), 79
plasmonic nanobubble, 218
plasticization, 325
platelet-derived growth factor (PDGF), 4
platinum (IV), 253
poly (2-azepane ethyl methacrylate) (PAEMA), 91
poly[2-(diisopropylamino)ethyl methacrylate]-b-poly[(ethylene glycol) methyl ether methacrylate], 25
poly (2methacryloyloxyethyl phosphorylcholine) (PMPC), 91
poly(2-nitroimidazole) (PNI), 342
poly (amino thioketal) (PATK), 103
poly(bis(carboxyphenoxy) phosphazene, 47
polycaprolactone (PCL), 331
polydispersity index (PDI), 323
polydopamine (PDA), 87, 123, 149
polyelectrolyte shell, 43
poly (ethylene glycol) (PEG), 106
polyethylene glycol (PEG), 91, 161, 197
polyethyleneimine (PEI), 129
poly (hydroxypropyl methacrylamidelactate) (p(HPMAm-Lacn), 149
polylactic-co-glycolic acid (PGLA), 179
poly (lactic-co-glycolic acid) (PLGA), 149, 197
poly(N-isopropylacrylamide) (PNIPAM), 124, 149
poly (nisopropylacrylamide) (PNIPAM), 342
poly (propylene sulphide) (PPS), 109
poly(styrene-co-maleic anhydride) (SMA), 32
polyvinylpyrrolidone (PVP), 83
poly (ε-caprolactone), 251
porphyrin microbubbles, 249
porphysomes, 265
positron emission tomography (PET), 78, 123, 196, 218
pro-apoptosis peptide, 256
pro-apoptotic drugs, 297
proinflammatory factors, 262
prostate specific membrane antigen (PSMA), 231
protein corona, 354
protein kinase c iota (PRKCI), 240
protein kinase (PKc), 240
protein tyrosine phosphatase receptor type J (PTPRJ), 258
proteoglycans, 303
protoporphyrin IX, 255
prussian blue nanoparticles (PBNPs), 196
prussian blue (PB), 251
pyridine endoperoxide (PE), 104

Q

quantum dots, 39, 194, 234

R

radical oxygen species (ROS), 156
radiosensitizer, 264
radiotherapy, 212, 225
rat serum (RSA), 247
reactive oxygen species (ROS), 13, 78, 124
Refsum's disease, 310
renal clearance, 262
reticuloendothelial system (RES), 180, 196, 302
rhizomelic chondrodysplasia punctata (RCDP), 310
rolling circle amplification (RCA), 237
rose bengal (RB), 251

S

scanning electron microscopy (SEM), 324
seed and soil theory of carcinogenesis, 3
selenium-containing copolymer (Se-polymer), 104
SELEX, 229
Se–Se cleavage mechanism, 33
short hairpin RNA (shRNA), 240
signal hypothesis, 297
silicon dioxide-methylene blue (Sio2-MB), 91
silk fibroin (SF), 330
single-chain variable fragment (scFv), 282
single-photon emission computed tomography (SPECT), 146, 196
single stranded oligonucleotide, 229
small-cell lung cancer (SCLC), 87
small interfering RNAs (siRNAs), 226
small unilamellar vesicles (SUVs), 200
solid lipid nanoparticles (SLNs), 327
soybean phosphatidylcholine (SPC), 86
spherical gold nanoparticles, 41
spiro pyrans (SP), 159
squaraine (SQ), 250
stealth liposomes, 42
stimuli responsive drug delivery system (SRDDS), 22
stimuli-responsive systems, 324
streptavidin, 281
subcellular organelle, 261
supercritical carbon dioxide (SC-CO2), 84
super-enhanced permeation and retention (SUPR) effect, 5
superparamagnetic iron oxide nanoparticles (SPIONs), 282
superparamagnetic iron oxide (SPIO), 217
surface plasmon resonance, 235

T

targetability, 288
TAT peptide, 255
taxological evaluation, 347
temperature-sensitive liposomes (TSLs), 199
tetraethyl orthosilicate (TEOS), 79
tetrakis[4-(2-mercaptoethoxy) phenyl] ethylene (TPE-4SH), 43
tetraphenyl porphyrin (TPP), 101
theranostic nanomedicine, 209, 288
theranostic nanoparticles, 43
theranostics, 208
therapeutic efficacy, 39
thermal imaging, 249
thermal multimodality imaging, 248
thermoanalytical technique, 325
thermo-chemo-radio therapy, 218
thermochemotherapy, 251
thermoelastic, 250
thermogravimetric (TG) analysis, 333

Index

thermo-responsive drug delivery, 332
toggling SELEX, 228
tragacanth gum (TG), 127
transcription factor receptors, 262
transferrin, 254
transferrin receptor (Tf-R), 258
transitional growth factor – β (TGF-β), 6
transmission electron microscopy (TEM), 324
Traut's reagent, 279
trigycerol monostearate (TGM), 134
triphenylphosphonium, 306
tumor-associated macrophages (TAMs), 6
tumor microenvironment (TME), 1
tumor necrosis factor – α (TNF-α), 6
tumor-related markers (TAMs), 343
tumor vasculature, 4
tumour microenvironment (TME), 121, 254, 340
tumour-specific antigens (TAAs), 343
two-photon excitation (TPE), 163
two-photon fluorophore (TP), 92
two-photon luminescence (TPL), 197
tyrosine protein kinase like 7 (PTK7), 231

U

ultrasound-assisted drug delivery, 334
ultrasound (US), 146
ultraviolet spectroscopy, 326
ultraviolet (UV), 128
untargeted radiotherapy, 41
up conversion nanoparticles (UCNPs), 122, 236, 251

V

vascular endothelial growth factor (VEGF), 115
vascular endothelial growth factor A (VEGFA), 234

W

Warburg effect, 8
wingless type (Wnt), 122
World Health Organization (WHO), 114

X

X-linked adrenoleukodystrophy, 310
X-ray crystallography, 248

Z

Zellweger spectrum of disease, 310
zeolite imidazolate frameworks (ZIFs), 24
zeta potential, 301
zwitterionic surface, 302

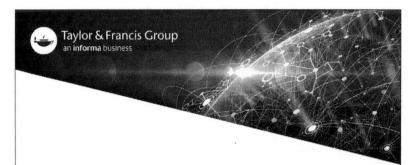